A Course in Theoretical Physics

A Course in Theoretical Physics

P. J. SHEPHERD

Central concepts, laws and results, with full mathematical derivations, in:

Quantum Mechanics
Thermal and Statistical Physics
Many-Body Theory
Classical Theory of Fields
Special and General Relativity
Relativistic Quantum Mechanics
Gauge Theories in Particle Physics

WILEY

A John Wiley & Sons, Ltd., Publication

This edition first published 2013

Registered office
John Wiley & Sons Ltd, The Atrium, Southern Gate, Chichester, West Sussex, PO19 8SQ, United Kingdom

For details of our global editorial offices, for customer services and for information about how to apply for permission to reuse the copyright material in this book please see our website at www.wiley.com.

Library of Congress Cataloging-in-Publication Data applied for.

A catalogue record for this book is available from the British Library.

HB ISBN: 9781118481349
PB ISBN: 9781118481424

Typeset in 10/12pt Times by Aptara Inc., New Delhi, India
Printed and bound in Malaysia by Vivar Printing Sdn Bhd

2 2013

To N.

Contents

Notation

The following types of alphabetic symbol are used in the mathematical expressions in this book:

Nonbold italic letters (both Latin-script and Greek-script) are used for quantities taking values (or for functions taking a range of values) in the form of single numbers (in some system of units). Examples are energy (E), wave function (ψ), magnitude of magnetic induction (B), polarizability (α), temperature (T), and the speed of light (c). Nonbold italic letters are also used for vector and tensor indices.

Bold italic letters (mostly Latin script) are used to denote three-component vectors and pseudovectors. Examples are the radius vector ($\boldsymbol{r}$), three-momentum ($\boldsymbol{p}$), magnetic induction ($\boldsymbol{B}$), and angular momentum ($\boldsymbol{L}$).

Regular (that is, nonbold and upright) Latin-script letters are used to denote standard functions (sin, cos, exp, etc.) and physical units (kg, m, s, J, W, etc.). They are also used (as are regular Greek-script letters) to denote matrices, including row matrices and column matrices. Examples are (3×3) rotation matrices λ, (2×2) Pauli matrices σ_x, σ_y, σ_z, (4×4) Dirac matrices γ^0, γ^1, γ^2, γ^3, and (4×1) Dirac column-matrix wave functions Ψ. Unit and null matrices are denoted by 1 and 0, respectively.

Bold upright letters (both Latin-script and Greek-script) are used to denote three-component vector matrices. An example is the vector Pauli matrix $\boldsymbol{\sigma}$ with components σ_x, σ_y, and σ_z.

Preface

The twentieth century saw the birth and development of two great theories of physics – quantum mechanics and relativity, leading to an extraordinarily accurate description of the world, on all scales from subatomic to cosmological.

Indeed, the transformation of our understanding of the physical world during the past century has been so profound and so wide in its scope that future histories of science will surely acknowledge that the twentieth century was the century of physics.

Should not the great intellectual achievements of twentieth-century physics by now be part of our culture and of our way of viewing the world? Perhaps they should, but this is still very far from being the case. To take an example, Einstein's general theory of relativity is one of the most beautiful products of the human mind, in any field of endeavor, providing us with a tool to understand our universe and its evolution. Yet, far from having become part of our culture, general relativity is barely studied, even in university physics degree courses, and the majority of graduate physicists cannot even write down Einstein's field equations, let alone understand them.

The reason for this is clear. There is a consensus amongst university physics professors (reinforced by the fact that as students they were mostly victims of the same consensus) that the mathematics of general relativity is "too difficult". The result is that the general theory of relativity rarely finds its way into the undergraduate physics syllabus.

Of course, any mathematical methods that have not been carefully explained first will always be "too difficult". This applies in all fields of physics – not just in general relativity. Careful, and appropriately placed, explanations of mathematical concepts are essential. Serious students are unwilling to take results on trust, but at the same time may not wish to spend valuable time on detailed and often routine mathematical derivations. It is better that the latter be provided, so that a student has at least the choice as to whether to read them or omit them. Claims that "it can be shown that..." can lead to relatively unproductive expenditure of time on the filling of gaps – time that students could better spend on gaining a deeper understanding of physics, or on improving their social life.

Several excellent first-year physics degree texts have recognized the likely mathematical limitations of their readers, and, while covering essentially the entire first-year syllabus, ensure that their readers are not left struggling to fill in the gaps in the derivations.

This book is the result of extending this approach to an account of theoretical topics covered in the second, third and fourth years of a degree course in physics in most universities. These include quantum mechanics, thermal and statistical physics, special relativity, and the classical theory of fields, including electromagnetism. Other topics covered here, such as general relativity, many-body theory, relativistic quantum mechanics and gauge theories, are not always all covered in such courses, even though they are amongst the central achievements of twentieth-century theoretical physics.

The book covers all these topics in a single volume. It is self-contained in the sense that a reader who has completed a higher-school mathematics course (for example, A-level mathematics in the United Kingdom) should find within the book everything he or she might need in order to understand every part of it. Details of all derivations are supplied.

The theoretical concepts and methods described in this book provide the basis for commencing doctoral research in a number of areas of theoretical physics, and indeed some are often covered only in first-year postgraduate modules designed for induction into particular research areas. To illustrate their power, the book also includes, with full derivations of all expressions, accounts of Nobel-prize-winning work such as the Bardeen–Cooper–Schrieffer theory of superconductivity and the Weinberg–Salam theory of the electroweak interaction.

The book is divided into five parts, each of which provides a mathematically detailed account of the material typically covered in a forty-lecture module in the corresponding subject area.

Each of the modules is, necessarily, shorter than a typical textbook specializing in the subject of that module. This is not only because many detailed applications of the theory are not included here (nor would they be included in most university lecture modules), but also because, in each of the modules beyond the first, the necessary background can all be found in preceding modules of the book.

The book is the product of over thirty years of discussing topics in theoretical physics with students on degree courses at Exeter University, and of identifying and overcoming the (often subtle) false preconceptions that can impede our understanding of even basic concepts. My deepest thanks go to these students.

Like most course textbooks in such a highly developed (some would say, almost completed) subject area as theoretical physics, this book lays no claim to originality of the basic ideas expounded. Its distinctness lies more in its scope and in the detail of its exposition. The ideas themselves have been described in very many more-specialized textbooks, and my inclusion of some of these books in the lists of recommended supplementary reading is also to be regarded as my grateful acknowledgment of their authors' influence on my chosen approaches to the relevant subject matter.

John Shepherd
Exeter, July 2012

Module I

Nonrelativistic Quantum Mechanics

1

Basic Concepts of Quantum Mechanics

1.1 Probability interpretation of the wave function

In quantum mechanics the **state** of some particle moving in time and space is described by a (complex) **wave function** $\psi(\boldsymbol{r}, t)$ of the space coordinate $\boldsymbol{r}$ $(= \boldsymbol{i}x + \boldsymbol{j}y + \boldsymbol{k}z)$ and time coordinate t. The state of the particle is "described" by $\psi(\boldsymbol{r}, t)$ in the sense that knowledge of the precise form of this function enables us to make precise predictions of the probabilities of the various possible outcomes of any given type of physical measurement on the particle. In particular, in a **position** measurement the probability of finding the particle to be in an infinitesimal box of volume $\mathrm{d}x\mathrm{d}y\mathrm{d}z \equiv \mathrm{d}^3\boldsymbol{r}$ with centre at $\boldsymbol{r}$ at time t is

$$P(\boldsymbol{r}, t)\mathrm{d}x\mathrm{d}y\mathrm{d}z = \psi^*(\boldsymbol{r}, t)\psi(\boldsymbol{r}, t)\mathrm{d}x\mathrm{d}y\mathrm{d}z, \tag{1.1}$$

where $\psi^*(\boldsymbol{r}, t)$ is the complex conjugate of the function $\psi(\boldsymbol{r}, t)$. Equation (1.1) may be written alternatively as

$$P(\boldsymbol{r}, t)\mathrm{d}^3\boldsymbol{r} = |\psi(\boldsymbol{r}, t)|^2\,\mathrm{d}^3\boldsymbol{r}. \tag{1.2}$$

Clearly, since any probability must be a dimensionless number, and $\mathrm{d}^3\boldsymbol{r}$ has the dimensions of volume $[L]^3$, the quantity $P(\boldsymbol{r}, t)$ is a **probability density**, with dimensions $[L]^{-3}$.

The wave function $\psi(\boldsymbol{r}, t)$ is also known as the **probability amplitude**, and has (in three-dimensional space) dimensions $[L]^{-3/2}$.

Since the particle must be found somewhere in space, the continuous sum (i.e., integral) of the probabilities (1.2) over infinitesimal boxes filling all space must be equal to unity:

$$\int_{\text{all space}} |\psi|^2\,\mathrm{d}^3\boldsymbol{r} = 1. \tag{1.3}$$

A wave function ψ satisfying (1.3) is said to be **normalized**.

A Course in Theoretical Physics, First Edition. P. J. Shepherd.

1.2 States of definite energy and states of definite momentum

If a particle is in a state of **definite energy** E, the corresponding wave function $\psi(\boldsymbol{r}, t)$ can be separated as a product of a space-dependent factor $\phi(\boldsymbol{r})$ and a **time-dependent** factor that takes the form of a pure harmonic wave $e^{-i\omega t}$ [i.e., there is only one Fourier component (one frequency) in the Fourier decomposition of the time dependence], where the frequency ω is given by $\omega = E/\hbar$ (the **Einstein relation** $E = \hbar\omega$), in which $\hbar$ ($= 1.055 \times 10^{-34}$ Js) is Planck's constant h divided by 2π.

Thus, the time dependence of the wave function of a particle (or, indeed, a system of many particles) with definite energy E is **always** of the form $e^{-iEt/\hbar}$.

If a particle is in a state of **definite momentum** $\boldsymbol{p} = \boldsymbol{i}p_x + \boldsymbol{j}p_y + \boldsymbol{k}p_z$ (so that the energy is also well defined and the above-mentioned factorization occurs), the **space-dependent** part $\phi(\boldsymbol{r})$ of the corresponding wave function $\psi(\boldsymbol{r}, t)$ is a harmonic plane wave $e^{i\boldsymbol{k}\cdot\boldsymbol{r}}$ [i.e., there is only one Fourier component (one wave vector $\boldsymbol{k}$) in its Fourier decomposition], where the wave vector $\boldsymbol{k} = \boldsymbol{i}k_x + \boldsymbol{j}k_y + \boldsymbol{k}k_z$ is related to the momentum $\boldsymbol{p}$ by $\boldsymbol{k} = \boldsymbol{p}/\hbar$, i.e., $k_x = p_x/\hbar$, and so on. [This is the well known **de Broglie relation** $\boldsymbol{p} = \hbar\boldsymbol{k}$, which, for the magnitudes, gives

$$p \equiv |\boldsymbol{p}| = \hbar\,|\boldsymbol{k}| \equiv \hbar k = \frac{h}{2\pi} \cdot \frac{2\pi}{\lambda} = \frac{h}{\lambda},$$

where λ is the wavelength associated with wave number k.]

Thus, the space dependence of the wave function of a particle with definite momentum $\boldsymbol{p}$ is the **plane wave**

$$\phi_{\boldsymbol{p}}(\boldsymbol{r}) = e^{i\boldsymbol{p}\cdot\boldsymbol{r}/\hbar}. \tag{1.4}$$

[To see the reason for the designation "plane wave", we choose the z axis to lie along the direction of the momentum $\boldsymbol{p}$. Then (1.4) takes the form $e^{ipz/\hbar}$, which clearly has the same phase (and so takes the same value) over any surface $z = \text{const}$ (a plane).]

Successive planes on which the function (1.4) has the same phase are separated by a distance equal to the wavelength λ and are called **wave fronts**.

To understand the motion of wave fronts, consider a particle of definite energy E and definite momentum $\boldsymbol{p}$ (along the z axis). Then the corresponding wave function is

$$\psi_{E,\boldsymbol{p}}(t) \propto e^{i(pz-Et)/\hbar},$$

and the equation of motion of a wave front, that is, of a plane $z = z_{\text{P}}$ with given phase of the wave function, is found by equating the phase of $\psi_{E,\boldsymbol{p}}(t)$ to a constant:

$$pz_{\text{P}} - Et = \text{const},$$

that is,

$$z_{\text{P}} = z_{\text{P}}(t) = \frac{E}{p}t + \text{const} \equiv v_{\text{P}}t + \text{const},$$

Thus, a wave front (surface of constant phase) moves with **phase velocity** $v_{\text{P}} = E/p$ (> 0) along the positive z axis, that is, in the direction of the momentum $\boldsymbol{p}$.

1.3 Observables and operators

With every physically measurable quantity ("**observable**") A we can associate an operator $\hat{A}$, such that, if a particle is in a state with a well defined value a of A [we denote the wave function corresponding to this state by $\psi_a(\boldsymbol{r}, t)$], then

$$\hat{A}\psi_a(\boldsymbol{r}, t) = a\psi_a(\boldsymbol{r}, t). \tag{1.5}$$

Thus, the action of the operator $\hat{A}$ on the function $\psi_a(\boldsymbol{r}, t)$ is to reproduce precisely the same function of $\boldsymbol{r}$ and t but scaled by a constant factor a equal to the well defined value of A in this state. We say that $\psi_a(\boldsymbol{r}, t)$ is an **eigenfunction** of the operator $\hat{A}$, and a is the corresponding **eigenvalue** of the operator $\hat{A}$.

1.4 Examples of operators

As follows from section 1.2, the state of a particle with definite energy E and definite momentum $\boldsymbol{p}$ is described by a wave function of the form

$$\psi_{E,\boldsymbol{p}}(\boldsymbol{r}, t) \propto \mathrm{e}^{i(\boldsymbol{k}\cdot\boldsymbol{r}-\omega t)} = \mathrm{e}^{i(\boldsymbol{p}\cdot\boldsymbol{r}-Et)/\hbar}.$$

Accordingly [see (1.5)], there must exist an **energy operator** $\hat{E}$ and a **momentum operator** $\hat{\boldsymbol{p}}$ such that

$$\hat{E}\mathrm{e}^{i(\boldsymbol{p}\cdot\boldsymbol{r}-Et)/\hbar} = E\mathrm{e}^{i(\boldsymbol{p}\cdot\boldsymbol{r}-Et)/\hbar}$$

and

$$\hat{\boldsymbol{p}}\mathrm{e}^{i(\boldsymbol{p}\cdot\boldsymbol{r}-Et)/\hbar} = \boldsymbol{p}\mathrm{e}^{i(\boldsymbol{p}\cdot\boldsymbol{r}-Et)/\hbar}.$$

The differential operators

$$\hat{E} = i\hbar\frac{\partial}{\partial t} \tag{1.6}$$

and

$$\hat{\boldsymbol{p}} = -i\hbar\left(\boldsymbol{i}\frac{\partial}{\partial x} + \boldsymbol{j}\frac{\partial}{\partial y} + \boldsymbol{k}\frac{\partial}{\partial z}\right) \equiv -i\hbar\boldsymbol{\nabla} \tag{1.7}$$

clearly have the required properties (if we bear in mind that $\boldsymbol{p}\cdot\boldsymbol{r} = p_x x + p_y y + p_z z$).

Just as we have found the energy and momentum operators by considering wave functions corresponding to definite energy and definite momentum, we can find the **position operator** by considering a wave function corresponding to definite position. We restrict ourselves, for the moment, to motion confined to one dimension. Then a particle with (at some given time) a definite position $x = x_1$ will be described by a wave function with the following spatial dependence (at that time):

$$\phi_{x_1}(x) = \delta(x - x_1),$$

where $\delta(x - x_1)$ is the **Dirac delta-function**, equal to zero when $x \neq x_1$ and with the properties

$$\int f(x)\delta(x - x_1)\mathrm{d}x = f(x_1) \quad \text{and} \quad \int \delta(x - x_1)\mathrm{d}x = 1,$$

(provided that the range of integration includes the point $x = x_1$). We need an operator $\hat{x}$ with the following effect:

$$\hat{x}\phi_{x_1}(x) = x_1\phi_{x_1}(x),$$

that is,

$$\hat{x}\delta(x - x_1) = x_1\delta(x - x_1). \tag{1.8}$$

The purely multiplicative operator

$$\hat{x} = x$$

has the required property, since (1.8) becomes

$$x\delta(x - x_1) = x_1\delta(x - x_1),$$

which is clearly true both for $x \neq x_1$ and for $x = x_1$. Similarly, we have $\hat{y} = y$ and $\hat{z} = z$, and so the operator of the three-dimensional position vector $\boldsymbol{r}$ is $\hat{\boldsymbol{r}} = \boldsymbol{r}$.

The operator corresponding to a general observable $A(\boldsymbol{r}, \boldsymbol{p})$ is obtained by the prescription

$$A \to \hat{A} = A(\hat{\boldsymbol{r}}, \hat{\boldsymbol{p}}) \tag{1.9}$$

with

$$\hat{\boldsymbol{r}} = \boldsymbol{r} \quad \text{and} \quad \hat{\boldsymbol{p}} = -i\hbar\boldsymbol{\nabla}. \tag{1.10}$$

For example, the angular momentum, about the coordinate origin, of a particle at a vector distance $\boldsymbol{r}$ from the origin and with momentum $\boldsymbol{p}$ is given classically by the vector product $\boldsymbol{L} = \boldsymbol{r} \times \boldsymbol{p}$, and the corresponding operator $\hat{\boldsymbol{L}}$ is given by:

$$\hat{\boldsymbol{L}} = \hat{\boldsymbol{r}} \times \hat{\boldsymbol{p}} = -i\hbar\boldsymbol{r} \times \boldsymbol{\nabla}. \tag{1.11}$$

1.5 The time-dependent Schrödinger equation

The total energy E of a particle at a given point $\boldsymbol{r}$ at time t is given by the sum of the kinetic energy $T(\boldsymbol{r}, t)$ and potential energy $V(\boldsymbol{r}, t)$:

$$\begin{aligned} E &= E(\boldsymbol{r}, t) = T(\boldsymbol{r}, t) + V(\boldsymbol{r}, t) \\ &= \frac{\boldsymbol{p} \cdot \boldsymbol{p}}{2m} + V(\boldsymbol{r}, t) \equiv \frac{p^2}{2m} + V(\boldsymbol{r}, t), \end{aligned} \tag{1.12}$$

where $\boldsymbol{p} = \boldsymbol{p}(\boldsymbol{r}, t)$ is the momentum of the particle (of mass m) at the point $\boldsymbol{r}$ at time t.

We make the operator replacements (section 1.4):

$$E \to \hat{E} = i\hbar\frac{\partial}{\partial t},$$
$$\boldsymbol{p} \to \hat{\boldsymbol{p}} = -i\hbar\boldsymbol{\nabla} \quad \text{(so that } p^2 \equiv \boldsymbol{p}\cdot\boldsymbol{p} \to -\hbar^2\boldsymbol{\nabla}\cdot\boldsymbol{\nabla} \equiv -\hbar^2\nabla^2\text{)}, \tag{1.13}$$
$$V(\boldsymbol{r}, t) \to \hat{V} = V(\hat{\boldsymbol{r}}, t) = V(\boldsymbol{r}, t).$$

Since the relation $E = T + V$ is a physical requirement, a physically acceptable wave function $\psi(\boldsymbol{r}, t)$ must satisfy the equation $\hat{E}\psi = \hat{T}\psi + \hat{V}\psi$, that is,

$$i\hbar\frac{\partial\psi(\boldsymbol{r}, t)}{\partial t} = -\frac{\hbar^2}{2m}\nabla^2\psi(\boldsymbol{r}, t) + V(\boldsymbol{r}, t)\psi(\boldsymbol{r}, t). \tag{1.14}$$

This is the **time-dependent Schrödinger equation**, and is (for the nonrelativistic case and in three dimensions) a **completely general** differential equation (first order in time and second order in space) for the evolution of the wave function $\psi(\boldsymbol{r}, t)$ of a particle whose energy may or may not be well defined, moving in a potential that may or may not be varying in time.

1.6 Stationary states and the time-independent Schrödinger equation

We now specialize to the case when the particle is in a state with definite energy E. For this case (section 1.2) the wave function is always of the form

$$\psi(\boldsymbol{r}, t) = \phi(\boldsymbol{r})\mathrm{e}^{-iEt/\hbar}, \tag{1.15}$$

where $\phi(\boldsymbol{r})$ specifies the space-dependent part of the function. Because, from (1.15), $|\psi(\boldsymbol{r}, t)|^2 = |\phi(\boldsymbol{r})|^2$, which is independent of the time t, such a state is called a **stationary state**. When (1.15) is substituted into the time-dependent Schrödinger equation (1.14) the left-hand side becomes $E\psi(\boldsymbol{r}, t)$, and so, after cancellation of the factor $\mathrm{e}^{-iEt/\hbar}$, we obtain

$$-\frac{\hbar^2}{2m}\nabla^2\phi(\boldsymbol{r}) + V(\boldsymbol{r})\phi(\boldsymbol{r}) = E\phi(\boldsymbol{r}), \tag{1.16}$$

where we have also replaced $V(\boldsymbol{r}, t)$ by $V(\boldsymbol{r})$, since for a particle to be in a state of well-defined energy E it is a necessary (though not sufficient) condition that the potential-energy function be time independent. [This is clearly seen from the fact that equation (1.16) with different potentials $V_t(\boldsymbol{r}) \equiv V(\boldsymbol{r}, t)$ at different times t will have different solutions $\phi_t(\boldsymbol{r})$ with different energy eigenvalues E_t, contradicting the assumption of a definite energy E.] The equation (1.16) is the **time-independent Schrödinger equation**, and is an eigenvalue equation [see (1.5)] of the form

$$\hat{H}\phi(\boldsymbol{r}) = E\phi(\boldsymbol{r}) \tag{1.17}$$

with

$$\hat{H} = -\frac{\hbar^2}{2m}\nabla^2 + V(\boldsymbol{r}) \tag{1.18}$$

(the hamiltonian operator, or **hamiltonian**).

1.7 Eigenvalue spectra and the results of measurements

The set of all possible eigenvalues a of an operator $\hat{A}$ is called the **spectrum** of the operator $\hat{A}$. **A measurement of A always yields a value belonging to this spectrum.**

Two cases are possible:

(a) The wave function $\psi(\boldsymbol{r}, t)$ describing the state of the particle is an eigenfunction $\psi_i(\boldsymbol{r}, t)$ of $\hat{A}$, that is,

$$\hat{A}\psi_i(\boldsymbol{r}, t) = a_i\psi_i(\boldsymbol{r}, t). \tag{1.19}$$

Then the result of measuring A will certainly be a_i.

(b) The wave function $\psi(\boldsymbol{r}, t)$ describing the state of the particle is not an eigenfunction of $\hat{A}$; that is, the action of $\hat{A}$ on $\psi(\boldsymbol{r}, t)$ gives a function that is not simply a scaled version of $\psi(\boldsymbol{r}, t)$. But since the eigenfunctions $\psi_i(\boldsymbol{r}, t)$ of $\hat{A}$ form a complete set, in the sense that any normalized function can be expanded in terms of them, we may write $\psi(\boldsymbol{r}, t)$ as such an expansion:

$$\psi(\boldsymbol{r}, t) = \sum_i c_i\psi_i(\boldsymbol{r}, t). \tag{1.20}$$

Then a measurement of A can yield any eigenvalue a_i for which the corresponding eigenfunction appears in the sum (1.20) with nonzero weight c_i. For example, a measurement of A yields the result a_j with probability $|c_j|^2$.

1.8 Hermitian operators

The operators corresponding to all physical observables are **hermitian**.

Definition. An operator $\hat{A}$ is said to be hermitian if, for any pair of normalizable wave functions $\psi(\boldsymbol{r}, t)$ and $\chi(\boldsymbol{r}, t)$, the relation

$$\int \chi^*\hat{A}\psi \mathrm{d}^3\boldsymbol{r} = \int (\hat{A}\chi)^*\psi \mathrm{d}^3\boldsymbol{r} \tag{1.21}$$

always holds.

The eigenvalues of hermitian operators are real. This is proved as follows. Choose χ and ψ to be the same eigenfunction ψ_i of the operator $\hat{A}$, with corresponding eigenvalue a_i; that is, choose

$$\chi = \psi = \psi_i \qquad (\hat{A}\psi_i = a_i\psi_i).$$

Then (1.21) becomes

$$a_i \int \psi_i^*\psi_i \mathrm{d}^3\boldsymbol{r} = a_i^* \int \psi_i^*\psi_i \mathrm{d}^3\boldsymbol{r},$$

that is, $a_i = a_i^*$, or, in other words, the eigenvalue a_i (and hence any eigenvalue of $\hat{A}$) is real.

Now choose χ and ψ to be two different eigenfunctions of $\hat{A}$: let χ be ψ_i with eigenvalue a_i, and let ψ be ψ_j with eigenvalue a_j. Then (1.21) becomes

$$\int \psi_i^* \hat{A} \psi_j \mathrm{d}^3 \boldsymbol{r} = \int (\hat{A} \psi_i)^* \psi_j \mathrm{d}^3 \boldsymbol{r},$$

whence

$$a_j \int \psi_i^* \psi_j \mathrm{d}^3 \boldsymbol{r} = a_i^* \int \psi_i^* \psi_j \mathrm{d}^3 \boldsymbol{r}.$$

But $a_i^* = a_i$ and, since $a_j \neq a_i$, we have:

$$\int \psi_i^* \psi_j \mathrm{d}^3 \boldsymbol{r} = 0. \tag{1.22}$$

This result states that eigenfunctions of a hermitian operator that correspond to different eigenvalues of that operator are **orthogonal**.

We now use the idea of orthogonality to prove an inequality

$$\left| \int \phi^* \chi \mathrm{d}^3 \boldsymbol{r} \right|^2 \leq \left(\int \phi^* \phi \mathrm{d}^3 \boldsymbol{r} \right) \left(\int \chi^* \chi \mathrm{d}^3 \boldsymbol{r} \right) \tag{1.23}$$

that will be used in section 1.11 to prove the **general uncertainty relation**. The case when the functions ϕ and χ are the same function clearly corresponds to the equality in (1.23). The inequality (1.23), if true, will be true for any normalization of ϕ and χ, since when we multiply ϕ and χ by arbitrary factors these factors cancel in the inequality (1.23). Therefore, for convenience in examining the left-hand side of (1.23), we choose unit normalization for ϕ and χ. But if in the left-hand side $|\int \phi^* \chi \mathrm{d}^3 \boldsymbol{r}|^2$ the function χ contains not only a part proportional to ϕ but also a part proportional to a normalized function ρ **orthogonal** to ϕ, so that $\chi = a\phi + b\rho$, with $|a|^2 + |b|^2 = 1$ (note that then, in particular, $|a| \leq 1$), we have:

$$\left| \int \phi^* \chi \mathrm{d}^3 \boldsymbol{r} \right| = \left| \int \phi^* (a\phi + b\rho) \mathrm{d}^3 \boldsymbol{r} \right| = |a| \int \phi^* \phi \mathrm{d}^3 \boldsymbol{r} \leq \int \phi^* \phi \mathrm{d}^3 \boldsymbol{r}. \tag{1.24}$$

Similarly, expressing ϕ in this case as $\phi = c\chi + d\eta$, with the function η orthogonal to χ and with both χ and η normalized to unity (so that $|c|^2 + |d|^2 = 1$ and, in particular, $|c| \leq 1$), we have

$$\left| \int \phi^* \chi \mathrm{d}^3 \boldsymbol{r} \right| = \left| \int \chi^* \phi \mathrm{d}^3 \boldsymbol{r} \right| = \left| \int \chi^* (c\chi + d\eta) \mathrm{d}^3 \boldsymbol{r} \right| = |c| \int \chi^* \chi \mathrm{d}^3 \boldsymbol{r} \leq \int \chi^* \chi \mathrm{d}^3 \boldsymbol{r}, \tag{1.25}$$

so that, taking the product of (1.24) and (1.25), we obtain (1.23).

1.9 Expectation values of observables

The **expectation value** of an observable A in a particle state described by a wave function $\psi(\boldsymbol{r}, t)$ is defined as the average result of $N\,(\to \infty)$ measurements of A, all carried out on a particle in the same state $\psi(\boldsymbol{r}, t)$. The prescription for calculating the expectation value when the (normalized) function $\psi(\boldsymbol{r}, t)$ is known is

$$\langle A\rangle_{\psi(\boldsymbol{r},t)} \equiv \int \psi^*(\boldsymbol{r}, t)\hat{A}\psi(\boldsymbol{r}, t)\mathrm{d}^3\boldsymbol{r}. \tag{1.26}$$

To show that this prescription is equivalent to the definition of the expectation value as an average, we substitute the expansion (1.20) into (1.26) and use the fact that the eigenfunctions $\psi_i(\boldsymbol{r}, t)$ of $\hat{A}$ are normalized and mutually orthogonal (an **orthonormal** set). (We showed in section 1.8 that eigenfunctions belonging to different eigenvalues of a hermitian operator are always mutually orthogonal. Also, although eigenfunctions belonging to the same eigenvalue need not be mutually orthogonal, it is always possible to form linear combinations of them that are.) The result is

$$\langle A\rangle_{\psi(\boldsymbol{r},t)} = \sum_i |c_i|^2 a_i. \tag{1.27}$$

This relation asserts (in precise accord with its definition as an average) that the expectation value of A in the state $\psi(\boldsymbol{r}, t)$ is the sum of all possible outcome a_i, each weighted by the probability $|c_i|^2$ that in this state a measurement of A will yield the result a_i.

1.10 Commuting observables and simultaneous observability

If the operators corresponding to two observables A and B have a common, complete set of eigenfunctions ψ_{ij} (i labels the corresponding eigenvalues a_i of $\hat{A}$, and j labels the corresponding eigenvalues b_j of $\hat{B}$), we say that A and B can be **simultaneously well defined**, or, equivalently, that they are simultaneously observable.

The **commutator** $[\hat{A}, \hat{B}] \equiv \hat{A}\hat{B} - \hat{B}\hat{A}$ of any pair of simultaneously well-definable observables A and B is identically zero, in the sense that its action on an arbitrary normalizable function ψ gives zero. To show this, we use the fact that any such function ψ can be written as a linear combination of the common eigenfunctions ψ_{ij} of $\hat{A}$ and $\hat{B}$, so that

$$[\hat{A}, \hat{B}]\psi = [\hat{A}, \hat{B}]\sum_{i,j} c_{ij}\psi_{ij} = \sum_{i,j} c_{ij}[\hat{A}, \hat{B}]\psi_{ij} = 0, \tag{1.28}$$

which follows from the fact that

$$\begin{aligned}[\hat{A}, \hat{B}]\psi_{ij} &\equiv \hat{A}\hat{B}\psi_{ij} - \hat{B}\hat{A}\psi_{ij} = \hat{A}b_j\psi_{ij} - \hat{B}a_i\psi_{ij}\\ &= b_j\hat{A}\psi_{ij} - a_i\hat{B}\psi_{ij} = b_ja_i\psi_{ij} - a_ib_j\psi_{ij} = 0.\end{aligned}$$

The converse is clearly also true, in that if two operators $\hat{A}$ and $\hat{B}$ commute, that is, $[\hat{A}, \hat{B}] = 0$, they can be simultaneously well defined.

1.11 Noncommuting observables and the uncertainty principle

If $[\hat{A}, \hat{B}] \neq 0$, the observables A and B cannot be simultaneously well defined. We define the **uncertainty** in A in the state ψ by

$$(\Delta A)_\psi = \sqrt{\left\langle \left(\hat{A} - \langle \hat{A} \rangle_\psi\right)^2 \right\rangle_\psi} \tag{1.29}$$

(which is clearly the standard deviation of A, i.e., the root mean square deviation of A from the mean of A, where the "mean" here is the expectation value in the state ψ). Then, as is proved below, the product of the uncertainties in two observables A and B is given by the inequality (**general uncertainty relation**)

$$(\Delta A)_\psi (\Delta B)_\psi \geq \frac{1}{2} \left| \langle [\hat{A}, \hat{B}] \rangle_\psi \right|. \tag{1.30}$$

Proof: With the definition (1.29) in mind, we set $\phi \equiv (\hat{A} - \langle \hat{A} \rangle_\psi)\psi$ and $\chi \equiv (\hat{B} - \langle \hat{B} \rangle_\psi)\psi$, so that

$$\begin{aligned}\int \phi^* \phi \mathrm{d}^3 \boldsymbol{r} &= \int \left[\left(\hat{A} - \langle \hat{A} \rangle_\psi\right)\psi\right]^* \left(\hat{A} - \langle \hat{A} \rangle_\psi\right)\psi \mathrm{d}^3 \boldsymbol{r} \\ &= \int \psi^* \left(\hat{A} - \langle \hat{A} \rangle_\psi\right)\left(\hat{A} - \langle \hat{A} \rangle_\psi\right)\psi \mathrm{d}^3 \boldsymbol{r} = (\Delta A)_\psi{}^2,\end{aligned}$$

where we have used the fact that $\hat{A}$ is a hermitian operator [see (1.21)]. Similarly, $\int \chi^* \chi \mathrm{d}^3 \boldsymbol{r} = (\Delta B)_\psi{}^2$. Then the inequality (1.23) becomes:

$$\begin{aligned}(\Delta A)_\psi{}^2 (\Delta B)_\psi{}^2 &= \left(\int \phi^* \phi \mathrm{d}^3 \boldsymbol{r}\right)\left(\int \chi^* \chi \mathrm{d}^3 \boldsymbol{r}\right) \geq \left|\int \phi^* \chi \mathrm{d}^3 \boldsymbol{r}\right|^2 = \left|\mathrm{Re}\left(\int \phi^* \chi \mathrm{d}^3 \boldsymbol{r}\right)\right|^2 + \left|\mathrm{Im}\left(\int \phi^* \chi \mathrm{d}^3 \boldsymbol{r}\right)\right|^2 \\ &\geq \left|\mathrm{Im}\left(\int \phi^* \chi \mathrm{d}^3 \boldsymbol{r}\right)\right|^2 = \left|\frac{\int \phi^* \chi \mathrm{d}^3 \boldsymbol{r} - \int \chi^* \phi \mathrm{d}^3 \boldsymbol{r}}{2i}\right|^2 \\ &= \frac{1}{4}\left|\int \left(\left(\hat{A} - \langle \hat{A} \rangle_\psi\right)\psi\right)^* \left(\hat{B} - \langle \hat{B} \rangle_\psi\right)\psi \mathrm{d}^3 \boldsymbol{r} - \int \left(\left(\hat{B} - \langle \hat{B} \rangle_\psi\right)\psi\right)^* \left(\hat{A} - \langle \hat{A} \rangle_\psi\right)\psi \mathrm{d}^3 \boldsymbol{r}\right|^2 \\ &= \frac{1}{4}\left|\int \psi^* \left(\hat{A} - \langle \hat{A} \rangle_\psi\right)\left(\hat{B} - \langle \hat{B} \rangle_\psi\right)\psi \mathrm{d}^3 \boldsymbol{r} - \int \psi^* \left(\hat{B} - \langle \hat{B} \rangle_\psi\right)\left(\hat{A} - \langle \hat{A} \rangle_\psi\right)\psi \mathrm{d}^3 \boldsymbol{r}\right|^2,\end{aligned}$$

where we have again used the fact that that $\hat{A}$ and $\hat{B}$ are hermitian operators. Thus, we have

$$\begin{aligned}(\Delta A)_\psi{}^2 (\Delta B)_\psi{}^2 &\geq \frac{1}{4}\left|\langle \hat{A}\hat{B} \rangle_\psi - 2\langle \hat{A} \rangle_\psi \langle \hat{B} \rangle_\psi + \langle \hat{A} \rangle_\psi \langle \hat{B} \rangle_\psi - \left(\langle \hat{B}\hat{A} \rangle_\psi - 2\langle \hat{B} \rangle_\psi \langle \hat{A} \rangle_\psi + \langle \hat{B} \rangle_\psi \langle \hat{A} \rangle_\psi\right)\right|^2 \\ &= \frac{1}{4}\left|\langle \hat{A}\hat{B} - \hat{B}\hat{A} \rangle_\psi\right|^2 \equiv \frac{1}{4}\left|\langle [\hat{A}, \hat{B}] \rangle_\psi\right|^2,\end{aligned}$$

from which (1.30 follows). For example, when $\hat{A} = \hat{x} = x$ and $\hat{B} = \hat{p}_x = -i\hbar\partial/\partial x$, we have $[\hat{x}, \hat{p}_x]\psi = -i\hbar\left(x\frac{\partial \psi}{\partial x} - \frac{\partial}{\partial x}(x\psi)\right) = i\hbar\psi$. From this we see that $[\hat{x}, \hat{p}_x] = i\hbar$, and so (1.30) yields $(\Delta x)_\psi (\Delta p_x)_\psi \geq \frac{\hbar}{2}$, which asserts that for a particle in one-dimensional motion no state ψ exists for which the product of the uncertainties in the position and momentum is smaller than $\hbar/2$. In fact, the equality $(\Delta x)_\psi(\Delta p_x)_\psi = \frac{\hbar}{2}$ (corresponding to the smallest possible product of the position and momentum uncertainties) is found in the case of a one-dimensional harmonic oscillator in its ground state, as can be seen by substituting the gaussian space-dependent part $\phi(x) \propto \exp(-\alpha x^2)$ of the latter [see (3.36)] into the expressions for $(\Delta x)_\psi$ and $(\Delta p_x)_\psi$.

1.12 Time dependence of expectation values

Using the expression (1.26) for the expectation value of an observable A, and also the time-dependent Schrödinger equation (1.14) in the form

$$i\hbar\frac{\partial\psi(\boldsymbol{r},t)}{\partial t} = \hat{H}\psi(\boldsymbol{r},t) \tag{1.31}$$

(together with the complex conjugate of this equation), we can immediately obtain an expression for the rate of change of the expectation value with time:

$$i\hbar\frac{\mathrm{d}}{\mathrm{d}t}\langle A\rangle_\psi = \langle[\hat{A},\hat{H}]\rangle_\psi + i\hbar\langle\partial\hat{A}/\partial t\rangle_\psi\,, \tag{1.32}$$

which is known as **Ehrenfest's theorem**.

1.13 The probability-current density

We now use the time-dependent Schrödinger equation (1.14) to find the time rate of change of the probability density $P(\boldsymbol{r},t) = \psi^*(\boldsymbol{r},t)\psi(\boldsymbol{r},t)$. We find

$$\frac{\partial P(\boldsymbol{r},t)}{\partial t} + \boldsymbol{\nabla}\cdot\boldsymbol{j}(\boldsymbol{r},t) = 0 \tag{1.33}$$

(the **continuity equation** for the probability density). Here,

$$\boldsymbol{j} = -\frac{i\hbar}{2m}\left(\psi^*\boldsymbol{\nabla}\psi - (\boldsymbol{\nabla}\psi^*)\psi\right) \tag{1.34}$$

is the **probability-current density**.

1.14 The general form of wave functions

In this section, for simplicity, we consider the case of **one-dimensional** motion of a particle of mass m and potential energy $V(x)$.

The space part $\phi_E(x)$ of the wave function of a stationary state with energy E satisfies the time-independent Schrödinger equation:

$$-\frac{\hbar^2}{2m}\frac{\mathrm{d}^2\phi_E}{\mathrm{d}x^2} + V(x)\phi_E = E\phi_E, \tag{1.35}$$

that is

$$\frac{1}{\phi_E}\frac{\mathrm{d}^2\phi_E}{\mathrm{d}x^2} = \frac{2m}{\hbar^2}[V(x) - E].$$

Case (a): Consider points x at which $E > V(x)$, that is, points at which the kinetic energy $T(x)$ is positive, corresponding to a classically allowed situation. At such points,

$$\frac{1}{\phi_E}\frac{\mathrm{d}^2\phi_E}{\mathrm{d}x^2} < 0,$$

that is, $\phi_E(x)$ is **concave** to the x axis.

Case (b): At points x at which $E < V(x)$, that is, points at which the kinetic energy $T(x)$ is negative, corresponding to a classically forbidden situation, we have:

$$\frac{1}{\phi_E}\frac{\mathrm{d}^2\phi_E}{\mathrm{d}x^2} > 0,$$

that is, $\phi_E(x)$ is **convex** to the x axis.

These features of $\phi_E(x)$ are made clear in figure 1.1, in which, as a convention, we plot schematically the real part (or it could be the imaginary part) of $\phi_E(x)$ at the level of E.

The kinetic energy (and hence the momentum $p = \hbar k$) is greatest at the deepest part of the well (the centre, in this case); that is, k is largest, and hence the "local wavelength" $\lambda = 2\pi/k$ is shortest, near the centre of the well.

Similarly, since higher values of E correspond to higher values of the kinetic energy T (and hence of the momentum p) at all points, the wave functions of the higher-energy states have more oscillations (and hence more nodes).

Suppose now that the potential energy V is piecewise-constant in space, for example, as in the case of the "potential barrier" shown in figure 1.2.

In regions in which V is constant and smaller than E (**classically allowed regions**, such as the regions to the left and to the right of the barrier in figure 1.2), the space-dependent part of the wave function always has the **oscillatory** form

$$\phi_E(x) = Ae^{ikx} + Be^{-ikx}, \tag{1.36}$$

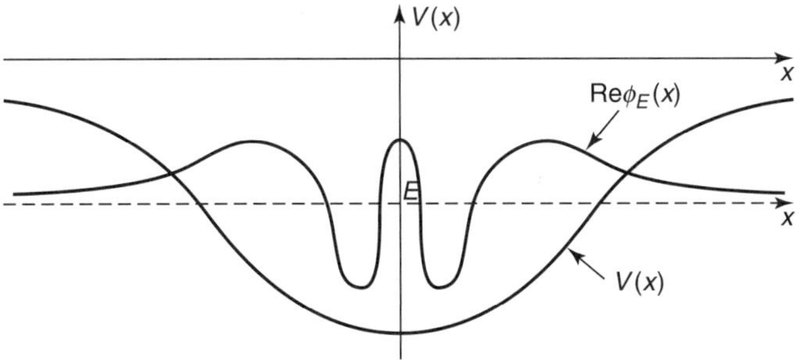

Figure 1.1 *Real or imaginary part (schematic) of the space part $\phi_E(x)$ of the wave function of a particle of definite energy E moving in a potential V(x).*

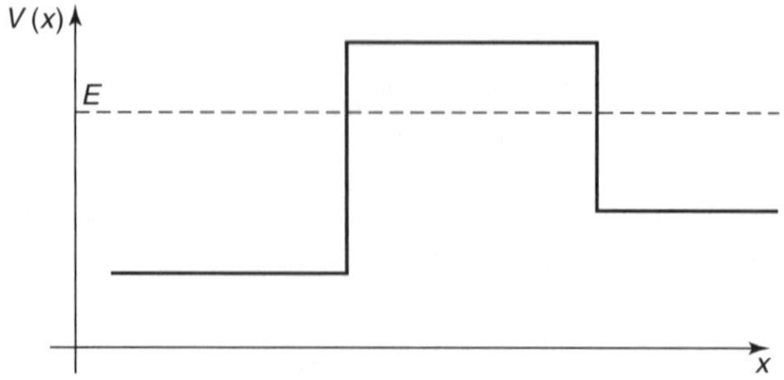

Figure 1.2 *A "potential barrier" V(x). The dashed line denotes the level of the total energy E of the particle.*

where k is found from the relation

$$E - V = T = \frac{\hbar^2 k^2}{2m},$$

in which V is the constant potential energy in the given region. Thus, the wave number k in (1.36) is given by

$$k = \left[\frac{2m}{\hbar^2}(E - V)\right]^{1/2}.$$

In regions in which V is constant and greater than E (**classically forbidden regions**, such as the region of the barrier in figure 1.2), the space-dependent part of the wave function has the general form of an **exponentially increasing** and an **exponentially decreasing** term:

$$\phi_E(x) = Ce^{\gamma x} + De^{-\gamma x}, \tag{1.37}$$

with

$$\frac{\hbar^2\gamma^2}{2m} = V - E,$$

where V is the constant potential energy in the given region. Thus, γ in (1.37) is given by

$$\gamma = \left[\frac{2m}{\hbar^2}(V - E)\right]^{1/2}.$$

To find the constants of the type A, B in the forms (1.36) or of the type C, D in the forms (1.37) in all the regions of constant V, we impose the following requirements on the wave function $\phi_E(x)$:

(i) The function $\phi_E(x)$ is continuous.
(ii) The first derivative $d\phi_E(x)/dx$ is continuous [except at singularities or infinite discontinuities of $V(x)$].
(iii) The function $\phi_E(x)$ is normalized.

These conditions are, in fact, **general physical requirements** on the space part of the wave function of a particle with a definite energy E. Moreover, it turns out that, in the general case, they cannot be met for **arbitrary** values of E, and the time-independent Schrödinger equation has solutions that satisfy these requirements only if the energy E is specified to have a value belonging to a specified spectrum. In certain regions of energy this spectrum may be discrete, and so the above requirements lead naturally to **energy quantization**.

1.15 Angular momentum

The vector angular-momentum operator $\hat{\boldsymbol{L}}$ (1.11) has the following cartesian components:

$$\hat{L}_x = -i\hbar\left(y\frac{\partial}{\partial z} - z\frac{\partial}{\partial y}\right) = i\hbar\left(\sin\varphi\frac{\partial}{\partial\theta} + \cot\theta\cos\varphi\frac{\partial}{\partial\varphi}\right),$$
$$\hat{L}_y = -i\hbar\left(z\frac{\partial}{\partial x} - x\frac{\partial}{\partial z}\right) = i\hbar\left(-\cos\varphi\frac{\partial}{\partial\theta} + \cot\theta\sin\varphi\frac{\partial}{\partial\varphi}\right),$$
$$\hat{L}_z = -i\hbar\left(x\frac{\partial}{\partial y} - y\frac{\partial}{\partial x}\right) = -i\hbar\frac{\partial}{\partial\varphi},$$

and, therefore,

$$\hat{L}^2 = \hat{L}_x^2 + \hat{L}_y^2 + \hat{L}_z^2 = -\hbar^2\left[\frac{1}{\sin\theta}\frac{\partial}{\partial\theta}\left(\sin\theta\frac{\partial}{\partial\theta}\right) + \frac{1}{\sin^2\theta}\frac{\partial^2}{\partial\varphi^2}\right]. \tag{1.38}$$

As will be proved in chapter 2, the eigenvalue spectrum of $\hat{L}^2$ is

$$0,\ 2\hbar^2,\ 6\hbar^2,\ 12\hbar^2,\ 20\hbar^2,\ \ldots,$$

that is,

$$l(l+1)\hbar^2, \text{ with } \quad l = 0, 1, 2, 3, 4, \ldots,$$

and the eigenvalue spectrum of $\hat{L}_z$ is

$$0,\ \pm\hbar,\ \pm 2\hbar,\ \pm 3\hbar,\ \ldots,\ \pm l\hbar,$$

that is,

$$m\hbar, \text{ with } \quad m = 0, \pm 1, \pm 2, \pm 3, \ldots, \pm l.$$

It is easily shown that $[\hat{L}^2, \hat{L}_z] = 0$, and so (see section 1.10) the operators $\hat{L}^2$ and $\hat{L}_z$ have a set of common eigenfunctions, labelled by the quantum numbers l and m [the **spherical harmonics** $Y_{lm}(\theta, \varphi)$]:

$$\begin{aligned} \hat{L}^2 Y_{lm}(\theta,\varphi) &= l(l+1)\hbar^2 Y_{lm}(\theta,\varphi), \\ \hat{L}_z Y_{lm}(\theta,\varphi) &= m\hbar Y_{lm}(\theta,\varphi). \end{aligned} \tag{1.39}$$

The spherical harmonics are given by the expression

$$Y_{lm}(\theta, \varphi) = N_{lm} P_l^m(\cos\theta) e^{im\varphi}, \tag{1.40}$$

where the $P_l^m(\cos\theta)$ are the **associated Legendre functions**, defined by

$$P_l^m(w) = \left(1 - w^2\right)^{|m|/2} \frac{d^{|m|} P_l(w)}{dw^{|m|}} \tag{1.41}$$

[$P_l(w)$ is a **Legendre polynomial** – see below], and the normalization constants N_{lm} are given by

$$N_{lm} = \varepsilon \left[\frac{2l+1}{4\pi} \cdot \frac{(l-|m|)!}{(l+|m|)!} \right]^{\frac{1}{2}}, \text{ with } \varepsilon = \begin{cases} (-1)^m & (m > 0) \\ 1 & (m \leq 0) \end{cases}$$

The $P_l(w)$ ($l = 0, 1, 2, \ldots$) are polynomials of order l (power series in w containing a finite number of powers of w, the highest being w^l), called **Legendre polynomials**, and are polynomial solutions of Legendre's equation

$$\frac{d}{dw}\left[(1-w^2)\frac{d}{dw}P_l(w)\right] + l(l+1)P_l(w) = 0.$$

The $P_l(w)$ are defined on the interval $|w| \leq 1$ and normalized by the requirement that

$$\int_{-1}^{1} P_l(w) P_{l'}(w) dx = \frac{2}{2l+1} \delta_{ll'}.$$

The first few Legendre polynomials $P_l(w)$ are

$$P_0(w) = 1,\ P_1(w) = w,\ P_2(w) = \frac{1}{2}\left(3w^2 - 1\right),\ P_3(w) = \frac{1}{2}\left(5w^3 - 3w\right), \cdots$$

The polynomials $P_l(w)$ have the parity of l, that is, they are even functions of w if the integer l is even, and odd functions of w if l is odd.

Since

$$[\hat{L}_x, \hat{L}_y] = i\hbar \hat{L}_z, \quad [\hat{L}_y, \hat{L}_z] = i\hbar L_x, \quad [\hat{L}_z, \hat{L}_x] = i\hbar \hat{L}_y, \tag{1.42}$$

that is, since the components of $\hat{\boldsymbol{L}}$ do not commute with one another, only one of the components of $\boldsymbol{L}$ can be well defined.

1.16 Particle in a three-dimensional spherically symmetric potential

For motion in a three-dimensional spherically symmetric potential $V(\boldsymbol{r}) = V(r)$ the time-independent Schrödinger equation for stationary states of energy E is [see (1.16)]

$$-\frac{\hbar^2}{2m}\nabla^2\phi(\boldsymbol{r}) + V(\boldsymbol{r})\phi(\boldsymbol{r}) = E\phi(\boldsymbol{r}). \tag{1.43}$$

In spherical polar coordinates this becomes

$$-\frac{\hbar^2}{2m}\left[\frac{1}{r^2}\frac{\partial}{\partial r}\left(r^2\frac{\partial}{\partial r}\right) + \frac{1}{r^2\sin\theta}\frac{\partial}{\partial\theta}\left(\sin\theta\frac{\partial}{\partial\theta}\right) + \frac{1}{r^2\sin^2\theta}\frac{\partial^2}{\partial\varphi^2}\right]\phi(\boldsymbol{r}) + V(r)\phi(\boldsymbol{r}) = E\phi(\boldsymbol{r}).$$

We now multiply this equation by $-2mr^2/\hbar^2$, try the substitution

$$\phi(\boldsymbol{r}) = \phi(r,\theta,\varphi) = R(r)Y(\theta,\varphi),$$

and divide the resulting equation by $\phi = RY$. The result is

$$\frac{1}{R}\frac{\mathrm{d}}{\mathrm{d}r}\left(r^2\frac{\mathrm{d}R}{\mathrm{d}r}\right) + \frac{2mr^2}{\hbar^2}(E - V(r)) = -\frac{1}{Y}\left[\frac{1}{\sin\theta}\frac{\partial}{\partial\theta}\left(\sin\theta\frac{\partial Y}{\partial\theta}\right) + \frac{1}{\sin^2\theta}\frac{\partial^2 Y}{\partial\varphi^2}\right]. \tag{1.44}$$

(Note that the first term on the left involves a total, not a partial, derivative with respect to r, since R depends only on r.) Since the left-hand side is a function of r, and the right-hand side is a function of θ and φ, and yet the two sides are equal, they must be equal to the same constant Λ. The right-hand side then becomes

$$-\left[\frac{1}{\sin\theta}\frac{\partial}{\partial\theta}\left(\sin\theta\frac{\partial Y}{\partial\theta}\right) + \frac{1}{\sin^2\theta}\frac{\partial^2 Y}{\partial\varphi^2}\right] = \Lambda Y. \tag{1.45}$$

Comparison with (1.38) shows that this can be written as

$$\frac{\hat{L}^2}{\hbar^2}Y = \Lambda Y, \quad \text{or} \quad \hat{L}^2 Y = \Lambda\hbar^2 Y.$$

This is none other than the first eigenvalue equation in (1.39), and so we can make the identifications

$$\Lambda = l(l+1),$$
$$Y(\theta,\varphi) = Y_{lm}(\theta,\varphi).$$

With this Λ the radial equation (obtained by setting the left-hand side of (1.44) equal to Λ) is, after division by r^2,

$$\frac{1}{r^2}\frac{\mathrm{d}}{\mathrm{d}r}\left(r^2\frac{\mathrm{d}R}{\mathrm{d}r}\right) + \left[\frac{2m}{\hbar^2}(E - V(r)) - \frac{l(l+1)}{r^2}\right]R = 0. \tag{1.46}$$

1.17 The hydrogen-like atom

In the case of a hydrogen-like atom (i.e., one-electron atom), with nuclear charge Ze (Z is the number of protons and e is the modulus of the electron charge $-e$) and nuclear mass m_{nuc}, equation (1.46) describes the **relative motion** of the electron–nucleus system if we identify r with the electron–nucleus distance $r = |\boldsymbol{r}_{\text{elec}} - \boldsymbol{r}_{\text{nuc}}|$, replace m by the **reduced mass** μ of the electron–nucleus system:

$$m = \mu = \frac{m_{\text{elec}} m_{\text{nuc}}}{m_{\text{elec}} + m_{\text{nuc}}},$$

and set $V(r)$ equal to the Coulomb potential energy of the electron–nucleus system:

$$V(r) = -\frac{Ze^2}{4\pi\varepsilon_0 r}. \tag{1.47}$$

The radial equation then has negative-energy (**bound-state**) solutions that satisfy the physical requirements of continuity, continuity of gradient (except at the singularity $r = 0$ of the Coulomb potential), and normalizability **only** if E belongs to the spectrum of values

$$E = E_n = -\frac{1}{2n^2} \cdot \frac{Ze^2}{4\pi\varepsilon_0 a_0(Z)}, \tag{1.48}$$

where the **principal quantum number** $n = 1, 2, 3, \ldots$, and

$$a_0(Z) = \frac{4\pi\varepsilon_0\hbar^2}{\mu Ze^2} = \frac{a_0(1)}{Z} \equiv \frac{a_0}{Z} \tag{1.49}$$

is the **Bohr radius** in the case when the nucleus has charge Ze.

The corresponding radial solutions are:

$$R_{nl}(r) = \mathrm{e}^{-\frac{1}{2}\alpha r}(\alpha r)^l L_{n+l}^{2l+1}(\alpha r),$$

with $\alpha = \left(\frac{8\mu|E_n|}{\hbar^2}\right)^{1/2}$ (i.e., α depends on the quantum number n), and $L_{n+l}^{2l+1}(\alpha r)$ is a polynomial (**Laguerre polynomial**) of degree $n' = n - l - 1$, that is, with $n - l - 1$ nodes.

For example,

$$R_{10}(r) = \left(\frac{Z}{a_0}\right)^{3/2} \cdot 2\mathrm{e}^{-r/a_0(Z)}, \tag{1.50}$$

$$R_{20}(r) = \left(\frac{Z}{2a_0}\right)^{3/2} \left(2 - \frac{Zr}{a_0}\right) \mathrm{e}^{-r/2a_0(Z)}, \tag{1.51}$$

$$R_{21}(r) = \left(\frac{Z}{2a_0}\right)^{3/2} \frac{r}{\sqrt{3}a_0(Z)} \mathrm{e}^{-r/2a_0(Z)}. \tag{1.52}$$

These radial functions are depicted schematically in figure 1.3.

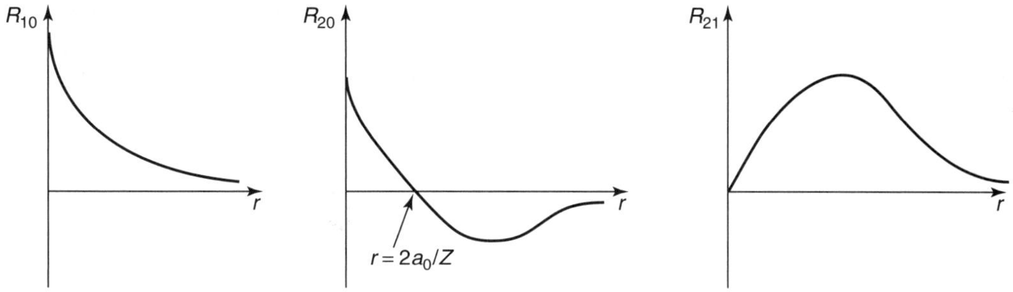

Figure 1.3 *The radial functions $R_{10}(r)$, $R_{20}(r)$ and $R_{21}(r)$.*

The full solutions have the form

$$\phi_{nlm}(r, \theta, \varphi) = R_{nl}(r) Y_{lm}(\theta, \varphi).$$

Consider the effect on these of **inversion**, that is, of replacing $\boldsymbol{r}$ by $-\boldsymbol{r}$, which means replacing (r, θ, φ) by $(r, \pi - \theta, \varphi + \pi)$. Under inversion,

$$e^{im\varphi} \rightarrow e^{im(\varphi+\pi)} = \begin{cases} e^{im\varphi} \text{ if } m \text{ is even,} \\ -e^{im\varphi} \text{ if } m \text{ is odd,} \end{cases}$$

that is, the factor $e^{im\varphi}$ is even under inversion if m is even, and odd under inversion if m is odd. We say that $e^{im\varphi}$ has the **parity** of m.

Similarly, since $P_l(\cos\theta)$ has the parity of l under inversion, and

$$P_l^m(w) = \left(1 - w^2\right)^{\frac{1}{2}|m|} \frac{d^{|m|} P_l(w)}{dw^{|m|}},$$

it follows that $P_l^m(\cos\theta)$ has the parity of $l - |m|$ under inversion.

Therefore, the spherical harmonic

$$Y_{lm}(\theta, \varphi) = N_{lm} P_l^m(\cos\theta) e^{im\varphi}$$

has the parity of l.

Thus, wave functions with even l are even under inversion, and wave functions with odd l are odd under inversion.

For example, the functions $\phi_{nlm} = \phi_{100}$, ϕ_{200} and ϕ_{300} are all even under inversion, and hence, as can be seen from the graphs in figure 1.3, have a cusp (a discontinuous gradient) at $r = 0$.

For example,

$$\frac{\partial}{\partial x} \frac{\partial \phi_{100}}{\partial x}\bigg|_{r=0} = -\infty. \tag{1.53}$$

This discontinuity of gradient is perfectly permissible in the case of the Coulomb potential, since the potential energy (1.47) is singular at $r = 0$ (it tends to $-\infty$ at $r = 0$), so that, for a given fixed total energy E, the kinetic energy tends to $+\infty$ at $r = 0$. Since, as can be seen from (1.12) and (1.13), the operator representing minus the kinetic energy contains second-derivative operators with respect to x, y and z, the relation (1.53) has a natural interpretation.

Functions with $l = 0, 1, 2, 3, 4, \ldots$ are called $s, p, d, f, g, \ldots$ functions, respectively. Since for $l = 0$ we must have $m = 0$ always, all s functions contain the factor

$$Y_{00} = 1/\sqrt{4\pi},$$

and so are independent of θ and φ, that is, are spherically symmetric functions.

For any given value of the principal quantum number n (except $n = 1$) there are three p functions ($l = 1$, $m = 1, 0, -1$). For these the radial part is $R_{n1}(r)$ and the three possible spherical harmonics are

$$Y_{10}(\theta, \varphi) = \left(\frac{3}{4\pi}\right)^{1/2} \cos\theta,$$
$$Y_{1,\pm1}(\theta, \varphi) = \mp\left(\frac{3}{8\pi}\right)^{1/2} \sin\theta \mathrm{e}^{\pm i\varphi}.$$

Any linear combination of the three functions ϕ_{n1m} ($m = 1, 0, -1$) will also be a solution of the Schrödinger equation for the hydrogen-like atom. In particular, consider the combinations

$$\begin{aligned} \phi_{n10} &= R_{n1}Y_{10} \propto R_{n1}\cos\theta, \\ \frac{1}{2}(\phi_{n11} + \phi_{n1,-1}) &= \frac{R_{n1}}{2}(Y_{11} + Y_{1,-1}) \propto R_{n1}\sin\theta\cos\varphi, \\ \frac{1}{2i}(\phi_{n11} - \phi_{n1,-1}) &= \frac{R_{n1}}{2i}(Y_{11} - Y_{1,-1}) \propto R_{n1}\sin\theta\sin\varphi. \end{aligned} \tag{1.54}$$

Since $\cos\theta = z/r$, $\sin\theta\cos\varphi = x/r$ and $\sin\theta\sin\varphi = y/r$, the three functions (1.54) are simply z, x and y, respectively, multiplied by a spherically symmetric function. Since spherically symmetric functions are invariant under coordinate rotations and reflections, the three functions (1.54) transform under coordinate rotations and reflections like z, x and y, respectively, and so are designated $\phi_{np_z}(\boldsymbol{r})$, $\phi_{np_x}(\boldsymbol{r})$ and $\phi_{np_y}(\boldsymbol{r})$. For example, we can depict the function $\phi_{2p_z}(\boldsymbol{r})$ by drawing surfaces of constant $|\phi_{2p_z}(\boldsymbol{r})|$, as in one of the diagrams in figure 1.4. The nodal plane $z = 0$ is one such surface and the proportionality to z is indicated by the opposite signs of the function in the half-spaces $z > 0$ and $z < 0$.

Similarly, from the five $l = 2$ spherical harmonics (corresponding to d functions) we can construct five linear combinations that transform, under coordinate rotations and reflections, like xy, yz, zx, $x^2 - y^2$, and $3z^2 - r^2$. For the example of $n = 3$ (the $3d$ functions) these are also depicted, with the corresponding designation of the respective functions, in figure 1.4.

For the $2p$ functions and for the $3d$ functions the number of nodes of the radial part is zero ($n - l - 1 = 0$ in both cases). Examples for which the radial part of the wave function has one node (the function $\phi_{2s}(\boldsymbol{r})$, for which $n - l - 1 = 1$) or two nodes (the function $\phi_{4p_z}(\boldsymbol{r})$, for which $n - l - 1 = 2$) are illustrated in figure 1.5 .

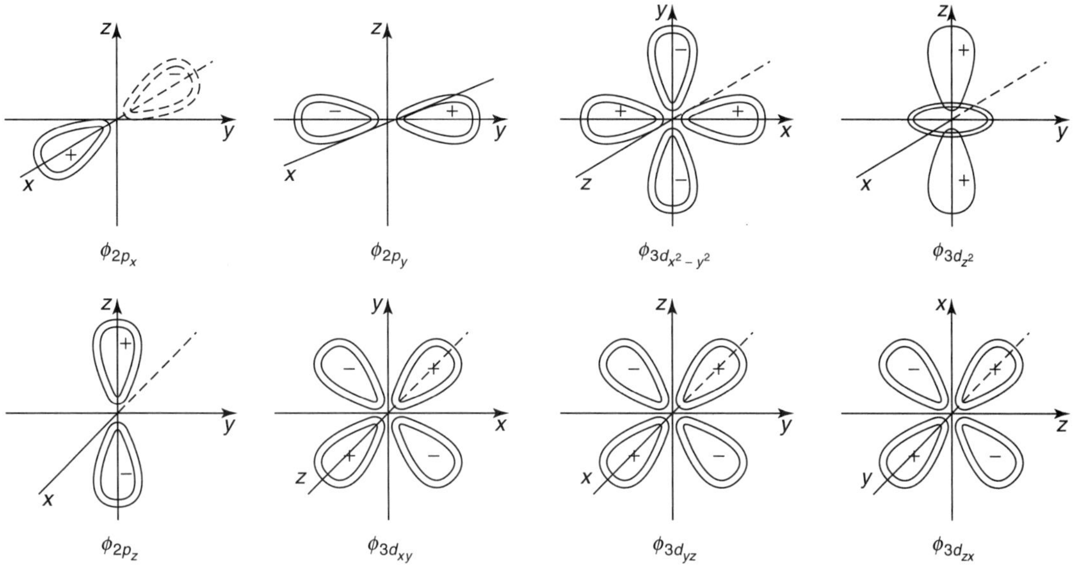

Figure 1.4 *Surfaces of constant modulus of the three 2p and five 3d functions.*

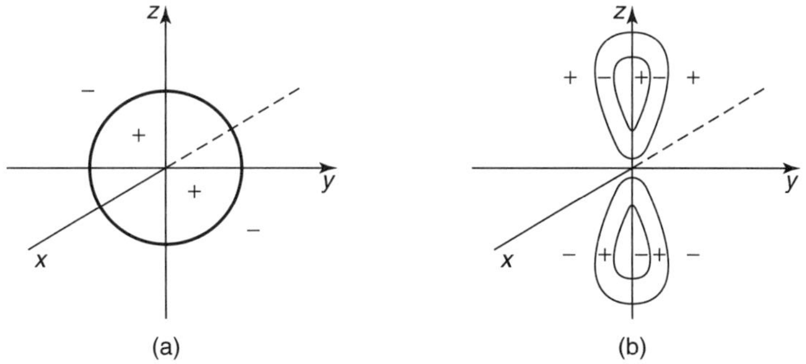

Figure 1.5 *Nodal surfaces and relative signs of (a) the function $\phi_{2s}(\mathbf{r})$ and (b) the function $\phi_{4p_z}(\mathbf{r})$.*

2

Representation Theory

2.1 Dirac representation of quantum mechanical states

Consider the representation of an ordinary three-component vector $\boldsymbol{a}$. In a specified orthogonal set of coordinate axes corresponding to coordinates x, y, z the vector $\boldsymbol{a}$ has components a_x, a_y, a_z, that is, $\boldsymbol{a} = a_x\boldsymbol{x} + a_y\boldsymbol{y} + a_z\boldsymbol{z}$, where we have denoted the unit vectors along the coordinate axes x, y, z by $\boldsymbol{x}, \boldsymbol{y}, \boldsymbol{z}$.

In any given representation, the corresponding components of $\boldsymbol{a}$ can be displayed in a column matrix. For example, in the representation corresponding to specified coordinate axes x, y, z the column matrix $\mathbf{a}$ representing the vector $\boldsymbol{a}$ is

$$\mathrm{a} \equiv \begin{pmatrix} a_x \\ a_y \\ a_z \end{pmatrix} = a_x\mathrm{x} + a_y\mathrm{y} + a_z\mathrm{z} \equiv a_x \begin{pmatrix} 1 \\ 0 \\ 0 \end{pmatrix} + a_y \begin{pmatrix} 0 \\ 1 \\ 0 \end{pmatrix} + a_z \begin{pmatrix} 0 \\ 0 \\ 1 \end{pmatrix},$$

where, in the x, y, z representation, the column matrices x, y, z corresponding to the unit vectors $\boldsymbol{x}, \boldsymbol{y}, \boldsymbol{z}$ each have one element equal to unity and the other two equal to zero. The matrix analog of the familiar expression $a_x = \boldsymbol{x} \cdot \boldsymbol{a}$ for the component a_x as the projection of the vector $\boldsymbol{a}$ on $\boldsymbol{x}$ is clearly

$$a_x = \mathrm{x}^\dagger \mathrm{a} \equiv (1\ 0\ 0) \begin{pmatrix} a_x \\ a_y \\ a_z \end{pmatrix}, \quad a_y = \mathrm{y}^\dagger \mathrm{a}, \quad a_z = \mathrm{z}^\dagger \mathrm{a} \tag{2.1}$$

where, for the purposes of the analogy that follows later in this section, we have written the unit row matrix (1 0 0) as the **adjoint** $\mathrm{x}^\dagger$ of the unit column matrix x, this being consistent with the (physicists', not mathematicians'!) definition of the adjoint of a matrix as the transpose of its complex conjugate.

A Course in Theoretical Physics, First Edition. P. J. Shepherd.

Of course, the unit vectors $\boldsymbol{x}, \boldsymbol{y}, \boldsymbol{z}$ specify only one possible cartesian coordinate system. In a rotated frame with axes corresponding to orthogonal unit vectors $\boldsymbol{x}', \boldsymbol{y}', \boldsymbol{z}'$ the **same** vector $\boldsymbol{a}$ has different components $a_{x'}, a_{y'}, a_{z'}$, and can be expressed as

$$\boldsymbol{a} = a_{x'}\boldsymbol{x}' + a_{y'}\boldsymbol{y}' + a_{z'}\boldsymbol{z}'.$$

As before, the components can be displayed in a column matrix, which we shall denote by a'. Thus, the vector $\boldsymbol{a}$ is now represented by the column matrix

$$\mathrm{a}' \equiv \begin{pmatrix} a_x' \\ a_y' \\ a_z' \end{pmatrix} = a_x'\mathrm{x}' + a_y'\mathrm{y}' + a_z'\mathrm{z}' \equiv a_x' \begin{pmatrix} 1 \\ 0 \\ 0 \end{pmatrix} + a_y' \begin{pmatrix} 0 \\ 1 \\ 0 \end{pmatrix} + a_z' \begin{pmatrix} 0 \\ 0 \\ 1 \end{pmatrix},$$

with components $a_{x'} = \boldsymbol{x}' \cdot \boldsymbol{a}$ given by the corresponding matrix products

$$a_{x'} = \mathrm{x}'^{\dagger}\mathrm{a}', \quad a_{y'} = \mathrm{y}'^{\dagger}\mathrm{a}', \quad a_{z'} = \mathrm{z}'^{\dagger}\mathrm{a}'. \tag{2.2}$$

The column-vector forms a and a' with components (2.1) and (2.2) are two different **representations** of the same vector $\boldsymbol{a}$ (corresponding to two choices of orientation, in this case, of the set of cartesian coordinate axes).

Guided by this, Dirac proposed that the value at point x (we assume a one-dimensional system for the moment) of the wave function ϕ_{a} describing a state specified by the set of quantum numbers a can be regarded as the projection of a **state vector** $|a\rangle$ on a **basis vector** $|x\rangle$. If, by analogy with the above dot-product expression $a_x = \boldsymbol{x} \cdot \boldsymbol{a}$ for a component of the vector $\boldsymbol{a}$, we write this projection as $\langle x \mid a\rangle$, the state vector $|a\rangle$ has components $\phi_a(x) \equiv \langle x \mid a\rangle$ (one component for every position x!). By analogy with the matrix-product representations (2.1) and (2.2) of the components of $\boldsymbol{a}$, we can, in the chosen representation, display the "components" $\phi_a(x) \equiv \langle x \mid a\rangle$ as matrix products too:

$$\phi_a(x) = |x\rangle^{\dagger}|a\rangle \equiv \langle x||a\rangle \equiv \langle x \mid a\rangle, \tag{2.3}$$

where (rather than introduce new notation) we are now using the symbols $|a\rangle$ and $|x\rangle$, not for the given state vector and basis vector, respectively, as above, but for the corresponding column matrices whose elements are their components in the given representation; we have also adopted the notation $\langle x|$ for the **adjoint** $|x\rangle^{\dagger}$ (a row matrix) of the basis "column matrix" $|x\rangle$. Since $\langle x \mid a\rangle$ forms an angular bracket, the row matrices $\langle\ldots|$ (and their antecedent "row vectors") will be called **bras**, and the column matrices $|\ldots\rangle$ (and their antecedent "column vectors") will be called **kets**. Whether a bra (ket) in any given expression is a row (column) vector or the corresponding row (column) matrix with elements equal to the components of the vector in some given representation will always be clear from the context.

What is the dimensionality of the vector $|a\rangle$? To answer this question, consider the example of a hydrogenic wave function with well defined ("good") quantum numbers n, l, m. The corresponding state vector is $|n\,l\,m\rangle$, and its projection on the bra $\langle \boldsymbol{r}| = \langle r\theta\varphi|$ gives the value of the wave function at the point $\boldsymbol{r}$:

$$\phi_{nlm}(\boldsymbol{r}) = \phi_{nlm}(r, \theta, \varphi) \equiv \langle r\theta\varphi \mid n\,l\,m\rangle.$$

Here, the possible values of the quantum numbers n, l, m are $n = 1, 2, 3, \ldots, l = 0, 1, \ldots, n-1$, and $m = -l, -l+1, \ldots, l-1, l$; that is, there are an infinite number of mutually orthogonal (and hence linearly independent) kets $|n\,l\,m\rangle$.

But if N vectors in the same space are linearly independent, they must have $\geq N$ components.

In fact, r, θ and φ can take infinitely many values, and so there are infinitely many basis bras, and hence infinitely many projections (components) of any given ket $|n\,l\,m\rangle$.

Thus, $|n\,l\,m\rangle$ is a vector in an infinite-dimensional space and the values of its (infinite number of) components constitute a square-integrable wave function. (Such a space, by definition, is a **Hilbert space**.)

We have discussed wave functions in the **coordinate representation**, that is, functions of spatial position $\boldsymbol{r}$. (It is only to such functions that the operator prescriptions given in section 1.4 for the position and momentum apply.) The same information about a system is contained in its wave function in the **momentum representation**:

$$\chi_a(\boldsymbol{p}) = \frac{1}{(2\pi\hbar)^{3/2}} \int \mathrm{e}^{-i\boldsymbol{p}\cdot\boldsymbol{r}/\hbar} \phi_a(\boldsymbol{r}) \mathrm{d}^3\boldsymbol{r}. \tag{2.4}$$

The function $\chi_a(\boldsymbol{p})$ is, of course, just the Fourier coefficient multiplying the plane wave $(2\pi\hbar)^{-3/2}\mathrm{e}^{i\boldsymbol{p}\cdot\boldsymbol{r}/\hbar}$ (momentum eigenfunction) in the Fourier expansion of $\phi_a(\boldsymbol{r})$ [the inverse of (2.4)]:

$$\phi_a(\boldsymbol{r}) = \frac{1}{(2\pi\hbar)^{3/2}} \int \mathrm{e}^{i\boldsymbol{p}\cdot\boldsymbol{r}/\hbar} \chi_a(\boldsymbol{p}) \mathrm{d}^3\boldsymbol{p}.$$

In Dirac bra-and-ket notation, the function (2.4) can be written as

$$\chi_a(\boldsymbol{p}) \equiv \langle \boldsymbol{p} \mid a \rangle \equiv \langle p_x\; p_y\; p_z \mid a \rangle. \tag{2.5}$$

Thus, just as for ordinary three-component vectors, we can write $|a\rangle$ in different frames, that is, in different representations. For example, $|a\rangle$ is represented by the column matrices

$$\begin{pmatrix} \langle \boldsymbol{r}_1 \mid a \rangle \\ \langle \boldsymbol{r}_2 \mid a \rangle \\ \cdot \\ \cdot \\ \cdot \end{pmatrix} \quad \text{and} \quad \begin{pmatrix} \langle \boldsymbol{p}_1 \mid a \rangle \\ \langle \boldsymbol{p}_2 \mid a \rangle \\ \cdot \\ \cdot \\ \cdot \end{pmatrix}$$

in the coordinate representation and momentum representation, respectively. These column-matrix representations are, of course, formal in the sense that $\boldsymbol{r}$ (and $\boldsymbol{p}$, in the case of free-particle states) takes a continuum of values, so that these column vectors must be regarded as having **continuously** rather than discretely labelled components!

There is a special representation for a ket $|a\rangle$, namely, the representation obtained by projecting the given ket $|a\rangle$ on to the basis vectors $\langle a|$ themselves. For example, a hydrogenic ket $|n\; l\; m\rangle$ can be written in the basis of the kets $|n_i\; l_j\; m_k\rangle$. In this representation (the **proper representation**, i.e., "own" representation), the ket $|2\ 1\ 1\rangle$ is represented by the column matrix

$$\begin{pmatrix} \langle 1\,0\,0 \mid 2\ 1\ 1\rangle \\ \langle 2\,0\,0 \mid 2\ 1\ 1\rangle \\ \langle 2\,1\,1 \mid 2\ 1\ 1\rangle \\ \langle 2\,1\,0 \mid 2\ 1\ 1\rangle \\ \langle 2\,1\,-1 \mid 2\ 1\ 1\rangle \\ \cdot \\ \cdot \\ \cdot \end{pmatrix} = \begin{pmatrix} 0 \\ 0 \\ 1 \\ 0 \\ 0 \\ \cdot \\ \cdot \\ \cdot \end{pmatrix}.$$

Thus, states in their own (proper) representation are column vectors with just one 1 and with 0s in all the other positions.

The proper representation is most useful in cases when the state space has a **finite** number of dimensions, for example, for **intrinsic** angular momentum (**spin**), isospin, and so on.

For example, for an electron, proton or neutron, we have

$$s \equiv j_{\text{intrinsic}} = 1/2, \quad m_s = \pm 1/2,$$

and so there are only two spin basis states $|s m_s\rangle$. In the proper representation these take the form

$$\left|\frac{1}{2}\frac{1}{2}\right\rangle = \begin{pmatrix} \left\langle \frac{1}{2}\frac{1}{2} \,\middle|\, \frac{1}{2}\frac{1}{2} \right\rangle \\ \left\langle \frac{1}{2}, -\frac{1}{2} \,\middle|\, \frac{1}{2}\frac{1}{2} \right\rangle \end{pmatrix} = \begin{pmatrix} 1 \\ 0 \end{pmatrix}$$

$$\left|\frac{1}{2}, -\frac{1}{2}\right\rangle = \begin{pmatrix} \left\langle \frac{1}{2}\frac{1}{2} \,\middle|\, \frac{1}{2}, -\frac{1}{2} \right\rangle \\ \left\langle \frac{1}{2}, -\frac{1}{2} \,\middle|\, \frac{1}{2}, -\frac{1}{2} \right\rangle \end{pmatrix} = \begin{pmatrix} 0 \\ 1 \end{pmatrix}.$$

A general state $|a\rangle$ is written in the b-representation as

$$|a\rangle = \begin{pmatrix} \langle b_1 \mid a\rangle \\ \langle b_2 \mid a\rangle \\ \cdot \\ \cdot \\ \cdot \end{pmatrix}.$$

Therefore,

$$\langle a| \equiv |a\rangle^\dagger = \left(\langle b_1 \mid a\rangle^* \ \langle b_2 \mid a\rangle^* \ \ldots \ldots \ldots\right).$$

But the row vector $\langle a|$ in the b-representation can be written as

$$\langle a| = \left(\langle a \mid b_1\rangle \ \langle a \mid b_2\rangle \ \ldots \ldots \ldots\right).$$

Therefore,

$$\langle b_1 \mid a\rangle^* = \langle a \mid b_1\rangle, \text{ etc.}$$

We always assume the bras and kets to have been normalized to unity; that is, for states with discrete quantum numbers we have

$$\langle n_i \mid n_j\rangle = \delta_{ij},$$

while for states with continuous labels we have, for example,

$$\langle \boldsymbol{r} \mid \boldsymbol{r}'\rangle = \delta^{(3)}(\boldsymbol{r} - \boldsymbol{r}') \equiv \delta(x - x')\delta(y - y')\delta(z - z').$$

2.2 Completeness and closure

Suppose that a set of ket vectors $|n_i\rangle$ forms a countable **complete** set; that is, any ket vector $|\phi\rangle$ in the same Hilbert space can be expanded as follows in terms of the $|n_i\rangle$:

$$|\phi\rangle = \sum_i c_i\, |n_i\rangle. \tag{2.6}$$

We now project (2.6) on to $\langle n_j|$:

$$\langle n_j \mid \phi\rangle = \sum_i c_i \langle n_j \mid n_i\rangle = \sum_i c_i \delta_{ij} = c_j,$$

that is,

$$c_j = \langle n_j \mid \phi\rangle.$$

We act on $|\phi\rangle$ in (2.6) with the operator

$$P_j \equiv |n_j\rangle\langle n_j|.$$

The result is

$$P_j|\phi\rangle = \sum_i c_i |n_j\rangle\langle n_j \mid n_i\rangle = \sum_i c_i |n_j\rangle\delta_{ij} = c_j|n_j\rangle;$$

that is, P_j projects out the j term in $|\phi\rangle$, and is called a **projection operator**. Therefore,

$$\left(\sum_j P_j\right)|\phi\rangle = \sum_j c_j|n_j\rangle = |\phi\rangle.$$

Thus, we have

$$\sum_j P_j = \sum_j |n_j\rangle\langle n_j| = 1, \tag{2.7}$$

where 1 is the unit matrix. The relation (2.7) is called the **closure relation**.

In the coordinate representation, the closure relation takes the form

$$\sum_j \langle \boldsymbol{r}' \mid n_j\rangle\langle n_j \mid \boldsymbol{r}\rangle = \langle \boldsymbol{r}' | \, 1 \, |\boldsymbol{r}\rangle = \langle \boldsymbol{r}' \mid \boldsymbol{r}\rangle = \delta^{(3)}(\boldsymbol{r} - \boldsymbol{r}'),$$

that is,

$$\sum_j \phi_{n_j}(\boldsymbol{r}')\phi^*_{n_j}(\boldsymbol{r}) = \delta^{(3)}(\boldsymbol{r} - \boldsymbol{r}').$$

The closure relation for a complete set of nondenumerable states $|k\rangle$ is

$$\int \mathrm{d}k \, |k\rangle \, \langle k| = 1, \tag{2.8}$$

that is,

$$\int \mathrm{d}k \phi_k(\boldsymbol{r}')\phi^*_k(\boldsymbol{r}) = \delta^{(3)}(\boldsymbol{r} - \boldsymbol{r}').$$

2.3 Changes of representation

Suppose we know the coordinate-space wave function $\phi_a(\boldsymbol{r}) \equiv \langle \boldsymbol{r} \mid a\rangle$ and wish to find the corresponding momentum-space wave function $\chi_a(\boldsymbol{p}) \equiv \langle \, \boldsymbol{p} \mid a\rangle$. We insert the closure relation

$$\int |\boldsymbol{r}\rangle \mathrm{d}^3\boldsymbol{r}\langle \boldsymbol{r}| = 1$$

into $\chi_a(\boldsymbol{p}) \equiv \langle \, \boldsymbol{p}|a\rangle$:

$$\chi_a(\boldsymbol{p}) \equiv \langle \, \boldsymbol{p} \mid a\rangle = \int \mathrm{d}^3\boldsymbol{r} \, \langle \, \boldsymbol{p} \mid \boldsymbol{r}\rangle \, \langle \boldsymbol{r} \mid a\rangle = \int \mathrm{d}^3\boldsymbol{r} \, \langle \boldsymbol{r} \mid \boldsymbol{p}\rangle^* \, \phi_a(\boldsymbol{r}) = \int \mathrm{d}^3\boldsymbol{r}\phi^*_{\boldsymbol{p}}(\boldsymbol{r})\phi_a(\boldsymbol{r}).$$

But $\phi_{\boldsymbol{p}}(\boldsymbol{r})$ is a normalized eigenfunction of the momentum operator $\hat{\boldsymbol{p}} = -i\hbar\boldsymbol{\nabla}$, that is, it is the plane wave

$$\phi_{\boldsymbol{p}}(\boldsymbol{r}) = \frac{1}{(2\pi\hbar)^{3/2}} \mathrm{e}^{i\,\boldsymbol{p}\cdot\boldsymbol{r}/\hbar}.$$

Therefore,

$$\chi_a(\boldsymbol{p}) = \frac{1}{(2\pi\hbar)^{3/2}} \int \mathrm{d}^3\boldsymbol{r}\mathrm{e}^{-i\boldsymbol{p}\cdot\boldsymbol{r}/\hbar}\phi_a(\boldsymbol{r}),$$

which is the **Fourier transform** of $\phi_a(\boldsymbol{r})$!

In general,

$$\langle j \mid a\rangle = \mathop{\mathrm{S}}_i \langle j \mid i\rangle \langle i \mid a\rangle ,$$

in which the generalized sum S . . . denotes a summation over the discrete values of i plus an integral over the continuous values of i. The quantities $\langle j \mid i\rangle$ are the elements of the **transformation matrix** from representation i to representation j.

2.4 Representation of operators

Let $\hat{F}$ be a local operator, for example, $-(\hbar^2/2m)(\mathrm{d}^2/\mathrm{d}x^2)$, and let the result of its action on a wave function $\phi_{a_i}(x)$ be

$$\hat{F}\phi_{a_i}(x) = \sum_j c_{ij}\phi_{a_j}(x), \tag{2.9}$$

where we have expressed the resulting function as a superposition of the functions $\phi_{a_j}(x)$ (which are assumed here to form a complete set). In the Dirac notation, (2.9) becomes

$$\hat{F}\langle x \mid a_i\rangle = \sum_j c_{ij}\langle x \mid a_j\rangle. \tag{2.10}$$

Now multiply both sides of (2.10) by $|x\rangle$ and integrate over x using the relation

$$\int \mathrm{d}x\, |x\rangle\, \langle x| = 1$$

in the right-hand side. Then (2.10) becomes

$$\hat{\hat{F}}|a_i\rangle = \sum_j c_{ij}\, |a_j\rangle, \tag{2.11}$$

which no longer contains any reference to the coordinate x. In (2.11), the operator

$$\hat{\hat{F}} \equiv \int \mathrm{d}x\, |x\rangle\, \hat{F}\, \langle x| \tag{2.12}$$

is an (x-independent) operator in the Hilbert space spanned by the kets $|a_i\rangle$.

The form (2.11) is the "abstract" (coordinate-free) form of (2.9). In fact, equation (2.9) [or (2.10)] is simply (2.11) written in the coordinate (x) representation, that is, it is the form (2.11) projected on to the basis bra $\langle x|$. To write (2.11) in the **energy representation**, we project it instead onto the basis bra $\langle E_m|$ and insert the closure relation for the energy eigenstates, in the form

$$\sum_n |E_n\rangle\langle E_n| = 1, \tag{2.13}$$

into the left-hand side of the resulting equation. We obtain

$$\sum_n \langle E_m| \hat{\hat{F}} |E_n\rangle \langle E_n \mid a_i\rangle \equiv \sum_n F_{mn} \langle E_n \mid a_i\rangle = \sum_j c_{ij}\langle E_m \mid a_j\rangle.$$

Here the F_{mn} are the elements of the matrix F multiplying the column vector in the left-hand side of the matrix form of this equation:

$$\begin{pmatrix} \langle E_1| \hat{\hat{F}} |E_1\rangle & \langle E_1| \hat{\hat{F}} |E_2\rangle & \dots \\ \langle E_2| \hat{\hat{F}} |E_1\rangle & \langle E_2| \hat{\hat{F}} |E_2\rangle & \dots \\ \dots & \dots & \dots \\ \dots & \dots & \dots \end{pmatrix} \begin{pmatrix} \langle E_1 \mid a_i\rangle \\ \langle E_2 \mid a_i\rangle \\ \dots \\ \dots \end{pmatrix} = \sum_j c_{ij} \begin{pmatrix} \langle E_1 \mid a_j\rangle \\ \langle E_2 \mid a_j\rangle \\ \dots \\ \dots \end{pmatrix}.$$

This is a matrix form (in the energy representation) of the operator equation (2.11).

For example, for stationary states the time-independent Schrödinger equation in the coordinate representation is [see equation (1.17)]

$$\hat{H}\phi_i(\boldsymbol{r}) = E_i\phi_i(\boldsymbol{r}).$$

The corresponding "representation-free" form in the Hilbert space of the kets $|E_i\rangle$ is

$$\hat{\hat{H}} \, |E_i\rangle = E_i \, |E_i\rangle \tag{2.14}$$

with

$$\hat{\hat{H}} \equiv \int \mathrm{d}^3\boldsymbol{r} \, |\boldsymbol{r}\rangle \, \hat{H} \, \langle \boldsymbol{r}|.$$

To obtain (2.14) in the energy representation [which is the "proper representation" for the kets $|E_i\rangle$ (see section 2.1)], we project (2.14) on to the basis bra $\langle E_m|$ and again use insertion of the energy-states closure relation (2.13). We obtain

$$\sum_n \langle E_m| \hat{\hat{H}} |E_n\rangle \langle E_n \mid E_i\rangle = E_i \, \langle E_m \mid E_i\rangle. \tag{2.15}$$

But the energy eigenkets form an orthonormal set, that is,

$$\langle E_n \mid E_i\rangle = \delta_{ni},$$

and so (2.15) becomes

$$\langle E_m| \hat{\hat{H}} |E_i\rangle = E_i\delta_{mi}.$$

Thus, in the energy representation, $\hat{\hat{H}}$ forms a **diagonal** matrix H, with the eigenvalues of $\hat{\hat{H}}$ on the diagonal.

2.5 Hermitian operators

Let $\hat{F}$ be a local hermitian operator [see (1.21)]. Then

$$\int \phi_n^*(\boldsymbol{r})\hat{F}\phi_m(\boldsymbol{r})\mathrm{d}^3\boldsymbol{r} = \int [\hat{F}\phi_n(\boldsymbol{r})]^*\phi_m(\boldsymbol{r})\mathrm{d}^3\boldsymbol{r} = \left(\int \phi_m^*(\boldsymbol{r})\hat{F}\phi_n(\boldsymbol{r})\mathrm{d}^3\boldsymbol{r}\right)^*. \quad (2.16)$$

If we write the wave functions in (2.16) in Dirac notation, this becomes

$$\int \mathrm{d}^3\boldsymbol{r}\,\langle n \mid \boldsymbol{r}\rangle\,\hat{F}\,\langle \boldsymbol{r} \mid m\rangle = \left(\int \mathrm{d}^3\boldsymbol{r}\,\langle m \mid \boldsymbol{r}\rangle\,\hat{F}\,\langle \boldsymbol{r} \mid n\rangle\right)^*,$$

which, if we use the appropriate analogue of (2.12), can be written as

$$\langle n|\,\hat{\hat{F}}\,|m\rangle = \langle m|\,\hat{\hat{F}}\,|n\rangle^*, \quad \text{i.e.,} \quad F_{nm} = (F_{mn})^*. \quad (2.17)$$

But, by definition of the **adjoint** $\mathrm{F}^\dagger$ of the matrix F, we have $(F_{mn})^* = (\mathrm{F}^\dagger)_{nm}$, and so the second equation (2.17) can be written as

$$F_{nm}\,[\equiv (\mathrm{F})_{nm}] = (\mathrm{F}^\dagger)_{nm},$$

which states that $\mathrm{F} = \mathrm{F}^\dagger$. Thus, hermitian operators correspond to **self-adjoint** matrices and, for this reason, are often referred to as **self-adjoint operators**. Indeed, if we **define** the adjoint $\hat{G}^\dagger$ of an operator $\hat{G}$ as the operator satisfying the relation

$$\int \phi_n^*(\boldsymbol{r})\hat{G}\phi_m(\boldsymbol{r})\mathrm{d}^3\boldsymbol{r} = \int [\hat{G}^\dagger\phi_n(\boldsymbol{r})]^*\phi_m(\boldsymbol{r})\mathrm{d}^3\boldsymbol{r}, \quad (2.18)$$

the first line of (2.16) is simply the statement that a hermitian operator is equal to its adjoint, and so is self-adjoint.

2.6 Products of operators

Consider the product of two operators $\hat{\hat{F}}$ and $\hat{\hat{G}}$ acting in a Hilbert space. Using the appropriate form of (2.12), we can write

$$\hat{\hat{F}}\hat{\hat{G}} = \int\!\!\int \mathrm{d}^3\boldsymbol{r}\mathrm{d}^3\boldsymbol{r}'|\boldsymbol{r}\rangle\hat{F}\langle\boldsymbol{r}\,|\,\boldsymbol{r}'\rangle\hat{G}\langle\boldsymbol{r}'| = \int\!\!\int \mathrm{d}^3\boldsymbol{r}\mathrm{d}^3\boldsymbol{r}'|\boldsymbol{r}\rangle\hat{F}\delta^{(3)}(\boldsymbol{r}-\boldsymbol{r}')\hat{G}\langle\boldsymbol{r}'| = \int \mathrm{d}^3\boldsymbol{r}|\boldsymbol{r}\rangle\hat{F}\hat{G}\langle\boldsymbol{r}|. \quad (2.19)$$

Thus, the rule relating a product of Hilbert-space operators to the product of the corresponding local operators is the same as that for single Hilbert-space and local operators. In particular, if two local operators $\hat{\hat{F}}$ and $\hat{\hat{G}}$ have commutator

$$[\hat{F}, \hat{G}] \equiv \hat{F}\hat{G} - \hat{G}\hat{F} = \hat{J}, \quad (2.20)$$

it follows from (2.19) that the commutator of the corresponding Hilbert-space operators is

$$\begin{aligned}[\hat{\hat{F}}, \hat{\hat{G}}] \equiv \hat{\hat{F}}\hat{\hat{G}} - \hat{\hat{G}}\hat{\hat{F}} &= \int \mathrm{d}^3\boldsymbol{r}\, |\boldsymbol{r}\rangle\, \hat{F}\hat{G}\, \langle\boldsymbol{r}| - \int \mathrm{d}^3\boldsymbol{r}\, |\boldsymbol{r}\rangle\, \hat{G}\hat{F}\, \langle\boldsymbol{r}| \\ &= \int \mathrm{d}^3\boldsymbol{r}\, |\boldsymbol{r}\rangle\, [\hat{F}, \hat{G}]\, \langle\boldsymbol{r}| = \int \mathrm{d}^3\boldsymbol{r}\, |\boldsymbol{r}\rangle\, \hat{J}\, \langle\boldsymbol{r}| = \hat{\hat{J}},\end{aligned}$$

that is, the commutation relations for the Hilbert-space operators take the same (canonical) form as those for the local operators to which they are related.

2.7 Formal theory of angular momentum

The eigenvalue equations for the operators corresponding to the square and z component of the angular momentum $\boldsymbol{L} = \boldsymbol{r} \times \boldsymbol{p}$ are [see (1.39)]

$$\begin{aligned}\hat{L}^2 Y_{lm}(\theta, \varphi) &= l(l+1)\hbar^2 Y_{lm}(\theta, \varphi), \\ \hat{L}_z Y_{lm}(\theta, \varphi) &= m\hbar Y_{lm}(\theta, \varphi),\end{aligned} \tag{2.21}$$

with

$$\begin{aligned}\hat{L}_z &= -i\hbar \frac{\partial}{\partial \varphi}, \\ \hat{L}^2 &= -\hbar^2 \left[\frac{1}{\sin\theta} \frac{\partial}{\partial\theta} \left(\sin\theta \frac{\partial}{\partial\theta} \right) + \frac{1}{\sin^2\theta} \frac{\partial^2}{\partial\varphi^2} \right].\end{aligned}$$

The analogues of equations (2.21) for the state vectors $|lm\rangle$, defined to be such that

$$Y_{lm}(\theta, \varphi) \equiv \langle \theta\varphi \mid lm \rangle,$$

are

$$\begin{aligned}\hat{\hat{l}}^2 |lm\rangle &= l(l+1)\hbar^2 |lm\rangle, \\ \hat{\hat{l}}_z |lm\rangle &= m\hbar |lm\rangle\end{aligned} \tag{2.22}$$

where $\hat{\hat{l}}^2$ and $\hat{\hat{l}}_z$ are operators in the Hilbert space of the kets $|lm\rangle$.

We now prove (2.22) without reference to angles or other coordinates, using a new (general algebraic) definition of angular momentum.

Definition: An **angular-momentum operator** $\hat{\boldsymbol{j}}$ is a vector operator

$$\hat{\boldsymbol{j}} = \boldsymbol{x}\hat{j}_x + \boldsymbol{y}\hat{j}_y + \boldsymbol{z}\hat{j}_z$$

whose components satisfy the commutation relations

$$[\hat{j}_i, \hat{j}_j] = i\hbar\varepsilon_{ijk}\hat{j}_k, \tag{2.23}$$

where ε_{ijk} is the Levi-Civita symbol, equal to $+1$ when ijk is an even permutation of xyz, equal to -1 when ijk is an odd permutation of xyz, and equal to 0 when two or more of the indices ijk refer to the same component.

For the three nontrivial choices of the pair $\{i, j\}$, equation (2.23) yields

$$[\hat{j}_x, \hat{j}_y] = i\hbar\hat{j}_z, \quad [\hat{j}_y, \hat{j}_z] = i\hbar\hat{j}_x, \quad [\hat{j}_z, \hat{j}_x] = i\hbar\hat{j}_y. \tag{2.24}$$

Our problem is to find the eigenvalues of the operators $\hat{j}^2$ and $\hat{j}_z$, or, in other words, to find the functions $F(j')$ and $G(m)$ in the relations

$$\begin{aligned} \hat{j}^2|j'm\rangle &= \hbar^2 F(j')|j'm\rangle, \\ \hat{j}_z|j'm\rangle &= \hbar G(m)|j'm\rangle, \end{aligned} \tag{2.25}$$

where the powers of $\hbar$ have been inserted in order to make F and G dimensionless, and the nature and possible values of the quantum numbers j' and m appearing in the eigenvalues F and G, respectively, remain to be determined.

We introduce new operators

$$\hat{j}^+ \equiv \hat{j}_x + i\hat{j}_y, \quad \hat{j}^- \equiv \hat{j}_x - i\hat{j}_y, \tag{2.26}$$

so that $\hat{j}_x = (\hat{j}^+ + \hat{j}^-)/2$ and $\hat{j}_y = (\hat{j}^+ - \hat{j}^-)/2i$. Since $\hat{j}_x$ and $\hat{j}_y$ are hermitian (self-adjoint), that is, since (see section 2.5)

$$\hat{j}_x^\dagger = \hat{j}_x \quad \text{and} \quad \hat{j}_y^\dagger = \hat{j}_y,$$

we have

$$(\hat{j}^+)^\dagger = \hat{j}^- \quad \text{and} \quad (\hat{j}^-)^\dagger = \hat{j}^+.$$

Thus, $\hat{j}^+$ and $\hat{j}^-$ are **mutually adjoint** operators.

From the commutation relations (2.24), we find

$$[\hat{j}^+, \hat{j}_z] = [\hat{j}_x, \hat{j}_z] + i[\hat{j}_y, \hat{j}_z] = -i\hbar\hat{j}_y + i.i\hbar\hat{j}_x = -\hbar\hat{j}^+.$$

Similarly,

$$\begin{aligned} [\hat{j}^-, \hat{j}_z] &= \hbar\hat{j}^-, \\ [\hat{j}^+, \hat{j}^-] &= 2\hbar\hat{j}_z, \\ [\hat{j}^+, \hat{j}^2] &= [\hat{j}^-, \hat{j}^2] = 0. \end{aligned}$$

Consider the ket $\hat{j}^+|j'm\rangle$. Operate on it first with $\hat{j}^2$, and then, instead, with $\hat{j}_z$.

The action of $\hat{j}^2$ yields

$$\hat{j}^2\hat{j}^+|j'm\rangle = \hat{j}^+\hat{j}^2|j'm\rangle = \hat{j}^+\hbar^2 F(j')|j'm\rangle = \hbar^2 F(j')\hat{j}^+|j'm\rangle.$$

Thus, like $|j'm\rangle$, the ket $\hat{j}^+|j'm\rangle$ is an eigenket of $\hat{j}^2$, with the same eigenvalue $\hbar^2 F(j')$. (By a similar argument, the same is also true of the ket $\hat{j}^-|j'm\rangle$.)

The action of $\hat{j}_z$ on $\hat{j}^+|j'm\rangle$ yields

$$\begin{aligned}\hat{j}_z\hat{j}^+|j'm\rangle &= \hat{j}^+\hat{j}_z|j'm\rangle + [\hat{j}_z, \hat{j}^+]|j'm\rangle \\ &= \hat{j}^+\hbar G(m)|j'm\rangle + \hbar\hat{j}^+|j'm\rangle \\ &= \hbar(G(m)+1)\,\hat{j}^+|j'm\rangle. \end{aligned} \tag{2.27}$$

Thus, the ket $\hat{j}^+|j'm\rangle$ is an eigenket of $\hat{j}_z$, with eigenvalue $\hbar[G(m)+1]$. (By a similar argument, the ket $\hat{j}^-|j'm\rangle$ is also an eigenket of $\hat{j}_z$, but with eigenvalue $\hbar[G(m)-1]$.)

We assign the quantum number $m+1$ to the ket $\hat{j}^+|j'm\rangle$, that is, we write

$$\hat{j}^+|j'm\rangle = c^+_{j'm}|j'm+1\rangle, \tag{2.28}$$

where $c^+_{j'm}$ is a numerical coefficient to be determined. Then

$$\hat{j}_z\hat{j}^+|j'm\rangle = c^+_{j'm}\hat{j}_z|j'm+1\rangle = c^+_{j'm}\hbar G(m+1)|j'm+1\rangle = \hbar G(m+1)\hat{j}^+|j'm\rangle. \tag{2.29}$$

Comparing (2.27) and (2.29), we find

$$G(m)+1 = G(m+1),$$

which is satisfied by $G(m) = m + \text{const}$, and therefore, in particular, by

$$G(m) = m.$$

Thus, the second relation of (2.25) becomes

$$\hat{j}_z|j'm\rangle = m\hbar|j'm\rangle. \tag{2.30}$$

We now consider the state

$$\left(\hat{j}_x^2 + \hat{j}_y^2\right)|j'm\rangle = \left(\hat{j}^2 - \hat{j}_z^2\right)|j'm\rangle = \hbar^2(F(j') - m^2)|j'm\rangle.$$

Thus, the ket $|j'm\rangle$ is an eigenket of $\hat{j}_x^2 + \hat{j}_y^2$, with eigenvalue $\hbar^2[F(j') - m^2]$. But any eigenvalue of $\hat{j}_x^2 + \hat{j}_y^2$ must be non-negative. Therefore, $m^2 \leq F(j')$, and so the quantum number m is bounded, that is, has a maximum value m_{max}.

By time-reversal invariance, if $\hat{j}_z$ has an eigenvalue $m\hbar$ it also has an eigenvalue $-m\hbar$ (corresponding to the time-reversed motion). Therefore, the minimum value of m is $-m_{\text{max}}$.

Since we can change m in steps of ± 1, the difference $m_{\text{max}} - (-m_{\text{max}}) = 2m_{\text{max}}$ must be an integer. Therefore, there are two possibilities: m_{max} (and hence every value of m) is an **integer**, or m_{max} (and hence every value of m) is a **half-integer**, that is, half an odd integer.

We now **define** a new number j by

$$j \equiv m_{\text{max}}. \tag{2.31}$$

Then m can take the values

$$m = j, j-1, \ldots, -(j-1), -j,$$

where j is an integer or half-integer.

We shall now find $F(j')$ and identify the quantum number j'. We use the relation

$$\hat{j}^{+}|j'm_{\max}\rangle \equiv \hat{j}^{+}|j'j\rangle = 0,$$

which follows from the fact that the ket $|j'm_{\max}+1\rangle$ does not exist. We now act on the ket $\hat{j}^{+}|j'j\rangle = 0$ with the operator $\hat{j}^{-}$, using the form

$$\begin{aligned}\hat{j}^{-}\hat{j}^{+} &= (\hat{j}_x - i\hat{j}_y)(\hat{j}_x + i\hat{j}_y) = \hat{j}_x^2 + \hat{j}_y^2 + i[\hat{j}_x, \hat{j}_y] \\ &= \hat{j}_x^2 + \hat{j}_y^2 + i.i\hbar\hat{j}_z = \hat{j}^2 - \hat{j}_z^2 - \hbar\hat{j}_z.\end{aligned}$$

Therefore,

$$\left(\hat{j}^2 - \hat{j}_z^2 - \hbar\hat{j}_z\right)|j'j\rangle = 0,$$

and so

$$\hat{j}^2|j'j\rangle = \left(\hat{j}_z^2 + \hbar\hat{j}_z\right)|j'j\rangle = \hbar^2(j^2+j)|j'j\rangle = \hbar^2 j(j+1)|j'j\rangle.$$

But we showed earlier in this section that the operator $\hat{j}^2$ has the same eigenvalue for the kets $\hat{j}^{\pm}|j'm\rangle$ as for the ket $|j'm\rangle$. Therefore,

$$\hat{j}^2|j'm\rangle = \hbar^2 j(j+1)|j'm\rangle.$$

Thus, we can identify j' with $j \equiv m_{\max}$, and finally obtain

$$\hat{j}^2|jm\rangle = \hbar^2 j(j+1)|jm\rangle, \tag{2.32}$$

so that in (2.25) we have

$$F(j') \equiv F(j) = j(j+1).$$

We now consider in more detail the effects of the so-called angular-momentum **raising and lowering** operators $\hat{j}^{+}$ and $\hat{j}^{-}$. We may write

$$\begin{aligned}\hat{j}^{+}|jm\rangle &= c_{jm}^{+}|jm+1\rangle, \\ \hat{j}^{-}|jm\rangle &= c_{jm}^{-}|jm-1\rangle.\end{aligned} \tag{2.33}$$

Also,

$$\hat{j}^{-}\hat{j}^{+}|jm\rangle = \left(\hat{j}^2 - \hat{j}_z^2 - \hbar\hat{j}_z\right)|jm\rangle = \hbar^2\left(j(j+1) - m(m+1)\right)|jm\rangle. \tag{2.34}$$

Alternatively, from (2.33) we have

$$\hat{j}^-\hat{j}^+|jm\rangle = \hat{j}^- c^+_{jm}|jm+1\rangle = c^+_{jm}\hat{j}^-|jm+1\rangle = c^+_{jm}c^-_{j,m+1}|jm\rangle. \tag{2.35}$$

Comparing (2.34) and (2.35), we have

$$c^+_{jm}c^-_{j,m+1} = \hbar^2\left(j(j+1) - m(m+1)\right),$$

which we can solve for c^+_{jm} and c^-_{jm} if we can find a relation connecting them. We use

$$\langle jm+1|\hat{j}^+|jm\rangle = \langle jm+1|c^+_{jm}|jm+1\rangle = c^+_{jm}.$$

Using the definition of the adjoint of an operator (see section 2.5), we have

$$\begin{aligned}(c^+_{jm})^* &= \langle jm+1|\hat{j}^+|jm\rangle^* = \langle jm|(\hat{j}^+)^\dagger|jm+1\rangle \\ &= \langle jm|\hat{j}^-|jm+1\rangle = \langle jm|c^-_{j,m+1}|jm\rangle = c^-_{j,m+1},\end{aligned}$$

so

$$c^+_{jm}c^-_{j,m+1} = c^+_{jm}(c^+_{jm})^* = |c^+_{jm}|^2 = \hbar^2\left(j(j+1) - m(m+1)\right).$$

Therefore, to within physically unimportant phase factors,

$$c^+_{jm} = \hbar\sqrt{j(j+1) - m(m+1)}$$

and

$$c^-_{jm} = (c^+_{j,m-1})^* = \hbar\sqrt{j(j+1) - m(m-1)}$$

Thus, the effects of the angular-momentum raising and lowering operators can be summarized as

$$\hat{j}^\pm|jm\rangle = \hbar\sqrt{j(j+1) - m(m\pm 1)}|jm\pm 1\rangle. \tag{2.36}$$

For example, we can use this formula to generate all the $2j+1$ kets $|jm\rangle$ for a given j by applying the operator $\hat{j}^-$ repeatedly to the ket $|jj\rangle$.

Finally, we consider the description of spin, for the example of spin-1/2 particles. The designation "spin-1/2" means $m_{\max}$ $(\equiv j) = 1/2$, and so the possible values of m are

$$m = 1/2, -1/2.$$

For spin, we denote j by s, and m by m_s. The angular-momentum kets are $|sm_s\rangle = \left|\frac{1}{2}m_s\right\rangle$. In commonly used notation, they are

$$\begin{aligned}|\alpha\rangle &\equiv |\uparrow\rangle = \left|\tfrac{1}{2}\tfrac{1}{2}\right\rangle, \\ |\beta\rangle &\equiv |\downarrow\rangle = \left|\tfrac{1}{2}, -\tfrac{1}{2}\right\rangle.\end{aligned}$$

We apply the operators $\hat{S}_z$ and $\hat{S}^2$ to these states:

$$\begin{aligned}
\hat{S}_z\,|\alpha\rangle &\equiv \hat{S}_z\left|\tfrac{1}{2}\tfrac{1}{2}\right\rangle = \tfrac{1}{2}\hbar\left|\tfrac{1}{2}\tfrac{1}{2}\right\rangle,\\
S_z\,|\beta\rangle &\equiv \hat{S}_z\left|\tfrac{1}{2},-\tfrac{1}{2}\right\rangle = -\tfrac{1}{2}\hbar\left|\tfrac{1}{2},-\tfrac{1}{2}\right\rangle,\\
\hat{S}^2\,|\alpha \text{ or } \beta\rangle &= \hbar^2\tfrac{1}{2}\left(\tfrac{1}{2}+1\right)|\alpha \text{ or } \beta\rangle = \tfrac{3}{4}\hbar^2\,|\alpha \text{ or } \beta\rangle\,.
\end{aligned}$$

We now use (2.36) to find the effects of the spin raising and lowering operators $\hat{S}^+$ and $\hat{S}^-$:

$$\begin{aligned}
\hat{S}^+\,|\beta\rangle &\equiv \hat{S}^+\left|\tfrac{1}{2},-\tfrac{1}{2}\right\rangle = \hbar\sqrt{\tfrac{1}{2}\left(\tfrac{1}{2}+1\right)-\left(-\tfrac{1}{2}\right)\left(-\tfrac{1}{2}+1\right)}\left|\tfrac{1}{2}\tfrac{1}{2}\right\rangle = \hbar\left|\tfrac{1}{2}\tfrac{1}{2}\right\rangle \equiv \hbar\,|\alpha\rangle\,,\\
\hat{S}^+\,|\alpha\rangle &\equiv \hat{S}^+\left|\tfrac{1}{2}\tfrac{1}{2}\right\rangle = \hbar\sqrt{\tfrac{1}{2}\left(\tfrac{1}{2}+1\right)-\tfrac{1}{2}\left(\tfrac{1}{2}+1\right)}\,|?\rangle = 0.
\end{aligned}$$

Clearly, as expected from the definition of j as $m_{\max}$, the ket $\hat{S}^+\,|\alpha\rangle$ does not exist. Similarly, we have

$$\begin{aligned}
\hat{S}^-|\alpha\rangle &= \hbar|\beta\rangle,\\
\hat{S}^-|\beta\rangle &= 0.
\end{aligned}$$

In the "proper representation" (see section 2.1), the spin kets are

$$\begin{aligned}
|\alpha\rangle \equiv \left|\tfrac{1}{2}\tfrac{1}{2}\right\rangle &= \begin{pmatrix}\left\langle\tfrac{1}{2}\tfrac{1}{2}\,\middle|\,\tfrac{1}{2}\tfrac{1}{2}\right\rangle\\ \left\langle\tfrac{1}{2},-\tfrac{1}{2}\,\middle|\,\tfrac{1}{2}\tfrac{1}{2}\right\rangle\end{pmatrix} = \begin{pmatrix}1\\0\end{pmatrix},\\
|\beta\rangle \equiv \left|\tfrac{1}{2},-\tfrac{1}{2}\right\rangle &= \begin{pmatrix}\left\langle\tfrac{1}{2}\tfrac{1}{2}\,\middle|\,\tfrac{1}{2},-\tfrac{1}{2}\right\rangle\\ \left\langle\tfrac{1}{2},-\tfrac{1}{2}\,\middle|\,\tfrac{1}{2},-\tfrac{1}{2}\right\rangle\end{pmatrix} = \begin{pmatrix}0\\1\end{pmatrix},
\end{aligned}$$

and the matrix of the operator $\hat{S}_z$ (consisting of elements $\left\langle\tfrac{1}{2}m_s\right|\hat{S}_z\left|\tfrac{1}{2}m_s'\right\rangle$) is

$$\mathrm{S}_z = \begin{pmatrix}\tfrac{1}{2}\hbar & 0\\ 0 & -\tfrac{1}{2}\hbar\end{pmatrix} = \tfrac{1}{2}\hbar\begin{pmatrix}1 & 0\\ 0 & -1\end{pmatrix}.$$

Similarly, using the above results for the effects of the operators $\hat{S}^+$ and $\hat{S}^-$ on the kets $|\alpha\rangle$ and $|\beta\rangle$, we find the corresponding matrices:

$$\mathrm{S}^+ = \begin{pmatrix}0 & \hbar\\ 0 & 0\end{pmatrix} = \hbar\begin{pmatrix}0 & 1\\ 0 & 0\end{pmatrix} \quad\text{and}\quad \mathrm{S}^- = \begin{pmatrix}0 & 0\\ \hbar & 0\end{pmatrix} = \hbar\begin{pmatrix}0 & 0\\ 1 & 0\end{pmatrix}.$$

The matrix of the operator $\hat{S}_x$ is that of $(\hat{S}^+ + \hat{S}^-)/2$ [see (2.26)]:

$$\mathrm{S}_x = \tfrac{1}{2}\hbar\begin{pmatrix}0 & 1\\ 1 & 0\end{pmatrix}.$$

Similarly, the matrix of the operator $\hat{S}_y$ is that of the operator $(\hat{S}^+ - \hat{S}^-)/2i$:

$$\mathrm{S}_y = \tfrac{1}{2}\hbar\begin{pmatrix}0 & -i\\ i & 0\end{pmatrix}.$$

Thus, we can write the vector-spin matrix **S** as

$$\mathbf{S} = \tfrac{1}{2}\hbar\boldsymbol{\sigma},$$

where the vector matrix $\boldsymbol{\sigma}$ is given by

$$\boldsymbol{\sigma} = \boldsymbol{x}\sigma_x + \boldsymbol{y}\sigma_y + \boldsymbol{z}\sigma_z,$$

in which

$$\sigma_x = \begin{pmatrix} 0 & 1 \\ 1 & 0 \end{pmatrix}, \quad \sigma_y = \begin{pmatrix} 0 & -i \\ i & 0 \end{pmatrix}, \quad \sigma_z = \begin{pmatrix} 1 & 0 \\ 0 & -1 \end{pmatrix} \tag{2.37}$$

are known as the **Pauli matrices**.

3

Approximation Methods

3.1 Time-independent perturbation theory for nondegenerate states

For only very few physical problems (e.g., the hydrogen-like atom and the harmonic oscillator) can the Schrödinger equation be solved exactly, and so methods of approximate solution must be used.

In some cases, the hamiltonian $\hat{H}$ of the problem differs only little from the hamiltonian $\hat{H}_0$ of an exactly solvable problem, so that $\hat{H}$ can be written as

$$\hat{H} = \hat{H}_0 + \lambda \hat{V},$$

where the extra term $\lambda \hat{V}$ is the "perturbation", λ being any **constant** that can be factored out of it. The method is to find formal expansions of the perturbed energies and perturbed wave functions in powers of the perturbation, and hence in powers of λ. Clearly, for this approach to be fruitful, the perturbation must be "small" in the sense that the resulting expansions converge, and converge fast enough to ensure that only the first few terms are needed to give results accurate enough for practical purposes.

Our problem is to solve the full time-independent Schrödinger equation

$$\hat{H}\phi_m(\boldsymbol{r}) \equiv (\hat{H}_0 + \lambda \hat{V})\phi_m(\boldsymbol{r}) = E_m \phi_m(\boldsymbol{r}) \tag{3.1}$$

for the "perturbed" stationary-state energies E_m and the corresponding energy eigenfunctions $\phi_m(\boldsymbol{r})$, given that we know the solution of the time-independent Schrödinger equation

$$\hat{H}_0 \phi_m^{(0)}(\boldsymbol{r}) = E_m^{(0)} \phi_m^{(0)}(\boldsymbol{r}) \tag{3.2}$$

for the "unperturbed" stationary-state energies $E_m^{(0)}$ and corresponding energy eigenfunctions $\phi_m^{(0)}(\boldsymbol{r})$.

The analogues of (3.1) and (3.2) in Dirac notation are

$$\hat{H}|m\rangle \equiv (\hat{H}_0 + \lambda \hat{V})|m\rangle = E_m |m\rangle \tag{3.3}$$

A Course in Theoretical Physics, First Edition. P. J. Shepherd.

and

$$\hat{H}_0|m\rangle^{(0)} = E_m^{(0)}|m\rangle^{(0)}, \tag{3.4}$$

that is, we denote the unperturbed kets by $|m\rangle^{(0)}$ and the corresponding perturbed kets by $|m\rangle$:

$$\phi_m^{(0)}(\boldsymbol{r}) \equiv \langle \boldsymbol{r}|m\rangle^{(0)}, \quad \phi_m(\boldsymbol{r}) \equiv \langle \boldsymbol{r}|m\rangle.$$

We assume that $|m\rangle$ can be expressed in powers of λ:

$$|m\rangle = |m\rangle^{(0)} + \lambda|m\rangle^{(1)} + \lambda^2|m\rangle^{(2)} + \cdots. \tag{3.5}$$

Clearly, if $\lambda = 0$ (no perturbation is present), then $|m\rangle = |m\rangle^{(0)}$ as required.

Analogously, for the energy we have the power series

$$E_m = E_m^{(0)} + \lambda E_m^{(1)} + \lambda^2 E_m^{(2)} + \cdots. \tag{3.6}$$

The aim of perturbation theory is to find the coefficients of the powers of λ in (3.5) and (3.6). If the perturbation is small enough for the series (3.5) and (3.6) to converge well, it will be sufficient to calculate just the terms proportional to λ and λ^2 in these series expansions.

Putting (3.5) and (3.6) into (3.3), we have

$$\begin{aligned}(\hat{H}_0 + \lambda\hat{V})&\left(|m\rangle^{(0)} + \lambda|m\rangle^{(1)} + \lambda^2|m\rangle^{(2)} + \cdots\right)\\ &= \left(E_m^{(0)} + \lambda E_m^{(1)} + \lambda^2 E_m^{(2)} + \cdots\right)\left(|m\rangle^{(0)} + \lambda|m\rangle^{(1)} + \lambda^2|m\rangle^{(2)} + \cdots\right).\end{aligned}$$

Since the choice of the constant λ is arbitrary, this equation must be true for all λ, and so we can extract all the equations we need by equating coefficients, first of λ^0 ($= 1$), then of λ^1 ($= \lambda$), then of λ^2, and so on.

Equating coefficients of λ^0, we get the Schrödinger equation of the unperturbed problem:

$$\hat{H}_0|m\rangle^{(0)} = E_m^{(0)}|m\rangle^{(0)},$$

that is,

$$\left(\hat{H}_0 - E_m^{(0)}\right)|m\rangle^{(0)} = 0. \tag{3.7}$$

Equating coefficients of λ^1, we have

$$\left(\hat{H}_0 - E_m^{(0)}\right)|m\rangle^{(1)} = \left(E_m^{(1)} - \hat{V}\right)|m\rangle^{(0)}. \tag{3.8}$$

Equating coefficients of λ^2, we have

$$\left(\hat{H}_0 - E_m^{(0)}\right)|m\rangle^{(2)} = \left(E_m^{(1)} - \hat{V}\right)|m\rangle^{(1)} + E_m^{(2)}|m\rangle^{(0)}, \tag{3.9}$$

while the terms proportional to λ^3 give

$$\left(\hat{H}_0 - E_m^{(0)}\right)|m\rangle^{(3)} = \left(E_m^{(1)} - \hat{V}\right)|m\rangle^{(2)} + E_m^{(2)}|m\rangle^{(1)} + E_m^{(3)}|m\rangle^{(0)}, \tag{3.10}$$

and so on.

We find the coefficient $E_m^{(1)}$ immediately by projecting both sides of (3.8) on to the bra ${}^{(0)}\langle m|$:

$$^{(0)}\langle m|\big(\hat{H}_0 - E_m^{(0)}\big)|m\rangle^{(1)} = {}^{(0)}\langle m|\big(E_m^{(1)} - \hat{V}\big)|m\rangle^{(0)}.$$

But, since $\hat{H}_0$ is self-adjoint (hermitian), that is, $\hat{H}_0{}^{\dagger} = \hat{H}_0$, the left-hand side of this can be written as

$$\big[{}^{(1)}\langle m|\big(\hat{H}_0 - E_m^{(0)}\big)|m\rangle^{(0)}\big]^*,$$

which is equal to zero, since

$$\big(\hat{H}_0 - E_m^{(0)}\big)|m\rangle^{(0)} = 0.$$

Thus,

$$E_m^{(1)}\,{}^{(0)}\langle m\mid m\rangle^{(0)} - {}^{(0)}\langle m|\hat{V}|m\rangle^{(0)} = 0,$$

or, since the kets $|m\rangle^{(0)}$ are normalized to unity,

$$E_m^{(1)} = {}^{(0)}\langle m|\hat{V}|m\rangle^{(0)}. \tag{3.11}$$

To obtain the first-order correction to the state vector $|m\rangle^{(0)}$ in (3.5), we expand the coefficient ket $|m\rangle^{(1)}$ as a sum over the unperturbed eigenkets $|n\rangle^{(0)}$ with $n \neq m$ [the contribution of the unperturbed eigenket $|n\rangle^{(0)} = |m\rangle^{(0)}$ to $|m\rangle$ is already taken into account in the first term in the expansion (3.5)]:

$$|m\rangle^{(1)} = \sum_{\substack{n \\ (\neq m)}} c_{mn}^{(1)}|n\rangle^{(0)}. \tag{3.12}$$

Putting this into equation (3.8) for $|m\rangle^{(1)}$ we find

$$\sum_{\substack{n \\ (\neq m)}} c_{mn}^{(1)}\big(\hat{H}_0 - E_m^{(0)}\big)\,|n\rangle^{(0)} = \big(E_m^{(1)} - \hat{V}\big)|m\rangle^{(0)},$$

and so

$$\sum_{\substack{n \\ (\neq m)}} c_{mn}^{(1)}\big(E_n^{(0)} - E_m^{(0)}\big)\,|n\rangle^{(0)} = \big(E_m^{(1)} - \hat{V}\big)|m\rangle^{(0)}.$$

To extract an expression for the coefficient $c_{mk}^{(1)}$ we project both sides of this equation on to the bra ${}^{(0)}\langle k|$ and use the orthonormality property

$$^{(0)}\langle k|n\rangle^{(0)} = \delta_{kn}.$$

Then

$$\sum_{\substack{n \\ (\neq m)}} c_{mn}^{(1)}\big(E_n^{(0)} - E_m^{(0)}\big)\,\delta_{kn} = {}^{(0)}\langle k|\big(E_m^{(1)} - \hat{V}\big)|m\rangle^{(0)} = -{}^{(0)}\langle k|\hat{V}|m\rangle^{(0)}$$

(since $k \neq m$), and, therefore,

$$c_{mk}^{(1)} = \frac{{}^{(0)}\langle k|\hat{V}|m\rangle^{(0)}}{E_m^{(0)} - E_k^{(0)}} \quad (k \neq m). \tag{3.13}$$

To find the second-order correction to the energy we first project equation (3.9) on to the bra ${}^{(0)}\langle m|$. Using the same procedure as before, that is, using the fact that $\hat{H}_0$ is self-adjoint and applying (3.7), we see that the left-hand side of the resulting equation vanishes, so that we obtain

$$0 = {}^{(0)}\langle m| \left(E_m^{(1)} - \hat{V}\right) |m\rangle^{(1)} + E_m^{(2)}\,{}^{(0)}\langle m|m\rangle^{(0)}.$$

Therefore, expressing $|m\rangle^{(1)}$ by means of (3.12) with (3.13), we have

$$\begin{aligned} E_m^{(2)} &= {}^{(0)}\langle m|\left(\hat{V} - E_m^{(1)}\right) \sum_{\substack{k \\ (\neq m)}} c_{mk}^{(1)}|k\rangle^{(0)} = \sum_{\substack{k \\ (\neq m)}} {}^{(0)}\langle m|\hat{V}|k\rangle^{(0)} c_{mk}^{(1)} \\ &= \sum_{\substack{k \\ (\neq m)}} \frac{{}^{(0)}\langle m|\hat{V}|k\rangle^{(0)(0)}\langle k|\hat{V}|m\rangle^{(0)}}{E_m^{(0)} - E_k^{(0)}}. \end{aligned}$$

Alternatively, since $\hat{V}$ is self-adjoint, that is, since

$${}^{(0)}\langle k|\hat{V}|m\rangle^{(0)} = \left[{}^{(0)}\langle m|\hat{V}|k\rangle^{(0)}\right]^*,$$

we have

$$E_m^{(2)} = \sum_{\substack{k \\ (\neq m)}} \frac{|{}^{(0)}\langle m|\hat{V}|k\rangle^{(0)}|^2}{E_m^{(0)} - E_k^{(0)}}. \tag{3.14}$$

Substituting (3.12) with (3.13) into the perturbation expansion (3.5) for the energy eigenket $|m\rangle$, and the results (3.11) and (3.14) into the perturbation expansion (3.6) for the energy E_m, we have

$$|m\rangle = |m\rangle^{(0)} + \lambda \sum_{\substack{k \\ (\neq m)}} \frac{{}^{(0)}\langle k|\hat{V}|m\rangle^{(0)}}{E_m^{(0)} - E_k^{(0)}} |k\rangle^{(0)} + \cdots, \tag{3.15}$$

$$E_m = E_m^{(0)} + \lambda\,{}^{(0)}\langle m|\hat{V}|m\rangle^{(0)} + \lambda^2 \sum_{\substack{k \\ (\neq m)}} \frac{|{}^{(0)}\langle m|\hat{V}|k\rangle^{(0)}|^2}{E_m^{(0)} - E_k^{(0)}} + \cdots. \tag{3.16}$$

The corresponding expressions in terms of wave functions and local operators are

$$\phi_m(\boldsymbol{r}) = \phi_m^{(0)}(\boldsymbol{r}) + \lambda \sum_{\substack{k \\ (\neq m)}} \left[\frac{\int \phi_k^{(0)\,*}(\boldsymbol{r}')V(\boldsymbol{r}')\phi_m^{(0)}(\boldsymbol{r}')\mathrm{d}^3\boldsymbol{r}'}{E_m^{(0)} - E_k^{(0)}}\right] \phi_k^{(0)}(\boldsymbol{r}) + \cdots, \tag{3.17}$$

$$E_m = E_m^{(0)} + \lambda \int \phi_m^{(0)\,*}(\boldsymbol{r})V(\boldsymbol{r})\phi_m^{(0)}(\boldsymbol{r})\mathrm{d}^3\boldsymbol{r} + \lambda^2 \sum_{\substack{k \\ (\neq m)}} \frac{\left|\int \phi_k^{(0)\,*}(\boldsymbol{r})V(\boldsymbol{r})\phi_m^{(0)}(\boldsymbol{r})\mathrm{d}^3\boldsymbol{r}\right|^2}{E_m^{(0)} - E_k^{(0)}} + \cdots. \tag{3.18}$$

As an example, we shall consider a hydrogen atom in its ground state in a uniform electric field $\boldsymbol{\mathcal{E}}$ pointing along the z axis. The full hamiltonian of the relative motion of the electron–proton system is

$$\hat{H} = \hat{H}_0 + \lambda\hat{V}.$$

Here, the unperturbed hamiltonian is

$$\hat{H}_0 = -\frac{\hbar^2}{2\mu}\nabla^2 - \frac{e^2}{4\pi\varepsilon_0 r},$$

where $r = |\boldsymbol{r}| = |\boldsymbol{r}_\mathrm{e} - \boldsymbol{r}_\mathrm{p}|$ and μ is the reduced mass of the electron–proton system (see section 1.17). The perturbation $\lambda\hat{V}$, equal to the potential energy $-\boldsymbol{\mathcal{E}}\cdot\boldsymbol{d}$ of the interaction of the electron–proton dipole moment $\boldsymbol{d} = e(\boldsymbol{r}_\mathrm{p} - \boldsymbol{r}_\mathrm{e}) \equiv -e\boldsymbol{r}$ (e is the proton charge) with the electric field of intensity $\boldsymbol{\mathcal{E}}$, is given by

$$\lambda\hat{V} = -\boldsymbol{\mathcal{E}}\cdot\boldsymbol{d} = e\boldsymbol{\mathcal{E}}\cdot\boldsymbol{r} = e\mathcal{E}z,$$

where z is the component of $\boldsymbol{r} = \boldsymbol{r}_\mathrm{e} - \boldsymbol{r}_\mathrm{p}$ along the direction of the field (the z direction). Since the field is assumed to be uniform, that is, constant in space, we treat $e\mathcal{E}$ as λ, and z as $\hat{V}$.

The perturbed wave function $\phi_{1s}(\boldsymbol{r})$ and perturbed energy E_{1s} corresponding to the unperturbed ground-state wave function $\phi_{1s}^{(0)}(\boldsymbol{r})$ and unperturbed energy $E_{1s}^{(0)}$ are given by (3.17) and (3.18) with $m \equiv 1s$, $\lambda \equiv e\mathcal{E}$, and $\hat{V} \equiv \hat{z} = z$.

The resulting equations are

$$\phi_{1s}(\boldsymbol{r}) = \phi_{1s}^{(0)}(\boldsymbol{r}) + e\mathcal{E}\sum_{\substack{k\\(\neq 1s)}}\left[\frac{\int \phi_k^{(0)*}(\boldsymbol{r}')z'\phi_{1s}^{(0)}(\boldsymbol{r}')\mathrm{d}^3\boldsymbol{r}'}{E_{1s}^{(0)} - E_k^{(0)}}\right]\phi_k^{(0)}(\boldsymbol{r}) + \cdots, \tag{3.19}$$

$$E_{1s} = E_{1s}^{(0)} + e\mathcal{E}\int \phi_{1s}^{(0)*}z\phi_{1s}^{(0)}\mathrm{d}^3\boldsymbol{r} + (e\mathcal{E})^2\sum_{\substack{k\\(\neq 1s)}}\frac{\left|\int \phi_k^{(0)*}z\phi_{1s}^{(0)}\mathrm{d}^3\boldsymbol{r}\right|^2}{E_{1s}^{(0)} - E_k^{(0)}} + \cdots, \tag{3.20}$$

where the sums over k are over all but one ($1s$) of a complete set of unperturbed hydrogen-atom states, for example, $2s$, $2p_x$, $2p_y$, $2p_z$, $3s$, $3p_x$, $3p_y$, $3p_z$, $3d_{xy}$, $3d_{yz}$, $3d_{zx}$, $3d_{x^2-y^2}$, $3d_{3z^2-r^2}$, and so on.

Because, as shown in section 1.17, $\phi_{np_x}^{(0)}(\boldsymbol{r})$ can be expressed as x times a function of $r \equiv |\boldsymbol{r}|$ only, while $\phi_{nd_{xy}}^{(0)}(\boldsymbol{r})$ is xy times a function of r only, and so on, it is clear that, except for the terms with $k = 2p_z, 3p_z, 4p_z$, and so on, every integral in (3.19) and (3.20) is equal to zero by symmetry.

Thus, the effect of an electric field along the z direction is to introduce into the original, unperturbed ground-state wave function $\phi_{1s}^{(0)}(\boldsymbol{r})$ an **admixture** of the unperturbed wave functions $\phi_{2p_z}^{(0)}(\boldsymbol{r})$, $\phi_{3p_z}^{(0)}(\boldsymbol{r})$, $\phi_{4p_z}^{(0)}(\boldsymbol{r})$, and so on.

This result is entirely to be expected on physical grounds. In an electric field in the z direction the electron and proton will be pulled in opposite directions along the z axis, and the electron-charge distribution around the nucleus will lose its spherical symmetry and acquire cylindrical symmetry, as shown schematically in figure 3.1.

Since the electron-charge density is proportional to the ground-state probability density $|\phi_{1s}(\boldsymbol{r})|^2$, this distortion of the charge density must result from an admixture of unperturbed hydrogen-atom wave functions that have cylindrical symmetry about the z axis (and only such functions), that is, p_z functions. For example, the

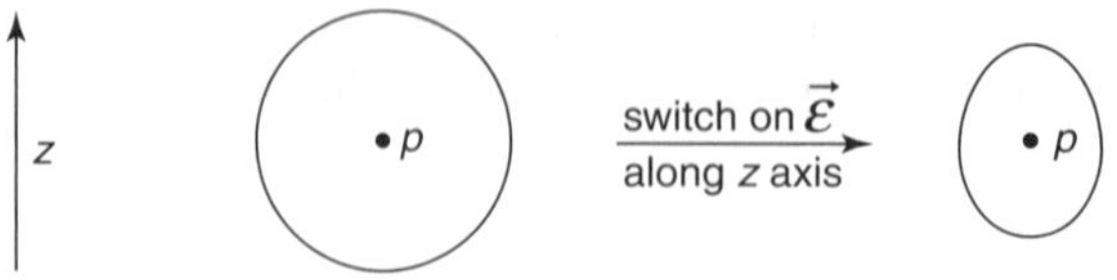

Figure 3.1 *Effect (schematic) of an electric field on the electron charge distribution of a hydrogen atom originally in the 1s state.*

superposition $\phi_{1s}^{(0)}(\boldsymbol{r}) + a\phi_{2p_z}^{(0)}(\boldsymbol{r})$ corresponding to (3.19) with just the largest first-order correction [the coefficient a here is negative, as can be seen by inspection of (3.19)] can be depicted schematically as in figure 3.2, in which the "contours" represent surfaces of constant modulus of the wave function (see section 1.17).

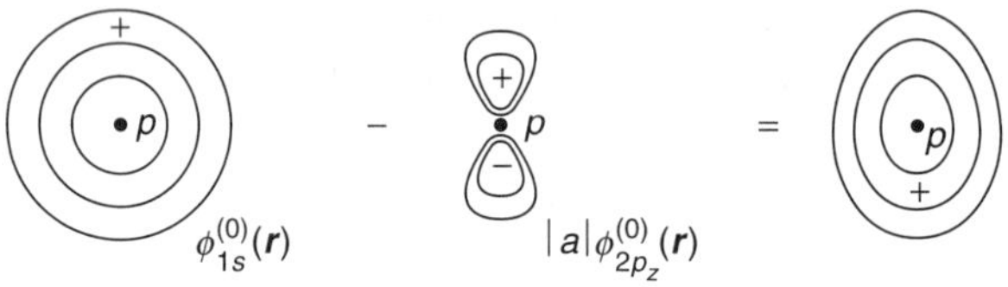

Figure 3.2 *Effect (schematic) of a small ($|a| << 1$) admixture of $\phi_{2p_z}^{(0)}(\boldsymbol{r})$ into $\phi_{1s}^{(0)}(\boldsymbol{r})$.*

The integrand in the first-order correction to the **energy** in (3.20) is clearly odd in z, and so this integral vanishes. The leading term in the energy shift in the electric field is thus the second-order term, with only the np_z states contributing to the sum over k.

Because the energy denominators in (3.19) and (3.20) get larger as the principal quantum number n of the states $\phi_{np_z}^{(0)}(\boldsymbol{r})$ increases, we may expect that the greatest contribution to the corrections to the wave function and energy of the ground state of the hydrogen atom will be made by the term with $k = 2p_z$.

This shift in the energy of a hydrogen atom in an electric field is called the **Stark effect**. The absence of an energy shift proportional to the electric field intensity $\mathcal{E}$ is not surprising. A molecule with a permanent dipole moment $\boldsymbol{d}$ would have an energy term proportional to $\mathcal{E}$ (namely, $-\boldsymbol{d} \cdot \mathcal{E}$) in an electric field. A free hydrogen atom is spherically symmetric and so does not have a permanent dipole moment. Consequently, there is no first-order energy shift. The term proportional to $\mathcal{E}^2$ (the second-order term $\lambda^2 E_{1s}^{(2)}$) in the perturbed energy of the hydrogen atom in a field $\boldsymbol{\mathcal{E}}$ may be regarded as the energy of interaction of the **induced dipole** $\boldsymbol{d} = \alpha\boldsymbol{\mathcal{E}}$ of the atom with the field $\boldsymbol{\mathcal{E}}$ (α is the **polarizability** of the atom).

3.2 Time-independent perturbation theory for degenerate states

The derivation of equations (3.15) and (3.16) for the perturbed energy eigenket $|m\rangle$ and the corresponding perturbed energy eigenvalue E_m required the assumption (in the derivation of (3.13) for $c_{mk}^{(1)}$ for all $k \neq m$ from the equation preceding it) that $E_n^{(0)} - E_m^{(0)} \neq 0$ for all $n \neq m$, that is, that the unperturbed energy eigenket $|m\rangle^{(0)}$ is **nondegenerate**. We now consider the problem of the calculation of the perturbed energy eigenkets and energy eigenvalues in the case when the corresponding unperturbed energy eigenkets correspond to the same energy, that is, are **degenerate**.

Consider, for example, the $n = 2$ eigenkets $|2s\rangle^{(0)}$, $|2p_x\rangle^{(0)}$, $|2p_y\rangle^{(0)}$ and $|2p_z\rangle^{(0)}$ of the unperturbed hamiltonian $\hat{H}_0$ of the relative motion of the electron–proton system (the hydrogen atom). All four eigenkets correspond to the same eigenvalue $E_2^{(0)}$, that is, they are degenerate. Moreover, any linear combination of these four eigenkets is also an eigenket of $\hat{H}_0$, with precisely the same energy $E_2^{(0)}$.

However, when a general perturbation $\lambda\hat{V}$ is added to $\hat{H}_0$ the structure of the expansions in λ of the four eigenkets of $\hat{H} = \hat{H}_0 + \lambda\hat{V}$ that correspond to the four $n = 2$ eigenkets of $\hat{H}_0$ is such that, if the limit $\lambda \to 0$ is taken, what remains is a set of four **specific**, mutually orthogonal, linear combinations of the four unperturbed $n = 2$ eigenkets.

Moreover, the four "$n = 2$ type" eigenkets of $\hat{H}$ correspond, in general, to **different** eigenvalues of $\hat{H}$, that is, the perturbation "lifts" (i.e., removes) the degeneracy.

For simplicity, and to illustrate the general method, we consider the first-order energy shifts and the structures of the perturbed energy eigenkets for a level that has **two-fold** degeneracy in the absence of the perturbation.

Two linearly independent (and mutually orthogonal) kets $|m_1\rangle^{(0)}$ and $|m_2\rangle^{(0)}$ are eigenkets of $\hat{H}_0$ with the same eigenvalue $E_m^{(0)}$:

$$\begin{aligned} \hat{H}_0|m_1\rangle^{(0)} &= E_m^{(0)}|m_1\rangle^{(0)}, \\ \hat{H}_0|m_2\rangle^{(0)} &= E_m^{(0)}|m_2\rangle^{(0)}, \end{aligned} \tag{3.21}$$

and, hence, so too is an arbitrary superposition

$$|m\rangle^{(0)} = a_1|m_1\rangle^{(0)} + a_2|m_2\rangle^{(0)} \tag{3.22}$$

of these two eigenkets. Here, to ensure that the ket (3.22) is normalized, we must have

$$|a_1|^2 + |a_2|^2 = 1. \tag{3.23}$$

We now put equation (3.22) into the equation (3.8) derived earlier by equating coefficients of λ^1 in the time-independent Schrödinger equation (3.3) containing the expansions (3.5) and (3.6).

The result is

$$\left(\hat{H}_0 - E_m^{(0)}\right)|m\rangle^{(1)} = \left(E_m^{(1)} - \hat{V}\right)\left(a_1|m_1\rangle^{(0)} + a_2|m_2\rangle^{(0)}\right).$$

Projecting this first on to ${}^{(0)}\langle m_1|$ and secondly on to ${}^{(0)}\langle m_2|$, we find that in both cases the left-hand sides vanish for the same reasons as before [the self-adjointness of $\hat{H}_0$, followed by the use of (3.21)], and, since $|m_1\rangle^{(0)}$ and $|m_2\rangle^{(0)}$ are orthonormal, we obtain

$$\begin{aligned} 0 &= \left(E_m^{(1)} - {}^{(0)}\langle m_1|\hat{V}|m_1\rangle^{(0)}\right)a_1 - {}^{(0)}\langle m_1|\hat{V}|m_2\rangle^{(0)}a_2, \\ 0 &= -{}^{(0)}\langle m_2|\hat{V}|m_1\rangle^{(0)}a_1 + \left(E_m^{(1)} - {}^{(0)}\langle m_2|\hat{V}|m_2\rangle^{(0)}\right)a_2. \end{aligned} \tag{3.24}$$

In matrix form, this reads

$$\begin{pmatrix} {}^{(0)}\langle m_1|\hat{V}|m_1\rangle^{(0)} - E_m^{(1)} & {}^{(0)}\langle m_1|\hat{V}|m_2\rangle^{(0)} \\ {}^{(0)}\langle m_2|\hat{V}|m_1\rangle^{(0)} & {}^{(0)}\langle m_2|\hat{V}|m_2\rangle^{(0)} - E_m^{(1)} \end{pmatrix}\begin{pmatrix} a_1 \\ a_2 \end{pmatrix} = \begin{pmatrix} 0 \\ 0 \end{pmatrix}$$

(a secular equation). This equation has the trivial solution $a_1 = a_2 = 0$, which is of no physical interest. It has a nontrivial solution for a_1 and a_2 only if

$$\det\begin{pmatrix} {}^{(0)}\langle m_1|\hat{V}|m_1\rangle^{(0)} - E_m^{(1)} & {}^{(0)}\langle m_1|\hat{V}|m_2\rangle^{(0)} \\ {}^{(0)}\langle m_2|\hat{V}|m_1\rangle^{(0)} & {}^{(0)}\langle m_2|\hat{V}|m_2\rangle^{(0)} - E_m^{(1)} \end{pmatrix} = 0,$$

which is a quadratic equation for $E_m^{(1)}$.

Denote the two solutions of this quadratic equation by $E_{m_\alpha}^{(1)}$ and $E_{m_\beta}^{(1)}$.

Substitute the first of these into the pair of equations (3.24) and solve the resulting equations [together with (3.23)] for the coefficients a_1 and a_2. Call the results a_1^α and a_2^α, and call the corresponding unperturbed linear superposition (3.22) $|m_\alpha\rangle^{(0)}$. In other words, the unperturbed superposition "corresponding" to the first-order energy shift $E_{m_\alpha}^{(1)}$ has the form

$$|m_\alpha\rangle^{(0)} = a_1^\alpha |m_1\rangle^{(0)} + a_2^\alpha\, |m_2\rangle^{(0)}\,. \tag{3.25}$$

The unperturbed superposition "corresponding" to the first-order energy shift $E_{m_\beta}^{(1)}$ is found similarly:

$$|m_\beta\rangle^{(0)} = a_1^\beta |m_1\rangle^{(0)} + a_2^\beta\, |m_2\rangle^{(0)}\,. \tag{3.26}$$

The perturbed eigenkets that have (3.25) and (3.26) as their limits (when $\lambda \to 0$) can be written, to first order in λ, as

$$|m_\alpha\rangle = |m_\alpha\rangle^{(0)} + \lambda |m_\alpha\rangle^{(1)} \tag{3.27}$$

and

$$|m_\beta\rangle = |m_\beta\rangle^{(0)} + \lambda |m_\beta\rangle^{(1)}. \tag{3.28}$$

We find the coefficient ket $|m_\alpha\rangle^{(1)}$ in (3.27) in the usual way, by first expanding it in the unperturbed eigenkets $|n\rangle^{(0)}$ of $\hat{H}_0$:

$$|m_\alpha\rangle^{(1)} = \sum_n c_{m_\alpha n}^{(1)}\, |n\rangle^{(0)},$$

and putting this expansion into (3.8) with $E_m^{(1)} = E_{m_\alpha}^{(1)}$ and $|m\rangle^{(0)} = |m_\alpha\rangle^{(0)}$. We then find

$$\sum_n c_{m_\alpha n}^{(1)} \left(\hat{H}_0 - E_m^{(0)}\right) |n\rangle^{(0)} = \left(E_{m_\alpha}^{(1)} - \hat{V}\right) |m_\alpha\rangle^{(0)}.$$

The two terms in the left-hand side with $n = m_1$ and $n = m_2$ clearly vanish (since the kets $|m_1\rangle^{(0)}$ and $|m_2\rangle^{(0)}$ are eigenkets of $\hat{H}_0$ with eigenvalue $E_m^{(0)}$), and so may be excluded from the sum.

The procedure is then as before, and we find that $|m_\alpha\rangle^{(1)}$ is given by the expression

$$|m_\alpha\rangle^{(1)} = \sum_{\substack{k \\ (\neq m_1, m_2)}} \frac{{}^{(0)}\langle k|\hat{V}|m_\alpha\rangle^{(0)}}{E_m^{(0)} - E_k^{(0)}}\, |k\rangle^{(0)}\,, \tag{3.29}$$

in which $|m_\alpha\rangle^{(0)}$ is given by (3.25). Similarly, $|m_\beta\rangle^{(1)}$ in (3.28) is given by

$$|m_\beta\rangle^{(1)} = \sum_{\substack{k \\ (\neq m_1, m_2)}} \frac{{}^{(0)}\langle k|\hat{V}|m_\beta\rangle^{(0)}}{E_m^{(0)} - E_k^{(0)}}\, |k\rangle^{(0)}\,, \tag{3.30}$$

in which $|m_\beta\rangle^{(0)}$ is given by (3.26).

If this procedure does not lift the degeneracy, that is, if it turns out that $E^{(1)}_{m_\alpha} = E^{(1)}_{m_\beta}$, the coefficients a_1 and a_2 in (3.22) cannot be determined from the secular equation derived from (3.8) and one must try substituting the linear superposition (3.22) into (3.9) instead, that is, into

$$\left(\hat{H}_0 - E^{(0)}_m\right)|m\rangle^{(2)} = \left(E^{(1)}_m - \hat{V}\right)|m\rangle^{(1)} + E^{(2)}_m|m\rangle^{(0)}. \tag{3.9}$$

For $|m\rangle^{(1)}$ here we use

$$|m\rangle^{(1)} = \sum_{\substack{k \\ (\neq m_1, m_2)}} \frac{^{(0)}\langle k|\hat{V}|m\rangle^{(0)}}{E^{(0)}_m - E^{(0)}_k}|k\rangle^{(0)},$$

which is the general form exemplified by (3.29) and (3.30). We then apply the same method as before (projecting (3.9) in turn on to $^{(0)}\langle m_1|$ and $^{(0)}\langle m_2|$ and showing that the left-hand sides of the resulting equations vanish) and again obtain a secular equation, this time with a determinant that yields a quadratic equation for $E^{(2)}_m$. If this equation has two distinct roots $E^{(2)}_{m_\alpha}$ and $E^{(2)}_{m_\beta}$ we can solve the secular equation for a_1 and a_2 for each choice of root, giving, via (3.22), two distinct linear combinations $|m_\alpha\rangle^{(0)}$ and $|m_\beta\rangle^{(0)}$ of the original unperturbed eigenkets $|m_1\rangle^{(0)}$ and $|m_2\rangle^{(0)}$.

If the degeneracy is not lifted in second-order perturbation theory, one may repeat the above procedure starting from equation (3.10) instead of (3.9), and so on until the perturbation lifts the degeneracy (if this happens at all).

As an example, we consider the Stark effect for the $n = 2$ states of the hydrogen atom. Let the applied uniform electric field $\boldsymbol{\mathcal{E}}$ be in the z direction, so that the perturbation is

$$\lambda\hat{V} = e\mathcal{E}\hat{z}$$

with $\lambda \equiv e\mathcal{E}$ and $\hat{V} \equiv \hat{z}$.

In the absence of the field, the four kets $|2s\rangle^{(0)}$, $|2p_x\rangle^{(0)}$, $|2p_y\rangle^{(0)}$ and $|2p_z\rangle^{(0)}$ are degenerate eigenkets of $\hat{H}_0$, with common eigenvalue $E^{(0)}_2$. In the presence of the field, the coefficient $E^{(1)}_2$ of λ ($= e\mathcal{E}$) in the perturbation expansion for the energy E_2 is found by solving the following quartic determinantal equation for $E^{(1)}_2$:

$$\begin{vmatrix} ^{(0)}\langle 2s|\hat{z}|2s\rangle^{(0)} - E^{(1)}_2 & ^{(0)}\langle 2s|\hat{z}|2p_x\rangle^{(0)} & ^{(0)}\langle 2s|\hat{z}|2p_y\rangle^{(0)} & ^{(0)}\langle 2s|\hat{z}|2p_z\rangle^{(0)} \\ ^{(0)}\langle 2p_x|\hat{z}|2s\rangle^{(0)} & ^{(0)}\langle 2p_x|\hat{z}|2p_x\rangle^{(0)} - E^{(1)}_2 & ^{(0)}\langle 2p_x|\hat{z}|2p_y\rangle^{(0)} & ^{(0)}\langle 2p_x|\hat{z}|2p_z\rangle^{(0)} \\ ^{(0)}\langle 2p_y|\hat{z}|2s\rangle^{(0)} & ^{(0)}\langle 2p_y|\hat{z}|2p_x\rangle^{(0)} & ^{(0)}\langle 2p_y|\hat{z}|2p_y\rangle^{(0)} - E^{(1)}_2 & ^{(0)}\langle 2p_y|\hat{z}|2p_z\rangle^{(0)} \\ ^{(0)}\langle 2p_z|\hat{z}|2s\rangle^{(0)} & ^{(0)}\langle 2p_z|\hat{z}|2p_x\rangle^{(0)} & ^{(0)}\langle 2p_z|\hat{z}|2p_y\rangle^{(0)} & ^{(0)}\langle 2p_z|\hat{z}|2p_z\rangle^{(0)} - E^{(1)}_2 \end{vmatrix} = 0$$

All matrix elements in this determinant vanish by symmetry, except for the elements $^{(0)}\langle 2s|\hat{z}\,|2p_z\rangle^{(0)}$ and $^{(0)}\langle 2p_z|\hat{z}\,|2s\rangle^{(0)}$ (which are mutually complex-conjugate):

$$^{(0)}\langle 2s|\hat{z}\,|2p_z\rangle^{(0)} \equiv \int \phi^{(0)*}_{2s}(\boldsymbol{r})z\phi^{(0)}_{2p_z}(\boldsymbol{r})\mathrm{d}^3\boldsymbol{r}.$$

To evaluate this element we use the wave functions

$$\phi^{(0)}_{2s}(\boldsymbol{r}) = \frac{1}{\sqrt{4\pi}}\left(\frac{1}{2a_0}\right)^{3/2}\left(2 - \frac{r}{a_0}\right)\mathrm{e}^{-r/2a_0}$$

and

$$\phi^{(0)}_{2p_z}(\boldsymbol{r}) = \frac{1}{\sqrt{4\pi}}\left(\frac{1}{2a_0}\right)^{3/2}\frac{r\cos\theta}{a_0}\mathrm{e}^{-r/2a_0},$$

and set $z = r\cos\theta$. The above integral then becomes:

$$\begin{aligned}
{}^{(0)}\langle 2s|\hat{z}\,|2p_z\rangle^{(0)} &\equiv \int \phi^{(0)*}_{2s}(\boldsymbol{r}) r\cos\theta\phi^{(0)}_{2p_z}(\boldsymbol{r})\mathrm{d}^3\boldsymbol{r} \\
&= \frac{1}{4\pi}\cdot\frac{1}{8a_0^4}\int_0^\infty r^2\mathrm{d}r\int_0^\pi \sin\theta\mathrm{d}\theta\int_0^{2\pi}\mathrm{d}\varphi\left(2-\frac{r}{a_0}\right)\mathrm{e}^{-r/a_0}r^2\cos^2\theta \\
&= \frac{1}{32\pi a_0^4}\int_0^\infty r^4\left(2-\frac{r}{a_0}\right)\mathrm{e}^{-r/a_0}\mathrm{d}r\int_0^\pi \sin\theta\cos^2\theta\mathrm{d}\theta.2\pi \\
&= \frac{1}{16a_0^4}\cdot a_0^5\int_0^\infty u^4(2-u)\mathrm{e}^{-u}\mathrm{d}u\int_{-1}^{1} w^2\mathrm{d}w \\
&= \frac{a_0}{16}\cdot\frac{2}{3}\left[2\int_0^\infty u^4\mathrm{e}^{-u}\mathrm{d}u - \int_0^\infty u^5\mathrm{e}^{-u}\mathrm{d}u\right] = \frac{a_0}{24}\left[(2\times 4!) - 5!\right] = -3a_0.
\end{aligned} \tag{3.31}$$

The determinantal equation then reads

$$\begin{vmatrix} -E_2^{(1)} & 0 & 0 & -3a_0 \\ 0 & -E_2^{(1)} & 0 & 0 \\ 0 & 0 & -E_2^{(1)} & 0 \\ -3a_0 & 0 & 0 & -E_2^{(1)} \end{vmatrix} = 0,$$

which has four roots

$$\begin{array}{lcccc} E_2^{(1)} = & 0, & 0, & -3a_0, & 3a_0 \\ & (E_{2\alpha}^{(1)}) & (E_{2\beta}^{(1)}) & (E_{2\gamma}^{(1)}) & (E_{2\delta}^{(1)}) \end{array}$$

that is, the first-order energy shifts are

$$\lambda E_2^{(1)} \equiv e\mathcal{E}E_2^{(1)} = 0, \quad 0, \quad -3e\mathcal{E}a_0, \quad 3e\mathcal{E}a_0\,.$$

The superposition of $n = 2$ kets that corresponds to the first of these roots is obtained by solving the secular equation

$$\begin{pmatrix} -E_2^{(1)} & 0 & 0 & -3a_0 \\ 0 & -E_2^{(1)} & 0 & 0 \\ 0 & 0 & -E_2^{(1)} & 0 \\ -3a_0 & 0 & 0 & -E_2^{(1)} \end{pmatrix}\begin{pmatrix} a_{2s} \\ a_{2p_x} \\ a_{2p_y} \\ a_{2p_z} \end{pmatrix} = 0$$

with $E_2^{(1)} = E_{2_\alpha}^{(1)} = 0$.

The result is

$$-3a_0 a^{\alpha}_{2p_z} = 0, \quad -3a_0 a^{\alpha}_{2s} = 0;$$

that is,

$$a^{\alpha}_{2p_z} = a^{\alpha}_{2s} = 0,$$

and so

$$|2_{\alpha}\rangle^{(0)} = a^{\alpha}_{2p_x}|2p_x\rangle^{(0)} + a^{\alpha}_{2p_y}|2p_y\rangle^{(0)},$$

with undetermined coefficients $a^{\alpha}_{2p_x}$ and $a^{\alpha}_{2p_y}$.

Similarly, for the second (β) root we find

$$|2_{\beta}\rangle^{(0)} = a^{\beta}_{2p_x}|2p_x\rangle^{(0)} + a^{\beta}_{2p_y}|2p_y\rangle^{(0)}.$$

The secular equation with the third root

$$E^{(1)}_2 = E^{(1)}_{2_{\gamma}} = -3a_0$$

gives

$$3a_0 a^{\gamma}_{2s} - 3a_0 a^{\gamma}_{2p_z} = 0, \quad 3a_0 a^{\gamma}_{2p_x} = 0, \quad 3a_0 a^{\gamma}_{2p_y} = 0;$$

that is,

$$a^{\gamma}_{2s} = a^{\gamma}_{2p_z} \quad \text{and} \quad a^{\gamma}_{2p_x} = a^{\gamma}_{2p_y} = 0.$$

Thus, the superposition of unperturbed kets that corresponds to the third root is

$$|2_{\gamma}\rangle^{(0)} = \frac{1}{\sqrt{2}}\left(|2s\rangle^{(0)} + |2p_z\rangle^{(0)}\right),$$

where the factor $1/\sqrt{2}$ ensures that $|2_{\gamma}\rangle^{(0)}$ is normalized.

Similarly, the superposition of unperturbed kets that corresponds to the fourth root is

$$|2_{\delta}\rangle^{(0)} = \frac{1}{\sqrt{2}}\left(|2s\rangle^{(0)} - |2p_z\rangle^{(0)}\right).$$

Thus, the wave functions corresponding to the third and fourth roots of the secular equation are

$$\phi^{(0)}_{2_{\gamma}}(\boldsymbol{r}) = \frac{1}{\sqrt{2}}\left(\phi^{(0)}_{2s}(\boldsymbol{r}) + \phi^{(0)}_{2p_z}(\boldsymbol{r})\right) \tag{3.32}$$

and

$$\phi_{2_\delta}^{(0)}(\boldsymbol{r}) = \frac{1}{\sqrt{2}}\left(\phi_{2s}^{(0)}(\boldsymbol{r}) - \phi_{2p_z}^{(0)}(\boldsymbol{r})\right), \tag{3.33}$$

with equal and opposite energy shifts $-3e\mathcal{E}a_0$ and $3e\mathcal{E}a_0$, respectively, in the field $\boldsymbol{\mathcal{E}}$. These energy shifts may be interpreted as shifts $-\boldsymbol{\mathcal{E}}\cdot\langle\boldsymbol{d}\rangle_\phi$ arising from the interaction of the field $\boldsymbol{\mathcal{E}}$ with the expectation value (in the appropriate state ϕ) of the dipole moment

$$\boldsymbol{d} = e(\boldsymbol{r}_\mathrm{p} - \boldsymbol{r}_\mathrm{e}) = -e(\boldsymbol{r}_\mathrm{e} - \boldsymbol{r}_\mathrm{p}) \equiv -e\boldsymbol{r} = -e(\boldsymbol{i}x + \boldsymbol{j}y + \boldsymbol{k}z)$$

(recall that e is the **proton** charge). Indeed, the expectation values of both x and y in both the states (3.32) and (3.33) are zero, as can easily be seen from the parity properties of the appropriate integrands, while the expectation value of z in the state (3.32) is calculated as follows:

$$\langle z\rangle_{\phi_{2_\gamma}^{(0)}} = \int \phi_{2_\gamma}^{(0)*}(\boldsymbol{r})z\phi_{2_\gamma}^{(0)}(\boldsymbol{r})\,\mathrm{d}^3\boldsymbol{r} = \frac{1}{2}\int\left(\phi_{2s}^{(0)}(\boldsymbol{r}) + \phi_{2p_z}^{(0)}(\boldsymbol{r})\right)^* z\left(\phi_{2s}^{(0)}(\boldsymbol{r}) + \phi_{2p_z}^{(0)}(\boldsymbol{r})\right)\mathrm{d}^3\boldsymbol{r}.$$

Here,

$$\int \phi_{2s}^{(0)*}(\boldsymbol{r})z\phi_{2s}^{(0)}(\boldsymbol{r})\mathrm{d}^3\boldsymbol{r} = 0 \text{ and } \int \phi_{2p_z}^{(0)*}(\boldsymbol{r})z\phi_{2p_z}^{(0)}(\boldsymbol{r})\mathrm{d}^3\boldsymbol{r} = 0$$

by symmetry, while from (3.31) we have

$$\int \phi_{2s}^{(0)*}(\boldsymbol{r})z\phi_{2p_z}^{(0)}(\boldsymbol{r})\mathrm{d}^3\boldsymbol{r} = \int \phi_{2p_z}^{(0)*}(\boldsymbol{r})z\phi_{2s}^{(0)}(\boldsymbol{r})\mathrm{d}^3\boldsymbol{r} = -3a_0.$$

Thus,

$$\langle\boldsymbol{d}\rangle_{\phi_{2_\gamma}} = -e\boldsymbol{k}\langle z\rangle_{\phi_{2_\gamma}} = 3\boldsymbol{k}ea_0,$$

and the energy $-\boldsymbol{\mathcal{E}}\cdot\langle\boldsymbol{d}\rangle_\phi$ of interaction of this dipole moment with the field $\boldsymbol{\mathcal{E}} = \mathcal{E}\boldsymbol{k}$ is equal to $-3e\mathcal{E}a_0$, as stated above. Repeating the above arguments with the wave function $\phi_{2_\delta}^{(0)}(\boldsymbol{r})$ (3.33) instead of $\phi_{2_\gamma}^{(0)}(\boldsymbol{r})$ (3.32), we find that the expectation value of the dipole moment has the same magnitude as before but the opposite sign, so that there is a corresponding equal but opposite energy shift in the electric field.

3.3 The variational method

Perturbation theory is useful if there is a closely related problem that has been solved exactly. If this is not so, the variational method, which is especially useful for finding the ground-state energy and ground-state wave function of a system, may be used.

According to the variational principle (proved below), the ground-state energy E_0 of a system with hamiltonian $\hat{H}$ satisfies

$$E_0 \leq \int \phi^*(\boldsymbol{r})\hat{H}\phi(\boldsymbol{r})\mathrm{d}^3\boldsymbol{r} \tag{3.34}$$

for any normalized wave function $\phi(\boldsymbol{r})$.

The "equals" sign applies, of course, only when $\phi(\boldsymbol{r})$ is the exact ground-state eigenfunction of the hamiltonian $\hat{H}$.

Thus, if we guess the functional form of $\phi(\boldsymbol{r})$ and let it depend on some parameter α (or a set of parameters), and then calculate $\int \phi^*(\boldsymbol{r})\hat{H}\phi(\boldsymbol{r})\mathrm{d}^3\boldsymbol{r}$ and find its minimum with respect to α (or with respect to the entire set of parameters), this minimum is, by (3.34), an upper bound on the ground-state energy E_0, and, if we have made an intelligent choice of the functional form of $\phi(\boldsymbol{r})$, will be close to E_0.

We now give a proof of (3.34). We expand the chosen function $\phi(\boldsymbol{r})$ in the exact (but unknown) orthonormal eigenfunctions $\phi_n(\boldsymbol{r})$ of the hamiltonian $\hat{H}$:

$$\phi(\boldsymbol{r}) = \sum_n c_n \phi_n(\boldsymbol{r}),$$

where the $\phi_n(\boldsymbol{r})$ satisfy

$$\hat{H}\phi_n(\boldsymbol{r}) = E_n \phi_n(\boldsymbol{r}).$$

Then

$$\begin{aligned}
\int \phi^*(\boldsymbol{r})\hat{H}\phi(\boldsymbol{r})d^3\boldsymbol{r} &= \sum_n c_n{}^* \sum_{n'} c_{n'} \iint \phi_n{}^*(\boldsymbol{r})\hat{H}\phi_{n'}(\boldsymbol{r})\mathrm{d}^3\boldsymbol{r} \\
&= \sum_n \sum_{n'} c_n{}^* c_{n'} \int \phi_n{}^*(\boldsymbol{r}) E_{n'} \phi_{n'}(\boldsymbol{r})\mathrm{d}^3\boldsymbol{r} \\
&= \sum_n \sum_{n'} c_{n'}{}^* c_n E_{n'} \delta_{nn'} = \sum_n |c_n|^2 E_n \geq E_0 \sum_n |c_n|^2,
\end{aligned}$$

since $E_n \geq E_0$, by definition of the ground-state energy. But, since $\phi(\boldsymbol{r})$ is normalized,

$$\begin{aligned}
\int \phi^*(\boldsymbol{r})\phi(\boldsymbol{r})\mathrm{d}^3\boldsymbol{r} &= \int \left(\sum_n c_n \phi_n(\boldsymbol{r})\right)^* \left(\sum_{n'} c_{n'} \phi_{n'}(\boldsymbol{r})\right) \mathrm{d}^3\boldsymbol{r} \\
&= \sum_n \sum_{n'} c_n{}^* c_{n'} \int \phi_n(\boldsymbol{r})^* \phi_{n'}(\boldsymbol{r})\mathrm{d}^3\boldsymbol{r} \\
&= \sum_n \sum_{n'} c_n{}^* c_{n'} \delta_{nn'} = \sum_n |c_n|^2 = 1
\end{aligned}$$

and, therefore,

$$E_0 \leq \int \phi^*(\boldsymbol{r})\hat{H}\phi(\boldsymbol{r})\mathrm{d}^3\boldsymbol{r},$$

as was to be proved.

As an example, consider the determination of the ground state of a particle of mass m executing one-dimensional simple harmonic oscillation with classical frequency ω. The hamiltonian is

$$\hat{H} = -\frac{\hbar^2}{2m}\frac{\mathrm{d}^2}{\mathrm{d}x^2} + \frac{1}{2}m\omega^2 x^2.$$

Now choose a trial wave function for the ground state. Since the potential is nonsingular everywhere, we choose a function that is continuous everywhere. The function should also go to zero as $x \to \pm\infty$ (since we expect the probability of finding the particle to vanish in regions where the potential energy tends to infinity). Moreover, as the wave function describing the ground state, it should have no nodes.

We therefore choose

$$\phi(x) = A\mathrm{e}^{-\alpha x^2},$$

where A is the normalization constant, found from the condition

$$1 = \int_{-\infty}^{\infty} |\phi|^2 \,\mathrm{d}x = A^2 \int_{-\infty}^{\infty} \mathrm{e}^{-2\alpha x^2} \mathrm{d}x = \frac{A^2}{\sqrt{2a}} \int_{-\infty}^{\infty} \mathrm{e}^{-y^2} \mathrm{d}y = A^2 \sqrt{\frac{\pi}{2a}},$$

that is,

$$A = \left(\frac{2\alpha}{\pi}\right)^{1/4}.$$

The quantity we wish to minimize is

$$\int \phi^*(x)\hat{H}\phi(x)\mathrm{d}x = \left(\frac{2\alpha}{\pi}\right)^{1/2} \int_{-\infty}^{\infty} \mathrm{e}^{-\alpha x^2} \left(-\frac{\hbar^2}{2m}\right) \frac{\mathrm{d}^2}{\mathrm{d}x^2} \mathrm{e}^{-\alpha x^2} \mathrm{d}x$$
$$+ \left(\frac{2\alpha}{\pi}\right)^{1/2} \int_{-\infty}^{\infty} \mathrm{e}^{-\alpha x^2} \frac{1}{2} m\omega^2 x^2 \mathrm{e}^{-\alpha x^2} \mathrm{d}x.$$

But

$$\frac{\mathrm{d}^2}{\mathrm{d}x^2} \mathrm{e}^{-\alpha x^2} = \frac{\mathrm{d}}{\mathrm{d}x}\left[-2\alpha x \mathrm{e}^{-\alpha x^2}\right] = -2\alpha \mathrm{e}^{-\alpha x^2} + 4\alpha^2 x^2 \mathrm{e}^{-\alpha x^2}.$$

Therefore,

$$\int_{-\infty}^{\infty} \phi^*(x)\hat{H}\phi(x)\mathrm{d}x = \left(\frac{2\alpha}{\pi}\right)^{1/2} \left[\frac{2\alpha\hbar^2}{2m} \int_{-\infty}^{\infty} \mathrm{e}^{-2\alpha x^2} \mathrm{d}x - \left(\frac{4\alpha^2\hbar^2}{2m} - \frac{1}{2} m\omega^2\right) \int_{-\infty}^{\infty} x^2 \mathrm{e}^{-2\alpha x^2} \mathrm{d}x\right].$$

But

$$\int_{-\infty}^{\infty} x^2 \mathrm{e}^{-2\alpha x^2} \mathrm{d}x = \frac{1}{(2\alpha)^{3/2}} \int_{-\infty}^{\infty} y^2 \mathrm{e}^{-y^2} \mathrm{d}y,$$

in which

$$\int_{-\infty}^{\infty} y^2 e^{-y^2} dy = \int_{y=-\infty}^{y=\infty} \left(-\frac{y}{2}\right) d\left(e^{-y^2}\right) = -\frac{y}{2} e^{-y^2}\Big|_{y=-\infty}^{y=\infty} + \frac{1}{2}\int_{-\infty}^{\infty} e^{-y^2} dy = \frac{\sqrt{\pi}}{2}.$$

Therefore,

$$\int_{-\infty}^{\infty} x^2 e^{-2\alpha x^2} dx = \frac{1}{(2\alpha)^{3/2}} \frac{\sqrt{\pi}}{2},$$

and so

$$\int \phi^*(x)\hat{H}\phi(x)dx = \left(\frac{2\alpha}{\pi}\right)^{1/2} \left\{ \frac{2\alpha\hbar^2}{2m}\left(\frac{\pi}{2\alpha}\right)^{1/2} - \left[\frac{2\alpha^2\hbar^2}{m} - \frac{1}{2}m\omega^2\right]\left(\frac{1}{2\alpha}\right)^{3/2}\frac{\sqrt{\pi}}{2}\right\}$$

$$= \frac{\alpha\hbar^2}{m} - \left[\frac{\alpha\hbar^2}{2m} - \frac{m\omega^2}{2\cdot 2\cdot 2\alpha}\right] = \frac{1}{2}\left(\frac{\alpha\hbar^2}{m} + \frac{m\omega^2}{4\alpha}\right).$$

Minimizing this with respect to α, that is, using the condition

$$\frac{d}{d\alpha}\int \phi^*(x)\hat{H}\phi(x)dx = \frac{\hbar^2}{2m} - \frac{m\omega^2}{8\alpha^2} = 0,$$

we find for the "best" value of α:

$$\alpha = \frac{m\omega}{2\hbar},$$

and so

$$\int \phi^*(x)\hat{H}\phi(x)dx\Big|_{\min} = \frac{1}{2}\left(\frac{m\omega\hbar^2}{2\hbar m} + \frac{m\omega^2}{4m\omega}\cdot 2\hbar\right) = \frac{1}{4}\hbar\omega + \frac{1}{4}\hbar\omega = \frac{1}{2}\hbar\omega, \tag{3.35}$$

while the "best approximation" to the ground-state wave function is

$$\phi(x) = \left(\frac{2\alpha}{\pi}\right)^{1/4} e^{-\alpha x^2} = \left(\frac{m\omega}{\pi\hbar}\right)^{1/4} e^{-m\omega x^2/2\hbar}. \tag{3.36}$$

Because we happened to make the correct choice of the functional form of $\phi(x)$, (3.35) and (3.36) are, in fact, the **exact** results for the ground state of the harmonic oscillator, as can be checked by showing that $\hat{H}\phi(x)$ with $\phi(x)$ given by (3.36) yields $E_0\phi(x)$ with $E_0 = \frac{1}{2}\hbar\omega$.

3.4 Time-dependent perturbation theory

As in time-independent perturbation theory, we consider a hamiltonian

$$\hat{H} = \hat{H}_0 + \hat{H}',$$

in which $\hat{H}_0$ is the hamiltonian of a solvable problem and $\hat{H}'$ is the "perturbation" to $\hat{H}_0$, except that, now, $\hat{H}' = \hat{H}'(t)$, that is, $\hat{H}'$ depends on the time t.

An example is the interaction of a hydrogen atom with light. Here, $\hat{H}'(t)$ is the energy of interaction of the instantaneous electron–proton dipole moment with the oscillating (and hence time- and space-dependent) electric and magnetic fields of the electromagnetic wave constituting the light.

Thus, time-dependent perturbation theory forms the basis of the theory of atomic, nuclear and solid state **spectroscopy**, and also of the theory of the dielectric permittivities and refractive indices of materials.

Let $\hat{H}'(t)$ be "switched on" at time $t = 0$ and "switched off" at time $t = \tau$:

$$\hat{H}'(t) = \begin{cases} \hat{V}(t) & (0 \le t \le \tau) \\ 0 & (t < 0 \text{ and } t > \tau) \end{cases}$$

where $\hat{V}(t)$ varies arbitrarily with t.

Suppose that the system, for all times $t < 0$, was in a stationary state $\psi_m^{(0)}(\boldsymbol{r}, t)$ of the unperturbed hamiltonian $\hat{H}_0$:

$$\psi_{\text{initial}} = \psi_m^{(0)}(\boldsymbol{r}, t) = \phi_m^{(0)}(\boldsymbol{r})\,\mathrm{e}^{-iE_m^{(0)}t/\hbar},$$

satisfying the time-dependent Schrödinger equation

$$i\hbar\frac{\partial\psi_m^{(0)}(\boldsymbol{r}, t)}{\partial t} = \hat{H}_0\psi_m^{(0)}(\boldsymbol{r}, t).$$

Here, of course,

$$\hat{H}_0\phi_m^{(0)}(\boldsymbol{r}) = E_m^{(0)}\phi_m^{(0)}(\boldsymbol{r}).$$

The wave function $\Psi(\boldsymbol{r}, t)$ describing the state of a system always evolves in accordance with the time-dependent Schrödinger equation

$$i\hbar\frac{\partial\Psi(\boldsymbol{r}, t)}{\partial t} = \hat{H}\Psi(\boldsymbol{r}, t), \tag{3.37}$$

where $\hat{H}$ is the full hamiltonian at the given time. For $0 \le t \le \tau$, the full hamiltonian contains the time-dependent term $\hat{V}(t)$, and so the state $\Psi(\boldsymbol{r}, t)$ of the system during this period will no longer be a stationary state of the unperturbed hamiltonian $\hat{H}_0$.

However, since the stationary states of $\hat{H}_0$ form a complete set, we can always write $\Psi(\boldsymbol{r}, t)$ as a linear combination of them:

$$\Psi(\boldsymbol{r}, t) = \sum_j a_j(t)\phi_j^{(0)}(\boldsymbol{r})\mathrm{e}^{-iE_j^{(0)}t/\hbar}, \tag{3.38}$$

where the coefficients $a_j(t)$ are determined by the state at $t = 0$ and the requirement that, for $0 \leq t \leq \tau$, the wave function satisfies

$$i\hbar\frac{\partial \Psi(\boldsymbol{r}, t)}{\partial t} = (\hat{H}_0 + \hat{V}(t))\Psi(\boldsymbol{r}, t). \tag{3.39}$$

Putting (3.38) into (3.39), we obtain

$$\sum_j \left[i\hbar\frac{\mathrm{d}a_j(t)}{\mathrm{d}t}\phi_j^{(0)}(\boldsymbol{r})\mathrm{e}^{-iE_j^{(0)}t/\hbar} + E_j^{(0)}a_j(t)\phi_j^{(0)}(\boldsymbol{r})\mathrm{e}^{-iE_j^{(0)}t/\hbar}\right]$$
$$= \sum_j \left[a_j(t)E_j^{(0)}\phi_j^{(0)}(\boldsymbol{r})\mathrm{e}^{-iE_j^{(0)}t/\hbar} + a_j(t)\hat{V}(t)\phi_j^{(0)}(\boldsymbol{r})\mathrm{e}^{-iE_j^{(0)}t/\hbar}\right].$$

The second sum on the left cancels the first sum on the right. We then multiply the resulting equation by $\phi_n^{(0)*}(\boldsymbol{r})$ and integrate over all space. Then, since

$$\int \phi_n^{(0)*}(\boldsymbol{r})\phi_j^{(0)}(\boldsymbol{r})\mathrm{d}^3\boldsymbol{r} = \delta_{nj},$$

we have

$$\frac{\mathrm{d}a_n(t)}{\mathrm{d}t} = -\frac{i}{\hbar}\sum_j {}^{(0)}\langle n|\hat{V}(t)\,|j\rangle^{(0)}\,\mathrm{e}^{i\omega_{nj}t}a_j(t), \tag{3.40}$$

where

$${}^{(0)}\langle n|\hat{V}(t)\,|j\rangle^{(0)} \equiv \int \phi_n^{(0)*}(\boldsymbol{r})\hat{V}(t)\phi_j^{(0)}(\boldsymbol{r})\mathrm{d}^3\boldsymbol{r}$$

and

$$\omega_{nj} = (E_n^{(0)} - E_j^{(0)})/\hbar.$$

To integrate (3.40) for $a_n(t)$ we need to know all the coefficients $a_j(t)$ already! But we **do** know $a_j(t)$ for times $t < 0$, namely,

$$a_j(t) = \delta_{jm} \text{ for } t < 0$$

[we may confirm that putting this into (3.38) yields the initial ($t < 0$) state], and so we use this as our **zeroth** approximation $a_j^{(0)}(t)$ to $a_j(t)$ for times $t > 0$ as well. Putting it into (3.40), we obtain the equation for the **first** approximation $a_n^{(1)}(t)$ to $a_n(t)$ (for $n \neq m$):

$$\frac{\mathrm{d}a_n^{(1)}(t)}{\mathrm{d}t} = -\frac{i}{\hbar}\sum_j {}^{(0)}\langle n|\hat{V}(t)\,|j\rangle^{(0)}\,\mathrm{e}^{i\omega_{nj}t}\delta_{jm} = -\frac{i}{\hbar}{}^{(0)}\langle n|\hat{V}(t)|m\rangle^{(0)}\mathrm{e}^{i\omega_{nm}t},$$

which is easily integrated to give

$$a_n^{(1)}(t) = -\frac{i}{\hbar}\int_0^t {}^{(0)}\langle n|\hat{V}(t')|m\rangle^{(0)} e^{i\omega_{nm}t'}\,dt'. \tag{3.41}$$

[Substituting (3.41) into the right-hand side of (3.40) then gives the equation for the second approximation, and so on.]

Since $\hat{V}(t) = 0$ for $t > \tau$, the quantity $a_n(t)$ as given by (3.41) is "frozen" at its $t = \tau$ value for all times $t > \tau$:

$$a_n^{(1)}(t > \tau) = a_n^{(1)}(\tau) = -\frac{i}{\hbar}\int_0^\tau {}^{(0)}\langle n|\hat{V}(t')|m\rangle^{(0)} e^{i\omega_{nm}t'}\,dt'. \tag{3.42}$$

Since measurement of any observable can only yield a value belonging to the eigenvalue spectrum of that observable, an energy measurement at any time $t > \tau$ must yield one or other of the eigenvalues of the prevailing hamiltonian (namely, of $\hat{H}_0$), corresponding to a particular unperturbed eigenstate, say, $\phi_n^{(0)}(\boldsymbol{r})$.

Since $a_n(t)$ is the weight of this state in $\Psi(\boldsymbol{r}, t)$ (3.38), and since, for all times $t > \tau$, we have $a_n(t) = a_n(\tau)$, the quantity $|a_n(\tau)|^2$ gives the probability of finding the system to be in the state $\psi_n^{(0)}(\boldsymbol{r}, t)$ at any time $t > \tau$ if the perturbation acted for a time τ. But since the system was originally (i.e., at times $t < \tau$) in the state $\psi_m^{(0)}(\boldsymbol{r}, t)$, this is the probability of a transition from the state $\psi_m^{(0)}(\boldsymbol{r}, t)$ to the state $\psi_n^{(0)}(\boldsymbol{r}, t)$ when the perturbation acts for a time τ. Therefore, this **transition probability** is given in the first approximation by the expression

$$w_{n\leftarrow m}^{(1)}(\tau) = \left|a_n^{(1)}(\tau)\right|^2 = \frac{1}{\hbar^2}\left|\int_0^\tau {}^{(0)}\langle n|\hat{V}(t)|m\rangle^{(0)} e^{i\omega_{nm}t}\,dt\right|^2. \tag{3.43}$$

We now consider two examples of $\hat{V}(t)$. In the first, we shall assume that $\hat{V}(t)$ is constant in time over the interval $0 \le t \le \tau$, and in the second we shall assume that $\hat{V}(t)$ is harmonic, with nonzero frequency ω in the interval $0 \le t \le \tau$.

The first case is indicated schematically in figure 3.3, which is schematic in the sense that, in the coordinate representation, the operator $\hat{V}(t)$ depends on $\boldsymbol{r}$ as well as t.

In this case,

$$\int_0^\tau {}^{(0)}\langle n|\hat{V}(t)|m\rangle^{(0)} e^{i\omega_{nm}t}\,dt = {}^{(0)}\langle n|\hat{V}|m\rangle^{(0)}\int_0^\tau e^{i\omega_{nm}t}\,dt = {}^{(0)}\langle n|\hat{V}|m\rangle^{(0)}\frac{e^{i\omega_{nm}\tau} - 1}{i\omega_{nm}}.$$

Figure 3.3 *Schematic representation of a perturbation $\hat{V}$ that is constant in time during the interval for which it is switched on.*

Therefore,

$$
\begin{aligned}
w_{n\leftarrow m}^{(1)}(\tau) &= \frac{1}{\hbar^2}\left|{}^{(0)}\langle n|\hat{V}|m\rangle^{(0)}\right|^2 \frac{(\mathrm{e}^{i\omega_{nm}\tau}-1)(\mathrm{e}^{-i\omega_{nm}\tau}-1)}{\omega_{nm}^2} \\
&= \frac{2}{\hbar^2}\left|{}^{(0)}\langle n|\hat{V}|m\rangle^{(0)}\right|^2 \left(\frac{1-\cos\omega_{nm}\tau}{\omega_{nm}^2}\right) \\
&= \frac{4}{\hbar^2}\left|{}^{(0)}\langle n|\hat{V}|m\rangle^{(0)}\right|^2 \left(\frac{\sin\frac{1}{2}\omega_{nm}\tau}{\omega_{nm}}\right)^2.
\end{aligned}
\tag{3.44}
$$

But the function

$$
\left(\frac{\sin\frac{1}{2}\omega_{nm}\tau}{\omega_{nm}}\right)^2 = \frac{\tau^2}{4}\left(\frac{\sin\frac{1}{2}\omega_{nm}\tau}{\frac{1}{2}\omega_{nm}\tau}\right)^2 \to \frac{\tau^2}{4} \quad \text{as} \quad \omega_{nm} \to 0,
$$

and is zero when $\frac{1}{2}\omega_{nm}\tau = n\pi$ with n an integer, and so is zero when $\omega_{nm} = 2n\pi/\tau$. Thus, this function has the form depicted in figure 3.4.

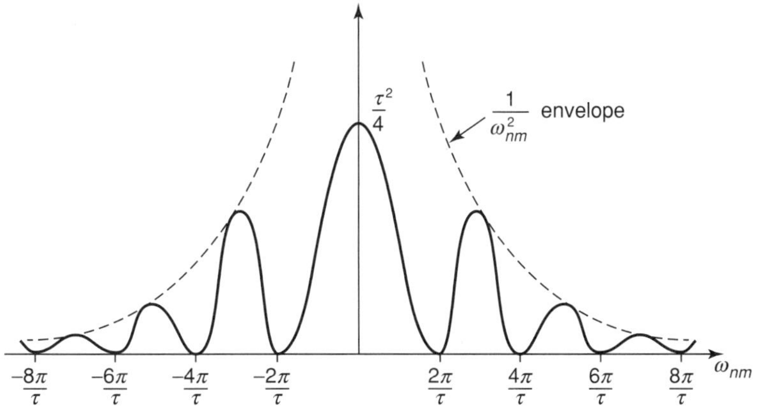

Figure 3.4 *Graph of* $\sin^2\frac{1}{2}\omega_{nm}\tau/\omega_{nm}^2$ *as a function of* ω_{nm}.

Thus, for final states $|n\rangle^{(0)}$ for which the matrix element ${}^{(0)}\langle n|\hat{V}|m\rangle^{(0)}$ does not vanish by symmetry, the graph in figure 3.4, which is the graph for a particular value of τ, shows that for this value of τ the transition probability is a maximum for transitions to states $|n\rangle^{(0)}$ for which $\omega_{nm} = 0$, that is, for which $E_n^{(0)} = E_m^{(0)}$. (For shorter times τ, the noncentral maxima at $\omega_{nm} = \pm 2\pi/\tau$, rather than the central maximum, may have the greatest height.) In the limit $\tau \to \infty$ the area under the central maximum goes to infinity linearly with τ, since the height of this maximum is proportional to τ^2 and its width is proportional to τ^{-1}. The area of any noncentral maximum (i.e., a maximum at any nonzero value of ω_{nm}) goes to zero in this limit, since the height is bounded by the $1/\omega_{nm}^2$ envelope and the width is proportional to τ^{-1}.

For finite τ, however, there is clearly an appreciable probability of **apparently** energy-nonconserving transitions, for example, to states with energies $E_n^{(0)}$ corresponding to nonzero values of ω_{nm} within the middle part of the central maximum, for which

$$|\omega_{nm}| \leq \frac{2\pi}{\tau},$$

that is, for which

$$\left|E_n^{(0)} - E_m^{(0)}\right| \tau \leq 2\pi\hbar = h. \tag{3.45}$$

It might appear that this is related in some way to the energy–time uncertainty relation

$$\Delta E \Delta\tau \geq \hbar, \tag{3.46}$$

which states that, if $\Delta\tau$ is the lifetime of a state (or is the time available to carry out a measurement of the energy), the energy of the state cannot be determined to within a smaller uncertainty than $\Delta E = \hbar/\Delta\tau$ (so that energy nonconservation within this uncertainty is permitted).

In fact, (3.45) has nothing to do with energy uncertainty, as can be seen from the fact that the final state $\psi_n^{(0)}$ could persist for ever, with a **definite** energy $E_n^{(0)}$ not equal to the original energy $E_m^{(0)}$. Nor, contrary to appearances, does (3.45) involve energy nonconservation (which only ever occurs within the limits of the quantum-mechanical uncertainty of the energy), since the gain or loss of energy by our system is **real** and arises from energy exchange with the "external" system with which it is interacting via the interaction potential $\hat{V}(t)$. (Our hamiltonian omits the intrinsic dynamics of this external system.)

Returning to the graph in figure 3.4, we see that as $\tau \to \infty$ the entire area under the curve becomes concentrated under the central maximum, which becomes infinitely narrow and infinitely high, with area

$$\frac{\tau^2}{4} \times \frac{2\pi}{\tau} = \frac{\pi\tau}{2},$$

so that

$$\left(\frac{\sin\frac{1}{2}\omega_{nm}\tau}{\omega_{nm}}\right)^2 \to \frac{\pi\hbar\tau}{2}\delta\left(E_n^{(0)} - E_m^{(0)}\right) \quad \text{as} \quad \tau \to \infty,$$

as can be verified by integration of this relation over ω_{nm}:

$$\int_{-\infty}^{\infty}\left(\frac{\sin\frac{1}{2}\omega_{nm}\tau}{\omega_{nm}}\right)^2 \mathrm{d}\omega_{nm} \to \frac{\pi\tau}{2}\int_{-\infty}^{\infty}\delta\left(E_n^{(0)} - E_m^{(0)}\right)\mathrm{d}\left(E_n^{(0)} - E_m^{(0)}\right) = \frac{\pi\tau}{2}.$$

Therefore, from (3.44), for $\tau \to \infty$,

$$w_{n\leftarrow m}^{(1)}(\tau) \to \frac{2\pi\tau}{\hbar}\left|{}^{(0)}\langle n|\hat{V}|m\rangle^{(0)}\right|^2 \delta\left(E_n^{(0)} - E_m^{(0)}\right), \tag{3.47}$$

which expresses the fact that our system neither gains nor loses energy if its interaction with the external system is constant in time over a long period τ [so that the Fourier expansion of $\hat{V}(t)$ in harmonic functions of t contains essentially only the zero-frequency component].

Therefore, in this case, the **transition probability per unit time** from state $\psi_m^{(0)}$ to state $\psi_n^{(0)}$ is:

$$\tilde{P}_{n\leftarrow m} \approx \tilde{P}^{(1)}_{n\leftarrow m} \equiv \frac{w^{(1)}_{n\leftarrow m}(\tau)}{\tau} = \frac{2\pi}{\hbar} \left|{}^{(0)}\langle n|V|m\rangle^{(0)}\right|^2 \delta\left(E_n^{(0)} - E_m^{(0)}\right). \tag{3.48}$$

If the transitions induced by $\hat{V}(t)$ are from a particular state $\psi_m^{(0)}$ to states $\psi_n^{(0)}$ in a continuum of states, all of which are of the same symmetry type (call it "type n") and give rise to nonzero matrix elements ${}^{(0)}\langle n|V|m\rangle^{(0)}$ that vary continuously with the energy of the state $\psi_n^{(0)}$ (an example is a continuum of plane-wave states), and if in this continuum there are $\rho_n(E_n^{(0)})\mathrm{d}E_n^{(0)}$ states in the interval $(E_n^{(0)}, E_n^{(0)} + \mathrm{d}E_n^{(0)})$ [i.e., $\rho_n(E)$ is the density of states of type n], then the total transition probability per unit time to states of type n is, in the long-τ limit in which equation (3.48) applies,

$$P_{n\leftarrow m} = \int \tilde{P}_{n\leftarrow m}\rho_n\left(E_n^{(0)}\right)\mathrm{d}E_n^{(0)} = \frac{2\pi}{\hbar}\left|{}^{(0)}\langle n|V|m\rangle^{(0)}\right|^2 \rho_n\left(E_m^{(0)}\right), \tag{3.49}$$

in which $\rho_n(E_m^{(0)})$ is the density of n-type states, evaluated at energy $E_m^{(0)}$.

The expression (3.49) is called **Fermi's golden rule**.

We now consider the second case, when the perturbation $\hat{V}(t)$ has the form of a harmonic wave:

$$\hat{V}(t) = 2\hat{v}\cos\omega t = \hat{v}\left(\mathrm{e}^{i\omega t} + \mathrm{e}^{-i\omega t}\right) \tag{3.50}$$

depicted schematically in figure 3.5.

Figure 3.5 *Schematic representation of a perturbation $\hat{V}$ that is harmonic in time during the interval for which it is switched on.*

(The case dealt with above, with a perturbation that is constant in time over the interval for which it is switched on, is just a special case of this, with $\omega = 0$.)

Putting (3.50) into the expression (3.42) found earlier, we find, for $t > \tau$,

$$a_n^{(1)}(t) = a_n^{(1)}(\tau) = \frac{1}{i\hbar}{}^{(0)}\langle n|\hat{v}|m\rangle^{(0)}\int_0^\tau \mathrm{e}^{i(\omega_{nm}+\omega)t}\mathrm{d}t + \frac{1}{i\hbar}{}^{(0)}\langle n|\hat{v}|m\rangle^{(0)}\int_0^\tau \mathrm{e}^{i(\omega_{nm}-\omega)t}\mathrm{d}t,$$

each term of which has the same form as in the constant-perturbation case, except that ω_{nm} is replaced by $\omega_{nm} + \omega$ in the first term and by $\omega_{nm} - \omega$ in the second.

Considering the contributions of the two terms separately, in the large-τ limit we have

$$w^{(1)\pm}_{n\leftarrow m}(\tau) = \frac{2\pi\tau}{\hbar}\left|{}^{(0)}\langle n|\hat{v}|m\rangle^{(0)}\right|^2 \delta\left(E_n^{(0)} - E_m^{(0)} \pm \hbar\omega\right),$$

so the transition probabilities per unit time that result from the first and second terms in (3.50) are

$$\tilde{P}^{(1)+}_{n\leftarrow m} = \frac{2\pi}{\hbar} \left| {}^{(0)}\langle n|\hat{v}|m\rangle^{(0)} \right|^2 \delta\left(E_n^{(0)} - E_m^{(0)} + \hbar\omega\right)$$

and

$$\tilde{P}^{(1)-}_{n\leftarrow m} = \frac{2\pi}{\hbar} \left| {}^{(0)}\langle n|\hat{v}|m\rangle^{(0)} \right|^2 \delta\left(E_n^{(0)} - E_m^{(0)} - \hbar\omega\right),$$

respectively.

The first of these is nonzero only when

$$E_n^{(0)} = E_m^{(0)} - \hbar\omega,$$

corresponding to **emission** of energy $\hbar\omega$, that is, to energy transfer to the external system (e.g., the oscillating electromagnetic field of a light wave) with which our system is interacting via (3.50).

The second contribution is nonzero only when

$$E_n^{(0)} = E_m^{(0)} + \hbar\omega,$$

corresponding to **absorption** of a quantum $\hbar\omega$ (a **photon**) from the external system (light wave).

When the transition is to final states in a continuum, the expression

$$P_{n\leftarrow m} = \int \tilde{P}_{n\leftarrow m}\rho_n\left(E_n^{(0)}\right) \mathrm{d}E_n^{(0)}$$

yields the corresponding total transition probabilities per unit time:

$$P^{+}_{n\leftarrow m} = \frac{2\pi}{\hbar} \left| {}^{(0)}\langle n|\hat{v}|m\rangle^{(0)} \right|^2 \rho_n\left(E_m^{(0)} - \hbar\omega\right), \tag{3.51}$$

corresponding to emission of radiation, and

$$P^{-}_{n\leftarrow m} = \frac{2\pi}{\hbar} \left| {}^{(0)}\langle n|\hat{v}|m\rangle^{(0)} \right|^2 \rho_n\left(E_m^{(0)} + \hbar\omega\right), \tag{3.52}$$

corresponding to absorption of radiation. The formulas (3.51) and (3.52) are further instances of the Fermi golden rule.

In both the above examples (the case of a constant perturbation and the case of a harmonic perturbation) we considered transitions to states obtaining after the perturbation was switched off, that is, transitions to eigenstates of the unperturbed hamiltonian $\hat{H}_0$. In the former case, at times t at which the perturbation is still acting, that is, for $0 \le t \le \tau$, instead of expanding $\Psi(\boldsymbol{r},t)$ in the stationary-state eigenfunctions of $\hat{H}_0$ we could equally validly expand it in the stationary-state eigenfunctions of the (constant) perturbed hamiltonian $\hat{H}_0 + \hat{V}$. In other words, instead of

$$\Psi(\boldsymbol{r},t) = \sum_n a_n(t)\,\phi_n^{(0)}(\boldsymbol{r})\,\mathrm{e}^{-iE_n^{(0)}t/\hbar}, \tag{3.53}$$

we could write

$$\Psi(\boldsymbol{r}, t) = \sum_n c_n \phi_n(\boldsymbol{r}) \, \mathrm{e}^{-iE_n t/\hbar}, \tag{3.54}$$

where

$$(\hat{H}_0 + \hat{V})\phi_n(\boldsymbol{r}) = E_n \phi_n(\boldsymbol{r}),$$

and, for $0 \leq t \leq \tau$, the coefficients c_n, unlike the a_n in (3.53), are **independent of** t, as may be confirmed by observing that (3.54) with time-independent coefficients c_n satisfies the relevant time-dependent Schrödinger equation:

$$i\hbar \frac{\partial \Psi(\boldsymbol{r}, t)}{\partial t} = (\hat{H}_0 + \hat{V})\Psi(\boldsymbol{r}, t).$$

The coefficients c_n are constant in time for so long as $\hat{V}$ is constant in time. Therefore, if the latter were true right back to time $t = 0$, that is, if $\hat{V}(t)$ were a true step function of t (an assumption that constitutes the **sudden approximation**), we could calculate the coefficients c_n as **frozen** at their values at time $t = 0$. Using the fact that, at all times before and including $t = 0$, we have

$$\Psi(\boldsymbol{r}, t) = \psi_{\text{initial}}(\boldsymbol{r}, t) = \psi_m^{(0)}(\boldsymbol{r}, t) = \phi_m^{(0)}(\boldsymbol{r}) \, \mathrm{e}^{-iE_m^{(0)} t/\hbar},$$

so that

$$\Psi(\boldsymbol{r}, 0) = \phi_m^{(0)}(\boldsymbol{r}),$$

and, equating the latter to the right-hand side of (3.54) with $t = 0$, we get

$$\phi_m^{(0)}(\boldsymbol{r}) = \sum_n c_n \phi_n(\boldsymbol{r}).$$

Therefore,

$$c_k = \int \phi_k^*(\boldsymbol{r}) \, \phi_m^{(0)}(\boldsymbol{r}) \, \mathrm{d}^3\boldsymbol{r},$$

which is the sudden approximation to the constant values taken by the coefficients c_k while the constant perturbation remains switched on.

4

Scattering Theory

4.1 Evolution operators and Møller operators

Consider a spinless particle being scattered by some fixed potential V. The hamiltonian is given by

$$\hat{H} = \hat{H}_0 + \hat{V}, \tag{4.1}$$

where $\hat{H}_0$ is the operator $\hat{p}^2/2m$ of the kinetic energy of the particle.

The state vector $|\psi\rangle_t$ of the particle satisfies

$$i\hbar\frac{\mathrm{d}}{\mathrm{dt}}|\psi\rangle_t = \hat{H}|\psi\rangle_t \tag{4.2}$$

(we drop the hat from now on), which has the formal solution

$$|\psi\rangle_t = \mathrm{e}^{-iHt/\hbar}|\psi\rangle_{t=0} \equiv \mathrm{e}^{-iHt/\hbar}|\psi\rangle \equiv U(t)|\psi\rangle, \tag{4.3}$$

where $U(t)$ is the **evolution operator**. If the scattering occurs at time $t = 0$, and $|\psi\rangle$ is not a bound state, then at times $t \to \pm\infty$ the particle is far from the scatterer and is moving as a free particle. The evolution operator for a free particle is

$$U_0(t) \equiv \mathrm{e}^{-iH_0t/\hbar}. \tag{4.4}$$

We therefore expect that a state vector $|\psi_{\mathrm{in}}\rangle \equiv |\psi_{\mathrm{in}}\rangle_{t=0}$ (the "in asymptote" of $|\psi\rangle$) exists such that

$$U(t)|\psi\rangle - U_0(t)|\psi_{\mathrm{in}}\rangle \to 0 \quad \text{as}\, t \to -\infty, \tag{4.5}$$

as indicated schematically in figure 4.1.

The operators $U(t)$ and $U_0(t)$ are both unitary, that is,

$$U^\dagger U = UU^\dagger = 1 \quad \text{and} \quad U_0^\dagger U_0 = U_0U_0^\dagger = 1, \tag{4.6}$$

as follows from the fact that both H and H_0 are self-adjoint (i.e., $H^\dagger = H$ and $H_0^\dagger = H_0$).

A Course in Theoretical Physics, First Edition. P. J. Shepherd.

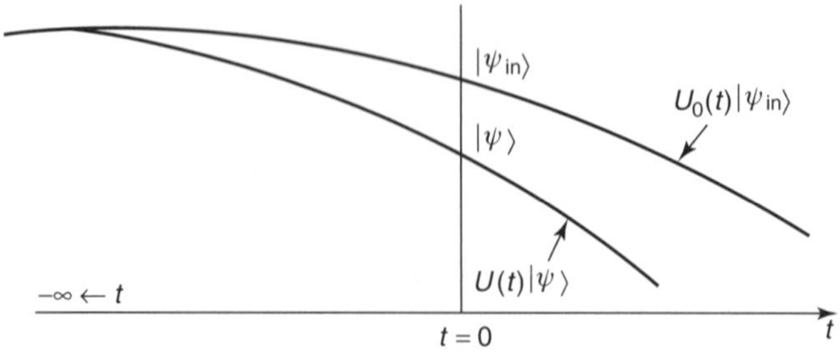

Figure 4.1 *Schematic diagram illustrating the meaning of the "in asymptote" of* $|\psi\rangle$.

We now let the operator $U^\dagger(t)$ act from the left on (4.5), and obtain

$$|\psi\rangle - U^\dagger(t)U_0(t)\,|\psi_{\text{in}}\rangle \to 0 \quad \text{as}\, t \to -\infty. \tag{4.7}$$

This may be written as

$$|\psi\rangle = \Omega_+\,|\psi_{\text{in}}\rangle\,, \tag{4.8}$$

in which

$$\Omega_+ \equiv \lim_{t\to-\infty} U^\dagger(t)U_0(t) \tag{4.9}$$

is a **Møller operator**.

Similarly, by considering the $t \to \infty$ behavior, depicted schematically in figure 4.2, we see that there exists a ket $|\psi_{\text{out}}\rangle$ (the "out asymptote" of $|\psi\rangle$) such that

$$|\psi\rangle = \Omega_-\,|\psi_{\text{out}}\rangle\,, \tag{4.10}$$

with

$$\Omega_- \equiv \lim_{t\to+\infty} U^\dagger(t)U_0(t). \tag{4.11}$$

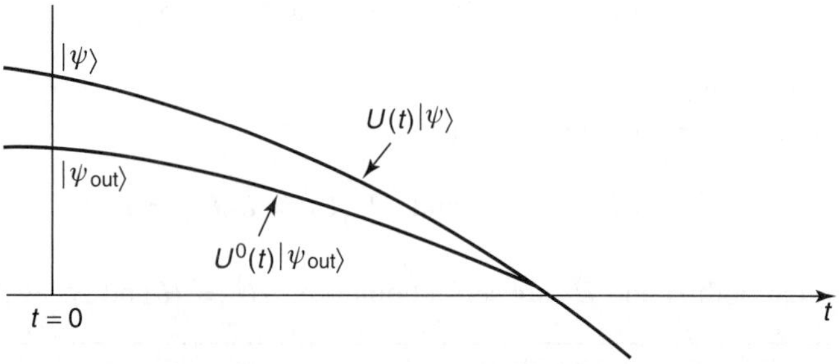

Figure 4.2 *Schematic diagram illustrating the meaning of the "out asymptote" of* $|\psi\rangle$.

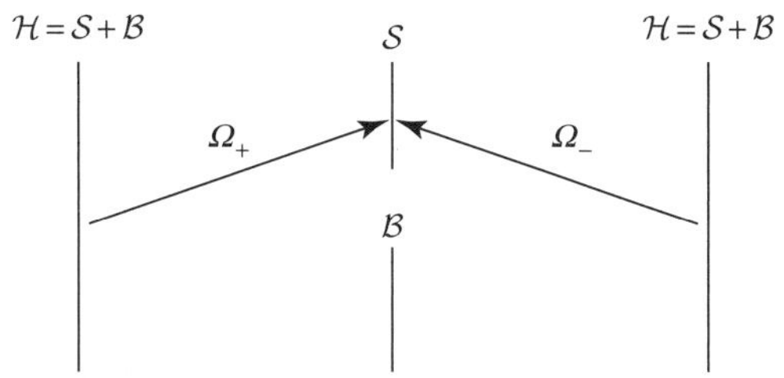

Figure 4.3 *Schematic diagram illustrating the "asymptotic condition".*

Although **bound** states (unlike $|\psi\rangle$ above) do not possess in and out asymptotes, **every** state $|\phi\rangle$ in the Hilbert space $\mathcal{H}$ (including the bound states) is the in asymptote of some state, denoted by $|\phi+\rangle$, and the out asymptote of another, denoted by $|\phi-\rangle$. Thus, the states

$$|\phi+\rangle \equiv \Omega_+ |\phi\rangle \tag{4.12}$$

and

$$|\phi-\rangle \equiv \Omega_- |\phi\rangle \tag{4.13}$$

exist for **all** states $|\phi\rangle$ in $\mathcal{H}$. This is called the **asymptotic condition**, a proof of which is given in Appendix A1.

The asymptotic condition is represented schematically in figure 4.3, in which $\mathcal{S}$ is the subspace of $\mathcal{H}$ spanned by the states $|\phi+\rangle \equiv \Omega_+ |\phi\rangle$ and the states $|\phi-\rangle \equiv \Omega_- |\phi\rangle$ (the so-called **scattering states**), and $\mathcal{B}$ is the subspace of $\mathcal{H}$ spanned by the bound states.

[The reason that a bound state develops into a scattering state under the action of $\Omega_\pm$ given by (4.9) and (4.11) is that the bound states are eigenstates of U and $U^\dagger$ (because they are eigenstates of H) but not of U_0, which consequently "spreads" the bound states.]

The Møller operators Ω_+ and Ω_- given by (4.9) and (4.11), being limits of products of evolution operators that do not change the norm of the kets on which they act, are themselves norm-preserving (isometric), that is, the norm

$$\|\Omega_\pm \psi\|^2 \equiv (\Omega_\pm \psi, \Omega_\pm \psi) = (\psi, \Omega_\pm^\dagger \Omega_\pm \psi) \equiv \langle \psi | \Omega_\pm^\dagger \Omega_\pm | \psi \rangle \tag{4.14}$$

is equal to the norm

$$\|\psi\|^2 \equiv (\psi, \psi) \equiv \langle \psi | \psi \rangle, \tag{4.15}$$

and this implies that

$$\Omega_\pm^\dagger \Omega_\pm |\psi\rangle = |\psi\rangle. \tag{4.16}$$

But, by the asymptotic condition, the states $\Omega_\pm |\psi\rangle$ in (4.14) exist for any $|\psi\rangle$ in $\mathcal{H}$, and so (4.16) is true for any $|\psi\rangle$ in $\mathcal{H}$. This implies the operator identity

$$\Omega_\pm^\dagger \Omega_\pm = 1. \tag{4.17}$$

To show that the operators $\Omega_\pm$ are unitary, we should have to show also that $\Omega_\pm \Omega_\pm^\dagger = 1$. If we act on the ket (4.16) with the operator $\Omega_\pm$ we obtain

$$\Omega_\pm \Omega_\pm^\dagger (\Omega_\pm |\psi\rangle) = \Omega_\pm |\psi\rangle ,$$

from which we **cannot** deduce that $\Omega_\pm \Omega_\pm^\dagger = 1$, since the ket $\Omega_\pm |\psi\rangle$ is **not** an arbitrary ket from $\mathcal{H}$ but belongs to the **subspace** $\mathcal{S}$ of scattering states (see figure 4.3). Thus, the operator $\Omega_\pm \Omega_\pm^\dagger = 1$ when it acts on states in $\mathcal{S}$, but $\Omega_\pm \Omega_\pm^\dagger \neq 1$ when it acts on states in $\mathcal{B}$.

Thus, if $\mathcal{B} \neq 0$, that is, if the hamiltonian H has bound states, the operators $\Omega_\pm$, despite being limits of a product of unitary operators, are not themselves unitary, but only isometric. Clearly, if H has no bound states, that is, if $\mathcal{B} = 0$, the operators Ω_0 are also unitary.

Note that Ω_+ and Ω_- each represent a **one-to-one** mapping of the states of $\mathcal{H}$ onto the states of a **subspace** $\mathcal{S}$ of $\mathcal{H}$! Clearly, this is possible only if $\mathcal{H}$ and $\mathcal{S}$ are infinite-dimensional vector spaces. Thus, an isometric operator fails to be unitary only on an infinite-dimensional space.

The Møller operators $\Omega_\pm$ satisfy the following useful relation, known as the **intertwining relation**:

$$H\Omega_\pm = \Omega_\pm H_0. \tag{4.18}$$

To prove this relation, we write

$$\begin{aligned} e^{iH\tau/\hbar}\Omega_\pm &= e^{iH\tau/\hbar} \lim_{t\to\mp\infty} e^{iHt/\hbar} e^{-iH_0t/\hbar} \\ &= \lim_{\substack{t\to\mp\infty \\ \text{(i.e. } t+\tau\to\mp\infty)}} e^{iH(t+\tau)/\hbar} e^{-iH_0(t+\tau)/\hbar} e^{iH_0\tau/\hbar} = \Omega_\pm e^{iH_0\tau/\hbar} \end{aligned}$$

Differentiating this with respect to τ and then setting $\tau = 0$, we obtain the intertwining relation (4.18).

Note that, since $\Omega_\pm^\dagger \Omega_\pm = 1$, it follows from the intertwining relation (4.18) that

$$\Omega_\pm^\dagger H \Omega_\pm = H_0.$$

This, if $\Omega_\pm$ were unitary, would imply that H and H_0 have the same eigenvalue spectrum, which is not possible if H has bound states and H_0 does not. Thus, $\Omega_\pm$ cannot be unitary if H has bound states, as shown above.

4.2 The scattering operator and scattering matrix

Acting on the ket $|\psi\rangle = \Omega_- |\psi_{\text{out}}\rangle$ with $\Omega_-^\dagger$ and using the property $\Omega_-^\dagger \Omega_- = 1$, we find

$$|\psi_{\text{out}}\rangle = \Omega_-^\dagger |\psi\rangle = \Omega_-^\dagger \Omega_+ |\psi_{\text{in}}\rangle \equiv S |\psi_{\text{in}}\rangle , \tag{4.19}$$

where

$$S = \Omega_-^\dagger \Omega_+ \tag{4.20}$$

is the **scattering operator**.

The scattering operator S is a norm-preserving (i.e., isometric) operator that maps any state $|\psi_{\rm in}\rangle$ in $\mathcal{H}$ into a state $|\psi_{\rm out}\rangle$ in $\mathcal{H}$, where $|\psi_{\rm out}\rangle$ is **not** restricted to any subspace of $\mathcal{H}$ (see figure 4.4). Therefore, the operator S is unitary.

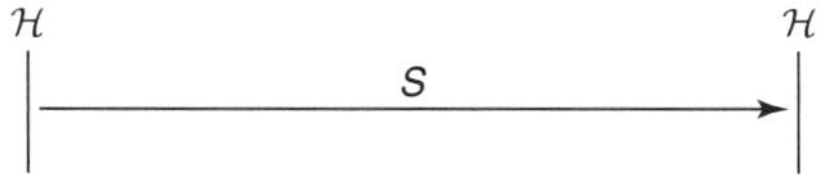

Figure 4.4 *Effect of the scattering operator S.*

Thus, as indicated schematically in figure 4.5, the true evolution of the state of the particle being scattered can be simulated by a three-stage process: (i) the ket describing the state of the particle at times $t \to -\infty$ evolves **freely**, that is, under the action of the free-evolution operator $U_0(t)$, into the $t = 0$ ket $|\psi_{\rm in}\rangle$; (ii) the scattering operator S acts on the $t = 0$ ket $|\psi_{\rm in}\rangle$, converting it into the $t = 0$ ket $|\psi_{\rm out}\rangle$; (iii) the $t = 0$ ket $|\psi_{\rm out}\rangle$ evolves **freely**, as $t \to \infty$, into the ket describing the state of the scattered particle. Since, as can be seen from figure 4.5, the initial state in the free-evolution stage (i) and the final state in the free-evolution stage (iii) are indistinguishable from the initial and final states, respectively, in the true evolution of the particle being scattered, this three-stage process provides an accurate simulation of the scattering process except at times close to $t = 0$, when the particle is in the vicinity of the scatterer. But since in real scattering experiments we prepare the incident particles far from the scatterer, and detect the scattered particles far from the scatterer, this simulation should be entirely adequate for the interpretation of such experiments.

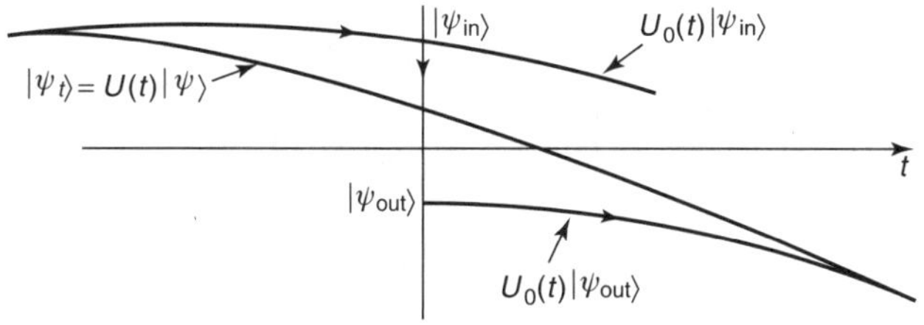

Figure 4.5 *Schematic diagram illustrating the way in which the true evolution of the state of a particle being scattered can be simulated by a three-stage process involving free evolution.*

The scattering operator S commutes with the free-particle hamiltonian H_0, as is demonstrated by the following simple proof:

$$SH_0 = \Omega_-^\dagger \Omega_+ H_0 = \Omega_-^\dagger H \Omega_+ = H_0 \Omega_-^\dagger \Omega_+ = H_0 S, \tag{4.21}$$

in which the intertwining is evident!

It follows from (4.21) and the unitarity of S ($S^\dagger S = SS^\dagger = 1$) that

$$H_0 = S^\dagger H_0 S = S H_0 S^\dagger, \tag{4.22}$$

whence follows the relation

$$\begin{aligned}
\text{“out energy”} &= \langle \psi_{{\rm out},t=\infty} | H_0 | \psi_{{\rm out},t=\infty} \rangle = \langle \psi_{\rm out} | U_0^\dagger(t=\infty) H_0 U_0(t=\infty) | \psi_{\rm out} \rangle \\
&= \langle \psi_{\rm out} | H_0 \, | \psi_{\rm out} \rangle \; (\text{since } [H_0, U_0] = 0 \text{ and } U_0^\dagger U_0 = 1) \\
&= \langle \psi_{\rm in} | \, S^\dagger H_0 S \, | \psi_{\rm in} \rangle = \langle \psi_{\rm in} | \, H_0 \, | \psi_{\rm in} \rangle \\
&= \langle \psi_{\rm in} | \, U_0^\dagger(t=-\infty) H_0 U_0(t=-\infty) \, | \psi_{\rm in} \rangle \\
&= \langle \psi_{{\rm in},t=-\infty} | H_0 | \psi_{{\rm in},t=-\infty} \rangle = \text{“in energy”}.
\end{aligned} \tag{4.23}$$

In other words, **energy is conserved in the scattering**.

It is convenient, therefore, to expand $|\psi_{\text{in}}\rangle$ and $|\psi_{\text{out}}\rangle$ in eigenkets of H_0. We use the momentum eigenkets $|\boldsymbol{p}\rangle$, satisfying

$$H_0 \,|\, \boldsymbol{p}\rangle = \frac{p^2}{2m}\,|\boldsymbol{p}\rangle \equiv E_p|\boldsymbol{p}\rangle. \tag{4.24}$$

The corresponding wave function of the coordinate $\boldsymbol{r}$ is

$$\langle \boldsymbol{r} \,|\, \boldsymbol{p}\rangle = \psi_{\boldsymbol{p}}(\boldsymbol{r}) = (2\pi\hbar)^{-3/2} \mathrm{e}^{i\,\boldsymbol{p}\cdot\boldsymbol{r}/\hbar}, \tag{4.25}$$

which is not normalizable (except to a delta function), and so cannot correspond to a physical state belonging to a Hilbert space. However, physical states belonging to $\mathcal{H}$ can be written as "packets" of the "improper" states $|\boldsymbol{p}\rangle$.

Thus, we write

$$|\psi_{\text{in}}\rangle = \int \mathrm{d}^3\boldsymbol{p}\,|\boldsymbol{p}\rangle\,\langle \boldsymbol{p} \mid \psi_{\text{in}}\rangle \equiv \int \mathrm{d}^3\boldsymbol{p}\,|\boldsymbol{p}\rangle\,\psi_{\text{in}}(\boldsymbol{p}), \tag{4.26}$$

where we have used the closure relation (see section 2.2)

$$\int \mathrm{d}^3\boldsymbol{p}\,|\boldsymbol{p}\rangle\,\langle \boldsymbol{p}| = 1. \tag{4.27}$$

Of course, if the coefficients $\psi_{\text{in}}(\boldsymbol{p})$ in the packet (4.26) are strongly peaked near $\boldsymbol{p} \approx \boldsymbol{p}_0$, then $|\psi_{\text{in}}\rangle$ **approximates** to the momentum eigenstate $|\boldsymbol{p}_0\rangle$.

Acting with S on the packet (4.26), we find the ket $|\psi_{\text{out}}\rangle$ [see (4.19)]:

$$|\psi_{\text{out}}\rangle = S\,|\psi_{\text{in}}\rangle = \int \mathrm{d}^3\boldsymbol{p} S\,|\boldsymbol{p}\rangle\,\psi_{\text{in}}(\boldsymbol{p}). \tag{4.28}$$

Thus, the probability amplitude that the particle in the state $|\psi_{\text{out}}\rangle$ has momentum $\boldsymbol{p}'$ is

$$\langle \boldsymbol{p}' \mid \psi_{\text{out}}\rangle \equiv \psi_{\text{out}}(\boldsymbol{p}') = \int \mathrm{d}^3\boldsymbol{p}\langle \boldsymbol{p}'|S|\boldsymbol{p}\rangle\psi_{\text{in}}(\boldsymbol{p}). \tag{4.29}$$

The matrix with elements $\langle \boldsymbol{p}'|S|\boldsymbol{p}\rangle$ is the ***S*-matrix** (in the momentum representation).

Let $\psi_{\text{in}}(\boldsymbol{p}) = \delta^{(3)}(\boldsymbol{p} - \boldsymbol{p}_0)$, that is, the in-state has momentum $\boldsymbol{p}_0$. Then

$$\langle \boldsymbol{p}'|\psi_{\text{out}}\rangle = \langle \boldsymbol{p}'|S|\boldsymbol{p}_0\rangle. \tag{4.30}$$

Thus, $\langle \boldsymbol{p}'|S|\boldsymbol{p}\rangle$ is interpreted as the probability amplitude that the particle in the state $|\psi_{\text{out}}\rangle$ is found to have momentum $\boldsymbol{p}'$ if it had momentum $\boldsymbol{p}$ in the state $|\psi_{\text{in}}\rangle$.

In general, if the system is prepared in the in-state $|\psi_{\text{in}}\rangle = |\chi_1\rangle$, then $\langle\chi_2|S|\chi_1\rangle$ is the probability amplitude that measurements of the appropriate observables on the out-state $|\psi_{\text{out}}\rangle$ will yield values characteristic of the state $|\chi_2\rangle$. This probability amplitude may be written as

$$\begin{aligned}\langle\chi_2|\,S\,|\chi_1\rangle &= \langle\chi_2|\,\Omega_-^\dagger\Omega_+\,|\chi_1\rangle \equiv (\chi_2, \Omega_-^\dagger\Omega_+\chi_1)\\ &= (\Omega_-\chi_2, \Omega_+\chi_1) \equiv (\chi_2-, \chi_1+) \equiv \langle\chi_2- \mid \chi_1+\rangle,\end{aligned} \tag{4.31}$$

which is equal to the overlap of the scattering state that corresponds to the in-asymptote $|\chi_1\rangle$ with the scattering state that corresponds to the out-asymptote $|\chi_2\rangle$. Analogously,

$$\langle \boldsymbol{p}'|S|\boldsymbol{p}\rangle = \langle \boldsymbol{p}'-|\boldsymbol{p}+\rangle, \tag{4.32}$$

where

$$|\boldsymbol{p}\pm\rangle = \Omega_\pm|\boldsymbol{p}\rangle. \tag{4.33}$$

The states $|\boldsymbol{p}\pm\rangle$ are eigenstates of the **full** hamiltonian H, with eigenvalues $E_p = p^2/2m$, as is shown by the following simple proof:

$$\begin{aligned} H\,|\boldsymbol{p}\pm\rangle &= H\Omega_\pm\,|\boldsymbol{p}\rangle = \Omega_\pm H_0\,|\boldsymbol{p}\rangle \\ &= \Omega_\pm E_p\,|\boldsymbol{p}\rangle = E_p\Omega_\pm\,|\boldsymbol{p}\rangle = E_p\,|\boldsymbol{p}\pm\rangle\,. \end{aligned} \tag{4.34}$$

In coordinate space, the corresponding wave functions $\psi_{\boldsymbol{p}}^{\pm}(\boldsymbol{r}) \equiv \langle \boldsymbol{r}\,|\,\boldsymbol{p}\pm\rangle$ satisfy

$$H\psi_{\boldsymbol{p}}^{\pm}(\boldsymbol{r}) = E_p\psi_{\boldsymbol{p}}^{\pm}(\boldsymbol{r})\,. \tag{4.35}$$

The result (4.34) implies that

$$U(t)\,|\boldsymbol{p}+\rangle = \mathrm{e}^{-iE_pt/\hbar}\,|\boldsymbol{p}+\rangle\,. \tag{4.36}$$

But,

$$U_0(t)\,|\boldsymbol{p}\rangle = \mathrm{e}^{-iE_pt/\hbar}\,|\boldsymbol{p}\rangle\,, \tag{4.37}$$

from which it is clear that

$$U(t)\,|\boldsymbol{p}+\rangle - U_0(t)\,|\boldsymbol{p}\rangle \not\to 0 \quad \text{as} \quad t\to\infty! \tag{4.38}$$

Thus, the asymptotic condition (see section 4.1) is not obeyed for the kets $|\boldsymbol{p}\rangle$ and $|\boldsymbol{p}+\rangle$. This is another manifestation of the improper character of the kets $|\boldsymbol{p}\rangle$ and $|\boldsymbol{p}+\rangle$. [Note that if, instead, we consider a **wave packet**

$$|\chi\rangle = \int \mathrm{d}^3\boldsymbol{p}\,\chi(\boldsymbol{p})\,|\boldsymbol{p}\rangle, \tag{4.39}$$

then

$$|\chi+\rangle = \Omega_+\,|\chi\rangle = \int \mathrm{d}^3\boldsymbol{p}\,\chi(\boldsymbol{p})\,\Omega_+\,|\boldsymbol{p}\rangle = \int \mathrm{d}^3\boldsymbol{p}\,\chi(\boldsymbol{p})\,|\boldsymbol{p}+\rangle \tag{4.40}$$

and

$$U(t)\,|\chi+\rangle - U_0(t)\,|\chi\rangle = \int \mathrm{d}^3\boldsymbol{p}\,\chi(\boldsymbol{p})\,\mathrm{e}^{-iE_pt/\hbar}\,(|\boldsymbol{p}+\rangle - |\boldsymbol{p}\rangle), \tag{4.41}$$

which **does** tend to zero as $t\to\infty$, by the Riemann–Lebesgue lemma.]

We now consider the general structure of the S-matrix in the momentum representation. Since $[H_0, S] = 0$ [see (4.21)], we have

$$0 = \langle \boldsymbol{p}'|[H_0, S]|\boldsymbol{p}\rangle = (E_{p'} - E_p)\langle \boldsymbol{p}'|S|\boldsymbol{p}\rangle. \tag{4.42}$$

Therefore, $\langle \boldsymbol{p}'|S|\boldsymbol{p}\rangle = 0$ except when $E_{p'} = E_p$, that is, when $p'^2 = p^2$. Thus, $\langle \boldsymbol{p}'|S|\boldsymbol{p}\rangle$ contains the Dirac delta function $\delta(E_{p'} - E_p)$ as a factor. We now write S in the form $S = 1 + R$ (which implies that $[H_0, R] = 0$, too). Then

$$\langle \boldsymbol{p}'|S|\boldsymbol{p}\rangle = \delta^{(3)}(\boldsymbol{p}' - \boldsymbol{p}) + \langle \boldsymbol{p}'|R|\boldsymbol{p}\rangle, \tag{4.43}$$

which may be written as

$$\langle \boldsymbol{p}'|S\,|\boldsymbol{p}\rangle = \delta^{(3)}(\boldsymbol{p}' - \boldsymbol{p}) - 2\pi i\delta(E_{p'} - E_p)t(\boldsymbol{p}' \leftarrow \boldsymbol{p}), \tag{4.44}$$

where the first part gives no scattering, and $t\left(\boldsymbol{p}' \leftarrow \boldsymbol{p}\right)$ contains no further δ-functions and may be assumed to be a smooth function of its arguments.

Since $t\left(\boldsymbol{p}' \leftarrow \boldsymbol{p}\right)$ is defined only for $\boldsymbol{p}'$ and $\boldsymbol{p}$ such that $p'^2 = p^2$, it does not define an operator. (To do that, we should need matrix elements for **all** values of $\boldsymbol{p}'$ and $\boldsymbol{p}$.) We shall show in the following, however, that we **can** define an operator $T(z)$ independently such that, for $z = E_p + i0$, its element $\langle \boldsymbol{p}'|T\,(z)\,|\boldsymbol{p}\rangle$ is just $t(\boldsymbol{p}' \leftarrow \boldsymbol{p})$ when $p'^2 = p^2$. The quantity $t\left(\boldsymbol{p}' \leftarrow \boldsymbol{p}\right)$ may thus be called the T-matrix on the energy shell $p'^2 = p^2$, or, simply, the **on-shell T-matrix**.

4.3 The Green operator and T operator

If $H = H_0 + V$, with $H_0 = p^2/2m$, the **Green operators** corresponding to H_0 and H are defined by

$$G^0\,(z) \equiv (z - H_0)^{-1} \tag{4.45}$$

and

$$G\,(z) \equiv (z - H)^{-1} \tag{4.46}$$

for any z, real or complex, for which the inverse of $z - H_0$ $(z - H)$ exists.

For such z,

$$(z - H_0)G^0\,(z) = 1. \tag{4.47}$$

In the coordinate representation,

$$\langle \boldsymbol{r}|(z - H_0)G^0(z)|\boldsymbol{r}'\rangle = \langle \boldsymbol{r}\mid \boldsymbol{r}'\rangle = \delta^{(3)}(\boldsymbol{r} - \boldsymbol{r}'). \tag{4.48}$$

But, for any $|\psi\rangle$ in $\mathcal{H}$, including $G^0(z)\,|\boldsymbol{r}'\rangle$, we have

$$\begin{aligned}\langle \boldsymbol{r}\,|H_0|\psi\rangle &\equiv \langle \boldsymbol{r}|\left(\int \mathrm{d}^3\boldsymbol{r}'|\boldsymbol{r}'\rangle\left(-\frac{\hbar^2}{2m}\nabla^2_{\boldsymbol{r}'}\right)\langle \boldsymbol{r}'|\right)|\psi\rangle \\ &= \int \mathrm{d}^3\boldsymbol{r}'\delta^{(3)}(\boldsymbol{r}-\boldsymbol{r}')\left(-\frac{\hbar^2}{2m}\nabla^2_{\boldsymbol{r}'}\right)\psi(\boldsymbol{r}') \\ &= -\frac{\hbar^2}{2m}\nabla^2\psi(\boldsymbol{r}) = -\frac{\hbar^2}{2m}\nabla^2\langle \boldsymbol{r}|\psi\rangle.\end{aligned}$$

Therefore,

$$\left(\frac{\hbar^2\nabla^2}{2m}+z\right)\langle \boldsymbol{r}|G^0(z)|\boldsymbol{r}'\rangle = \delta^{(3)}(\boldsymbol{r}-\boldsymbol{r}'), \tag{4.49}$$

which defines the function $G^0(\boldsymbol{r},\boldsymbol{r}';z) \equiv \langle \boldsymbol{r}|G^0(z)|\boldsymbol{r}'\rangle$ as the **Green function** of the differential operator $\hbar^2\nabla^2/2m + z$.

Exactly similarly, $\langle \boldsymbol{r}|\,G(z)\,|\boldsymbol{r}'\rangle$ is the Green function of the operator $\hbar^2\nabla^2/2m - V(r) + z$.

Let $z = E_n$, where E_n is an eigenvalue of the full hamiltonian H:

$$(E_n - H)\,|n\rangle = 0. \tag{4.50}$$

Then the operator $G(z)$ (4.46) is undefined for $z = E_n$.

If the spectrum $\{E_n\}$ of H is purely discrete (this will not be the case, of course, in scattering problems), completeness implies that

$$\sum_n |n\rangle\,\langle n| = 1. \tag{4.51}$$

Applying the operator $G(z) = (z-H)^{-1}$ to both sides of this, we have

$$G(z) = (z-H)^{-1}\sum_n |n\rangle\,\langle n| = \sum_n \frac{|n\rangle\,\langle n|}{z-E_n}, \tag{4.52}$$

which is well defined for all z not equal to an eigenvalue E_n.

The matrix element

$$\langle\chi|\,G(z)\,|\psi\rangle = \sum_n \frac{\langle\chi\mid n\rangle\,\langle n\mid\psi\rangle}{z-E_n} \equiv \sum_n \frac{\chi(n)^*\,\psi(n)}{z-E_n} \tag{4.53}$$

has poles at $z = E_n$ with residues $\langle\chi\mid n\rangle\,\langle n\mid\psi\rangle$, and so we may say that the operator $G(z)$ has poles at $z = E_n$ with residues equal to the projection operators $|n\rangle\,\langle n|$. [$G(z)$ is analytic except on the spectrum of H, and knowledge of $G(z)$ is equivalent to knowledge of the complete eigenvalue spectrum.]

To see what happens in the case of a **continuous** spectrum, consider the free Green function $G^0(z)$. The eigenkets of H_0 form an energy continuum and may be chosen in the form $|E, l, m\rangle$. Thus, we can write

$$G^0(z) = \sum_{lm} \int_0^\infty \mathrm{d}E \frac{|E, l, m\rangle\langle E, l, m|}{z - E}, \tag{4.54}$$

and, therefore,

$$\langle\chi|G^0(z)|\psi\rangle = \int_0^\infty \frac{\mathrm{d}E}{z - E} \sum_{lm} \langle\chi|E, l, m\rangle\langle E, l, m|\psi\rangle, \tag{4.55}$$

in which, if z has any real positive value, the denominator vanishes at a value of E in the range of integration; that is, $G^0(z)$ is analytic for all z except the **branch cut** running from 0 to ∞ along the real axis.

We use (4.55) to calculate the **discontinuity** across the branch cut (at $z = E_0$, say). The result is:

$$\begin{aligned} \operatorname{disc}\langle\chi|G^0(E_0)|\psi\rangle &\equiv \lim_{\varepsilon\to+0} \left\{\langle\chi|G^0(E_0 + i\varepsilon)|\psi\rangle - \langle\chi|G^0(E_0 - i\varepsilon)|\psi\rangle\right\} \\ &= \lim_{\varepsilon\to+0} \int \mathrm{d}E \left(\frac{1}{E_0 - E + i\varepsilon} - \frac{1}{E_0 - E - i\varepsilon}\right) \sum_{lm} \langle\chi|E, l, m\rangle\langle E, l, m|\psi\rangle. \end{aligned} \tag{4.56}$$

But

$$\lim_{\varepsilon\to+0} \frac{1}{E_0 - E \pm i\varepsilon} = \frac{\wp}{E_0 - E} \mp i\pi\delta(E - E_0), \tag{4.57}$$

where $\wp$ denotes the principal-value integral when the expression is integrated. Therefore,

$$\operatorname{disc}\langle\chi|G^0(E_0)|\psi\rangle = -2\pi i \sum_{lm} \langle\chi|E_0, l, m\rangle\langle E_0, l, m|\psi\rangle. \tag{4.58}$$

Thus, formally, disc $G^0(E_0)$ is $-2\pi i$ times the sum of the projection operators for the states $|E_0, l, m\rangle$ with energy E_0.

Since, in general, $\mathcal{H}$ has both bound and continuum states, the **full** Green function has both poles and a branch cut (see figure 4.6).

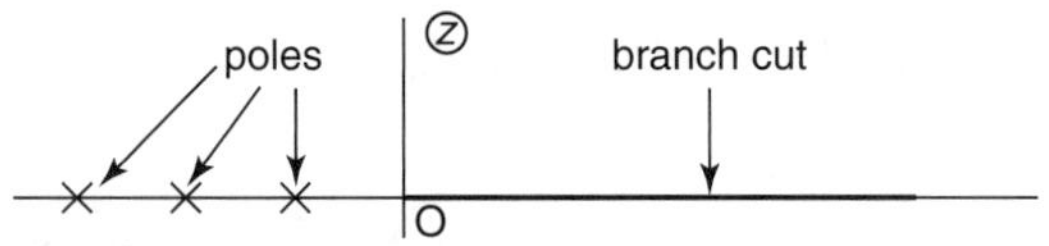

Figure 4.6 *Analytical structure of the Green function G(z) in the z plane.*

We can obtain a useful formula relating G and G^0 by using the operator identity

$$A^{-1} = B^{-1} + B^{-1}(B - A)A^{-1}. \tag{4.59}$$

We set $A = z - H$ and $B = z - H_0$. Then

$$G(z) = G^0(z) + G^0(z)VG(z), \tag{4.60}$$

or, setting $A = z - H_0$ and $B = z - H$, we get

$$G(z) = G^0(z) + G(z)VG^0(z). \tag{4.61}$$

Equations (4.60) and (4.61) are the two forms of the **Lippmann–Schwinger equation** for $G(z)$ in terms of $G^0(z)$, which is known via the equation

$$G^0 |\boldsymbol{p}\rangle = (z - H_0)^{-1} |\boldsymbol{p}\rangle = \frac{1}{z - E_p} |\boldsymbol{p}\rangle . \tag{4.62}$$

Note that, since $H_0^\dagger = H_0$ and $H^\dagger = H$, we have:

$$G^0(z)^\dagger = G^0(z^*) \quad \text{and} \quad G(z)^\dagger = G(z^*). \tag{4.63}$$

We now introduce a new operator $T(z)$, defined by

$$T(z) = V + VG(z)V, \tag{4.64}$$

known as the ***T*-operator**. By virtue of this definition, this operator must have the same analyticity properties (poles and branch cut) as the Green operator $G(z)$.

We now act on the left of $T(z)$ (4.64) with the operator $G^0(z)$:

$$G^0(z)T(z) = \left(G^0(z) + G^0(z)VG(z)\right) V = G(z)V \tag{4.65}$$

where the last equality follows from the Lippmann–Schwinger equation (4.60). Similarly, we find

$$T(z)G^0(z) = VG(z). \tag{4.66}$$

We can use these simple relations to find $G(z)$ in terms of $T(z)$. From the Lippmann–Schwinger equation (4.60) we have:

$$G(z) = G^0(z) + G^0(z)VG(z) = G^0(z) + G^0(z)T(z)G^0(z). \tag{4.67}$$

Also, putting $G(z)V = G^0(z)T(z)$ [see (4.65)] into the definition (4.64) of the T-operator, we obtain the Lippmann–Schwinger equation for $T(z)$:

$$T(z) = V + VG^0(z)T(z), \tag{4.68}$$

which one might hope to solve for $T(z)$ by iteration, if the scattering potential V is sufficiently weak. Substituting the first approximation $T^{(1)} = V$ into the right-hand side of (4.68) gives the second approximation

$$T^{(2)}(z) = V + VG^0(z)V, \tag{4.69}$$

which can then be substituted into the right-hand side of (4.68) to give the third approximation, and so on.

The resulting series

$$T(z) = V + VG^0(z)V + VG^0(z)VG^0(z)V + \cdots \tag{4.70}$$

is called the **Born series** for $T(z)$, and, for any given problem, may or may not converge. The first approximation $T = T^{(1)} = V$ is called the **Born approximation**.

Note also that, from the definition of $T(z)$, we have

$$T(z)^\dagger = T(z^*). \tag{4.71}$$

The action of the Møller operators $\Omega_\pm$ on any ket $|\phi\rangle$ in $\mathcal{H}$ may be expressed in terms of expressions involving the action of the Green operator $G(z)$ or T-operator $T(z)$. Consider, for example, the scattering state $|\phi-\rangle$ [see equations (4.13) and (4.11)]. We have

$$|\phi-\rangle \equiv \Omega_- |\phi\rangle = \lim_{t\to\infty} U^\dagger(t)U_0(t)|\phi\rangle = |\phi\rangle + \frac{i}{\hbar}\int_0^\infty \mathrm{d}\tau U^\dagger(\tau)VU_0(\tau)|\phi\rangle, \tag{4.72}$$

where, as in Appendix A1, we have written Ω_- as the time integral of its time derivative. The integral converges, as is shown in Appendix A1. We do not change $|\phi-\rangle$ by writing it in the form

$$|\phi-\rangle = |\phi\rangle + \frac{i}{\hbar}\lim_{\varepsilon\to+0}\int_0^\infty \mathrm{d}\tau \mathrm{e}^{-\varepsilon\tau/\hbar}U^\dagger(\tau)VU_0(\tau)|\phi\rangle \tag{4.73}$$

(in fact, the factor $\mathrm{e}^{-\varepsilon\tau/\hbar}$ clearly improves the convergence of the integral). Similarly, since the expression for $|\phi+\rangle$ involves an integral over τ from $-\infty$ to 0, we insert the factor $e^{\varepsilon\tau/\hbar}$ and again take the limit $\varepsilon \to +0$. This procedure is equivalent to replacing the potential V by $V(\tau) = Ve^{\mp\varepsilon\tau/\hbar}$ [but neglecting the effect of this replacement on $U(\tau)$]. That this procedure is valid constitutes what is known as the **adiabatic theorem** and is not surprising, since, for small ε, $V(\tau) = V\mathrm{e}^{\mp\varepsilon\tau/\hbar} = V$ near $\tau = 0$ but $V(\tau) \to 0$ as $\tau \to \pm\infty$, as we should expect, since as $\tau \to \pm\infty$ the particle will be outside the range of the potential.

Using the closure relation

$$\int \mathrm{d}^3\boldsymbol{p}\,|\boldsymbol{p}\rangle\langle\boldsymbol{p}| = 1,$$

we now insert a complete set of states $|\boldsymbol{p}\rangle$, with the result

$$|\phi-\rangle \equiv \Omega_- |\phi\rangle = |\phi\rangle + \frac{i}{\hbar}\lim_{\varepsilon\to+0}\int \mathrm{d}^3\boldsymbol{p}\int_0^\infty \mathrm{d}\tau \left[\mathrm{e}^{-\varepsilon\tau/\hbar}U^\dagger(\tau)VU_0(\tau)\right]|\boldsymbol{p}\rangle\langle\boldsymbol{p}\mid\phi\rangle. \tag{4.74}$$

But

$$U_0(\tau)|\boldsymbol{p}\rangle = \mathrm{e}^{-iE_p\tau/\hbar}|\boldsymbol{p}\rangle, \tag{4.75}$$

so the expression in the square brackets in (4.74) can be replaced by

$$[\cdots] \equiv \mathrm{e}^{-i(E_p - i\varepsilon - H)\tau/\hbar} V. \tag{4.76}$$

Therefore,

$$\begin{aligned}\int_0^\infty \mathrm{d}\tau\,[\ldots] &= \left.\frac{\mathrm{e}^{-i(E_p-H)\tau/\hbar}\mathrm{e}^{-\varepsilon\tau/\hbar}}{-i(E_p - i\varepsilon - H)/\hbar} V\right|_0^\infty \\ &= -i\hbar(E_p - i\varepsilon - H)^{-1} V = -i\hbar G(E_p - i\varepsilon)V,\end{aligned} \tag{4.77}$$

so, from (4.74), we have

$$|\phi-\rangle \equiv \Omega_- \,|\phi\rangle = |\phi\rangle + \lim_{\varepsilon\to+0} \int \mathrm{d}^3 \boldsymbol{p} G(E_p - i\varepsilon)V\,|\boldsymbol{p}\rangle\,\langle \boldsymbol{p} \mid \phi\rangle. \tag{4.78}$$

Similarly, in the expression for $|\phi+\rangle$, under the integral $\int_0^{-\infty} \mathrm{d}\tau \ldots$, we make the replacement $V \to \lim_{\varepsilon\to+0} V\mathrm{e}^{+\varepsilon\tau/\hbar}$ and obtain

$$|\phi+\rangle \equiv \Omega_+ \,|\phi\rangle = |\phi\rangle + \lim_{\varepsilon\to+0} \int \mathrm{d}^3 \boldsymbol{p} G(E_p + i\varepsilon)V\,|\boldsymbol{p}\rangle\,\langle \boldsymbol{p} \mid \phi\rangle. \tag{4.79}$$

The signs in the term $\pm i\varepsilon$ in equation (4.79) for $|\phi+\rangle$ and equation (4.78) for $|\phi-\rangle$ may be taken as the reasons for the choice of subscripts on the Møller operators $\Omega_\pm$, which are related by equations (4.79) and (4.78) to the Green operator $G(z)$.

Since $G(z)V = G^0(z)T(z)$ [see (4.65)], we can also express the action of the Møller operators in terms of the action of the T-operator:

$$|\phi\pm\rangle \equiv \Omega_\pm \,|\phi\rangle = |\phi\rangle + \lim_{\varepsilon\to+0} \int \mathrm{d}^3 \boldsymbol{p} G^0(E_p \pm i\varepsilon)T(E_p \pm i\varepsilon)\,|\boldsymbol{p}\rangle\,\langle \boldsymbol{p} \mid \phi\rangle. \tag{4.80}$$

Finally, we may express the action of the scattering operator S in terms of the action of the T-operator. The result (see Appendix A2) is

$$\langle \boldsymbol{p}'|S\,|\boldsymbol{p}\rangle = \delta^{(3)}(\boldsymbol{p}' - \boldsymbol{p}) - 2\pi i\delta(E_{p'} - E_p) \lim_{\varepsilon\to+0} \langle \boldsymbol{p}'|T(E_p + i\varepsilon)|\boldsymbol{p}\rangle. \tag{4.81}$$

Comparing this with equation (4.44), we see that the quantity $t(\boldsymbol{p}' \leftarrow \boldsymbol{p})$ in the latter expression can be identified as

$$t(\boldsymbol{p}' \leftarrow \boldsymbol{p}) = \lim_{\varepsilon\to+0} \langle \boldsymbol{p}'|T(E_p + i\varepsilon)|\boldsymbol{p}\rangle \equiv \langle \boldsymbol{p}'|T(E_p + i0)|\boldsymbol{p}\rangle \quad (E_{p'} = E_p). \tag{4.82}$$

The expression (4.82) is clearly an **on-shell** ($E_{p'} = E_p$) matrix element (evaluated at $z = E_p + i0$) of the T-operator $T(z)$, whose **general** elements $\langle \boldsymbol{p}'|T(z)\,|\boldsymbol{p}\rangle$ constitute the **off-shell T-matrix**. Thus, the statements following equation (4.44) are confirmed.

4.4 The stationary scattering states

In the last section we found [see (4.79) and (4.78)]

$$|\phi\pm\rangle \equiv \Omega_{\pm}\,|\phi\rangle = |\phi\rangle + \lim_{\varepsilon\to+0}\int \mathrm{d}^3\boldsymbol{p}G(E_p \pm i\varepsilon)V\,|\boldsymbol{p}\rangle\,\langle\boldsymbol{p}\mid\phi\rangle$$
$$\equiv |\phi\rangle + \int \mathrm{d}^3\boldsymbol{p}G(E_p \pm i0)V\,|\boldsymbol{p}\rangle\,\phi(\boldsymbol{p}). \tag{4.83}$$

If the ket $|\phi\rangle$ is now expanded in the momentum eigenkets $|\boldsymbol{p}\rangle$, that is,

$$|\phi\rangle = \int \mathrm{d}^3\boldsymbol{p}\phi(\boldsymbol{p})\,|\boldsymbol{p}\rangle \quad (\text{with } \phi(\boldsymbol{p}) \equiv \langle\boldsymbol{p}\,|\,\phi\rangle), \tag{4.84}$$

so that

$$|\phi\pm\rangle \equiv \Omega_{\pm}|\phi\rangle = \int \mathrm{d}^3\boldsymbol{p}\phi(\boldsymbol{p})\Omega_{\pm}\,|\boldsymbol{p}\rangle = \int \mathrm{d}^3\boldsymbol{p}\phi(\boldsymbol{p})\,|\boldsymbol{p}\pm\rangle, \tag{4.85}$$

then (4.83) becomes

$$\int \mathrm{d}^3\boldsymbol{p}\phi(\boldsymbol{p})|\boldsymbol{p}\pm\rangle = \int \mathrm{d}^3\boldsymbol{p}\phi(\boldsymbol{p})[|\boldsymbol{p}\rangle + G(E_p \pm i0)V|\boldsymbol{p}\rangle]. \tag{4.86}$$

Since this holds for arbitrary $\phi(\boldsymbol{p})$, we have

$$|\boldsymbol{p}\pm\rangle = |\boldsymbol{p}\rangle + G(E_p \pm i0)V\,|\boldsymbol{p}\rangle\,. \tag{4.87}$$

To convert this into an **equation** for $|\boldsymbol{p}\pm\rangle$, we note that $G(z)V = G^0(z)T(z)$ [see equation (4.65)], and that, by the definition (4.64) of $T(z)$, we have

$$\begin{aligned} T(E_p \pm i0)\,|\boldsymbol{p}\rangle &= [V + VG(E_p \pm i0)V]\,|\boldsymbol{p}\rangle \\ &= V[|\boldsymbol{p}\rangle + G(E_p \pm i0)V\,|\boldsymbol{p}\rangle] = V\,|\boldsymbol{p}\pm\rangle\,, \end{aligned} \tag{4.88}$$

where in the last equality we have used (4.87). Therefore,

$$|\boldsymbol{p}\pm\rangle = |\boldsymbol{p}\rangle + G^0(E_p \pm i0)V\,|\boldsymbol{p}\pm\rangle\,, \tag{4.89}$$

which is the desired equation for $|\boldsymbol{p}\pm\rangle$ (the Lippmann–Schwinger equation for $|\boldsymbol{p}\pm\rangle$).

We note also that, from (4.88), for $|\boldsymbol{p}'| = |\boldsymbol{p}|$, we have

$$t(\boldsymbol{p}' \leftarrow \boldsymbol{p}) \equiv \langle\boldsymbol{p}'|T(E_p + i0)|\boldsymbol{p}\rangle = \langle\boldsymbol{p}'|V|\boldsymbol{p}+\rangle. \tag{4.90}$$

Alternatively, taking the adjoint of the second version (–) of (4.88), and using the relation (4.71), we obtain

$$t(\boldsymbol{p}' \leftarrow \boldsymbol{p}) = \langle\boldsymbol{p}'|T(E_p + i0)|\boldsymbol{p}\rangle = \langle\boldsymbol{p}'-|V|\boldsymbol{p}\rangle, \tag{4.91}$$

so that $t(\boldsymbol{p}' \leftarrow \boldsymbol{p})$ can be calculated from knowledge of either $|\boldsymbol{p}+\rangle$ or $|\boldsymbol{p}-\rangle$.

We now write (4.89) in the coordinate representation, by projecting it on to the bra $\langle \boldsymbol{r}|$ and replacing the Hilbert-space operator V by

$$V = \int \mathrm{d}^3 \boldsymbol{r}' |\boldsymbol{r}'\rangle V(\boldsymbol{r}')\langle \boldsymbol{r}'| \tag{4.92}$$

[see (2.12)]. As a result, the **scattering-state wave function** $\psi_{\boldsymbol{p}}^{\pm}(\boldsymbol{r}) \equiv \langle \boldsymbol{r}|\boldsymbol{p}\pm\rangle$ is expressed as

$$\psi_{\boldsymbol{p}}^{\pm}(\boldsymbol{r}) \equiv \langle \boldsymbol{r} \mid \boldsymbol{p}\pm\rangle = \langle \boldsymbol{r} \mid \boldsymbol{p}\rangle + \int \mathrm{d}^3 \boldsymbol{r}' \, \langle \boldsymbol{r}| \, G^0(E_p \pm i0)|\, \boldsymbol{r}'\rangle V(\boldsymbol{r}')\langle \boldsymbol{r}' \mid \boldsymbol{p}\pm\rangle. \tag{4.93}$$

We now show that, in the limit $r \to \infty$, that is, far from the scatterer, the "+" version of this wave function has the form of **a plane wave and an outgoing spherical wave**, that is, that

$$\psi_{\boldsymbol{p}}^{+}(\boldsymbol{r}) \underset{(r\to\infty)}{\to} \text{const} \times \left[\mathrm{e}^{i\boldsymbol{p}\cdot\boldsymbol{r}/\hbar} + f(p\hat{\boldsymbol{r}} \leftarrow \boldsymbol{p}) \frac{\mathrm{e}^{ipr/\hbar}}{r} \right], \tag{4.94}$$

where $\hat{\boldsymbol{r}}$ is the unit vector in the direction of $\boldsymbol{r}$ and

$$f(\boldsymbol{p}' \leftarrow \boldsymbol{p}) = -(2\pi)^2 m\hbar \langle \boldsymbol{p}'|V|\boldsymbol{p}+\rangle = -(2\pi)^2 m\hbar t(\boldsymbol{p}' \leftarrow \boldsymbol{p}). \tag{4.95}$$

[The wave $\mathrm{e}^{ipr/\hbar}/r$ is called "outgoing" because, when the time-dependent part $\mathrm{e}^{-iE_pt/\hbar}$ is included (see section 1.2), giving $\mathrm{e}^{i(pr-E_pt)/\hbar}/r$, a point of constant phase (i.e., a point with $pr - E_pt = \text{const}$) in the wave moves with speed (phase velocity) E_p/p (>0); similarly, the wave $\mathrm{e}^{-ipr/\hbar}/r$, which appears in the "–" version, is an "incoming" wave with phase velocity $-E_p/p$ (<0).]

The quantity $f(p\hat{\boldsymbol{r}} \leftarrow \boldsymbol{p})$ is clearly the relative amplitude of the scattered spherical wave and, in general, $f(\boldsymbol{p}' \leftarrow \boldsymbol{p})$ is known as the **scattering amplitude**.

We now give a proof of (4.94) and (4.95).

We first need an expression for $\langle \boldsymbol{r}| \, G^0(z)|\, \boldsymbol{r}'\rangle$. Inserting $\int \mathrm{d}^3 p \, |\boldsymbol{p}\rangle \, \langle \boldsymbol{p}| = 1$, we find:

$$\langle \boldsymbol{r}|G^0(z)|\, \boldsymbol{r}'\rangle = \int \mathrm{d}^3 \boldsymbol{p} \langle \boldsymbol{r}|G^0(z)|\boldsymbol{p}\rangle \langle \boldsymbol{p}|\, \boldsymbol{r}'\rangle. \tag{4.96}$$

But

$$G^0(z)\, |\boldsymbol{p}\rangle = \frac{1}{z - E_p} \, |\boldsymbol{p}\rangle\,, \tag{4.97}$$

where $E_p = p^2/2m$, and

$$\langle \boldsymbol{p} \mid \boldsymbol{r}'\rangle = \langle \boldsymbol{r}' \mid \boldsymbol{p}\rangle^* = (2\pi\hbar)^{-3/2} \mathrm{e}^{-i\boldsymbol{p}\cdot\boldsymbol{r}'/\hbar}. \tag{4.98}$$

Therefore,

$$\langle \boldsymbol{r}|G^0(z)|\, \boldsymbol{r}'\rangle = \frac{1}{(2\pi\hbar)^3} \int \mathrm{d}^3 \boldsymbol{p} \frac{\mathrm{e}^{i\boldsymbol{p}\cdot(\boldsymbol{r}-\boldsymbol{r}')/\hbar}}{z - p^2/2m}. \tag{4.99}$$

In Appendix A3 this integral is evaluated and the value $z = E_p \pm i0$ is substituted into the result, giving

$$\langle \boldsymbol{r}|G^0(E_p \pm i0)|\boldsymbol{r}'\rangle = -\frac{m}{2\pi\hbar^2}\frac{\mathrm{e}^{\pm ip|\boldsymbol{r}-\boldsymbol{r}'|/\hbar}}{|\boldsymbol{r}-\boldsymbol{r}'|}. \tag{4.100}$$

Thus, (4.93) becomes

$$\langle \boldsymbol{r} \mid \boldsymbol{p}\pm\rangle = \langle \boldsymbol{r} \mid \boldsymbol{p}\rangle - \frac{m}{2\pi\hbar^2}\int \mathrm{d}^3\boldsymbol{r}'\frac{\mathrm{e}^{\pm ip|\boldsymbol{r}-\boldsymbol{r}'|/\hbar}}{|\boldsymbol{r}-\boldsymbol{r}'|}V(\boldsymbol{r}')\langle \boldsymbol{r}'|\boldsymbol{p}\pm\rangle. \tag{4.101}$$

If $V(\boldsymbol{r}')$ falls off sufficiently fast with $|\boldsymbol{r}'|$ we make no error by restricting the range of integration over $\boldsymbol{r}'$. We then consider the limit when r is greater than all the r' that contribute to the integral ($r \to \infty$). We can write

$$\begin{aligned}|\boldsymbol{r}-\boldsymbol{r}'| &= |r\hat{\boldsymbol{r}}-\boldsymbol{r}'| = (r^2 - 2r\hat{\boldsymbol{r}}\cdot\boldsymbol{r}' + r'^2)^{1/2} = r\left(1 - \frac{2\hat{\boldsymbol{r}}\cdot\boldsymbol{r}'}{r} + \frac{r'^2}{r^2}\right)^{1/2} \\ &= r - \hat{\boldsymbol{r}}\cdot\boldsymbol{r}' + \frac{1}{2r}(r'^2 + (\hat{\boldsymbol{r}}\cdot\boldsymbol{r}')^2) + \cdots \to r - \hat{\boldsymbol{r}}\cdot\boldsymbol{r}' \text{ as } r \to \infty \ \ (r' \text{ restricted}).\end{aligned} \tag{4.102}$$

Then

$$\mathrm{e}^{\pm ip|\boldsymbol{r}-\boldsymbol{r}'|/\hbar} \to \mathrm{e}^{\pm ipr/\hbar}\mathrm{e}^{\mp ip\hat{\boldsymbol{r}}\cdot\boldsymbol{r}'/\hbar} \tag{4.103}$$

and

$$\frac{1}{|\boldsymbol{r}-\boldsymbol{r}'|} = \frac{1}{r\left(1-\dfrac{\hat{\boldsymbol{r}}\cdot\boldsymbol{r}'}{r}\right)} \to \frac{1}{r}\left(1+\frac{\hat{\boldsymbol{r}}\cdot\boldsymbol{r}'}{r}\right) \to \frac{1}{r} \quad \text{as } r \to \infty. \tag{4.104}$$

Therefore, (4.101) becomes

$$\langle \boldsymbol{r} \mid \boldsymbol{p}\pm\rangle \underset{(r\to\infty)}{\to} \langle \boldsymbol{r} \mid \boldsymbol{p}\rangle - \frac{m}{2\pi\hbar^2}\frac{\mathrm{e}^{\pm ipr/\hbar}}{r}\int \mathrm{d}^3\boldsymbol{r}'\mathrm{e}^{\mp ip\hat{\boldsymbol{r}}\cdot\boldsymbol{r}'/\hbar}V(\boldsymbol{r}')\langle \boldsymbol{r}'|\boldsymbol{p}\pm\rangle. \tag{4.105}$$

But

$$\mathrm{e}^{\mp ip\hat{\boldsymbol{r}}\cdot\boldsymbol{r}'/\hbar} = (\mathrm{e}^{\pm ip\hat{\boldsymbol{r}}\cdot\boldsymbol{r}'/\hbar})^* = (2\pi\hbar)^{3/2}\langle \boldsymbol{r}'|\pm p\hat{\boldsymbol{r}}\rangle^* = (2\pi\hbar)^{3/2}\langle \pm p\hat{\boldsymbol{r}} \mid \boldsymbol{r}'\rangle \tag{4.106}$$

and

$$\langle \boldsymbol{r} \mid \boldsymbol{p}\rangle = (2\pi\hbar)^{-3/2}\mathrm{e}^{i\boldsymbol{p}\cdot\boldsymbol{r}/\hbar}. \tag{4.107}$$

Therefore,

$$\langle \boldsymbol{r}|\boldsymbol{p}\pm\rangle \underset{(r\to\infty)}{\to} (2\pi\hbar)^{-3/2}\left[\mathrm{e}^{i\boldsymbol{p}\cdot\boldsymbol{r}/\hbar} - \frac{\mathrm{e}^{\pm ipr/\hbar}}{r}(2\pi)^2 m\hbar\int \mathrm{d}^3\boldsymbol{r}'\langle \pm p\hat{\boldsymbol{r}}|\boldsymbol{r}'\rangle V(\boldsymbol{r}')\langle \boldsymbol{r}'|\boldsymbol{p}\pm\rangle\right], \tag{4.108}$$

that is,

$$\langle \boldsymbol{r} \mid \boldsymbol{p}\pm\rangle \underset{(r\to\infty)}{\to} (2\pi\hbar)^{-3/2}\left[e^{i\boldsymbol{p}\cdot\boldsymbol{r}/\hbar} - (2\pi)^2 m\hbar \langle \pm p\hat{\boldsymbol{r}} | V | \boldsymbol{p}\pm\rangle \frac{e^{\pm ipr/\hbar}}{r}\right]. \tag{4.109}$$

Consider the function $\langle \boldsymbol{r}|\boldsymbol{p}+\rangle$, which consists of the incident plane wave and an outgoing spherical wave, whose coefficient is $-(2\pi)^2 m\hbar \langle p\hat{\boldsymbol{r}}| V |\boldsymbol{p}+\rangle$. We showed above [see (4.90)] that

$$\langle \boldsymbol{p}'|V|\boldsymbol{p}+\rangle = t(\boldsymbol{p}' \leftarrow \boldsymbol{p}). \tag{4.110}$$

It is therefore convenient to define a **scattering amplitude**

$$f(\boldsymbol{p}' \leftarrow \boldsymbol{p}) = -(2\pi)^2 m\hbar t(\boldsymbol{p}' \leftarrow \boldsymbol{p}), \tag{4.111}$$

so that, asymptotically as $r \to \infty$,

$$\psi_{\boldsymbol{p}}^{+}(\boldsymbol{r}) \equiv \langle \boldsymbol{r} \mid \boldsymbol{p}+\rangle \underset{(r\to\infty)}{\to} (2\pi\hbar)^{-3/2}\left[e^{i\boldsymbol{p}\cdot\boldsymbol{r}/\hbar} + f(p\hat{\boldsymbol{r}} \leftarrow \boldsymbol{p})\frac{e^{ipr/\hbar}}{r}\right]. \tag{4.112}$$

From this formula it can be seen that the scattering amplitude f has the dimensions of length (as can also be seen from the definition of f, if we bear in mind that $t(\boldsymbol{p}' \leftarrow \boldsymbol{p})$ has the dimensions [energy] × [momentum]$^{-3}$).

Having found this expression for $|\boldsymbol{p}+\rangle = \Omega_+|\boldsymbol{p}\rangle$ for a given plane-wave state $|\boldsymbol{p}\rangle$, which is, of course, an improper state, we now show how the result can be used to study the scattering of a real particle.

The motion of a real particle **at all times** is described by the "orbit" $U(t)|\psi\rangle$, but, as $t \to -\infty$, we have $U(t)|\psi\rangle \to U_0(t)|\psi_{\text{in}}\rangle$. We can expand the in-state as a packet of plane waves:

$$|\psi_{\text{in}}\rangle = \int d^3\boldsymbol{p}\phi(\boldsymbol{p})|\boldsymbol{p}\rangle. \tag{4.113}$$

If $\phi(\boldsymbol{p})$ is sharply peaked about a momentum $\boldsymbol{p}_0$, the packet represents a particle with rather well defined momentum $\boldsymbol{p} \approx \boldsymbol{p}_0$. For example, if $\phi(\boldsymbol{p}) \approx \delta^{(3)}(\boldsymbol{p}-\boldsymbol{p}_0)$, then

$$|\psi_{\text{in}}\rangle \approx |\boldsymbol{p}_0\rangle, \tag{4.114}$$

which develops into

$$|\psi+\rangle = \Omega_+|\psi_{\text{in}}\rangle = \Omega_+|\boldsymbol{p}_0\rangle = |\boldsymbol{p}_0+\rangle, \tag{4.115}$$

and this is the reason one can use to justify the customary simple approach based on the improper states $|\boldsymbol{p}\rangle$ and $|\boldsymbol{p}+\rangle$. In other words, $|\boldsymbol{p}+\rangle$ is an approximate representation of the scattering state if the incident momentum is well defined. The improper state $|\boldsymbol{p}-\rangle$ (in terms of which the scattering state can also be expanded) has no such approximate relation to any physical state, since the momentum after the collision (as $t \to \infty$), unlike the momentum of the incident particle, is not well defined here.

As stated above, **at earlier times** ($t \to -\infty$) the orbit is described by

$$U_0(t)|\psi_{\text{in}}\rangle = \int d^3\boldsymbol{p}\phi(\boldsymbol{p})U_0(t)|\boldsymbol{p}\rangle = \int d^3\boldsymbol{p}\phi(\boldsymbol{p})e^{-iE_pt/\hbar}|\boldsymbol{p}\rangle, \tag{4.116}$$

and the corresponding wave function is

$$\begin{aligned}\langle \boldsymbol{r}|U_0(t)\,|\psi_{\rm in}\rangle &= \int \mathrm{d}^3\boldsymbol{p}\phi(\boldsymbol{p})\mathrm{e}^{-iE_pt/\hbar}\,\langle \boldsymbol{r}\mid \boldsymbol{p}\rangle \\ &= (2\pi\hbar)^{-3/2}\int \mathrm{d}^3\boldsymbol{p}\mathrm{e}^{i(\boldsymbol{p}\cdot\boldsymbol{r}-E_pt)/\hbar} \equiv \psi_{\rm in}(\boldsymbol{r},t).\end{aligned} \tag{4.117}$$

Shortly before $t = 0$, which is the time at which the centre of the wave packet $\psi_{\rm in}(\boldsymbol{r}, t)$ would reach the plane perpendicular to its direction of motion and containing the target responsible for the scattering potential, the orbit is

$$\begin{aligned}U(t)\,|\psi\rangle &= U(t)\Omega_+\,|\psi_{\rm in}\rangle = U(t)\int \mathrm{d}^3\boldsymbol{p}\phi(\boldsymbol{p})\Omega_+\,|\boldsymbol{p}\rangle \\ &= U(t)\int \mathrm{d}^3\boldsymbol{p}\phi(\boldsymbol{p})\,|\boldsymbol{p}+\rangle = \int \mathrm{d}^3\boldsymbol{p}\phi(\boldsymbol{p})\mathrm{e}^{-iE_pt/\hbar}\,|\boldsymbol{p}+\rangle\end{aligned} \tag{4.118}$$

(since $H|\boldsymbol{p}\pm\rangle = E_p\,|\boldsymbol{p}\pm\rangle$), so that

$$\langle \boldsymbol{r}|\,U(t)\,|\psi\rangle \equiv \psi(\boldsymbol{r},t) = \int \mathrm{d}^3\boldsymbol{p}\phi(\boldsymbol{p})\mathrm{e}^{-iE_pt/\hbar}\,\langle \boldsymbol{r}\mid \boldsymbol{p}+\rangle. \tag{4.119}$$

For $r \to \infty$ we can use the asymptotic form (4.112) of $\langle \boldsymbol{r} \mid \boldsymbol{p}+\rangle$:

$$\begin{aligned}\psi(\boldsymbol{r},t) \underset{(r\to\infty)}{\to} & (2\pi\hbar)^{-3/2}\int \mathrm{d}^3\boldsymbol{p}\phi(\boldsymbol{p})e^{-iE_pt/\hbar}\left[\mathrm{e}^{i\boldsymbol{p}\cdot\boldsymbol{r}/\hbar} + f(p\hat{\boldsymbol{r}} \leftarrow \boldsymbol{p})\frac{\mathrm{e}^{ipr/\hbar}}{r}\right] \\ &= \psi_{\rm in}(\boldsymbol{r},t) + \psi_{\rm sc}(\boldsymbol{r},t),\end{aligned} \tag{4.120}$$

where we have used the definition (4.117) of $\psi_{\rm in}(\boldsymbol{r}, t)$ and have introduced the notation $\psi_{\rm sc}(\boldsymbol{r}, t)$ for the second term. From our discussion of $\psi_{\rm in}(\boldsymbol{r}, t)$ it is clear that $\psi_{\rm in}(\boldsymbol{r}, t)$ is precisely the wave function that would have resulted if the target had had no effect on the incident particle, that is, it is the "**unscattered wave**". Correspondingly, $\psi_{\rm sc}(\boldsymbol{r}, t)$ is the "**scattered wave**".

If $\phi(\boldsymbol{p})$ is strongly peaked near $\boldsymbol{p} = \boldsymbol{p}_0$ and $f(p\hat{\boldsymbol{r}} \leftarrow \boldsymbol{p})$ is a slowly varying function of $\boldsymbol{p}$ in the vicinity of $\boldsymbol{p}_0$ (the latter is not the case if there is a so-called **resonance** near $\boldsymbol{p}_0$), we can take f evaluated at $\boldsymbol{p} = \boldsymbol{p}_0$ outside the integral, to give

$$\psi_{\rm sc}(\boldsymbol{r},t) = \text{const}\frac{f(p_0\hat{\boldsymbol{r}} \leftarrow \boldsymbol{p}_0)}{r}\int \mathrm{d}^3\boldsymbol{p}\phi(\boldsymbol{p})\mathrm{e}^{i(pr-E_pt)/\hbar}. \tag{4.121}$$

Since $\phi(\boldsymbol{p})$ is strongly peaked near $\boldsymbol{p}_0$, the direction of which we take to define the direction of the unit vector $\hat{z}$, we can replace $\mathrm{e}^{ipr/\hbar}$ in the integrand by $\mathrm{e}^{i\boldsymbol{p}\cdot\hat{z}r/\hbar}$, since the latter takes the value $\mathrm{e}^{ip_0r/\hbar}$ when $\boldsymbol{p} = \boldsymbol{p}_0$, as required.

Hence, using the definition of $\psi_{\rm in}(\boldsymbol{r}, t)$, we find

$$\psi_{\rm sc}(\boldsymbol{r},t) = \frac{f(p_0\hat{\boldsymbol{r}} \leftarrow \boldsymbol{p}_0)}{r}\psi_{\rm in}(\hat{z}r,t), \tag{4.122}$$

the meaning of which is summarized in figure 4.7. Thus, the scattered wave at vector position $\boldsymbol{r}$ at time t is just the product of $f(p_0\hat{\boldsymbol{r}} \leftarrow \boldsymbol{p}_0)/r$ with the wave ψ_{in} evaluated at time t and at a distance $r = |\boldsymbol{r}|$ past the target on the z axis.

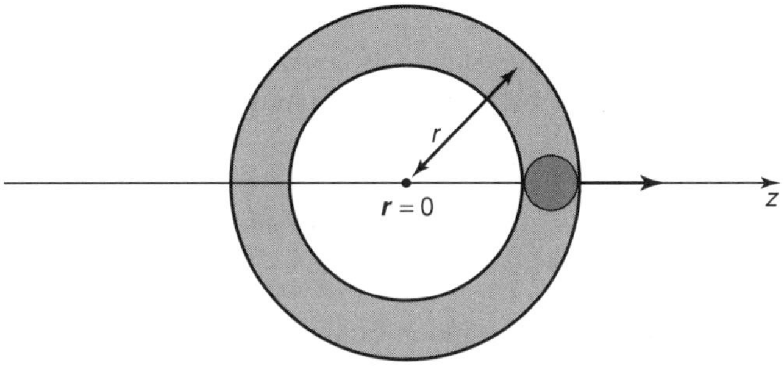

Figure 4.7 *The region of dark shading represents the region of nonzero $\psi_{\text{in}}(\boldsymbol{r}, t)$ (the unscattered packet at time t), moving in the positive z direction. The lightly shaded region (a shell) represents the region of nonzero $\psi_{\text{sc}}(\boldsymbol{r}, t)$ (the scattered wave at time t).*

Note that figure 4.7 corresponds to zero **impact parameter** $\boldsymbol{\rho} = x\hat{\boldsymbol{x}} + y\hat{\boldsymbol{y}}$, defined as the vector distance of the centre of the packet ψ_{in} from the z axis. If the impact parameter $\boldsymbol{\rho}$ has large enough magnitude, as in figure 4.8, then, as can be seen, $\psi_{\text{in}}(\hat{z}r, t) = 0$ (zero value on the z axis), and so $\psi_{\text{sc}}(\boldsymbol{r}, t) = 0$ (there is no scattering, because the projectile has missed the target).

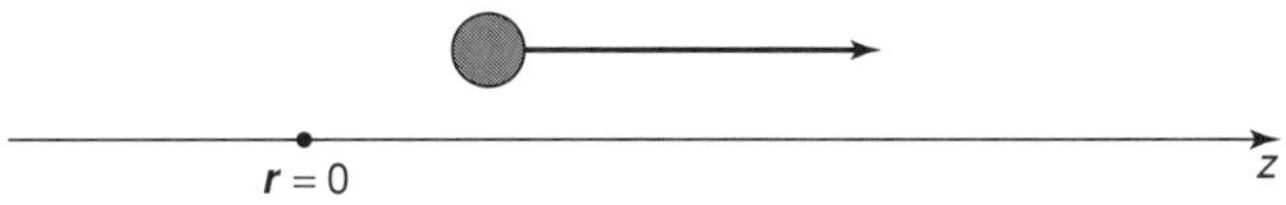

Figure 4.8 *An incident wave packet with nonzero impact parameter. The shaded region represents the region of nonzero $\psi_{\text{in}}(\boldsymbol{r}, t)$ (nonzero unscattered packet at time t).*

In a scattering experiment, we count the numbers of particles scattered in various directions (see figure 4.9), and we are interested in the probability that an incident particle is scattered into a given solid angle $\mathrm{d}\Omega$ in a given direction (θ, φ).

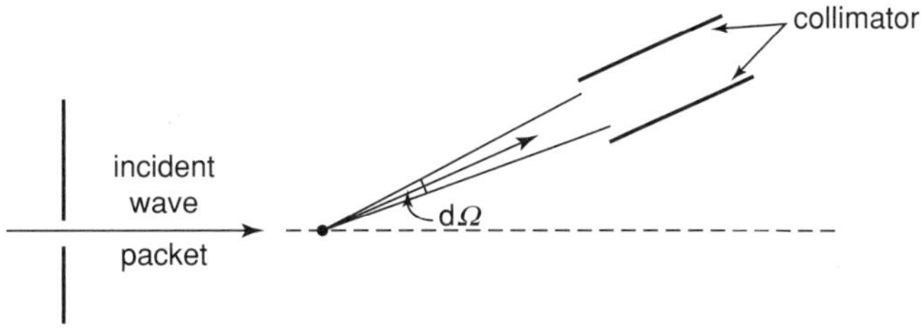

Figure 4.9 *Scheme of a particle-scattering experiment.*

For a given incident wave packet $\psi_{\rm in}(\boldsymbol{r},t)=\phi(\boldsymbol{r},t)$ the probability of finding the scattered particle in the volume element $\mathrm{d}^3\boldsymbol{r}=r^2\mathrm{d}r\,\sin\theta\mathrm{d}\theta\mathrm{d}\varphi$ at vector position $\boldsymbol{r}$ is

$$|\psi_{\rm sc}(\boldsymbol{r},t)|^2\,r^2\mathrm{d}r\,\sin\theta\mathrm{d}\theta\mathrm{d}\varphi=\mathrm{d}\Omega r^2\mathrm{d}r\frac{\left|f(p_0\hat{\boldsymbol{r}}\leftarrow\boldsymbol{p}_0)\right|^2}{r^2}\,|\psi_{\rm in}(\hat{z}r,t)|^2\,. \tag{4.123}$$

The probability $w_\phi(\hat{\boldsymbol{r}}\leftarrow\hat{z})\mathrm{d}\Omega$ that the scattered particle is scattered into the solid angle $\mathrm{d}\Omega$ in the direction of $\boldsymbol{r}$ when the incident wave packet is ϕ is obtained by integrating (4.123) over all r:

$$w_\phi(\hat{\boldsymbol{r}}\leftarrow\hat{z})\mathrm{d}\Omega=\mathrm{d}\Omega|f(p_0\hat{\boldsymbol{r}}\leftarrow\boldsymbol{p}_0)|^2\int_{-\infty}^{\infty}\mathrm{d}r\,|\psi_{\rm in}(\hat{z}r,t)|^2, \tag{4.124}$$

where we have extended the lower limit of integration from 0 to $-\infty$, since the incident packet has certainly vanished on the negative z axis by the time that the scattered particle is detected in the collimator.

The expression (4.124) is appropriate for a **particular** incident packet, that is, a packet with a particular impact parameter. In a scattering experiment we bombard the target(s) with very many particles and study the angular distribution of the scattered particles. Our aim is not good enough to ensure that the projectile particles have the same impact parameter every time, but it is usually possible to "prepare" the incident particles so that they are described by wave packets of the **same shape**.

For definiteness, let $\phi(\boldsymbol{r},t)=\psi_{\rm in}(\boldsymbol{r},t)$ describe a packet with zero impact parameter, that is, a packet centred on the z axis. We now consider a packet $\phi_{\boldsymbol{\rho}}(\boldsymbol{r},t)$ with the same shape but with nonzero impact parameter $\boldsymbol{\rho}$ (see figure 4.10).

Clearly,

$$\phi_\rho(\hat{z}r,t)\equiv\phi_{\boldsymbol{\rho}}(x=0,\,y=0,\,z=r,t)=\psi_{\rm in}(-\rho_x,-\rho_y,r,t)=\psi_{\rm in}(\hat{z}r-\boldsymbol{\rho},t), \tag{4.125}$$

and, therefore,

$$w_{\phi_\rho}(\hat{\boldsymbol{r}}\leftarrow\hat{z})\mathrm{d}\Omega=\mathrm{d}\Omega|f(p_0\hat{\boldsymbol{r}}\leftarrow\boldsymbol{p}_0)|^2\int_{-\infty}^{\infty}\mathrm{d}r\,|\psi_{\rm in}(\hat{z}r-\boldsymbol{\rho},t)|^2. \tag{4.126}$$

We now divide the **whole** area of the (x,y) plane into area elements $\mathrm{d}x\mathrm{d}y=\mathrm{d}^2\boldsymbol{\rho}$ and define $\mathrm{d}\sigma(\theta,\varphi)$ as the (continuous) sum of all the area elements $\mathrm{d}^2\boldsymbol{\rho}$ weighted by the probability that a particle whose packet is

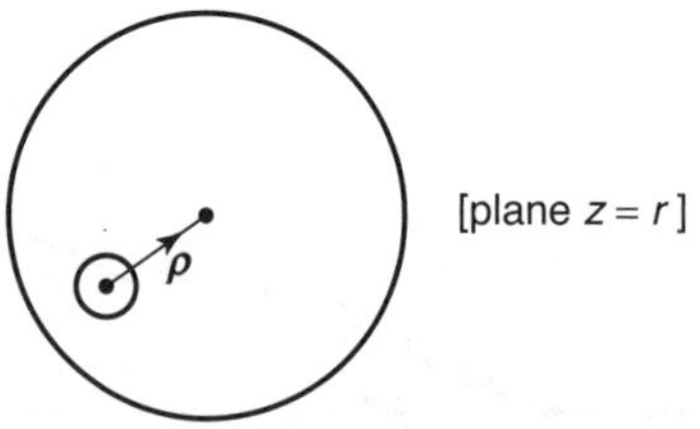

Figure 4.10 *Scattering with nonzero impact parameter $\boldsymbol{\rho}$. The interior of the circle represents the region of nonzero $\phi_{\boldsymbol{\rho}}(\boldsymbol{r},t)$ (nonzero incident packet at time t) in the plane $z=r$, and the dot inside the small circle represents the line of the z axis, on which the scattering centre lies.*

centred in $\mathrm{d}^2\boldsymbol{\rho}$ at $\boldsymbol{\rho}$ is scattered into $\mathrm{d}\Omega$ in the direction of (θ, φ):

$$\mathrm{d}\sigma(\theta, \varphi) = \int \mathrm{d}^2\boldsymbol{\rho}\, w_{\phi_\rho}(\hat{\boldsymbol{r}} \leftarrow \hat{z})\mathrm{d}\Omega = \mathrm{d}\Omega |f(p_0\hat{\boldsymbol{r}} \leftarrow \boldsymbol{p}_0)|^2 \int \mathrm{d}^2\boldsymbol{\rho} \int_{-\infty}^{\infty} \mathrm{d}r\, |\psi_{\text{in}}(\hat{z}r - \boldsymbol{\rho}, t)|^2$$
$$= \mathrm{d}\Omega |f(p_0\hat{\boldsymbol{r}} \leftarrow \boldsymbol{p}_0)|^2 \int \mathrm{d}^3\boldsymbol{r}\, |\psi_{\text{in}}(\boldsymbol{r}, t)|^2 = \mathrm{d}\Omega \left|f(p_0\hat{\boldsymbol{r}} \leftarrow \boldsymbol{p}_0)\right|^2. \tag{4.127}$$

Therefore,

$$\frac{\mathrm{d}\sigma(\theta, \varphi)}{\mathrm{d}\Omega} = |f(p_0\hat{\boldsymbol{r}} \leftarrow \boldsymbol{p}_0)|^2. \tag{4.128}$$

The quantity $\mathrm{d}\sigma(\theta, \varphi)/\mathrm{d}\Omega$ is called the **differential scattering cross-section**.

It is clear that $\mathrm{d}\sigma(\theta, \varphi) = |f(p_0\,\hat{\boldsymbol{r}} \leftarrow \boldsymbol{p}_0)|^2\mathrm{d}\Omega$ is the **effective area** presented by the target for scattering into solid angle $\mathrm{d}\Omega$ in the direction of $\boldsymbol{r}$. The effective area for scattering in **all** directions is clearly

$$\sigma(\boldsymbol{p}_0) \equiv \int \frac{\mathrm{d}\sigma(\theta, \varphi)}{\mathrm{d}\Omega}\mathrm{d}\Omega = \int |f(p_0\hat{\boldsymbol{r}} \leftarrow \boldsymbol{p}_0)|^2\mathrm{d}\Omega \tag{4.129}$$

and is called the **total scattering cross-section**.

4.5 The optical theorem

From the unitarity of the S-matrix, we have

$$S^\dagger S = 1. \tag{4.130}$$

We write $S = 1 + R$, as before. Then

$$R + R^\dagger = -R^\dagger R. \tag{4.131}$$

Therefore,

$$\langle \boldsymbol{p}'|R|\boldsymbol{p}\rangle + \langle \boldsymbol{p}'|R^\dagger|\boldsymbol{p}\rangle = -\langle \boldsymbol{p}'|R^\dagger R|\boldsymbol{p}\rangle, \tag{4.132}$$

which can be rewritten as

$$\langle \boldsymbol{p}'|R|\boldsymbol{p}\rangle + \langle \boldsymbol{p}|R|\boldsymbol{p}'\rangle^* = -\int \mathrm{d}^3\boldsymbol{p}''\langle \boldsymbol{p}'|R^\dagger|\boldsymbol{p}''\rangle\langle \boldsymbol{p}''|R|\boldsymbol{p}\rangle = -\int \mathrm{d}^3\boldsymbol{p}''\langle \boldsymbol{p}''|R|\boldsymbol{p}'\rangle^*\langle \boldsymbol{p}''|R|\boldsymbol{p}\rangle. \tag{4.133}$$

But, as follows from (4.43), (4.44) and (4.95),

$$\langle \boldsymbol{p}'|R\,|\boldsymbol{p}\rangle = -2\pi i\delta(E_{p'} - E_p)t(\boldsymbol{p}' \leftarrow \boldsymbol{p}) = \frac{i}{2\pi m\hbar}\delta(E_{p'} - E_p)f(\boldsymbol{p}' \leftarrow \boldsymbol{p}), \tag{4.134}$$

and so

$$\frac{i}{2\pi m\hbar}\delta(E_{p'}-E_p)\left[f(\boldsymbol{p}'\leftarrow\boldsymbol{p})-f(\boldsymbol{p}\leftarrow\boldsymbol{p}')^*\right]$$
$$=-\frac{1}{(2\pi m\hbar)^2}\int \mathrm{d}^3\boldsymbol{p}''\delta(E_{p''}-E_{p'})f(\boldsymbol{p}''\leftarrow\boldsymbol{p}')^*\delta(E_{p''}-E_p)f(\boldsymbol{p}''\leftarrow\boldsymbol{p}). \tag{4.135}$$

Using the presence of the factor $\delta(E_{p''}-E_{p'})$ to set $\delta(E_{p''}-E_p)$ equal to $\delta(E_{p'}-E_p)$, and cancelling the latter δ-function, we obtain

$$f(\boldsymbol{p}'\leftarrow\boldsymbol{p})-f(\boldsymbol{p}\leftarrow\boldsymbol{p}')^*=\frac{i}{2\pi m\hbar}\int \mathrm{d}^3\boldsymbol{p}''\delta(E_{p''}-E_{p'})f(\boldsymbol{p}''\leftarrow\boldsymbol{p}')^*f(\boldsymbol{p}''\leftarrow\boldsymbol{p}). \tag{4.136}$$

In (4.136) we let $\boldsymbol{p}'=\boldsymbol{p}$ [$f(\boldsymbol{p}\leftarrow\boldsymbol{p})$ is the forward-scattering amplitude], and divide the result by $2i$:

$$\mathrm{Im} f(\boldsymbol{p}\leftarrow\boldsymbol{p})=\frac{1}{4\pi m\hbar}\int \mathrm{d}\Omega_{p''}\int \mathrm{d}p''\,p''^2\delta(E_{p''}-E_p)f(\boldsymbol{p}''\leftarrow\boldsymbol{p})^*f(\boldsymbol{p}''\leftarrow\boldsymbol{p}). \tag{4.137}$$

But $E_{p''}=p''^2/2m$, and so $\mathrm{d}p''=\left(m/p''\right)\mathrm{d}E_{p''}$. Integrating over $E_{p''}$, we obtain

$$\mathrm{Im} f(\boldsymbol{p}\leftarrow\boldsymbol{p})=\frac{p}{4\pi\hbar}\int \mathrm{d}\Omega_{p''}f(\boldsymbol{p}''\leftarrow\boldsymbol{p})^*f(\boldsymbol{p}''\leftarrow\boldsymbol{p})\quad(|\boldsymbol{p}''|=|\boldsymbol{p}|)$$
$$=\frac{p}{4\pi\hbar}\int \mathrm{d}\Omega_{p''}|f(\boldsymbol{p}''\leftarrow\boldsymbol{p})|^2. \tag{4.138}$$

Therefore,

$$\mathrm{Im} f(\boldsymbol{p}\leftarrow\boldsymbol{p})=\frac{p}{4\pi\hbar}\sigma(\boldsymbol{p}), \tag{4.139}$$

which is known as the **optical theorem**.

Since $\sigma(p)>0$, the forward scattering amplitude $f(\boldsymbol{p}\leftarrow\boldsymbol{p})$ has a positive imaginary part. Clearly, if we can measure

$$|f(\boldsymbol{p}\leftarrow\boldsymbol{p})|=\sqrt{[\mathrm{Re} f(\boldsymbol{p}\leftarrow\boldsymbol{p})]^2+[\mathrm{Im} f(\boldsymbol{p}\leftarrow\boldsymbol{p})]^2}$$

by extrapolation of values of $|f(\boldsymbol{p}'\leftarrow\boldsymbol{p})|$ obtained from nonforward measurements of $[\mathrm{d}\sigma/\mathrm{d}\Omega](\boldsymbol{p}'\leftarrow\boldsymbol{p})$, and find $\mathrm{Im} f(\boldsymbol{p}\leftarrow\boldsymbol{p})$ from measurements of $\sigma(p)$, we can find $|\mathrm{Re} f(\boldsymbol{p}\leftarrow\boldsymbol{p})|$.

It is an experimental fact that $\sigma(p)$ for scattering of elementary particles becomes constant (i.e., independent of p) at high energies. The optical theorem (4.139) then predicts that, at high energies, $\mathrm{Im} f(\boldsymbol{p}\leftarrow\boldsymbol{p})\propto p$, so that the forward elastic-scattering peak of $\mathrm{d}\sigma/\mathrm{d}\Omega=|f|^2$ will rise at least as fast as p^2. This is well confirmed by experiment.

4.6 The Born series and Born approximation

If we can determine $T(z)$ we can find the quantity

$$t(\boldsymbol{p}' \leftarrow \boldsymbol{p}) = \langle \boldsymbol{p}'|T(E_p + i0)|\boldsymbol{p}\rangle \quad (|\boldsymbol{p}'| = |\boldsymbol{p}|), \tag{4.140}$$

and hence, from equation (4.95), the scattering amplitude $f(\boldsymbol{p}' \leftarrow \boldsymbol{p})$.

By iteration of the Lippmann–Schwinger equation $T(z) = V + VG^0(z)T(z)$ we earlier derived the **Born series**

$$T(z) = V + VG^0(z)V + VG^0(z)VG^0(z)V + \cdots, \tag{4.141}$$

of which the first approximation

$$T(z) \approx V \tag{4.142}$$

is the **Born approximation**.

Thus,

$$t(\boldsymbol{p}' \leftarrow \boldsymbol{p}) = \langle \boldsymbol{p}'|T(E_p + i0)|\boldsymbol{p}\rangle = \langle \boldsymbol{p}'|V|\boldsymbol{p}\rangle + \langle \boldsymbol{p}'|VG^0(E_p + i0)V|\boldsymbol{p}\rangle + \cdots \tag{4.143}$$

We may expect the series (4.143) to converge when $G^0(E_p + i0)V$ is small in some sense. Since $G^0(E_p + i0) = (E_p + i0 - H_0)^{-1}$, this means that we may expect good convergence when the incident-particle energy E_p is high and (or) the scattering potential V is small.

In other words, for a given incident energy E_p there is some V small enough for the series to converge, while for a given V there is some incident-particle energy E_p above which the series converges. (In some cases, however, this energy may be so high that the use of a potential becomes unrealistic, for example, because of relativistic effects.)

In the Born approximation we set $T(z) = V$, and so

$$t(\boldsymbol{p}' \leftarrow \boldsymbol{p}) \approx t^{(1)}(\boldsymbol{p}' \leftarrow \boldsymbol{p}) = \langle \boldsymbol{p}'|V|\boldsymbol{p}\rangle. \tag{4.144}$$

Therefore [see (4.95)],

$$\begin{aligned} f(\boldsymbol{p}' \leftarrow \boldsymbol{p}) \approx f^{(1)}(\boldsymbol{p}' \leftarrow \boldsymbol{p}) &= -(2\pi)^2 m\hbar t^{(1)}(\boldsymbol{p}' \leftarrow \boldsymbol{p}) \\ &= -\frac{m}{2\pi\hbar^2}\int \mathrm{d}^3\boldsymbol{r}\mathrm{e}^{-i\boldsymbol{p}'\cdot\boldsymbol{r}/\hbar}V(\boldsymbol{r})\mathrm{e}^{i\boldsymbol{p}\cdot\boldsymbol{r}/\hbar}, \end{aligned} \tag{4.145}$$

so that

$$f^{(1)}(\boldsymbol{p}' \leftarrow \boldsymbol{p}) = -\frac{m}{2\pi\hbar^2}\int \mathrm{d}^3\boldsymbol{r}\mathrm{e}^{-i\boldsymbol{q}\cdot\boldsymbol{r}}V(\boldsymbol{r}), \tag{4.146}$$

where $\hbar\boldsymbol{q} = \boldsymbol{p}' - \boldsymbol{p}$ is the **momentum transfer** in the scattering (see figure 4.11).

From figure 4.11, the momentum transfer $\hbar\boldsymbol{q}$ has magnitude

$$\hbar q = 2p\sin(\theta/2) \tag{4.147}$$

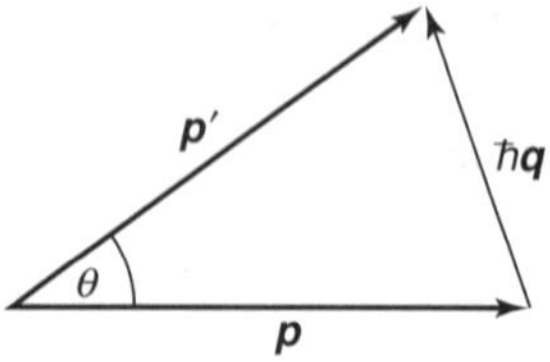

Figure 4.11 *Vector diagram illustrating the incident-particle momentum* $\mathbf{p}$*, scattered-particle momentum* $\mathbf{p}'$ *and momentum transfer* $\hbar\mathbf{q}$*.*

and clearly becomes large at high energies (high p).

Thus, all that we need for the first Born approximation to the scattering amplitude is the **Fourier transform** of $V(\mathbf{r})$.

For example, if $V(\mathbf{r})$ is spherically symmetric, that is, if $V(\mathbf{r}) = V(r)$, we can do the angular integration in (4.146), with the result

$$f^{(1)}(\mathbf{p}' \leftarrow \mathbf{p}) = -\frac{2m}{\hbar^2}\int_0^\infty r^2 \mathrm{d}r \frac{\sin qr}{qr} V(r). \tag{4.148}$$

We can also obtain from (4.146) a simple expression for the **forward** scattering amplitude. Setting $\mathbf{p}' = \mathbf{p}$ ($\mathbf{q} = 0$), we find

$$f^{(1)}(\mathbf{p} \leftarrow \mathbf{p}) = -\frac{m}{2\pi\hbar^2}\int \mathrm{d}^3\mathbf{r} V(r), \tag{4.149}$$

which is **independent of the energy** (and reduces to the result

$$f^{(1)}(\mathbf{p} \leftarrow \mathbf{p}) = -\frac{2m}{\hbar^2}\int \mathrm{d}r r^2 V(r) \tag{4.150}$$

when the scattering potential has spherical symmetry).

Consider the high-energy behavior of $f^{(1)}(\mathbf{p}' \leftarrow \mathbf{p})$. As p becomes large, $\hbar q = 2p\sin(\theta/2)$ becomes large (for $\theta \neq 0$), and $f^{(1)}(\mathbf{p}' \leftarrow \mathbf{p}) \to 0$ because of the rapid oscillation of the factor $e^{-i\mathbf{q}\cdot\mathbf{r}}$. Of course, this does not apply in the forward direction (where $q = 0$ always, independently of p), and so the scattering becomes concentrated in the forward direction (giving the "forward peak") as $p \to \infty$.

Note that, since V is hermitian,

$$\langle \mathbf{p}'|V|\mathbf{p}\rangle^* = \langle \mathbf{p}|V^\dagger|\mathbf{p}'\rangle = \langle \mathbf{p}|V|\mathbf{p}'\rangle. \tag{4.151}$$

Putting $\mathbf{p}' = \mathbf{p}$ into (4.151) shows that $\langle \mathbf{p}|\ V\ |\mathbf{p}\rangle$ (and hence $f^{(1)}(\mathbf{p} \leftarrow \mathbf{p})$) is real, that is, $\mathrm{Im} f^{(1)}(\mathbf{p} \leftarrow \mathbf{p}) = 0$. But, by the optical theorem (4.139), which is a consequence of the unitarity of the S-matrix, we have

$$\mathrm{Im} f(\mathbf{p} \leftarrow \mathbf{p}) = \frac{p}{4\pi\hbar}\int \mathrm{d}\Omega\ |f|^2. \tag{4.152}$$

which is **not** equal to zero if there is any scattering at all! Thus, we may say that **the Born approximation violates unitarity**. The reason is clear. The right-hand side of (4.152) is of order V^2, while the left-hand side is only of order V. Thus, by calculating $f(\boldsymbol{p} \leftarrow \boldsymbol{p})$ to order V only (the Born approximation) we completely miss its imaginary part.

4.7 Spherically symmetric potentials and the method of partial waves

Earlier, we showed that $[S, H_0] = 0$ and that, as a consequence, energy is conserved in a scattering experiment. Similarly, since

$$S = \Omega_-^\dagger \Omega_+ = \lim_{\substack{t\to\infty \\ t'\to-\infty}} \left(e^{iH_0t/\hbar}e^{-iHt/\hbar}\right)(e^{iHt'/\hbar}e^{-iH_0t'/\hbar}),$$

and since $[H, \boldsymbol{L}] = 0$ for a **spherically symmetric potential** (and, of course $[H_0, \boldsymbol{L}] = 0$), we have

$$[S, \boldsymbol{L}] = 0, \tag{4.153}$$

so that **angular momentum is conserved in scattering by a spherically symmetric potential**.

This means that the operator S is diagonal in the angular-momentum eigenstates, that is, in the eigenstates $Y_{lm}(\theta, \varphi) \equiv \langle \theta, \varphi \mid l, m\rangle$ of L^2 (with eigenvalues $l(l+1)\hbar^2$) and L_z (with eigenvalues $m\hbar$), as well as being diagonal in the eigenstates of H_0 (with eigenvalues E). Denoting the eigenstates of the complete set of commuting observables H_0, L^2 and L_z by $|E, l, m\rangle$, we have

$$\langle E', l', m'|S|E, l, m\rangle = \delta(E' - E)\delta_{l'l}\delta_{m'm}s_l(E). \tag{4.154}$$

The coordinate-space wave function $\psi_{Elm}(\boldsymbol{r}) \equiv \langle \boldsymbol{r}|E, l, m\rangle$ corresponding to the ket $|E, l, m\rangle$ is

$$\psi_{Elm}(\boldsymbol{r}) = i^l\left(\frac{2m}{\pi\hbar^2 k}\right)^{1/2}\frac{\hat{j}_l(kr)}{r}Y_{lm}(\theta, \varphi), \tag{4.155}$$

where $\hat{j}_l(kr)$ is the Riccati–Bessel function of order l.

[We prove this as follows: the coordinate-space wave function $\psi_{Elm}(\boldsymbol{r}) \equiv \langle \boldsymbol{r}|E, l, m\rangle$ corresponding to the ket $|E, l, m\rangle$ is a solution of the free time-independent Schrödinger equation, which takes the form of equation (1.43) with $V(r) = 0$:

$$-\frac{\hbar^2}{2m}\nabla^2\phi(\boldsymbol{r}) = E\phi(\boldsymbol{r}).$$

In spherical polar coordinates this equation takes the form

$$-\frac{\hbar^2}{2m}\left[\frac{1}{r^2}\frac{\partial}{\partial r}\left(r^2\frac{\partial}{\partial r}\right) + \frac{1}{r^2\sin\theta}\frac{\partial}{\partial\theta}\left(\sin\theta\frac{\partial}{\partial\theta}\right) + \frac{1}{r^2\sin^2\theta}\frac{\partial^2}{\partial\varphi^2}\right]\phi(\boldsymbol{r}) = E\phi(\boldsymbol{r}).$$

As in section 1.16, the substitution $\phi(\boldsymbol{r}) = \phi(r, \theta, \varphi) = R(r)Y(\theta, \varphi)$ leads to a solution of the form

$$\phi(\boldsymbol{r}) = R(r)Y_{lm}(\theta, \varphi)$$

where $Y_{lm}(\theta, \varphi)$ is a spherical harmonic and $R(r)$ satisfies the radial equation (1.46) with $V(r) = 0$, that is, satisfies the equation

$$\frac{1}{r^2}\frac{\mathrm{d}}{\mathrm{d}r}\left(r^2\frac{\mathrm{d}R}{\mathrm{d}r}\right) + \left[\frac{2mE}{\hbar^2} - \frac{l(l+1)}{r^2}\right]R = 0.$$

The substitution

$$R(r) = \frac{y(r)}{r}$$

then leads directly to the following equation for $y(r)$:

$$\left(\frac{\mathrm{d}^2}{\mathrm{d}r^2} - \frac{l(l+1)}{r^2} + k^2\right)y(r) = 0, \tag{4.156}$$

which is called the **free radial Schrödinger equation**. As an ordinary second-order linear differential equation, this has two linearly independent solutions. As $r \to 0$ the term k^2 can be neglected in comparison with the centrifugal term, and the solutions clearly behave like linear combinations of r^{l+1} and r^{-l}. The physically relevant solution $y(r)$ must vanish at the origin, and so must be that which behaves like r^{l+1}. This solution is the Riccati–Bessel function $\hat{j}_l(z)$. The latter, as may be checked by substitution into the above equation, takes the following series form

$$\hat{j}_l(z) \equiv z j_l(z) \equiv \left(\frac{\pi z}{2}\right)^{1/2} J_{l+\frac{1}{2}}(z) = z^{l+1}\sum_{n=0}^{\infty}\frac{(-z^2/2)^n}{n!(2l+2n+1)!!} \overset{(z\to 0)}{=} \frac{z^{l+1}}{(2l+1)!!}[1 + O(z^2)],$$

where $j_l(z)$ is the spherical Bessel function of order l, and $J_\lambda(z)$ is an ordinary Bessel function. The first three Riccati–Bessel functions are:

$$\hat{j}_0(z) = \sin z, \quad \hat{j}_1(z) = \frac{1}{z}\sin z - \cos z, \quad \hat{j}_2(z) = \left(\frac{3}{z^2} - 1\right)\sin z - \frac{3}{z}\cos z.$$

The Riccati–Bessel functions satisfy

$$\int_0^\infty \mathrm{d}r\,\hat{j}_l(k'r)\hat{j}_l(kr) = \frac{\pi}{2}\delta(k' - k).$$

Consequently, the coordinate-space wave function $\psi_{Elm}(\boldsymbol{r}) \equiv \langle \boldsymbol{r} \mid E, l, m\rangle$ satisfying the desired orthonormalization condition

$$\int \mathrm{d}^3\boldsymbol{r}\psi^*_{E'l'm'}(\boldsymbol{r})\psi_{Elm}(\boldsymbol{r}) = \int \mathrm{d}^3\boldsymbol{r}\langle E'l'm'|\boldsymbol{r}\rangle\langle\boldsymbol{r}|Elm\rangle = \langle E'l'm'|Elm\rangle = \delta(E' - E)\delta_{l'l}\delta_{m'm}$$

(the right-hand side of which clearly has dimensions [energy]$^{-1}$) is

$$\psi_{Elm}(\boldsymbol{r}) = i^l\left(\frac{2m}{\pi\hbar^2 k}\right)^{1/2}\frac{\hat{j}_l(kr)}{r}Y_{lm}(\theta, \varphi), \tag{4.157}$$

which, since $\hbar^2k^2/2m = E$, clearly has the necessary dimensions [energy]$^{-1/2}$ [length]$^{-3/2}$.]

Why, in (4.154), have we written $s_l(E)$, and not $s_{lm}(E)$? The answer is that, because $[S, \boldsymbol{L}] = 0$, we also have $[S, L^{\pm}] = 0$, where [see (2.36)]

$$L^{\pm}\,|l, m\rangle \equiv (L_x \pm i L_y)\,|l, m\rangle = \sqrt{l(l+1) - m(m \pm 1)}\,|l, m \pm 1\rangle\,.$$

Therefore,

$$\begin{aligned}\langle l, m+1|\, S\,|l, m+1\rangle &= \frac{\left\{L^+\,|l, m\rangle\right\}^{\dagger} SL^+\,|l, m\rangle}{l(l+1) - m(m+1)} = \frac{\langle l, m|\, L^- SL^+\,|l, m\rangle}{l(l+1) - m(m+1)} \\ &= \frac{\langle l, m|\, SL^- L^+\,|l, m\rangle}{l(l+1) - m(m+1)} = \langle l, m|\, S\,|l, m\rangle\,, \end{aligned} \tag{4.158}$$

which is therefore independent of m.

Clearly,

$$S\,|E, l, m\rangle = s_l(E)\,|E, l, m\rangle\,, \tag{4.159}$$

and since S is unitary (and hence norm-preserving) the eigenvalue $s_l(E)$ has modulus unity:

$$|s_l(E)| = 1. \tag{4.160}$$

We can thus define a quantity $\delta_l(E)$ by the relation

$$s_l(E) = \mathrm{e}^{2i\delta_l(E)}, \tag{4.161}$$

where the factor 2 is inserted in order that $\delta_l(E)$ coincide with the **phase shift** encountered later. Thus,

$$\langle E', l', m'|S\,|E, l, m\rangle = \delta(E' - E)\delta_{l'l}\delta_{m'm}\mathrm{e}^{2i\delta_l(E)}. \tag{4.162}$$

We now use (4.162) to find the relation between the scattering amplitude $f(\boldsymbol{p}' \leftarrow \boldsymbol{p})$ and the phase shifts $\delta_l(E)$. We showed earlier [see equation (4.44)] that

$$\begin{aligned}\langle \boldsymbol{p}'|S\,|\boldsymbol{p}\rangle &= \delta^{(3)}(\boldsymbol{p}' - \boldsymbol{p}) - 2\pi i\delta(E_{p'} - E_p)t(\boldsymbol{p}' \leftarrow \boldsymbol{p}) \\ &= \langle \boldsymbol{p}'|\boldsymbol{p}\rangle + \frac{i}{2\pi m\hbar}\delta(E_{p'} - E_p)f(\boldsymbol{p}' \leftarrow \boldsymbol{p}). \end{aligned} \tag{4.163}$$

Therefore,

$$\begin{aligned} i\frac{\delta(E_{p'} - E_p)}{2\pi m\hbar}f(\boldsymbol{p}' \leftarrow \boldsymbol{p}) &= \langle \boldsymbol{p}'|(S-1)\,|\boldsymbol{p}\rangle = \int \mathrm{d}E \sum_{l,m} \langle \boldsymbol{p}'|(S-1)\,|Elm\rangle\,\langle Elm|\boldsymbol{p}\rangle \\ &= \int \mathrm{d}E \sum_{l,m} \left(\mathrm{e}^{2i\delta_l(E)} - 1\right) \langle \boldsymbol{p}'|Elm\rangle\langle Elm|\boldsymbol{p}\rangle. \end{aligned} \tag{4.164}$$

Thus, we need the quantities $\langle Elm|\boldsymbol{p}\rangle$ defining the transformation from the $|Elm\rangle$ states to the $|\boldsymbol{p}\rangle$ states:

$$|\boldsymbol{p}\rangle = \int \mathrm{d}E \sum_{l,m} |Elm\rangle \langle Elm|\boldsymbol{p}\rangle. \tag{4.165}$$

The transformation coefficient $\langle Elm|\boldsymbol{p}\rangle$ is simply the complex conjugate of the momentum-space wave function $\langle \boldsymbol{p}|Elm\rangle$, and the latter is given by the expression

$$\langle \boldsymbol{p}|Elm\rangle = (mp)^{-1/2}\delta(E - E_p)Y_{lm}(\hat{\boldsymbol{p}}), \tag{4.166}$$

where $Y_{lm}(\hat{\boldsymbol{p}})$ is the spherical harmonic $Y_{lm}(\theta, \varphi)$ evaluated at angles (θ, φ) corresponding to the direction defined by the unit vector $\hat{\boldsymbol{p}}$.

Equation (4.166) follows from the fact that the ket $|Elm\rangle$ has nonzero projections only on to momentum eigenstates $|\boldsymbol{p}\rangle$ with momenta such that $p^2/2m$ $(\equiv E_p) = E$, while the factor $(mp)^{-1/2}$ gives $\langle \boldsymbol{p} \mid Elm\rangle$ the correct dimensions $[\text{energy}]^{-1/2}[\text{momentum}]^{-3/2}$ and ensures fulfilment of the orthonormalization condition

$$\int \mathrm{d}^3\boldsymbol{p}\psi^*_{E'l'm'}(\boldsymbol{p})\psi_{Elm}(\boldsymbol{p}) = \int \mathrm{d}^3\boldsymbol{p}\langle E'l'm'|\boldsymbol{p}\rangle\langle \boldsymbol{p}|Elm\rangle = \langle E'l'm'|Elm\rangle = \delta(E' - E)\delta_{l'l}\delta_{m'm},$$

as can be seen by writing $p^2\mathrm{d}p/mp = 2E_p\mathrm{d}p/p = 2E_p\mathrm{d}E_p/2E_p = \mathrm{d}E_p$ and using the integral $\int \mathrm{d}E_p\delta(E' - E_p)\delta(E - E_p) = \delta(E' - E)$.

Therefore, from (4.164) and (4.166),

$$\begin{aligned}\frac{i\delta(E_{p'} - E_p)}{2\pi m\hbar} f(\boldsymbol{p}' \leftarrow \boldsymbol{p}) &= \int \mathrm{d}E \sum_{l,m} \left(\mathrm{e}^{2i\delta_l(E)} - 1\right) \frac{\delta(E - E_{p'})\delta(E - E_p)Y_{lm}(\hat{\boldsymbol{p}}')Y^*_{lm}(\hat{\boldsymbol{p}})}{mp'^{1/2}p^{1/2}} \\ &= \frac{1}{mp}\delta(E_{p'} - E_p)\sum_{l,m} Y_{lm}(\hat{\boldsymbol{p}}')\left[s_l(E_p) - 1\right]Y^*_{lm}(\hat{\boldsymbol{p}}),\end{aligned} \tag{4.167}$$

or

$$f(\boldsymbol{p}' \leftarrow \boldsymbol{p}) = \frac{2\pi\hbar}{ip}\sum_{l,m} Y_{lm}(\hat{\boldsymbol{p}}')\left[s_l(E_p) - 1\right]Y^*_{lm}(\hat{\boldsymbol{p}}). \tag{4.168}$$

Using the spherical-harmonics addition theorem

$$\sum_{m=-l}^{l} Y^*_{lm}(\hat{\boldsymbol{p}}')Y_{lm}(\hat{\boldsymbol{p}}) = \frac{2l + 1}{4\pi}P_l(\cos\theta), \tag{4.169}$$

where θ is the angle between $\boldsymbol{p}'$ and $\boldsymbol{p}$ (the scattering angle), we obtain

$$f(\boldsymbol{p}' \leftarrow \boldsymbol{p}) \equiv f(E_p, \theta) = \frac{\hbar}{2ip}\sum_{l=0}^{\infty}(2l + 1)\left[s_l(E_p) - 1\right]P_l(\cos\theta). \tag{4.170}$$

This may be rewritten as

$$f(E,\theta) = \sum_{l=0}^{\infty}(2l+1)f_l(E)P_l(\cos\theta), \tag{4.171}$$

where

$$f_l(E) = \frac{s_l(E)-1}{2ik} = \frac{e^{2i\delta_l(E)}-1}{2ik} = \frac{e^{i\delta_l}\sin\delta_l}{k} \quad (k = p/\hbar). \tag{4.172}$$

Since

$$\int_{-1}^{1} d(\cos\theta)P_l(\cos\theta)P_{l'}(\cos\theta) = \frac{2\delta_{ll'}}{2l+1}, \tag{4.173}$$

we can invert (4.171) to obtain

$$f_l(E) = \frac{1}{2}\int_{-1}^{1} d(\cos\theta)f(E,\theta)P_l(\cos\theta). \tag{4.174}$$

Thus, knowledge of all the $f_l(E)$, that is, of all the $\delta_l(E)$, gives $f(E,\theta)$ via (4.171), while from $f(E,\theta)$ we can find each $f_l(E)$, and hence each $\delta_l(E)$, from (4.174).

The total cross-section $\sigma(p)$ is

$$\begin{aligned}\sigma(p) &= \int \frac{d\sigma}{d\Omega}d\Omega = \int_0^{2\pi} d\varphi \int_{-1}^{1} |f|^2\, d(\cos\theta) \\ &= 2\pi\sum_{l,l'}(2l+1)(2l'+1)f_l^*(E)f_{l'}(E)\int_{-1}^{1} d(\cos\theta)P_l(\cos\theta)P_{l'}(\cos\theta) \\ &\equiv \sum_l \sigma_l(p),\end{aligned} \tag{4.175}$$

where

$$\sigma_l(p) = 4\pi(2l+1)\,|f_l(E)|^2 = 4\pi(2l+1)\frac{\sin^2\delta_l}{k^2} \quad \left[E = \frac{p^2}{2m} = \frac{\hbar^2k^2}{2m}\right]. \tag{4.176}$$

Thus, we may write:

$$\sigma(p) = \frac{4\pi}{k^2}\sum_l (2l+1)\sin^2\delta_l(k). \tag{4.177}$$

The unitarity of S is equivalent to the reality of δ_l, which, in turn, implies that $|\sin\delta_l| \le 1$. Hence, we have the so-called **unitarity bound** on the partial cross-sections σ_l:

$$\sigma_l(k) \le 4\pi\frac{2l+1}{k^2}. \tag{4.178}$$

[The equality here applies when $\delta_l = (2n+1)\pi/2$, which is a condition often associated with a resonance.]

At low energies, as we shall see, only a few phase shifts δ_l are nonzero. For example, at very low energies differential scattering cross-sections become isotropic (i.e., f is independent of θ), indicating that only δ_0 is nonzero. In this case (pure s-wave scattering), we have

$$f(E,\theta) = f^{(s)}(E) = \frac{1}{k}\mathrm{e}^{i\delta_0(E)}\sin\delta_0(E), \tag{4.179}$$

and measurement of $\mathrm{d}\sigma/\mathrm{d}\Omega = |f|^2$ determines $\sin\delta_0$ to within the sign.

4.8 The partial-wave scattering states

Consider the state $\Omega_+\,|Elm\rangle$. Using the intertwining relation (4.18)

$$H\Omega_+ = \Omega_+ H_0 \quad (H = H_0 + V) \tag{4.180}$$

we have

$$H\Omega_+\,|Elm\rangle = \Omega_+ H_0\,|Elm\rangle = E\Omega_+\,|Elm\rangle\,, \tag{4.181}$$

that is, $\Omega_+\,|Elm\rangle$ is an eigenstate of H with eigenvalue E. Moreover, since Ω_+ commutes with $\boldsymbol{L}$, the state $\Omega_+\,|Elm\rangle$ is also an eigenstate of L^2 and L_z, with eigenvalues $l(l+1)\hbar^2$ and $m\hbar$, respectively. Thus, we can write

$$\Omega_+\,|Elm\rangle \equiv |Elm+\rangle \tag{4.182}$$

(by analogy with $\Omega_+\,|\boldsymbol{p}\rangle \equiv |\boldsymbol{p}+\rangle$).

With $V = 0$ we found [see (4.157)]

$$\langle \boldsymbol{r}\,|\,Elm\rangle = i^l\left(\frac{2m}{\pi\hbar^2 k}\right)^{1/2}\frac{\hat{j}_l(kr)}{r}Y_{lm}(\hat{\boldsymbol{r}}). \tag{4.183}$$

With a nonzero V we get

$$\langle \boldsymbol{r}\,|\,Elm+\rangle = i^l\left(\frac{2m}{\pi\hbar^2 k}\right)^{1/2}\frac{\psi^+_{l,k}(r)}{r}Y_{lm}(\hat{\boldsymbol{r}}), \tag{4.184}$$

where $\psi^+_{l,k}(r)$ satisfies the radial equation [compare with (4.156)]

$$\left[\frac{\mathrm{d}^2}{\mathrm{d}r^2} - \frac{l(l+1)}{r^2} - U(r) + k^2\right]\psi^+_{l,k}(r) = 0 \tag{4.185}$$

in which $U(r) \equiv 2mV(r)/\hbar^2$. Clearly,

$$\psi^+_{l,k}(r) \to \hat{j}_l(kr) \quad \text{as } V(r) \Rightarrow 0. \tag{4.186}$$

The solution $\psi^+_{l,k}(r)$ is chosen to satisfy $\psi^+_{l,k}(0) = 0$.

From the normalization property

$$\langle E'l'm'|Elm\rangle = \delta(E' - E)\delta_{l'l}\delta_{m'm} \tag{4.187}$$

it follows that

$$\int_0^\infty \mathrm{d}r\, \psi^+_{l,k'}(r)^* \psi^+_{l,k}(r) = \frac{\pi}{2}\delta(k' - k). \tag{4.188}$$

[We shall call $\psi^+_{l,k}(r)$ the **normalized** radial wave function.]

We now find the expression for the scattering states $|\boldsymbol{p}+\rangle$ in terms of the partial-wave scattering states $|Elm+\rangle$. We have

$$|\boldsymbol{p}\rangle = \int \mathrm{d}E \sum_{l,m} |Elm\rangle\, \langle Elm|\boldsymbol{p}\rangle. \tag{4.189}$$

Acting on this with Ω_+, we get

$$|\boldsymbol{p}+\rangle = \int \mathrm{d}E \sum_{l,m} \langle Elm|\boldsymbol{p}\rangle\, |Elm+\rangle, \tag{4.190}$$

and so

$$\langle \boldsymbol{r} \mid \boldsymbol{p}+\rangle = \int \mathrm{d}E \sum_{l,m} \langle Elm \mid \boldsymbol{p}\rangle\langle \boldsymbol{r} \mid Elm+\rangle. \tag{4.191}$$

Substituting for the transformation coefficient

$$\langle Elm \mid \boldsymbol{p}\rangle = (mp)^{-1/2}\delta(E - E_p)Y^*_{lm}(\hat{\boldsymbol{p}}) \tag{4.192}$$

found earlier [see (4.166)], and for $\langle \boldsymbol{r} \mid Elm+\rangle$ (4.184), we obtain

$$\begin{aligned}\langle \boldsymbol{r}|\boldsymbol{p}+\rangle &= \frac{1}{(mp)^{1/2}}\left(\frac{2m}{\pi\hbar p}\right)^{1/2} \sum_l i^l \frac{\psi^+_{l,k}(r)}{r} \sum_m Y^*_{lm}(\hat{\boldsymbol{p}})Y_{lm}(\hat{\boldsymbol{r}}) \\ &= (2\pi\hbar)^{-3/2}\frac{\hbar}{pr}\sum_l (2l+1)i^l\psi^+_{l,k}(r)P_l(\hat{\boldsymbol{p}}\cdot\hat{\boldsymbol{r}}) \quad (p = \hbar k),\end{aligned} \tag{4.193}$$

where we have used the spherical-harmonics addition theorem (4.169). But we know that, for large r,

$$\langle \boldsymbol{r} | \boldsymbol{p}+\rangle \rightarrow (2\pi\hbar)^{-3/2}\left[e^{i\boldsymbol{p}.\boldsymbol{r}/\hbar} + f(p\hat{r} \leftarrow \boldsymbol{p})\frac{e^{ipr/\hbar}}{r}\right]. \tag{4.194}$$

We now substitute Bauer's expansion

$$e^{i\boldsymbol{p}\cdot\boldsymbol{r}/\hbar} = e^{i\boldsymbol{k}\cdot\boldsymbol{r}} = e^{ikr\cos\theta} = \sum_{l=0}^{\infty} i^l(2l+1)\frac{\hat{j}_l(kr)}{kr}P_l(\hat{\boldsymbol{r}}\cdot\hat{\boldsymbol{k}}) \tag{4.195}$$

and

$$f(p\hat{\boldsymbol{r}} \leftarrow \boldsymbol{p}) = \sum_{l=0}^{\infty}(2l+1)f_l(p)P_l(\hat{\boldsymbol{r}}\cdot\hat{\boldsymbol{k}}) \quad (\boldsymbol{p} = \hbar\boldsymbol{k}). \tag{4.196}$$

The result is

$$\langle \boldsymbol{r} | \boldsymbol{p}+\rangle \rightarrow (2\pi\hbar)^{-3/2}\frac{\hbar}{pr}\sum_l (2l+1)\left[i^l\hat{j}_l(kr) + kf_l(p)e^{ipr/\hbar}\right]P_l(\hat{\boldsymbol{r}}\cdot\hat{\boldsymbol{p}}). \tag{4.197}$$

Comparing this with (4.193), we see that

$$\psi^+_{l,k}(r) \rightarrow \hat{j}_l(kr) + kf_l(k)e^{i(kr-l\pi/2)} \quad \text{as } r \rightarrow \infty. \tag{4.198}$$

The right-hand side of (4.198) is the combination of **partial waves** that corresponds to the full $r \rightarrow \infty$ form

$$e^{i\,\boldsymbol{p}\cdot\boldsymbol{r}/\hbar} + f(p\hat{\boldsymbol{r}} \leftarrow \boldsymbol{p})\frac{e^{ipr/\hbar}}{r}, \tag{4.199}$$

that is, $\hat{j}_l(kr)$ is the "incident-free l-wave" and $kf_l(k)e^{i(kr-l\pi/2)}$ is (for $r \rightarrow \infty$) the "scattered, outgoing l-wave".

We replace $\hat{j}_l(kr)$ by its asymptotic form $\sin(kr - l\pi/2)$ and put

$$kf_l(k) = e^{i\delta_l(k)}\sin\delta_l(k). \tag{4.200}$$

Then (4.198) becomes

$$\begin{aligned}
\psi^+_{l,k}(r) &\rightarrow \sin\left(kr - \frac{1}{2}l\pi\right) + e^{i\delta_l(k)}\sin\delta_l(k)e^{i\left(kr-\frac{1}{2}l\pi\right)} \\
&= e^{i\delta_l(k)}\left\{[\cos\delta_l(k) - i\sin\delta_l(k)]\sin\left(kr-\frac{1}{2}l\pi\right) + \sin\delta_l(k)\left[\cos\left(kr-\frac{1}{2}l\pi\right) + i\sin\left(kr-\frac{1}{2}l\pi\right)\right]\right\} \\
&= e^{i\delta_l(k)}\left\{\cos\delta_l(k)\sin\left(kr-\frac{1}{2}l\pi\right) + \sin\delta_l(k)\cos\left(kr-\frac{1}{2}l\pi\right)\right\} \\
&= e^{i\delta_l(k)}\sin\left(kr-\frac{1}{2}l\pi+\delta_l(k)\right).
\end{aligned} \tag{4.201}$$

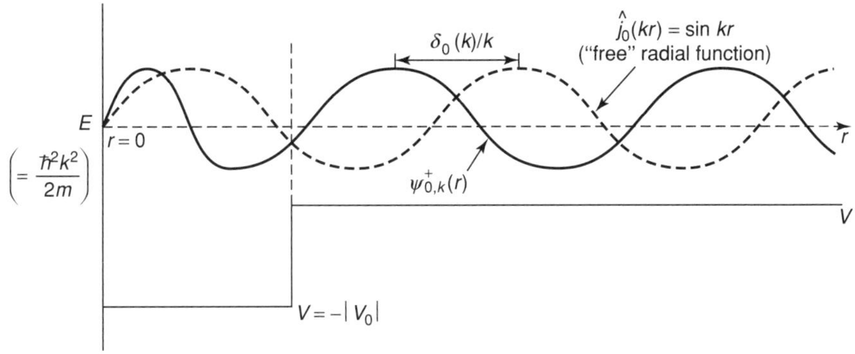

Figure 4.12 *Effect of an attractive potential on the phase of the $l = 0$ partial wave for a particle being scattered by the potential. The dashed curve represents the "free-particle" radial $l = 0$ function, and the solid curve represents the scattering-state $l = 0$ radial function.*

It can be seen from (4.201) that, as $r \to \infty$, the function $\psi^+_{l,k}(r)$ is proportional to the free-particle function $\hat{j}_l(kr)$ [$\to \sin(kr - \frac{1}{2}l\pi)$] **except that the phase of the oscillations is shifted by an amount** $\delta_l(k)$. This is the origin of the name **phase shift** for $\delta_l(k)$.

Consider, for example, the effect of an attractive potential on the $l = 0$ wave. In the region of the potential the kinetic momentum is larger (and hence the wavelength is smaller) than it is outside the region of the potential. In effect, the potential appears to "pull the wave in", relative to the wave of an unscattered particle (see figure 4.12).

It is clear from figure 4.12 that, although the **wavelengths** of the functions $\psi^+_{0,k}(r)$ and $\hat{j}_0(kr)$ are the same outside the region of the potential, the **phase** of $\psi^+_{0,k}(r)$ is in advance of that of $\hat{j}_0(kr)$ by $\delta_0(k)$. Moreover, the same mechanism operates for partial waves with all values of l. Hence, in general, **an attractive potential gives rise to a positive phase shift**.

Similarly, **a repulsive potential gives rise to a negative phase shift**.

A further interesting form of $\psi^+_{0,k}(r)$ is obtained by putting the $r \to \infty$ asymptotic form

$$\hat{j}_l(kr) = \sin(kr - l\pi/2) = \left(e^{i(kr-l\pi/2)} - e^{-i(kr-l\pi/2)}\right)/2i$$

and

$$kf_l(k) = [s_l(k) - 1]/2i$$

into (4.198). Then

$$\begin{aligned}\psi^+_{l,k}(r) &\to \frac{i}{2}\left(e^{-i(kr-l\pi/2)} - s_l(k)e^{i(kr-l\pi/2)}\right)\\ &= \frac{i e^{il\pi/2}}{2}\left(e^{-ikr} - s_l(k)e^{-il\pi}e^{ikr}\right) \propto e^{-ikr} - (-1)^l s_l(k)e^{ikr}.\end{aligned} \tag{4.202}$$

This form may be viewed purely conceptually as an "incident" incoming spherical wave e^{-ikr} and an outgoing spherical wave e^{ikr} produced by the target. The fact that the coefficient $(-1)^l s_l(k) = (-1)^l e^{2i\delta_l(k))}$ has modulus equal to unity is a reflection of the fact that the potential here does not create or destroy particles, that is, does not change the amplitude of the wave.

Module II

Thermal and Statistical Physics

5

Fundamentals of Thermodynamics

5.1 The nature of thermodynamics

In most physical problems we are interested not in properties of the whole universe but in those of some part of it. We call this part the "system", and the rest of the universe the "surroundings":

$$\text{universe} = \text{system} + \text{surroundings}$$

We study either large-scale (i.e., bulk or **macroscopic**) properties of the "system", or atomic-scale (**microscopic**) properties.

When **heat** (to be defined) is involved, the laws relating the macroscopic properties of a system, that is, the laws relating the pressure, volume, temperature, internal energy, entropy, and so on (again, to be defined), form the basis of the science of **thermodynamics**.

5.2 Walls and constraints

The system and surroundings are separated by a **boundary** or **wall**. Depending on the nature of the wall, the system and surroundings can exchange any or all of the following:

Energy   Matter   Volume

(In the latter case, one system grows in size at the expense of the other.)

The system may have **constraints**. In thermodynamics a constraint is a parameter or property of the system that can take a definite (**nonfluctuating**) value that can be fixed from outside the system.

Examples are:

- the volume V of a gas
- the number N of molecules in a system
- the net electric charge Q of a system.

A Course in Theoretical Physics, First Edition. P. J. Shepherd.

5.3 Energy

A system can also have a definite **energy** E. For example, the energy E of a gas consists of the rest energies and kinetic energies of the individual molecules, the potential energies of the molecules in any external fields, and the potential energy (sum of all the pair potential energies) of the mutual interaction of the molecules.

5.4 Microstates

In classical mechanics the microscopic state (**microstate**) of a system at a given time t is specified by giving the precise positions and velocities of every atom in the system at that time.

In quantum mechanics the microstate is specified by giving the exact wave function of the N-particle system (a function of the coordinates of all N particles of the system) at time t.

(Of course, both these tasks are impossible for a large system!)

In general, an unimaginably large number of microstates is possible for a given energy and a given set of values of the constraints.

5.5 Thermodynamic observables and thermal fluctuations

Besides the energy and the constraints, a system may have several other measurable macroscopic properties – the so-called **thermodynamic observables**. Examples are:

- pressure
- refractive index
- spectral absorption

The measured value of a thermodynamic observable depends on the microstate of the system at the time of the measurement. Since the microstate changes continuously, so do the values of the thermodynamic observables. We say that the thermodynamic observables are subject to **thermal fluctuations**.

For example, suppose that at time $t = 0$ we prepare many samples of a gas, all with the same energy and constraints (in particular, the same values of N and V). In general, each of these systems will be in a different microstate at $t = 0$.

Then, at some later time t (the same for all the samples), we measure the pressure p exerted on a given large area of wall of each container (a "large" area in the sense that very many molecules collide with it during the pressure measurement).

Let $n(p)$ be the number of measurements that yield a pressure value between 0 and p. Then the number that yield pressure values between 0 and $p + \delta p$ $(\delta p \to 0)$ will be

$$n(p + \delta p) = n(p) + \left.\frac{\mathrm{d}n}{\mathrm{d}p}\right|_p \delta p;$$

that is, the number of systems with pressures between p and $p + \delta p$ at the given time t is

$$\delta n(p) = n(p + \delta p) - n(p) = \left.\frac{\mathrm{d}n}{\mathrm{d}p}\right|_p \delta p.$$

If we measure the pressures of a total of n_{tot} systems, the fraction found to have pressures between p and $p + \delta p$ is

$$\frac{\delta n(p)}{n_{\mathrm{tot}}} = \frac{1}{n_{\mathrm{tot}}} \left.\frac{\mathrm{d}n}{\mathrm{d}p}\right|_p \delta p \equiv \tilde{f}(p)\delta p,$$

that is, $\tilde{f}(p)$ is defined as

$$\tilde{f}(p) = \frac{1}{n_{\mathrm{tot}}} \left.\frac{\mathrm{d}n}{\mathrm{d}p}\right|_p .$$

But in the limit of an infinite number n_{tot} of measurements, the **fraction** $\tilde{f}(p)\delta p$ of measurements yielding pressure values in a given interval $(p, p + \delta p)$ converges to a steady value $f(p)\delta p$, which is the **probability** of finding the pressure to lie in $(p, p + \delta p)$.

Thus, this probability is proportional to δp, as we should expect. The coefficient $f(p)$ of δp in this expression is called the **pressure distribution function**. It is a **probability density**, with the dimensions of $[p]^{-1}$ in order to make the probability dimensionless.

Thus, by measuring the fraction $\delta n(p)/n_{\mathrm{tot}}$ of systems with pressures in the interval $(p, p + \delta p)$ in the limit $n_{\mathrm{tot}} \to \infty$ we can find $f(p)$.

For example, such measurements might yield the curve shown in figure 5.1, in which the shaded area gives the probability that the value of the pressure lies in the range δp shown.

The area under this curve is:

$$\sum_{\delta p} f(p)\delta p \to \int_0^\infty f(p)\mathrm{d}p = \int_0^\infty \frac{\mathrm{d}n(p)}{n_{\mathrm{tot}}} = \frac{1}{n_{\mathrm{tot}}} [n(\infty) - n(0)] .$$

But $n(\infty)$ is the number of measurements yielding results between $p = 0$ and $p = \infty$. Since all n_{tot} measurements must yield results in this range, we have $n(\infty) = n_{\mathrm{tot}}$. Also, $n(0) = 0$, since no measurement can yield results in the "interval" between 0 and 0.

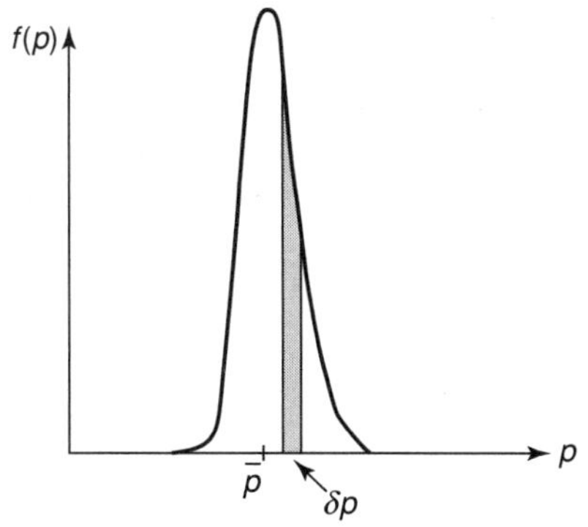

Figure 5.1 *Schematic pressure distribution function $f(p)$.*

Therefore,

$$\int_0^\infty f(p)\mathrm{d}p = \frac{n_{\mathrm{tot}}}{n_{\mathrm{tot}}} = 1,$$

that is, the continuous sum of the probabilities $f(p)\mathrm{d}p$ over the whole range of values of p is 1, as expected for the sum of the probabilities of all possible outcomes.

The distribution of measured pressures has a mean value $\bar{p}$, and a width characterizing the fluctuations from the mean.

If very many particles contribute to the result of a measurement, for example, if the area on which the pressure is measured is large, the distribution $f(p)$ will be extremely narrow, that is, the fluctuations from the mean pressure will be extremely small.

The different values of p in the distribution are associated with the different microstates of the system at the measurement time.

Definition: A **thermodynamic state** is one in which we know the probabilities of the microstates sufficiently accurately to be able to make good predictions of the distributions [such as $f(p)$] of the thermodynamic observables – in particular, of the mean values of the observables.

If we now make pressure measurements on all our macroscopically identical systems (i.e., with the same E, V and N) at a later time t', we obtain a pressure distribution $f_{t'}(p)$, which may differ from the one we found at time t.

If we wait long enough, however, the pressure distribution function may eventually settle to some final shape.

Definition: **Thermodynamic equilibrium** is a situation in which the distributions (and hence the mean values and widths) of all the thermodynamic observables have ceased to change with time.

For certain systems this may take a very long time indeed; for example, for a piece of glass it may take thousands of years for thermodynamic equilibrium to be reached (as manifested, for example, in the eventual constancy of the measured distribution, over a large number of samples, of the observable known as the refractive index).

The thermodynamic equilibrium state [and hence the definite mean values of all the observables (pressure, refractive index, X-ray diffraction pattern, etc.)] **is determined by the constraints and the value of the energy**.

Consequently, for a given system all "thermodynamic states" (see the definition) with given values of the constraints and a given energy E tend to the same thermodynamic equilibrium state.

A **state variable** is defined as a property of a system that takes a definite value in a thermodynamic equilibrium state of the system.

5.6 Thermodynamic degrees of freedom

Consider a gas mixture with two chemical species A and B.

We can find a thermodynamic equilibrium state with any N_A we like. Choose a particular value for N_A.

We can find a thermodynamic equilibrium state with this N_A and any N_B we like. Choose a particular value for N_B.
For the given N_A and N_B we can find a thermodynamic equilibrium state with any V we like. Choose a particular value for V.
For the given N_A, N_B and V we can find thermodynamic equilibrium states with infinitely many different values of E. Choose a value for E.
For the given values of N_A, N_B, V and E there is **only one** possible thermodynamic equilibrium state, that is, for this state we have **no freedom** to choose the value of any **other** state variable.

Definition: The **number of thermodynamic degrees of freedom** is the number of **state variables** (constraints, energy and thermodynamic observables) required to specify fully and uniquely a thermodynamic equilibrium state.

The number of thermodynamic degrees of freedom is **numerically** equal to the number of constraints plus one ($3 + 1 = 4$ in the above case), but the actual thermodynamic degrees of freedom chosen to specify a given thermodynamic equilibrium state can be **any** set of this number of state variables (four, in the above case).

In a given thermodynamic equilibrium state all other state variables beyond those chosen to specify it are completely determined by the values of the chosen ones.

5.7 Thermal contact and thermal equilibrium

Imagine a one-component gas enclosed in a rigid container of some given material, and with certain definite values of the energy and constraints (E, V and N). Let the gas be in the (unique) thermodynamic equilibrium state corresponding to these values of E, V and N, so that it exerts, for example, the appropriate corresponding (mean) pressure p on the walls.

A second rigid container, of the same material, containing a gas in the thermodynamic equilibrium state appropriate to **its** values E', V' and N', is now brought into contact with the first container (see figure 5.2).

If, after the contact, the pressures in the containers remain unchanged [$\tilde{p} = p$ and $\tilde{p}' = p'$] (whatever the initial thermodynamic equilibrium states), we say that the containers have **adiabatic walls** (***definition***).

If, on the other hand, in general, the pressures change [$\tilde{p} \neq p$ and $\tilde{p}' \neq p'$], we say that the walls are **diathermal** (***definition***) and that **thermal interaction** has taken place.

Two systems in contact via a diathermal wall are said to be in **thermal contact** (***definition***).

(Later, when we have defined **heat**, we shall see that diathermal walls conduct heat, while adiabatic walls do not.)

Two systems in thermodynamic equilibrium states **and** in thermal contact are said to be in **thermal equilibrium** with each other (**definition**).

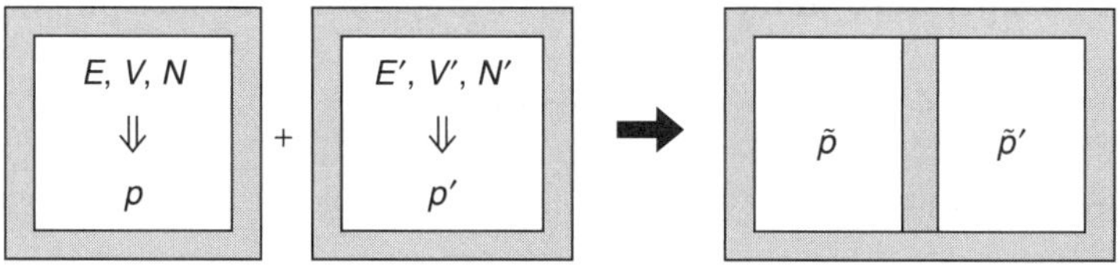

Figure 5.2 *Two systems, each in a thermodynamic equilibrium state, are brought into contact.*

5.8 The zeroth law of thermodynamics

Consider three bodies A, B and C. Suppose that both A and B are in thermal equilibrium with C, in the (wider) sense that if A and C are brought into thermal contact their thermodynamic states (as indicated, e.g., by their pressures) do not change, and if B and C are brought into thermal contact, **their** thermodynamic states do not change.

Then it is an **experimental fact** that if, instead, A and B are brought into thermal contact, their thermodynamic states also do not change.

Zeroth law: If A and B are separately in thermal equilibrium with C, they are also in thermal equilibrium with each other.

By bringing in further bodies D, E, F, and so on, one can see that this law applies to an arbitrarily large number of bodies.

Moreover, it is a further fact of experience that all bodies in thermal equilibrium in the sense described above have the same "hotness" to the touch.

We therefore postulate that they all have **the same value** of some new property, which we might call the "degree of hotness", or "**temperature**".

5.9 Temperature

It is natural to seek a measure of temperature in terms of those properties of a system that change when it is brought into thermal contact with another system and comes to thermal equilibrium with that system.

The most promising "thermometric properties" of this type would appear to be those which vary **monotonically** with the degree of hotness as assessed by the sense of touch. (If they were to increase and then decrease with increasing degree of hotness, equal values of the property would correspond to widely differing degrees of hotness, and this would clearly not be satisfactory.)

The system whose thermometric property X is to be measured when the system is brought into thermal contact with other systems is called the "thermometric system" or "**thermometer**".

Examples

Thermometric system	Thermometric property (X)
I Gas in a rigid (constant-volume) container	Pressure (p)
II A length of platinum wire	Electrical resistance (R)
III Mercury (Hg) in a glass capillary	Length of Hg column (l_{Hg})

We now ***define*** the "**temperature on the X-scale**" to be proportional to (linear in) X:

$$T_X = \text{const}.X.$$

For example, for thermometer I

$$T_p = ap, \tag{5.1}$$

for thermometer II

$$T_R = bR, \tag{5.2}$$

and for thermometer III

$$T_{l_{\mathrm{Hg}}} = c l_{\mathrm{Hg}}. \tag{5.3}$$

We find the constants of proportionality (a, b, c, etc.) for each thermometer by assigning (by international agreement) the value $T_{\mathrm{tr}} = 273.16$ K to the temperature of the (**unique**) thermodynamic equilibrium state of pure water in which **all three phases** coexist in equilibrium (see the phase diagram in figure 5.3).

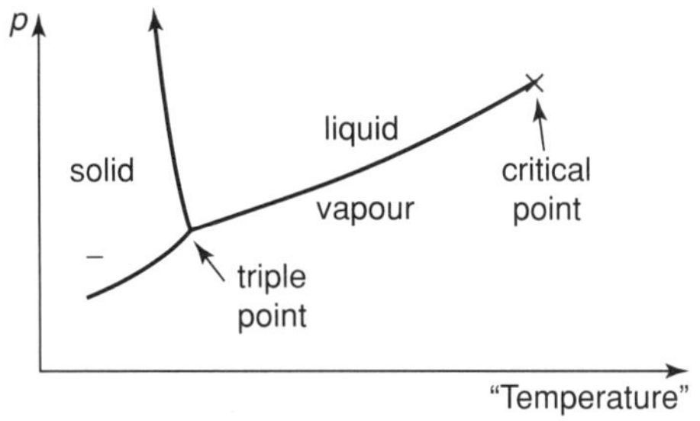

Figure 5.3 *Phase diagram of water.*

To calibrate the constant-volume gas thermometer I [i.e., to find a in (5.1)], we place it in **thermal contact** with water at its triple point and measure the eventual pressure p of the gas in the thermometer. Call the result p_{tr}. Then, from (5.1),

$$273.16 = a p_{\mathrm{tr}}, \quad \text{i.e.,} \quad a = 273.16/p_{\mathrm{tr}}.$$

Thus, (5.1) becomes

$$T_p = \frac{p}{p_{\mathrm{tr}}} \times 273.16\,\mathrm{K}, \tag{5.4}$$

which means that, if thermometer I is placed in thermal contact with another system and allowed to reach thermal equilibrium with it, and the pressure of the gas in the thermometer is measured to be p, the "temperature of the system [and thermometer] on the p-scale" is given by (5.4).

Similarly, for thermometer II,

$$b = \frac{273.16}{R_{\mathrm{tr}}},$$

so that the "temperature of a system on the resistance scale", that is, as found by measuring R, is

$$T_R = \frac{R}{R_{\mathrm{tr}}} \times 273.16\,\mathrm{K}. \tag{5.5}$$

For thermometer III,

$$c = \frac{273.16}{(l_{\mathrm{Hg}})_{\mathrm{tr}}}$$

and

$$T_{l_{\mathrm{Hg}}} = \frac{l_{\mathrm{Hg}}}{(l_{\mathrm{Hg}})_{\mathrm{tr}}} \times 273.16\,\mathrm{K}. \tag{5.6}$$

Except at the **one** fixed point (the triple point of water), at which $T_{p_{\mathrm{tr}}} = T_{R_{\mathrm{tr}}} = T_{(l_{\mathrm{Hg}})_{\mathrm{tr}}} = 273.16\,\mathrm{K}$, in general **the three scales do not coincide**.

To see this, suppose that thermometers I, II and III are put in thermal contact with a system whose temperature is to be measured. If the three thermometers always agreed, that is, if $T_p = T_R = T_{l_{\mathrm{Hg}}}$ always, we should have [from (5.4), (5.5) and (5.6)],

$$\frac{p}{p_{\mathrm{tr}}} = \frac{R}{R_{\mathrm{tr}}} = \frac{l_{\mathrm{Hg}}}{(l_{\mathrm{Hg}})_{\mathrm{tr}}}$$

always (!), that is, the **pressure** in I would be **exactly linearly** related to the **resistance** in II, and to the **length of the mercury column** in III, **always**. There is no reason whatever why this should be so, and experimentally it is not so!

In fact, even if we restrict our attention to constant-volume gas thermometers (i.e., measure the "temperature on the pressure scale", T_p), we find that thermometers with **different gases**, or even with the **same gases** but in **different amounts**, disagree in their values of T_p (except, of course, at the water triple point).

But, it **is** an experimental fact that **all** constant-volume gas thermometers, containing whatever type of gas, approach agreement on the measured value of T_p in the limit as the amount of gas tends to zero, that is, as the measured pressures, including p_{tr}, tend to zero. Therefore, we can **define** the **ideal-gas temperature** T as:

$$T = \lim_{p,\,p_{\mathrm{tr}} \to 0} \left[\frac{p}{p_{\mathrm{tr}}} \times 273.16 \right] \mathrm{K}$$

("ideal-gas", because in this limit the molecules are, on the average, sufficiently far apart to be treated accurately as noninteracting).

Later, by purely thermodynamic and statistical reasoning, we shall define the concept of **absolute temperature** T, and show that it is **identical to the ideal-gas temperature** (hence the use of the same symbol T for the latter). (As we shall see later, this scale has an **absolute zero** of 0 K (zero degrees Kelvin), i.e., the absolute temperature cannot take negative values.)

The **Celsius scale** is the same as the absolute scale except that its zero (its origin) is shifted. The Celsius temperature T^{C} of a system is related to its absolute temperature T by

$$T^{\mathrm{C}} = T - 273.15\ (^{\circ}\mathrm{C}).$$

The figure of 273.15 appears in this definition because, with T_{tr} **defined** as 273.16 K, the best experimental value of T for the thermodynamic equilibrium state of liquid water + ice at $p = 1$ atmosphere used to be $T = 273.15$ K, and the best experimental value of T for the thermodynamic equilibrium state of liquid water + water vapour at $p = 1$ atmosphere used to be $T = 373.15$ K.

Note that the values 273.15 K (= 0.00 °C) and 373.15 K (=100.00 °C) for these temperatures are **approximate, experimental** values [as are also the more accurate recent values 273.15 K (= 0.00 °C) and 373.12 K (= 99.97 °C)], but **the water triple-point temperature** T_{tr}^{C} = 273.16 – 273.15 °C = 0.01 °C is **exact by definition**.

The **units** of °C are the **same size** as the Kelvin (since the scales differ only by a defined shift of origin).

[It is also possible to define a "**centrigrade scale**", with **two** fixed points:

- T^{cent} (water + ice, $p = 1$ atm) = 0 degrees exactly;
- T^{cent} (water + vapour, $p = 1$ atm) = 100 degrees exactly.

Comparison with the best experimental results expressed in degrees Celsius (0.00 °C and 99.97 °C, respectively) shows that the centigrade unit differs in size from the degree Celsius (and hence from the Kelvin), and so the centigrade scale bears no relation to the absolute temperatures to which thermodynamic formulas apply. In such formulas, T is **always** the ideal-gas (absolute) temperature.]

5.10 The International Practical Temperature Scale

Gas thermometry is difficult. In practice we employ other methods and use the International Practical Temperature Scale of 1968, which assigns absolute temperatures to **ten** reference points as follows:

Triple point of H_2	13.81 K
Boiling point of water at 1 atm	373.15 K
Boiling point of H_2 at 1 atm	20.28 K
Boiling point of Ne at 1 atm	27.102 K
Triple point of O_2	54.361 K
Boiling point of O_2 at 1 atm	90.188 K
Boiling point of H_2 at 25/76 atm	17.042 K
Melting point of Zn at 1 atm	692.73 K
Melting point of Ag at 1 atm	1235.08 K
Melting point of Au at 1 atm	1337.58 K

plus, of course, the **exact** (by definition) triple point of water (= 273.16 K).

Three thermometers are used in three distinct ranges of temperature:

(i) Platinum resistance thermometer in the range 13.81 K to 903.89 K;
(ii) Pt/(90%Pt + 10%Rh) thermocouple in the range 903.89 K to 1337.58 K;
(iii) radiation pyrometer for temperatures higher than 1337.58 K.

These thermometers are calibrated to give the above values at the 11 reference points, and detailed formulas and numerical procedures are given for interpolating between the reference points, that is, for calculating a temperature from the reading on the instrument.

5.11 Equations of state

Recall that in a thermodynamic equilibrium state **all** the thermodynamic observables are determined by U, V and N (for a one-component system), or by U, V, N_1, ... , N_r (for an r-component system).

Thus, for a one-component system there must exist expressions for the thermodynamic observables p and T in terms of U, V and N:

$$p = p(U, V, N), \tag{5.7}$$

$$T = T(U, V, N), \tag{5.8}$$

where $p(U, V, N)$ is some function of U, V and N that enables p to be calculated.

If we know the function $T(U, V, N)$ we can solve

$$T = T(U, V, N),$$

for U in terms of T, V and N, that is, we get for U an expression

$$U = U(T, V, N).$$

If we substitute this expression into (5.7) we get p as a function of T, V and N:

$$p = p(T, V, N) \tag{5.9}$$

which can also be written as

$$f = f(p, T, V, N) = 0 \tag{5.9$'$}$$

where

$$f = p - p(T, V, N).$$

Equations (5.9) and (5.9′) are two forms of the **equation of state** of the system, that is, the equation relating p, T, V and N for the system.

For example, for an ideal gas (a gas of noninteracting molecules), we shall show later by simple quantum mechanics that the equation of state is the familiar

$$pV = Nk_{\text{B}}T \tag{5.10}$$

(k_{B} is **Boltzmann's constant**), or

$$pV = nRT$$

where n is the number of moles [1 mole $= M$ kg, where M is the molecular weight] and R is the gas constant.

5.12 Isotherms

Consider an ideal gas, and in the equation of state (5.10) fix N (by enclosing the gas in a container with an impermeable wall). Also, fix T to be T_1, by putting the container in thermal contact with a **large** system (a **heat reservoir**) of temperature T_1 (large enough for its temperature to be unchanged whatever happens to our gas).

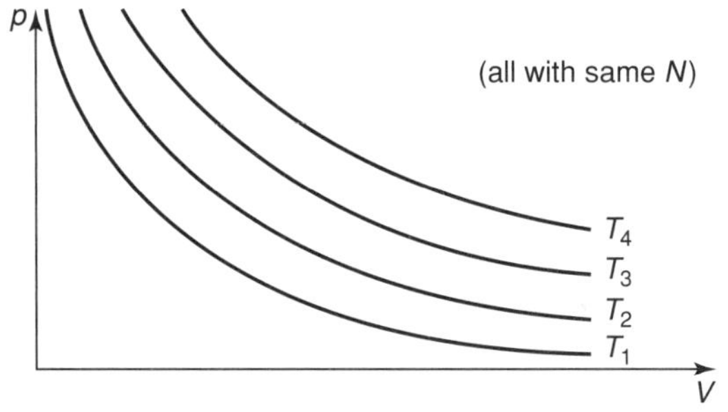

Figure 5.4 *Isotherms* ($T_4 > T_3 > T_2 > T_1$).

Then we can vary V and plot $p = Nk_BT_1/V = \text{const}/V$ against V. [Each state must be a **thermodynamic equilibrium state**, since otherwise equation (5.10) will not apply.]

We then repeat the measurements for other temperatures T_2, T_3, T_4, with $T_4 > T_3 > T_2 > T_1$ (see figure 5.4, in which each curve with constant T is an **isotherm**).

5.13 Processes

We shall consider processes that take a system from an initial **equilibrium** state to a final **equilibrium** state. Such processes may be **reversible** or **irreversible**.

Definition: A **reversible process** is one such that the "system" can subsequently be returned to its original equilibrium state in a manner that also returns the "surroundings" to **their** original state.

We now consider processes with transfer of energy to or from the system. We can identify **three** types of energy transfer:

(i) **Nondissipative work**, by slow, frictionless change of a constraint
(ii) **Dissipative work**, with or without change of a constraint
(iii) **Heat flow**

5.13.1 Nondissipative work

Example 1 Suppose we have a gas in a cylinder with a piston. Let the initial equilibrium state "be" (p_1, V_1).

Compress the gas to a new equilibrium state (p_2, V_2) by applying an external force F to the piston.

Perform the compression infinitely slowly, so that at every stage the pressure p is well defined and uniform, corresponding to an equilibrium state (p, V). Such a process is called a quasistatic process (that is, a quasistatic process can be regarded as a sequence of infinitesimally separated equilibrium states).

In an intermediate state in such a process, when the pressure is p, the balancing force on the piston is

$$F = pA,$$

where A is the area of the piston (see figure 5.5).

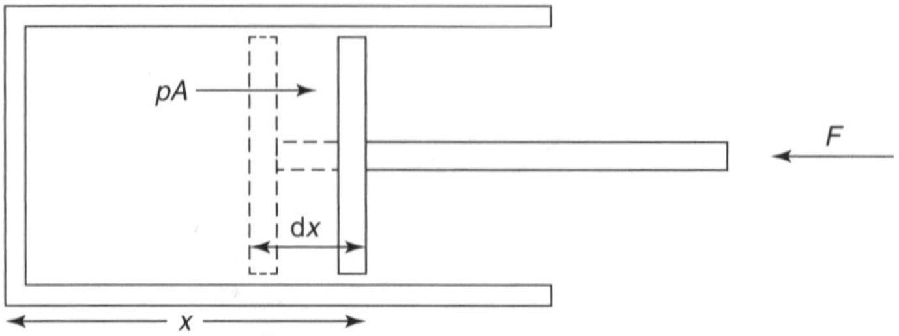

Figure 5.5 *Quasistatic compression of a gas.*

Now increase F infinitesimally so that the piston moves into the cylinder by dx. If the process is **frictionless** (no "dissipation" of energy), the work done on the gas is

$$đW = F\mathrm{d}x = pA\mathrm{d}x = -p\mathrm{d}V,$$

since the change in volume in the process is

$$\mathrm{d}V = A(x - \mathrm{d}x) - Ax = -A\mathrm{d}x.$$

The total work done on the gas in the quasistatic and frictionless compression from (p_1, V_1) to (p_2, V_2) is, therefore,

$$W = -\int_{V_1}^{V_2} p\mathrm{d}V$$

(> 0, if $V_2 < V_1$, as in compression). Of course, p here will be a function of V.

This is an example of **nondissipative work**, and involves the slow (quasistatic) frictionless change of the constraint V.

The process is also **reversible**, in the sense defined above, since upon slow reduction of the external force F the gas will expand quasistatically against the external force F, returning to its original equilibrium state (p_1, V_1) and doing work on the outside world.

[If some of this work was dissipated in friction, there would not be enough mechanical work available to return the external world to its original state, i.e., the process would be **irreversible**.]

Example 2 Consider the slow, quasistatic stretching of a rod, from length l_1 to length l_2 by a force f. When the length is l, the work done in stretching the rod to $l + \mathrm{d}l$ is $đW = f\mathrm{d}l$, that is,

$$W = \int_{l_1}^{l_2} f\mathrm{d}l$$

(> 0, if $l_2 > l_1$). Here, of course, f is a function of l.

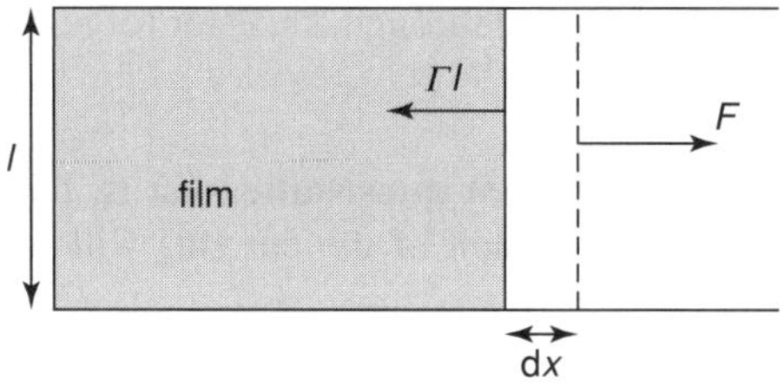

Figure 5.6 Stretching of a surface film.

Example 3 Consider the stretching of a surface film with surface tension Γ (see figure 5.6).

The work done on the film is

$$W = \int F\mathrm{d}x = \int \Gamma l \mathrm{d}x = \int \Gamma \mathrm{d}A,$$

where $\mathrm{d}A = l\mathrm{d}x$ is an infinitesimal change of the area of the film. (If the film is two sided, for example, for a soap film, then $W = 2\int \Gamma \mathrm{d}A$.)

5.13.2 Dissipative work

Example 1 Consider the situation illustrated in figure 5.7.

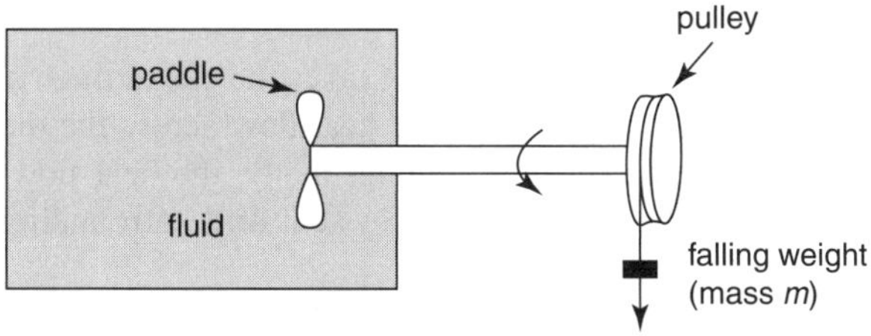

Figure 5.7 Stirring of a viscous fluid by the action of a falling weight.

This process, **even if carried out quasistatically** so that the weight is only just able to turn the paddle against the viscous drag ($m \approx 0$), is **irreversible** in the sense that if we reverse the torque (by suspending the small mass from the other side of the pulley), the paddle rotates in the opposite direction but the fluid does **not** return to its original state, that is, it does not return work to the mechanical system.

The work in this process is **dissipative work**.

This example of dissipative work involves no change of constraint. Dissipative work may also be done simultaneously **with** change of a constraint, for example, against frictional forces in the compression of a gas by a piston.

In fact, even if there is **no** friction between the piston and the wall of the cylinder, **rapid** (as opposed to quasistatic) compression involves dissipative work against the higher-than-equilibrium pressure that acts on the gas side of the piston when the gas does not have time to relax to equilibrium.

Example 2 Consider the passage of a current I through a resistor (of resistance R) immersed in a fluid. The dissipative work performed in time t is I^2Rt.

Again, this process is irreversible (**even when quasistatic**, that is, for $I \to 0$), since reversing the battery electromotive force (and hence the direction of the current) will not cause the fluid to return to its original state.

5.13.3 Heat flow

If two bodies (systems) with **different** temperatures are placed in **thermal contact** with each other, and no work [of either type (i) or type (ii)] is done on either system, the two systems change their thermodynamic states (as indicated, e.g., by measurements of their temperatures). The accompanying spontaneous flow of energy between them is called **heat flow**.

Heat flow between two bodies with **different** temperatures is **irreversible**. [It also necessarily involves temperature gradients, i.e., the absence of a single well-defined temperature.]

It is, however, possible to have heat flow between two bodies at the **same** temperature. This heat flow (**isothermal heat flow**) is **reversible**, as illustrated by the following example.

Let a cylinder containing a gas be in thermal contact with a heat reservoir at temperature T (see figure 5.8).

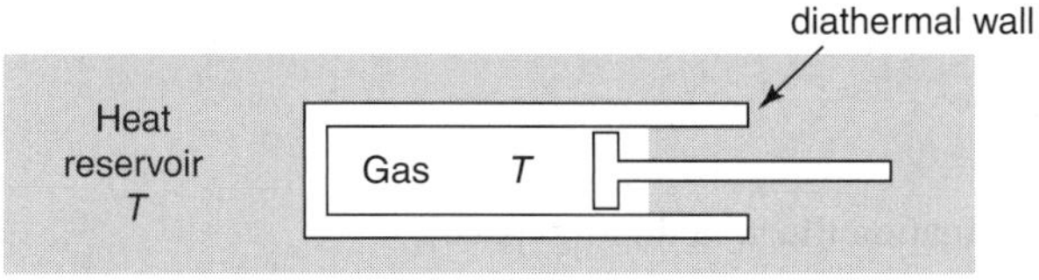

Figure 5.8 *Quasistatic compression of a gas in thermal contact with a heat reservoir.*

Quasistatic compression of the gas by a force $F = pA$ takes the gas through a succession of equilibrium states, in each of which the temperature adjusts to T by heat flow across the diathermal wall separating it from the reservoir. If the force F is then reduced infinitesimally, the heat flows required to maintain the gas temperature at T are reversed in direction, and the system **and** surroundings can be returned to their original state.

Isothermal (and hence reversible) heat flows are important for **thermodynamic engines** (discussed later).

5.14 Internal energy and heat

Having defined work (both dissipative and nondissipative) in precise mechanical terms, we now give more precise definitions of **internal energy** and **heat**.

5.14.1 Joule's experiments and internal energy

Joule took a tub of water with thermally insulating (adiabatic) walls. Using a paddle wheel to do dissipative work (equal to the torque times the total angle of rotation), he found that [in modern units] it required 4.2 kJ of work to raise the temperature of 1 kg of water by 1 °C (the "mechanical equivalent of heat").

More importantly, he found that, **no matter how** the adiabatic work was performed (e.g., by paddle wheel, by passage of an electric current, etc.), it needed the **same** amount of work to take water between the **same** two equilibrium states.

This is summarized in the following law:

If a thermally isolated system is brought from one equilibrium state to another, the work needed for this is independent of the process used.

In other words, $W_{\text{adiabatic}}$ depends only on the beginning and end equilibrium points, that is, it **is independent of the path between them**.

Thus, there must exist a **state function** (**state variable**) whose difference between two "end points" (2 and 1) is equal to the adiabatic work.

We call this state function the **internal energy** U, that is,

$$W_{\text{adiabatic}} = U_2 - U_1.$$

Since in Joule's experiment the tub was not given kinetic energy, or moved up or down to change its potential energy, the dissipative work goes into increasing the sum of the kinetic energies of the individual molecules and the potential energies of their mutual interaction.

5.14.2 Heat

If, instead, the system is **not** thermally isolated, the work W done in taking the system between equilibrium states 1 and 2 depends on the path (i.e., sequence of intermediate states) taken between them.

For a given change $1 \to 2$, $\Delta U = U_2 - U_1$ is fixed but $W \neq \Delta U$, that is,

$$W \neq W_{\text{adiabatic}}.$$

Definition: We call the difference between ΔU and W the **heat** Q, that is,

$$\Delta U = W + Q. \tag{5.11}$$

[When $Q = 0$ (adiabatic process), this gives $\Delta U = W_{\text{adiabatic}}$, as it should.] Both Q and W are defined to be positive when they increase the internal energy of the system.

The expression (5.11) is the **first law of thermodynamics**.

The work W can be nondissipative or dissipative work, or some combination of the two.

For an **infinitesimal process**, the first law takes the form

$$\mathrm{d}U = đW + đQ.$$

Here, $\mathrm{d}U$ means "a differential change of the state (state variable) U", and $đW$ and $đQ$ denote infinitesimal amounts of work and heat supplied to the system.

The quantities $đW$ and $đQ$ are **not** to be interpreted as the differential **changes** of **any** state function, **that is, there is no such state function as a work content or a heat content!**

For example, for infinitesimal (and quasistatic) compression of a gas,

$$đW = -p\mathrm{d}V \quad \text{(see earlier)}$$

that is,

$$dU = -p dV + đQ$$

or

$$đQ = dU + p dV \quad \text{(reversible)}$$

where the brackets indicate that, as argued earlier, this process is reversible.

For example, suppose state 2 of an ideal gas can be obtained from state 1 by **adiabatic** quasistatic compression (see figure 5.9).

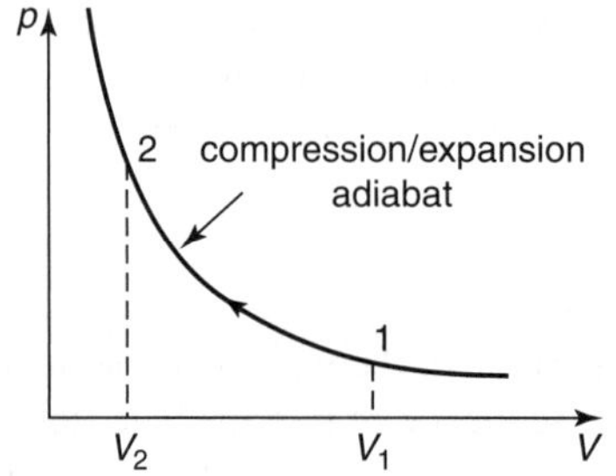

Figure 5.9 *Adiabatic compression of an ideal gas from equilibrium state 1 to equilibrium state 2.*

The change in the internal energy of the gas in this process is

$$\Delta U = U_2 - U_1 = W_{\text{adiabatic}} = -\int_{V_1}^{V_2} p dV = \int_{V_2}^{V_1} p dV = \text{area under curve} \quad (> 0).$$

It is also possible to get from 1 to 2 by other quasistatic (and reversible) processes (see figure 5.10).

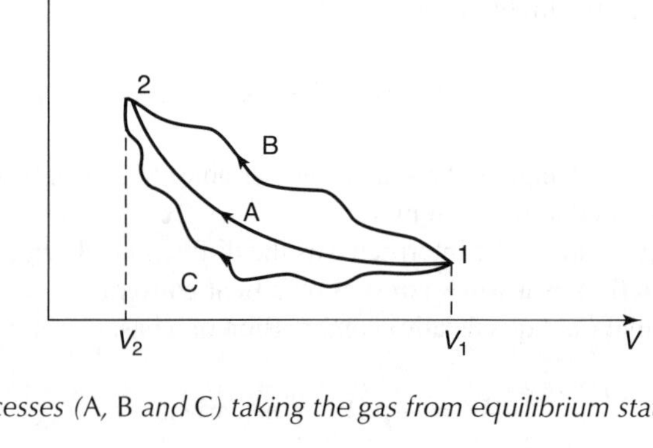

Figure 5.10 *Three processes (A, B and C) taking the gas from equilibrium state 1 to equilibrium state 2.*

The change in the internal energy is, of course, the same in all three processes:

$$\begin{aligned}\Delta U &= U_2 - U_1 = W_{\text{adiabatic}}\,(W_{\text{A}}) \\ &= W_{\text{A}}\,(\text{area under A}) + Q_{\text{A}}\,(\text{with } Q_{\text{A}} = 0) \\ &= W_{\text{B}}\,(\text{area under B}) + Q_{\text{B}}\,(\text{with } Q_{\text{B}} < 0) \\ &= W_{\text{C}}\,(\text{area under C}) + Q_{\text{C}}\,(\text{with } Q_{\text{C}} > 0)\end{aligned}$$

Thus, heat must be taken out of the system in process B and added to the system in process C.

In the opposite processes $2 \to 1$ (expansion), for which cases the Ws are negative, heat must be added in process B (to maintain the higher pressure) and taken out in process C.

Sometimes, the same energy source can be regarded as yielding either heat or work, depending on what constitutes the "system". For example, take an electric heating coil around a box containing a gas (see figure 5.11).

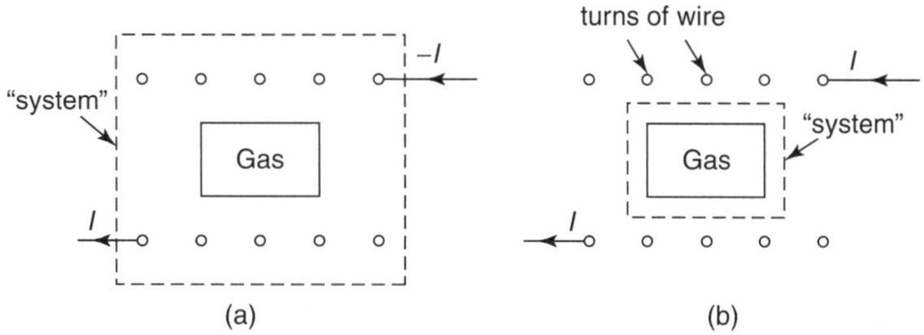

Figure 5.11 *Two ways of viewing energy transfer resulting from the flow of an electric current I.*

In figure 5.11a, where the coils are part of the "system", the energy supplied is **dissipative work** (by the battery on the coil).

In figure 5.11b it is **heat flow**, since it flows spontaneously into the system on account of the temperature difference between the coils and the "system".

5.15 Partial derivatives

For what follows, we give here a brief account of **partial derivatives**, and explain the meanings of the terms **exact** and **inexact differentials**.

Suppose that a function f depends on the variables x, y and z:

$$f = f(x, y, z).$$

How does f change when we replace

$$x \to x + \mathrm{d}x, \quad y \to y + \mathrm{d}y, \quad z \to z + \mathrm{d}z?$$

That is, what is

$$\mathrm{d}f \equiv f(x + \mathrm{d}x, y + \mathrm{d}y, z + \mathrm{d}z) - f(x, y, z)?$$

Answer: df will contain a term proportional to dx, a term proportional to dy, and a term proportional to dz.

The coefficients of dx, dy and dz in these terms in df are called the **partial derivatives of** f **(*definition*)**:

$$\mathrm{d}f = \left(\frac{\partial f}{\partial x}\right)_{y,z} \mathrm{d}x + \left(\frac{\partial f}{\partial y}\right)_{z,x} \mathrm{d}y + \left(\frac{\partial f}{\partial z}\right)_{x,y} \mathrm{d}z$$

Regard this as **notation** for the coefficients.

To justify this notation, fix y and z (so that dy = dz = 0), and divide the resulting expression for df by dx:

$$\frac{\mathrm{d}f}{\mathrm{d}x}[\mathrm{d}y = \mathrm{d}z = 0] = \left(\frac{\partial f}{\partial x}\right)_{y,z}.$$

Example: Let $f = f(x, y) = x^3 y + \text{const}$. Then

$$\mathrm{d}f = \left(\frac{\partial f}{\partial x}\right)_{y} \mathrm{d}x + \left(\frac{\partial f}{\partial y}\right)_{x} \mathrm{d}y$$
$$= 3x^2 y\mathrm{d}x + x^3 \mathrm{d}y$$

is an **exact differential** of the **point function** $f(x, y)$:

$$\left(\frac{\partial f}{\partial x}\right)_{y} = 3x^2 y \quad \text{and} \quad \left(\frac{\partial f}{\partial y}\right)_{x} = x^3.$$

Now consider an **infinitesimal quantity** $đg$, given in terms of the differentials dx and dy by

$$đg = 2x^2 y^5 \mathrm{d}x + 3x^3 y^4 \mathrm{d}y.$$

No point function $g(x, y)$ exists that changes by this amount when its arguments are shifted by dx and dy.

Thus, $đg$ is **not** the differential change of **any** point function $g(x, y)$. It just an infinitesimal quantity like $đW$ and $đQ$! We call it an **inexact differential**.

But, divide $đg$ by xy^2:

$$đg/xy^2 = 2xy^3 \mathrm{d}x + 3x^2 y^2 \mathrm{d}y \equiv \mathrm{d}h,$$

which **is** the differential change of the point function $h(x, y) = x^2 y^3$, that is, $1/xy^2$ is the **integrating factor** that converts the **inexact differential** $đg$ into the **exact differential** dh.

5.16 Heat capacity and specific heat

Let heat Q flow into a system, **let dissipative work be absent**, and let the temperature of the system change by ΔT. The **heat capacity** C of the system is **defined** as:

$$C = \lim_{\Delta T \to 0} \left(\frac{Q}{\Delta T}\right) = \frac{đQ}{\mathrm{d}T}.$$

The **specific heat** c is the heat capacity per unit mass:

$$c = \frac{1}{m}\frac{đQ}{dT}.$$

The **molar specific heat** c_{mol} is

$$c_{mol} = \frac{1}{n}\frac{đQ}{dT},$$

where n is the number of moles.

Since (even in the required absence of dissipative work) there are a large number of possible paths between the end points of the process, each with a different Q, there are a large number of possible heat capacities. We shall restrict ourselves to two.

5.16.1 Constant-volume heat capacity

Since $đQ = dU + pdV$ (if only nondissipative work occurs), heat added at constant volume (i.e., with $dV = 0$) is $đQ(dV = 0) \equiv đQ_V = dU$. Therefore,

$$C_V = \frac{đQ_V}{dT} = \left(\frac{\partial U}{\partial T}\right)_V,$$

where the **partial derivative** $(\partial U/\partial T)_V$ is the rate of change of U with T when V is held fixed.

5.16.2 Constant-pressure heat capacity

This is ***defined*** as

$$C_p = đQ_p/dT,$$

where $đQ_p$ denotes heat added at constant pressure (isobarically); again, dissipative work is absent.

Define a new state function

$$H = U + pV,$$

which is known as the **enthalpy**.

A differential change of H can be written as

$$\begin{aligned} dH &= dU + d(pV) \\ &= dU + pdV + Vdp \end{aligned}$$

But

$$dU + pdV = đQ$$

for nondissipative (reversible) expansion or compression. Therefore,

$$dH = đQ + Vdp,$$

and so

$$đQ_p = \mathrm{d}H$$

(since $\mathrm{d}p = 0$ for addition of heat at constant p).

Therefore,

$$C_p = đQ_p/\mathrm{d}T = \left(\frac{\partial H}{\partial T}\right)_p$$

The enthalpy H is useful in **flow processes** (e.g., in engineering) and in **chemistry** (ΔH in an isobaric chemical reaction is the "heat of reaction").

5.17 Applications of the first law to ideal gases

We shall show later that for an ideal gas we have

$$U = \frac{3}{2}Nk_{\mathrm{B}}T.$$

Thus, U for an ideal gas depends only on N and T (for a gas of interacting molecules, U also depends on V). Thus, for fixed N, for an ideal gas we have $U = U(T)$.

For example, consider the free expansion of a gas, as illustrated in figure 5.12.

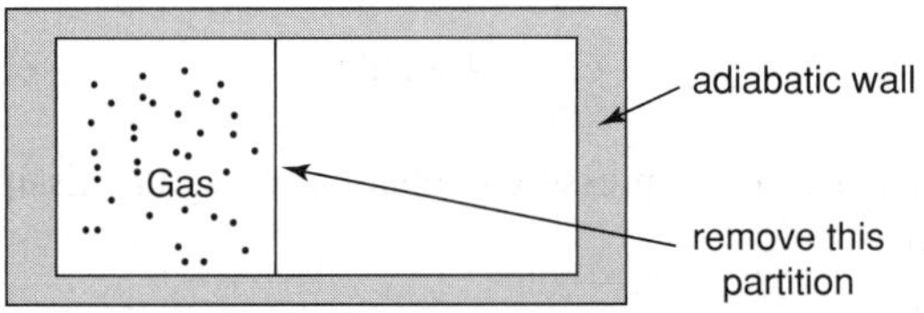

Figure 5.12 *Free expansion of a gas.*

In this sudden expansion the gas does no external work ($-W = 0$) and, since the walls are adiabatic, $Q = 0$. Therefore,

$$\Delta U = 0.$$

Since, for a fixed amount of an **ideal** gas, U is a single-valued function of T alone (so that T is a single-valued function of U alone), the result $\Delta U = 0$ means that

$$T_{\mathrm{final}} = T_{\mathrm{initial}} \quad \text{(for an ideal gas).}$$

[Note that $\Delta U = 0$ for the free expansion of **any** gas in a rigid adiabatic enclosure (since energy is always conserved), but only for an ideal gas do we expect that $T_{\mathrm{final}} = T_{\mathrm{initial}}$.]

For a **nonideal** gas (gas of noninteracting molecules), expansion increases the potential energy of their interaction (because the average intermolecular separation increases), and the kinetic energy must **decrease** to ensure that $\Delta U = 0$, that is, T must decrease:

$$\left(\frac{\partial T}{\partial V}\right)_U < 0,$$

so that, in free expansion from V_1 to V_2,

$$T_{\text{final}} - T_{\text{initial}} = \int_{V_1}^{V_2} \left(\frac{\partial T}{\partial V}\right)_U \mathrm{d}V < 0 \quad \text{(cooling)}$$

[$(\partial T/\partial V)_U$ is the **Joule coefficient**.]

5.18 Difference of constant-pressure and constant-volume heat capacities

For an infinitesimal process with no dissipative work we have

$$đQ = \mathrm{d}U + p\mathrm{d}V$$

(see earlier). But

$$\mathrm{d}U = \left(\frac{\partial U}{\partial T}\right)_V \mathrm{d}T + \left(\frac{\partial U}{\partial V}\right)_T \mathrm{d}V.$$

For an **ideal gas**, $U = U(T)$, that is,

$$(\partial U/\partial V)_T = 0.$$

Therefore,

$$\mathrm{d}U = \left(\frac{\partial U}{\partial T}\right)_V \mathrm{d}T = C_V \mathrm{d}T,$$

and so

$$đQ = C_V \mathrm{d}T + p\mathrm{d}V.$$

Therefore,

$$C_p = \frac{đQ_p}{\mathrm{d}T} = C_V + p\left(\frac{\partial V}{\partial T}\right)_p.$$

For an ideal gas, $pV = nRT$ (n is the number of moles), that is, $V = nRT/p$, and so

$$\left(\frac{\partial V}{\partial T}\right)_p = \frac{nR}{p}.$$

Therefore,

$$C_p = C_V + nR \tag{5.12}$$

for an ideal gas.

5.19 Nondissipative-compression/expansion adiabat of an ideal gas

Consider a nondissipative adiabatic expansion.

Because the work is nondissipative,

$$đQ = \mathrm{d}U + p\mathrm{d}V.$$

Because the process is adiabatic ($đQ = 0$),

$$0 = \mathrm{d}U + p\mathrm{d}V.$$

Because the gas is ideal [so that $U = U(T)$],

$$0 = C_V\mathrm{d}T + p\mathrm{d}V.$$

Because $p = nRT/V$ for an ideal gas, we have

$$0 = C_V\mathrm{d}T + nRT\frac{\mathrm{d}V}{V}, \quad \text{i.e.,} \quad 0 = C_V\frac{\mathrm{d}T}{T} + nR\frac{\mathrm{d}V}{V}$$

Integrating,

$$C_V \ln T + nR\ln V = \text{const.}$$

Therefore, with the aid of (5.12),

$$\ln T + \frac{C_p - C_V}{C_V}\ln V = \text{const},$$

that is, $\ln T + (\gamma - 1)\ln V = \text{const}$, where $\gamma \equiv C_p/C_V$ is the ratio of the heat capacities. Therefore,

$$TV^{\gamma-1} = \text{const.}$$

Put $T = pV/nR$. Then

$$pV^{\gamma} = \text{const.}$$

Using $U = 3Nk_BT/2 = 3nRT/2$, we find

$$C_V = \left(\frac{\partial U}{\partial T}\right)_{V,N} = \frac{3}{2}nR.$$

Therefore,

$$C_p = C_V + nR = \frac{5}{2}nR,$$

and so

$$\gamma = \frac{C_p}{C_V} = \frac{5}{3}.$$

Thus, reversible expansion/compression adiabats of an ideal gas have the form

$$pV^{5/3} = \text{const}, \quad \text{i.e.,} \quad p = \frac{\text{const.}}{V^{5/3}}.$$

A gas with a given number n of moles can have any pressure and volume, and every such point in the (p, V) plane lies on an adiabat.

For example, the point (p_1, V_1) lies (if the gas is ideal) on the adiabat

$$pV^{5/3} = \text{const.} = p_1V_1^{5/3}.$$

This adiabat is shown in figure 5.13.

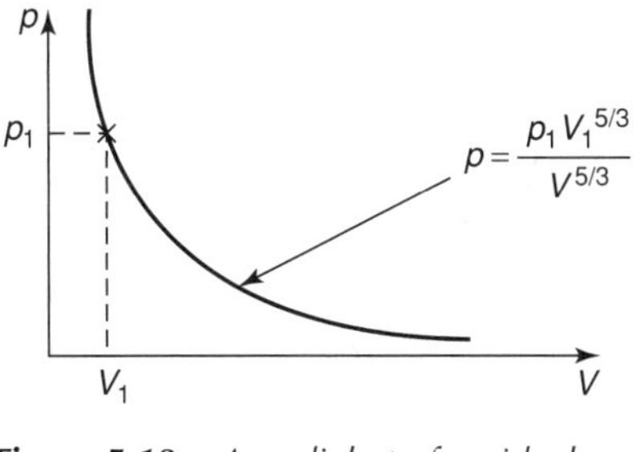

Figure 5.13 *An adiabat of an ideal gas.*

Every point (p, V) on the adiabat corresponds to a **different** temperature T, such that $pV = nRT$; for example, the point (p_1, V_1) corresponds to a temperature T_1 such that $p_1V_1 = nRT_1$.

Draw the **isotherm** $pV = nRT_1$ passing through (p_1, V_1), that is, corresponding to temperature T_1 (see figure 5.14).

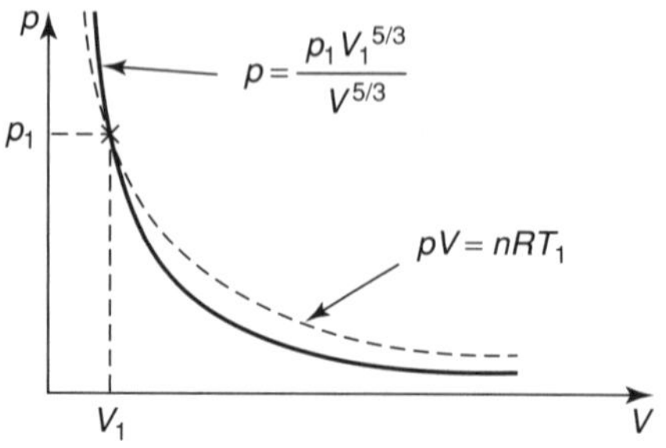

Figure 5.14 Intersection of an adiabat and an isotherm of an ideal gas.

At the point of intersection,

$$\text{slope of adiabat} = \left.\frac{\mathrm{d}p}{\mathrm{d}V}\right|_{\text{adiabat}} = \frac{\mathrm{d}}{\mathrm{d}V}\left[\frac{\text{const.}}{V^{\gamma}}\right] = -\gamma\frac{\text{const.}}{V^{\gamma+1}} = -\gamma\frac{p}{V};$$

$$\text{slope of isotherm} = \left.\frac{\mathrm{d}p}{\mathrm{d}V}\right|_{\text{isotherm}} = \frac{\mathrm{d}}{\mathrm{d}V}\left[\frac{nRT}{V}\right] = -\frac{nRT}{V^2} = -\frac{p}{V}.$$

Therefore,

$$\frac{\text{slope of adiabat}}{\text{slope of isotherm}} = \gamma.$$

We can fill the (p, V) plane with isotherms $pV = nRT$ (each with a different value of T) and adiabats $pV^{5/3} = \text{const}$ (each with a different constant α, β, δ, etc.) (see figure 5.15).

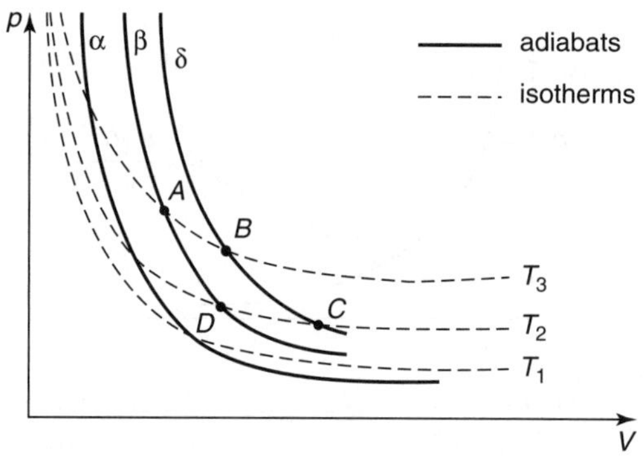

Figure 5.15 Isotherms and adiabats of an ideal gas.

Clearly, on adiabat β we have T_A ($= T_3$) $> T_D$ ($= T_2$), while on adiabat δ we have T_B ($= T_3$) $> T_C$ ($= T_2$), that is, T decreases as we move down an adiabat on the (p, V) diagram.

We now analyse some processes using the first law $\mathrm{d}U = đW + đQ$.

(i) $\mathrm{D} \to \mathrm{A}$
This is along an adiabat (the one with const $= \beta$), and so $đQ = 0$. Therefore,

$$U_\mathrm{A} - U_\mathrm{D} = \int đW = -\int_{V_\mathrm{D}}^{V_\mathrm{A}} p\mathrm{d}V = -\beta \int_{V_\mathrm{D}}^{V_\mathrm{A}} \frac{\mathrm{d}V}{V^{5/3}} = -\beta\left(-\frac{1}{2/3}\right)\frac{1}{V^{2/3}}\bigg|_{V_\mathrm{D}}^{V_\mathrm{A}}$$

$$= \frac{3\beta}{2}\frac{V}{V^{5/3}}\bigg|_{V_\mathrm{D}}^{V_\mathrm{A}} = \frac{3}{2}\,pV|_D^A = \frac{3}{2}\,nRT|_D^A = \frac{3}{2}\,nRT_3 - \frac{3}{2}\,nRT_2,$$

as expected, since $U = U(T) = 3nRT/2$ for an ideal gas.

(ii) $\mathrm{A} \to \mathrm{B}$

$$U_\mathrm{B} - U_\mathrm{A} = \int_\mathrm{A}^\mathrm{B} đW + \int_\mathrm{A}^\mathrm{B} đQ = -\int_{V_\mathrm{A}}^{V_\mathrm{B}} p\mathrm{d}V + Q.$$

But for an **ideal gas** we have $U_\mathrm{B} - U_\mathrm{A} = 0$ (since the states A and B have the same temperature). Therefore, for an ideal gas,

$$Q = \int_{V_\mathrm{A}}^{V_\mathrm{B}} p\mathrm{d}V = \text{area under curve},$$

that is, heat must be added for expansion along an isotherm.

6

Quantum States and Temperature

6.1 Quantum states

Consider **one** particle undergoing oscillation in a parabolic potential $V = \frac{1}{2}kx^2$ (see figure 6.1), so that the restoring force $f = -\mathrm{d}V/\mathrm{dx} = -kx$ is linear in x (simple harmonic motion).

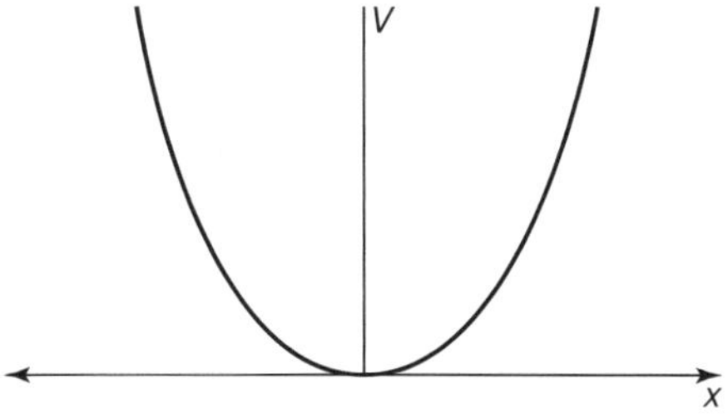

Figure 6.1 *Parabolic potential.*

The classical equation of motion is

$$ma = f,$$

that is,

$$m\frac{d^2x}{dt^2} = -kx,$$

which has the general solution

$$x(t) = A\cos\omega t + B\sin\omega t,$$

where $\omega = \sqrt{k/m}$ is the classical angular frequency $\omega = 2\pi\nu$ (A and B can be found by specifying two conditions, e.g., specifying the values of x and $\mathrm{d}x/\mathrm{d}t$ at $t = 0$).

A Course in Theoretical Physics, First Edition. P. J. Shepherd.

Whereas in classical mechanics the particle can have **any** amplitude of oscillation (and hence any total energy), in quantum mechanics the energy can only take the discrete values

$$\varepsilon_n = \left(n + \tfrac{1}{2}\right) h\nu \quad \text{(same } \nu \text{ as above)}$$

where h is Planck's constant and $n = 0, 1, 2, \ldots$. We say the oscillator has n **quanta of energy**. Thus, the possible energy values are:

$$\varepsilon_0 = \tfrac{1}{2}h\nu, \quad \varepsilon_1 = \tfrac{3}{2}h\nu, \quad \varepsilon_2 = \tfrac{5}{2}h\nu, \quad \text{etc.}$$

Suppose now that we have N such particles, each subject to the same linear restoring force $-kx$, and (hence) each oscillating with the same angular frequency $\omega = \sqrt{k/m}$, that is, we have N oscillators. Suppose also that we have n quanta of energy (i.e., energy $nh\nu$) to distribute among the N oscillators.

Treat the N oscillators as N boxes, separated by $N - 1$ partitions. Treat the n quanta as n balls to be placed in the boxes. If the balls (and the partititions, respectively) are distinguishable from each other, the number of ways of ordering the n balls and $N - 1$ partitions is $(n + N - 1)!$. (Starting from the left, there are $n + N - 1$ objects to select from for the first position, then $n + N - 2$ to select from for the second, and so on.)

But if the oscillators are indistinguishable (as they are), permuting boxes alone (and hence partitions) does not give a new state. Therefore, divide by $(N - 1)!$. And if, as is the case, the quanta are indistinguishable, permuting the balls alone does not give a new state. Therefore, divide by $n!$. The result is that the number of distinct states of N oscillators with n quanta in total is

$$W = \frac{(n + N - 1)!}{n!(N - 1)!}. \tag{6.1}$$

Check: For **one** oscillator ($N = 1$) and n quanta equation (6.1) gives

$$W = \frac{(n + 1 - 1)!}{n!(1 - 1)!} = \frac{n!}{n!} = 1 \;\; (0! = 1),$$

that is, there is **one** state for each value of n, as expected for $N = 1$.

This result can be represented as in figure 6.2.

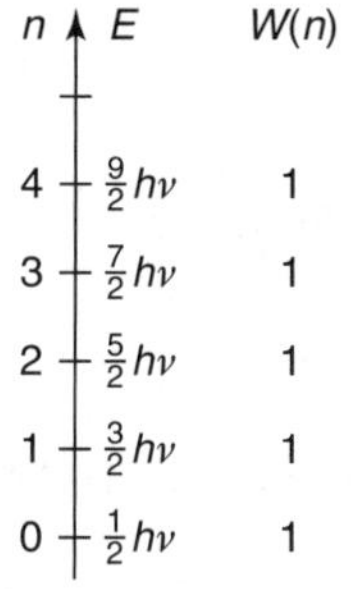

Figure 6.2 *Values of E and W(n) for the states of a single one-dimensional harmonic oscillator.*

We now introduce a function $g(E)$, such that the number of states in the energy range $(E, E + \mathrm{d}E)$ is $g(E)\mathrm{d}E$. The function $g(E)$ is called the **density of states**, and clearly has the dimensions of $[\text{energy}]^{-1}$.

It is clear that

$$g(E_n) = W(n) \times \frac{1}{h\nu} = (\text{for } N = 1)\ \frac{1}{h\nu},$$

which is independent of n. Thus, $g(E)$ is independent of E in the case $N = 1$.

If, however, $N \sim 10^{23}$ and $n \sim 10^{23}$ (typical values for a macroscopic system of particles [oscillators]), we can use Stirling's approximation for large factorials:

$$\ln(N!) \sim N \ln N - N$$

to show that (6.1) gives $W \sim 10^{10^{23}}$. This is true for each value of n, and so all the values of $W(n)$ in figure 6.3 are of this order.

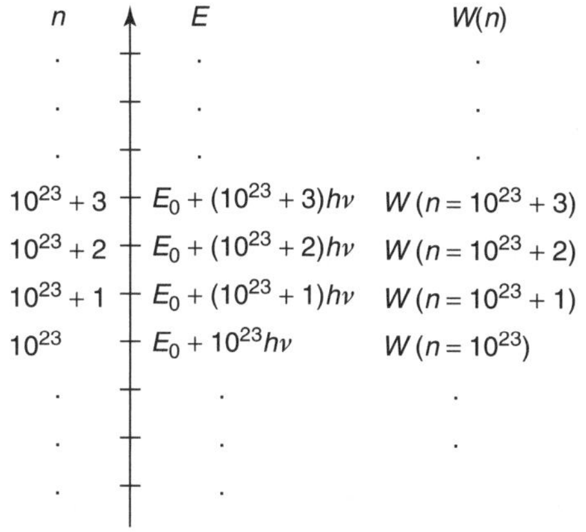

Figure 6.3 *Values of E and W(n) for the states of a system of N noninteracting one-dimensional harmonic oscillators.*

In figure 6.3, $E_0 = \frac{1}{2}Nh\nu$ is the total **zero-point energy** of the system. Each **level** is separated by energy $h\nu$, and, from the diagram, the number of states in the interval $h\nu$ containing the energy value $E_0 + 10^{23}h\nu$ is $W(n = 10^{23})$, that is, the density of states is, at energy E_n,

$$g(E_n) = \frac{W(n)}{h\nu}.$$

The number $W(n)$ increases with n. In fact,

$$W(n+1) \overset{(6.1)}{=} \frac{(n+1+N-1)!}{(n+1)!(N-1)!} = \frac{(n+N)!}{(n+1)!(N-1)!} = \frac{n+N}{n+1} W(n) = \left(1 + \frac{N-1}{n+1}\right) W(n).$$

Thus, each time we add 1 to n, $W(n)$ is multiplied by $1 + \frac{N-1}{n+1}$ (= 1 when $N = 1$, as already shown), that is, by essentially the **same factor** each time, since, to within a negligible error,

$$1 + \frac{N-1}{10^{23}+2} = 1 + \frac{N-1}{10^{23}+1}.$$

Call this constant factor $1 + K$, and write

$$W(n+1) - W(n) = \frac{dW}{dn} dn = \frac{dW}{dn}$$

(since $dn = 1$).

Therefore,

$$(1 + K)W(n) - W(n) = \frac{dW}{dn},$$

that is,

$$\frac{dW}{dn} = KW,$$

which has the solution

$$W = \text{const.e}^{Kn} = \text{const.e}^{(N-1)n/(n+1)} \approx \text{const.e}^{N}.$$

In other words, **the density of states of our *N*-oscillator system increases exponentially with *N*.**

6.2 Effects of interactions

Assume the system to be thermally isolated.

If there were no interactions between the oscillators, a system starting in a given state of the above type, with exact energy $E = \frac{1}{2}Nh\nu + nh\nu$, would remain in that state **forever** (a **stationary state**).

Interactions between the oscillators have **two** effects:

(i) They spread the states in energy; that is, instead of being concentrated at energy $E_0 + nh\nu$ ($E_0 = \frac{1}{2}Nh\nu$) the states are distributed (**in the same numbers**) more uniformly over the energy axis. [An energy interval of width $h\nu$ at E contains $W(E) = \text{const.}h\nu e^N$ states.] Therefore, the average energy spacing between the states is

$$\frac{h\nu}{W(E)} = \text{const.e}^{-N},$$

so that the average energy spacing when $N \sim 10^{23}$ is $\sim 10^{10^{23}}$ Joules (or ergs, etc. – the energy units are utterly irrelevant on this scale!).

(ii) The system no longer remains for ever in a given state of the type described, but undergoes **quantum jumps** from state to state.

For example, suppose a state i persists for a time of order t. Then **Heisenberg's energy–time uncertainty principle** shows that its energy is uncertain by an amount $\Delta E \sim h/t$; that is, the system can undergo transitions from state i to states j within an energy range ΔE about the energy of state i, as illustrated in figure 6.4.

Thus, the probability $\tilde{p}_i$ that the system is in state i changes with time.

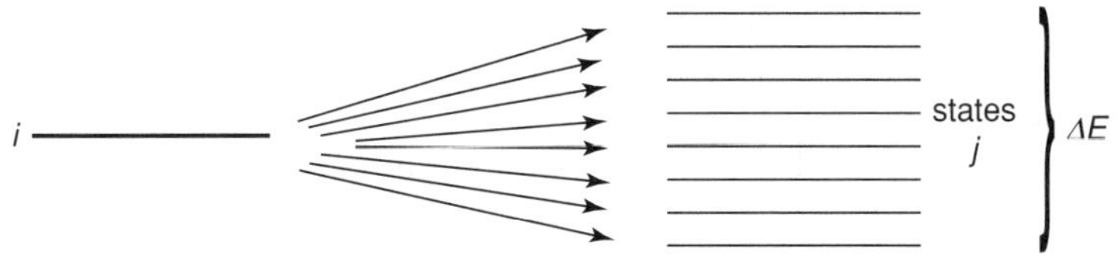

Figure 6.4 *Possible quantum jumps from the nonstationary state i.*

There are two contributions to $\mathrm{d}\tilde{p}_i/\mathrm{d}t$:

(i) The probability $\tilde{p}_i$ is **decreasing** as a result of transitions from i to states j in the range ΔE. This contribution to $\mathrm{d}\tilde{p}_i/\mathrm{d}t$ is proportional to $\tilde{p}_i$ itself, and to $\nu_{j\leftarrow i}$ (the one-to-one jump rate [number of jumps per second] from i to j):

$$\left.\frac{\mathrm{d}\tilde{p}_i}{\mathrm{d}t}\right|_{(\mathrm{i})} = -\sum_j \nu_{j\leftarrow i}\tilde{p}_i, \tag{i}$$

where the sum is over the states j in the range ΔE. But [contribution (ii)], $\tilde{p}_i$ is also **increasing** as a result of transitions from the states j to i. This contribution to $\mathrm{d}\tilde{p}_i/\mathrm{d}t$ is

$$\left.\frac{\mathrm{d}\tilde{p}_i}{\mathrm{d}t}\right|_{(\mathrm{ii})} = \sum_j \nu_{i\leftarrow j}\tilde{p}_j. \tag{ii}$$

But, for a given pair of states, the forward and reverse one-to-one jump rates depend on the modulus squared of the matrix element of the hamiltonian (see section 3.4), and so are equal, that is,

$$\nu_{j\leftarrow i} = \nu_{i\leftarrow j} \equiv \nu_{ij}.$$

Adding (i) and (ii), we get

$$\frac{\mathrm{d}\tilde{p}_i}{\mathrm{d}t} = \sum_j \nu_{ij}\left(\tilde{p}_j - \tilde{p}_i\right), \tag{6.2}$$

which is **Fermi's master equation**.

Thermodynamic equilibrium corresponds to **unchanging** probabilities of states:

$$\frac{\mathrm{d}\tilde{p}_i}{\mathrm{d}t} = 0.$$

From (6.2), this is satisfied if $\tilde{p}_j = \tilde{p}_i$ for all states j in the sum (6.2). Thus, when a **thermally isolated** system reaches thermodynamic equilibrium the state probabilities $\tilde{p}_i$ for all members of any set of **mutually accessible** states are equal.

This is the **principle of equal equilibrium probability** (sometimes called the **principle of equal *a priori* probability**).

In any real situation the **experimental** uncertainty δE in the energy is always **much** greater than the Heisenberg uncertainty ΔE (typically, $\delta E \sim 10^{-10}$ J and $\Delta E \sim 10^{-27}$ J).

We assume, in fact, that our chosen system has access to **all** the microstates in the range δE, that is, there exist nonzero jump rates ν_{ij} that link **every pair** of states in the range δE, either directly or indirectly. This is called the **ergodic assumption**.

Thus, whatever state we start from, the system will explore all the accessible states eventually.

Therefore, the time average of some quantity dependent on the microscopic state (for example, the pressure) over a long enough time will be the same as the average of the pressure over an **ensemble** of systems in which each accessible state is represented once (that is, with equal probability). Thus, the fluctuation distribution and ensemble distribution are identical.

The principle of equal equilibrium probability plus the ergodic hypothesis lead to the following result.

For an isolated (constant-energy) system, with "accessibility band" δE [between E and $E + \delta E$], all microstates **within** this band have **equal probabilities** and all microstates **outside** it have **zero probability** (see figure 6.5). (For the **equilibrium**, time-independent probabilities we use the notation p_i rather than $\tilde{p}_i$.)

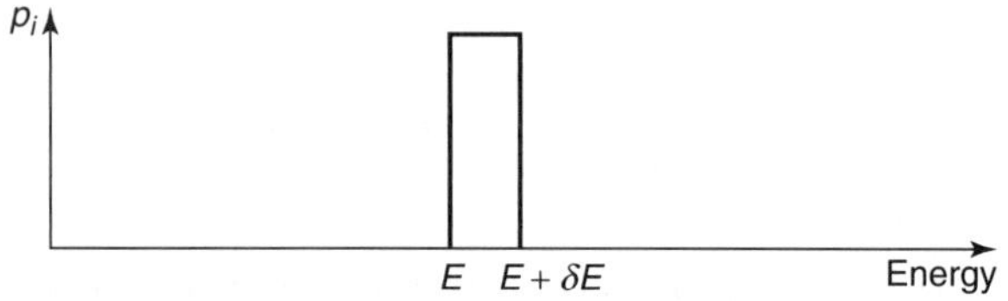

Figure 6.5 *Probability p_i of finding a closed system to be in quantum state i.*

The number of states in the "accessibility band" is

$$\int_E^{E+\delta E} g(E)\mathrm{d}E,$$

which can be approximated by $g(E)\delta E$ if δE is so small that the variation of $g(E)$ over it can be ignored.

Since all these states have equal probability, we have

$$p_i = \begin{cases} \dfrac{1}{g(E)\delta E} & \text{(for } E \text{ inside the accessibility band)} \\ 0 & \text{(for } E \text{ outside the accessibility band)} \end{cases} \tag{6.3}$$

Check:

$$\sum_{\text{states } i} p_i = \sum_{i=1}^{i=g(E)\delta E} \frac{1}{g(E)\delta E} = 1,$$

as required.

The distribution (6.3) (depicted in figure 6.5) is called a **microcanonical distribution**.

6.3 Statistical meaning of temperature

Bring two systems A and B into thermal contact (see figure 6.6).

Here, $U'_A + U'_B = U_A + U_B$ by energy conservation. What is the final, equilibrium sharing of the energy between A and B?

Thermal physics answer: It is the sharing that ensures that $T_B = T_A$.

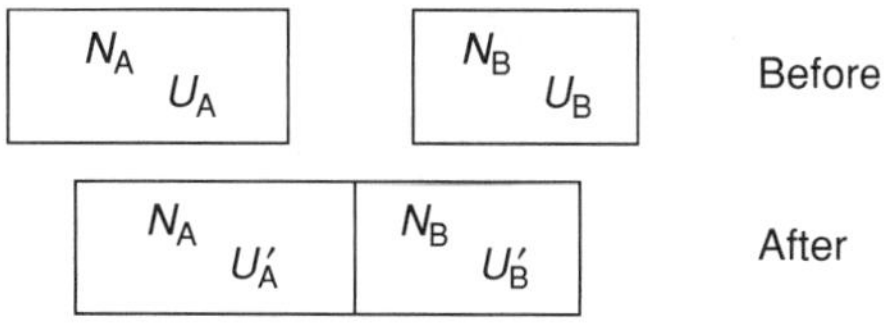

Figure 6.6 *Thermal contact of two systems.*

Quantum statistics answer: It is the most probable sharing, that is, the one that can be realized in the most ways.

Suppose the energy of system A before thermal contact is $E_{\mathrm{A}} = E_{\mathrm{A}}^{\text{before}} \pm (\delta E_{\mathrm{A}}/2)$, that is, the accessibility band of system A has width δE_{A}.

The number of accessible states of A is

$$\int_{E_{\mathrm{A}}^{\text{before}}-(\delta E_{\mathrm{A}}/2)}^{E_{\mathrm{A}}^{\text{before}}+(\delta E_{\mathrm{A}}/2)} g_{\mathrm{A}}(E_{\mathrm{A}})\mathrm{d}E_{\mathrm{A}} \approx g_{\mathrm{A}}\left(E_{\mathrm{A}}^{\text{before}}\right)\delta E_{\mathrm{A}}.$$

Similarly, if the energy of B before the thermal contact is $E_{\mathrm{B}} = E_{\mathrm{B}}^{\text{before}} \pm (\delta E_{\mathrm{B}}/2)$, the number of accessible states of B is approximately

$$g_{\mathrm{B}}\left(E_{\mathrm{B}}^{\text{before}}\right)\delta E_{\mathrm{B}}.$$

Therefore, the number of accessible states of the composite (but still disjoint) system "A plus B" is approximately

$$g_{\mathrm{A}}\left(E_{\mathrm{A}}^{\text{before}}\right) g_{\mathrm{B}}\left(E_{\mathrm{B}}^{\text{before}}\right)\delta E_{\mathrm{A}}\delta E_{\mathrm{B}}. \tag{6.4}$$

Now bring A and B into thermal contact, so that energy exchange occurs, and then quickly separate them again. If the energy of A is then measured to be $E_{\mathrm{A}} = E_{\mathrm{A}}^{\text{after}} \pm (\delta E_{\mathrm{A}}/2)$ (assume the same uncertainty as before) and that of B is found to be $E_{\mathrm{B}} = E_{\mathrm{B}}^{\text{after}} \pm (\delta E_{\mathrm{B}}/2)$, the number of accessible states of the composite (but disjoint) system is now

$$g_{\mathrm{A}}\left(E_{\mathrm{A}}^{\text{after}}\right) g_{\mathrm{B}}\left(E_{\mathrm{B}}^{\text{after}}\right)\delta E_{\mathrm{A}}\delta E_{\mathrm{B}}. \tag{6.5}$$

We expect the direction of the energy transfer to be such that the sharing ("partition") $(E_{\mathrm{A}}^{\text{after}}, E_{\mathrm{B}}^{\text{after}})$ is more probable than the partition $(E_{\mathrm{A}}^{\text{before}}, E_{\mathrm{B}}^{\text{before}})$, that is, to be such that the number of accessible microstates for the "after" partition is greater than the number for the "before" partition.

Thus, comparing (6.5) and (6.4), we expect

$$g_{\mathrm{A}}\left(E_{\mathrm{A}}^{\text{after}}\right) g_{\mathrm{B}}\left(E_{\mathrm{B}}^{\text{after}}\right) > g_{\mathrm{A}}\left(E_{\mathrm{A}}^{\text{before}}\right) g_{\mathrm{B}}\left(E_{\mathrm{B}}^{\text{before}}\right),$$

and, hence,

$$\ln\left[g_{\mathrm{A}}\left(E_{\mathrm{A}}^{\text{after}}\right) g_{\mathrm{B}}\left(E_{\mathrm{B}}^{\text{after}}\right)\right] > \ln\left[g_{\mathrm{A}}\left(E_{\mathrm{A}}^{\text{before}}\right) g_{\mathrm{B}}\left(E_{\mathrm{B}}^{\text{before}}\right)\right].$$

In general, if upon thermal contact we have spontaneous energy transfer such that

$$E_A \to E_A + dE_A$$

and

$$E_B \to E_B + dE_B \quad (dE_B = -dE_A),$$

the corresponding change in $\ln(g_A g_B)$ should be positive, that is, $d\ln(g_A g_B) > 0$.

But for this transfer we have

$$\begin{aligned} d\ln(g_A g_B) &= d(\ln g_A + \ln g_B) \\ &= \frac{\partial \ln g_A(E_A)}{\partial E_A} dE_A + \frac{\partial \ln g_B(E_B)}{\partial E_B} dE_B \\ &= \left[\frac{\partial \ln g_A(E_A)}{\partial E_A} - \frac{\partial \ln g_B(E_B)}{\partial E_B}\right] dE_A > 0. \end{aligned}$$

(The partial derivatives here mean $(\partial../\partial E)_{V,N}$, i.e., with the constraints held constant.)

Therefore, if $dE_A > 0$ (i.e., A is "heated" in the process), we must have

$$\frac{\partial \ln g_A(E_A)}{\partial E_A} > \frac{\partial \ln g_B(E_B)}{\partial E_B},$$

that is, $\partial \ln g_A/\partial E_A$must be a measure of the "coldness" of system A. (Since A is colder than B, it receives energy from B: $dE_A > 0$.)

So, define the **statistical temperature** $\tilde{T}$ (a measure of "hotness") by

$$\frac{1}{k\tilde{T}} \equiv \frac{\partial \ln g(E)}{\partial E},$$

where k is some constant, and the derivative is evaluated at the energy of the system.

We shall see later that if we choose k to be Boltzmann's constant k_B:

$$k_B \cong 1.38 \times 10^{-23}\,\mathrm{JK}^{-1},$$

then $\tilde{T}$ turns out to be identical with the ideal-gas temperature, that is,

$$\frac{1}{k_B T} = \frac{\partial \ln g(E)}{\partial E}. \tag{6.6}$$

If the two systems A and B are left in thermal contact, energy transfer will continue until that sharing with the maximum value of $g_A g_B$ is reached. The energies at which this occurs will be denoted by E_A^{eq} and $E_B^{eq}(= E - E_A^{eq})$ (see figure 6.7).

All other energy sharings (E'_A, E'_B) are **possible**, subject to $E'_A + E'_B = E$ (just as any distribution of the suits is possible in a hand of bridge), but **the energy sharing (E_A^{eq}, E_B^{eq}) can be realized in the most ways, and so is overwhelmingly the most probable.**

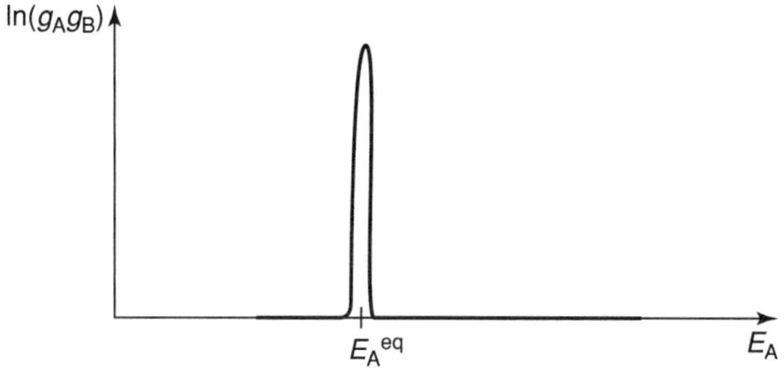

Figure 6.7 *Number of possible states of the combined system after contact, as a function of the energy of one of the subsystems.*

Since $\ln(g_A g_B)$ is maximum at $E_A = E_A^{eq}$, making the change from $E_A = E_A^{eq}$ to $E_A = E_A^{eq} + dE_A$ gives **zero** first-order differential change of $\ln(g_A g_B)$:

$$d[\ln(g_A g_B)] = 0 \quad \text{when} \quad E_A = E_A^{eq},$$

that is,

$$\left[\frac{\partial \ln g_A(E_A)}{\partial E_A} - \frac{\partial \ln g_B(E_B)}{\partial E_B}\right] dE_A = 0$$

for any infinitesimal change dE_A when $E_A = E_A^{eq}$. Thus,

$$T_A = T_B \quad \text{at equilibrium.}$$

Example: Let A and B be very large systems, with $T_A = 300.0$ K and $T_B = 299.9$ K.

Find the changes in $g_A g_B$ when A and B are brought into thermal contact and the energy transfers

$$\text{(i)}\, d\,E_A = -10^{-14}\,\text{J},$$
$$\text{(ii)}\, d\,E_A = +10^{-14}\,\text{J}$$

occur. (Transfer (i) is in the **expected** direction A → B, while (ii) is in the **opposite** direction B → A.)

We get the change

$$\ln(g_A g_B) \to \ln(g'_A g'_B) = \ln(g_A g_B) + d\ln(g_A g_B).$$

But

$$\begin{aligned} d\ln(g_A g_B) &= \left(\frac{1}{k_B T_A} - \frac{1}{k_B T_B}\right) dE_A \\ &= \frac{1}{1.38 \times 10^{-23}}\left(\frac{1}{300} - \frac{1}{299.9}\right) \times (\mp 10^{-14}) \\ &= \pm 8.2 \times 10^2 = \pm \ln e^{8.2\times 10^2} = \pm \ln 10^{351}. \end{aligned}$$

Therefore,

$$\ln(g'_A g'_B) = \ln(g_A g_B) \pm \ln 10^{351} = \ln(10^{\pm 351} g_A g_B),$$

so that

$$g'_A g'_B = 10^{\pm 351} g_A g_B.$$

Thus, there are $10^{+351}/10^{-351} = 10^{702}$ times more accessible states when A has lost energy 10^{-14}J to B than when A has gained energy 10^{-14}J from B.

Thus, an energy transfer of 10^{-14}J from B (at $T_B = 299.9$ K) to A (at $T_A = 300.0$ K) is **possible**, but **10^{702} times less likely** than the same energy transfer (heat transfer) in the opposite direction.

This leads to the **Clausius statement of the second law of thermodynamics (modified)**:

Spontaneous heat flow from a cooler to a hotter body is overwhelmingly improbable *[Clausius said "impossible"].*

6.4 The Boltzmann distribution

If a system is **thermally isolated**, that is, enclosed by adiabatic walls, the probability that it is in any given accessible microstate is the same for all accessible microstates.

Suppose, instead, that our system (A) is **not** thermally isolated, but is in thermal contact with a large reservoir (R) at temperature T.

Then, once equilibrium has been achieved, so that our system A is held at temperature T, what is the probability p_i of finding A to be in quantum state i?

To answer this question, we use the fact that the system **plus** reservoir "A plus R" **is** an isolated system, with constant **total** energy E.

Suppose that the system A is in quantum state i, with energy E_i.

The number of states of "A plus R" in which A is in quantum state i is

$$g_R(E_R)\delta E_R,$$

where

$$E_R = E - E_i, \tag{6.7}$$

that is, the probability p_i that **system** A is in quantum state i is proportional to the number of accessible states of the **reservoir** R when this is the case:

$$p_i \propto g_R(E_R). \tag{6.8}$$

Now suppose that the reservoir is so large that energy transfers of the order of the energy of the system A do not change the reservoir temperature $T_R = T$, that is,

$$\frac{1}{k_B T} = \frac{\partial \ln g_R(E_R)}{\partial E_R} = \text{const.}$$

Solving this, we find

$$g_R(E_R) \propto e^{E_R/k_B T}.$$

Therefore, from (6.8) and (6.7),

$$p_i \propto e^{E_R/k_B T} \propto e^{-E_i/k_B T}.$$

Call the proportionality constant $1/Z$. Then

$$p_i = \frac{e^{-E_i/k_B T}}{Z}, \tag{6.9}$$

which is the **Boltzmann distribution**.

We find Z from the condition that $\sum_i p_i = 1$:

$$Z = \sum_i e^{-E_i/k_B T}. \tag{6.10}$$

The quantity Z is the **partition function**. [**The sum over i is over N-particle quantum states i, and not a sum over energies** E_i. Although E_i is the energy of state i, it is the energy of a vast number of other states, too!]

When in thermal contact (see figure 6.8) with R, system A can have **any** energy and be in **any** of the quantum states corresponding to the chosen energy.

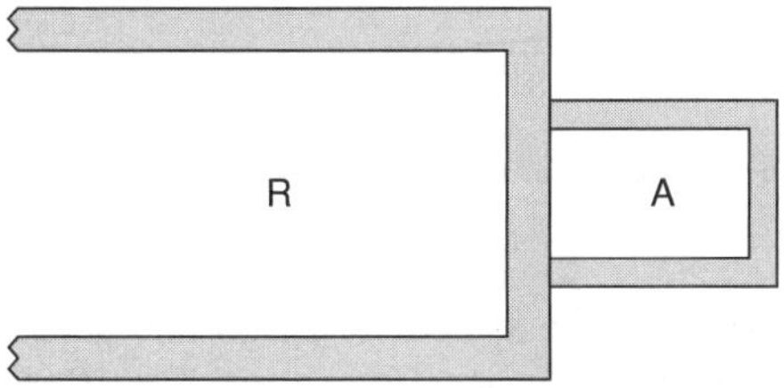

Figure 6.8 *System* A *in thermal contact with reservoir* R.

Suppose now that we have $n \to \infty$ systems A (all with the same N and V), such that the fraction n_i/n of them in state i is $n_i/n = p_i = (1/Z)\exp(-E_i/k_B T)$.

This "ensemble" of systems is called the **canonical ensemble** for the given N, V and T [just as an ensemble of systems with energies in a **limited** range $(E, E + dE)$ such that all states in this range have equal probability is called a **microcanonical ensemble** (see section 6.2)].

The **thermal average** $\langle \Omega \rangle$ of a physical quantity Ω (for example, energy, pressure, etc.) that takes the value Ω_i in quantum state i is **defined** as the average of Ω over the canonical ensemble, that is,

$$\langle \Omega \rangle = \sum_{\substack{\text{states} \\ i}} p_i \Omega_i = \frac{1}{Z} \sum_i \Omega_i e^{-E_i/k_B T} = \frac{\sum_i \Omega_i e^{-E_i/k_B T}}{\sum_i e^{-E_i/k_B T}}. \tag{6.11}$$

For example, the thermal average energy, to be identified with U, is

$$U = \langle E \rangle = \frac{\sum_i E_i e^{-E_i/k_B T}}{\sum_i e^{-E_i/k_B T}}, \tag{6.12}$$

where E_i is the energy of microstate (quantum state) i.

Similarly, the thermal average pressure P is

$$P = \langle P \rangle = \frac{\sum_i P_i e^{-E_i/k_B T}}{\sum_i e^{-E_i/k_B T}},$$

where P_i is the pressure in microstate i.

Plotted against E_i, the Boltzmann probability distribution (6.9) takes the form shown in figure 6.9, in which we have assigned the value $E_i = 0$ to the lowest-energy state (the **ground state**).

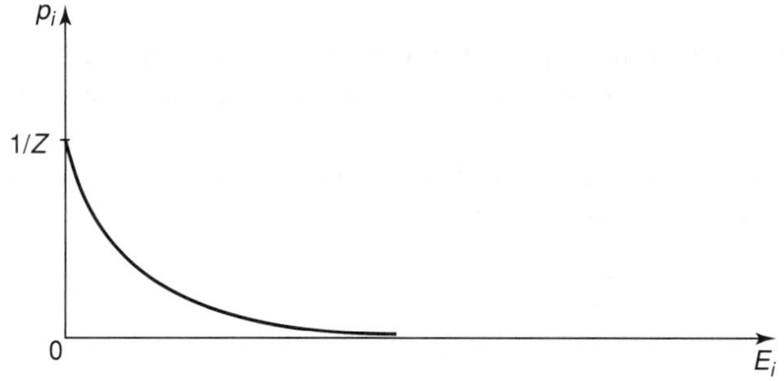

Figure 6.9 *Boltzmann probability distribution.*

Thus, the most probable **microstate** of the system is the ground state, with energy $E_i = 0$ (**even if the system is in thermal contact with a reservoir at temperature** $T = 10{,}000$ K).

Of course, this does **not** mean that $E = 0$ is the most probable **energy**! As we already know, the number $g(E)dE$ of states in an energy range $(E, E + dE)$ of given width dE increases very rapidly with E, and this counteracts the rapid fall of the function $p(E) = (1/Z)\exp(-E/k_B T)$ with energy [the function $p(E)$ is **defined** here to be such that $p(E_i) = p_i$].

Thus, the probability that a system in thermal contact with a heat reservoir of temperature T has energy in the range $(E, E + dE)$ is

$$f(E)dE = p(E)g(E)dE, \tag{6.13}$$

[the probability of **one** state in the range, times the **number** of states in the range $(E, E + dE)$].

The function $f(E)$ is the **energy distribution function**. Its maximum occurs at the energy at which

$$\frac{\partial f(E)}{\partial E} = 0,$$

that is, when

$$\frac{\partial \ln f(E)}{\partial E} = 0,$$

that is, when

$$\frac{\partial \ln p(E)}{\partial E} + \frac{\partial \ln g(E)}{\partial E} = 0.$$

If the system is in thermal contact with a heat reservoir at temperature T_R, then

$$p(E) = \frac{e^{-E/k_B T_R}}{Z}.$$

Therefore,

$$\frac{\partial \ln p(E)}{\partial E} = \frac{\partial \ln e^{-E/k_B T_R}}{\partial E} - \frac{\partial \ln Z}{\partial E}(\leftarrow = 0) = \frac{\partial(-E/k_B T_R)}{\partial E} = -\frac{1}{k_B T_R}.$$

Also,

$$\frac{\partial \ln g(E)}{\partial E} = \frac{1}{k_B T}$$

by definition of T, where T is the temperature of the **system**. Therefore, $f(E)$ is a maximum at the energy for which

$$-\frac{1}{k_B T_R} + \frac{1}{k_B T} = 0,$$

that is, at the energy for which

$$T = T_R.$$

Thus, the most probable energy of the system is that for which $T = T_R$, as expected. Denote this energy by E_0.

We can also find how $f(E)$ varies near E_0 [that is, near the maximum of $f(E)$], using the Taylor expansion of $\ln f(E)$ about E_0:

$$\ln f(E) = \ln f(E_0) + (E - E_0) \left.\frac{\partial \ln f}{\partial E}\right|_{E=E_0} + \frac{(E-E_0)^2}{2!} \left.\frac{\partial^2 \ln f}{\partial E^2}\right|_{E=E_0} + \cdots.$$

But

$$\left.\frac{\partial \ln f}{\partial E}\right|_{E=E_0} = 0$$

(the above condition for a maximum), and, therefore,

$$\ln f(E) = \ln f(E_0) + \frac{(E-E_0)^2}{2!} \left.\frac{\partial^2 \ln f}{\partial E^2}\right|_{E=E_0} + \cdots .$$

But, since $f(E) = p(E)g(E)$,

$$\begin{aligned}
\left.\frac{\partial^2 \ln f}{\partial E^2}\right|_{E=E_0} &= \frac{\partial}{\partial E}\left(\frac{\partial \ln p(E)}{\partial E} + \frac{\partial \ln g(E)}{\partial E}\right)\bigg|_{E=E_0} \\
&= \frac{\partial}{\partial E}\left(-\frac{1}{k_B T_R} + \frac{1}{k_B T}\right)\bigg|_{E=E_0} = \frac{\partial}{\partial E}\left(\frac{1}{k_B T}\right)\bigg|_{E=E_0} \\
&= -\frac{1}{k_B T^2} \left.\frac{\partial T}{\partial E}\right|_{E=E_0} = -\frac{1}{k_B T^2 C_V},
\end{aligned}$$

since the partial derivative here is at constant volume and $(\partial E/\partial T)_V$ evaluated at $E = E_0 \equiv U$ is $(\partial U/\partial T)_V \equiv C_V$ (see section 5.16.1). Therefore,

$$\ln f(E) = \ln f(E_0) - \frac{(E-E_0)^2}{2k_B T^2 C_V},$$

and so

$$f(E) = f(E_0) e^{-(E-E_0)^2/2k_B T^2 C_V}. \tag{6.14}$$

This is a **gaussian (normal) distribution**.

[All gaussian distributions take the standard form

$$f(x) = f(x_0) e^{-(x-x_0)^2/2(\Delta x)^2},$$

where Δx is the root-mean-square (rms) deviation of x from the mean value x_0:

$$\Delta x \equiv \sqrt{\langle (x-x_0)^2 \rangle}$$

(the "standard deviation" from x_0).]

Thus, the rms deviation of the energy from E_0 is

$$\Delta E \equiv \sqrt{\langle (E-E_0)^2 \rangle} = \sqrt{k_B T^2 C_V}. \tag{6.15}$$

For many systems, $E_0 \sim N k_B T$, and so $C_V \sim N k_B$. Therefore,

$$\Delta E \sim N^{1/2} k_B T,$$

while the **fractional fluctuation** is

$$\frac{\Delta E}{E_0} \sim \frac{N^{1/2} k_B T}{N k_B T} \sim \frac{1}{N^{1/2}}. \tag{6.16}$$

For example, for a system of 10^{26} atoms, $\Delta E/E_0 \sim 10^{-13}$. Thus, **the fluctuations of the energy away from the value E_0 for which $T = T_R$ are negligibly small for large systems**. As an example, if a 10^{26}-atom system has thermal average energy 10^6 J at $T = T_R$, the energy fluctuation (standard deviation) will be approximately $10^{-13} \times 10^6\,\text{J} = 10^{-7}$ J.

7

Microstate Probabilities and Entropy

7.1 Definition of general entropy

For an N-particle system, let the probability (not necessarily equilibrium) that the system is in the N-particle quantum state (microstate) i be $\tilde{p}_i$.

We then **define** the **general entropy** $\tilde{S}$ as

$$\tilde{S} = -k_B \sum_i \tilde{p}_i \ln \tilde{p}_i \tag{7.1}$$

(an average, over all states i, of $-k_B \ln \tilde{p}_i$).

Example 1 Let $\tilde{p}_i$ be the **equilibrium** probabilities for a **closed** system, that is (see section 6.2),

$$\tilde{p}_i = p_i = \begin{cases} \dfrac{1}{M} & \text{for } E_0 - \dfrac{\delta E}{2} \le E_i \le E_0 + \dfrac{\delta E}{2}, \\ 0 & \text{otherwise,} \end{cases}$$

where $M = g(E_0)\delta E$ is the number of states in the band $(E_0 - \delta E/2,\ E_0 + \delta E/2)$ (the **accessibility band**).

In (7.1) this gives

$$\tilde{S} = S_{\text{closed}} = -k_B \sum_{i=1}^{M} \left(\frac{1}{M}\right) \ln\left(\frac{1}{M}\right) = k_B \sum_{i=1}^{M} \frac{1}{M} \ln M = k_B \ln M$$

(we drop the tilde for the equilibrium S). Thus, S in this case is a logarithmic measure of the number of states accessible to the system.

A Course in Theoretical Physics, First Edition. P. J. Shepherd.

Example 2 Let $\tilde{p}_i$ be the **equilibrium** probabilities for an **open** system, that is (see section 6.4),

$$\tilde{p}_i = p_i = \frac{\mathrm{e}^{-E_i/k_\mathrm{B}T}}{Z}.$$

Then

$$\ln \tilde{p}_i = -\frac{E_i}{k_\mathrm{B}T} - \ln Z,$$

and so, from equation (7.1),

$$\tilde{S} = S_\mathrm{open} = -k_\mathrm{B}\sum_i p_i\left[-\frac{E_i}{k_\mathrm{B}T} - \ln Z\right] = \frac{\langle E\rangle}{T} + k_\mathrm{B}\ln Z = \frac{U}{T} + k_\mathrm{B}\ln Z, \tag{7.2}$$

where we have used the fact that $\sum_i p_i = 1$ and have identified the thermal average N-particle energy as U.

[Classical thermodynamics defines the **Helmholtz free energy** F as

$$F = U - TS, \tag{7.3}$$

so equation (7.2) implies that

$$F = -k_\mathrm{B}T\ln Z, \tag{7.4}$$

which provides **the central link between thermodynamics and statistical mechanics**.]

7.2 Law of increase of entropy

For general, nonequilibrium states of a closed system, $\tilde{S}$ increases with time until equilibrium is reached, that is, until the probabilities $\tilde{p}_i$ become equal to the equilibrium probabilities p_i.

Proof From (7.1),

$$\frac{\mathrm{d}\tilde{S}}{\mathrm{d}t} = -k_\mathrm{B}\sum_i \ln \tilde{p}_i\frac{\mathrm{d}\tilde{p}_i}{\mathrm{d}t} - k_\mathrm{B}\sum_i\frac{\mathrm{d}\tilde{p}_i}{\mathrm{d}t}.$$

But, since $\sum_i \tilde{p}_i = 1$, the last term is zero. Write the first term as

$$-\frac{k_\mathrm{B}}{2}\left[\sum_i \ln \tilde{p}_i\frac{\mathrm{d}\tilde{p}_i}{\mathrm{d}t} + \sum_j \ln \tilde{p}_j\frac{\mathrm{d}\tilde{p}_j}{\mathrm{d}t}\right]$$

and use Fermi's master equation (6.2) for the two derivatives:

$$\frac{\mathrm{d}\tilde{p}_i}{\mathrm{d}t} = \sum_j \nu_{ij} \left(\tilde{p}_j - \tilde{p}_i\right),$$

$$\frac{\mathrm{d}\tilde{p}_j}{\mathrm{d}t} = \sum_i \nu_{ji} \left(\tilde{p}_i - \tilde{p}_j\right) = -\sum_i \nu_{ij} \left(\tilde{p}_j - \tilde{p}_i\right)$$

(since $\nu_{ji} = \nu_{ij}$). Therefore,

$$\frac{\mathrm{d}\tilde{S}}{\mathrm{d}t} = -\frac{k_{\mathrm{B}}}{2} \sum_{i,j} \nu_{ij} \left(\ln \tilde{p}_i - \ln \tilde{p}_j\right) \left(\tilde{p}_j - \tilde{p}_i\right) = \frac{k_{\mathrm{B}}}{2} \sum_{i,j} \nu_{ij} \left(\ln \tilde{p}_j - \ln \tilde{p}_i\right) \left(\tilde{p}_j - \tilde{p}_i\right).$$

But

$$\left(\ln \tilde{p}_j - \ln \tilde{p}_i\right) \left(\tilde{p}_j - \tilde{p}_i\right) \geq 0$$

for **all** choices of j and i. Therefore,

$$\frac{\mathrm{d}\tilde{S}}{\mathrm{d}t} \geq 0. \tag{7.5}$$

Clearly, only when $\tilde{p}_i = \tilde{p}_j$ for all accessible states i, j (equal equilibrium probability) do we have $\mathrm{d}\tilde{S}/\mathrm{d}t = 0$ (equilibrium of the **closed** system).

It follows from (7.5) that, in any process,

$$\tilde{S}_{\text{later}} \geq \tilde{S}_{\text{earlier}}.$$

For example, if a process takes a closed system from thermodynamic state A at time t_1 to thermodynamic state B at time $t_2 > t_1$, we have

$$\tilde{S}_{\mathrm{B}} \geq \tilde{S}_{\mathrm{A}}. \tag{7.6}$$

If the process is reversible, that is, if we can also take the system from B at time t_1' to A at some later time $t_2' > t_1'$, the law of increase of entropy would imply

$$\tilde{S}_{\mathrm{A}} \geq \tilde{S}_{\mathrm{B}}. \tag{7.7}$$

But (7.6) and (7.7) are compatible only if

$$\tilde{S}_{\mathrm{A}} = \tilde{S}_{\mathrm{B}},$$

that is, for reversible processes in a closed system the change in entropy is zero:

$$\Delta\tilde{S} = 0.$$

The law $\mathrm{d}\tilde{S}/\mathrm{d}t \geq 0$ is one of the most significant laws in physics – it can be said to determine the direction of the **arrow of time**.

7.3 Equilibrium entropy S

We now show that, for a system in thermodynamic equilibrium, the entropy S takes the simple form

$$S = k_B \ln g(E_0) \tag{7.8}$$

for both closed and open systems, provided that E_0 is understood as follows:

For a **closed system**, E_0 is the central energy of the accessibility band.
For an **open system**, E_0 is the energy of the system in that sharing of energy with the surroundings that equalizes the temperatures of the system and the surroundings.

Proof For a **closed system**, we showed that

$$S = k_B \ln M \quad \text{with} \quad M = g(E_0)\delta E$$

that is,

$$S = k_B \ln g(E_0) + k_B \ln \delta E.$$

For example, suppose (typically) that

$$g(E_0) = 10^{10^{23}}\,\mathrm{J}^{-1} \quad \text{and} \quad \delta E = 10^{-10}\,\mathrm{J}.$$

Then, $\ln \delta E <<<< \ln g(E_0)$ and so

$$S = k_B \ln g(E_0). \tag{7.9}$$

[The units are utterly unimportant here. For example, using energy units μJ instead of J we have

$$g(E_0) = 10^{10^{23}} \times 10^{-6}(\mu\mathrm{J})^{-1} = 10^{(10^{23}-6)}\,(\mu\mathrm{J})^{-1}$$
$$\delta E = 10^{-10} \times 10^{6}\,\mu\mathrm{J} = 10^{-4}\,\mu\mathrm{J}.$$

Again, $\ln \delta E <<<< \ln g(E_0)$, so that

$$S = k_B \ln g(E_0). \tag{7.10}$$

Here, (7.9) and (7.10) give essentially the same numerical values, since $10^{23} - 6 \approx 10^{23}$.]

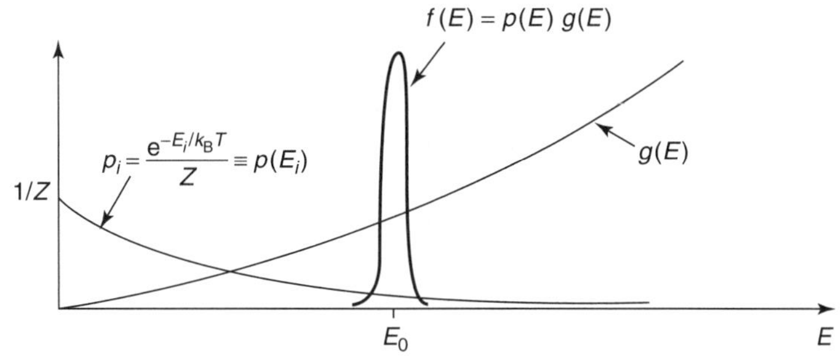

Figure 7.1 *Microstate probability p_i (Boltzmann distribution) and corresponding energy distribution function f(E) for an open system.*

Thus, we can use

$$S = k_B \ln g(E_0)$$

whatever the units (within reason!) one uses for $g(E_0)$.

For an **open system** we found [see equations (6.9), (6.10) and (6.14)–(6.16)] the results depicted in figure 7.1, in which $g(E)$ is the density-of-states function for the system. That is, we found that the **energy** distribution function $f(E)$ is concentrated in a narrow gaussian peak of width

$$\Delta E = \sqrt{k_B T^2 C_V}$$

such that

$$\frac{\Delta E}{E_0} \sim \frac{1}{N^{1/2}},$$

where E_0 is the "system" energy for which the temperature of the system equals the temperature of the surroundings.

Thus, whereas a **closed** system has **zero** probability of having an energy E outside the accessibility band ($E_0 - \delta E/2 < E < E_0 + \delta E/2$), an **open** system with a large number N of particles has a **negligible** (but nonzero) probability of having an energy E outside the range ΔE of fluctuations about the thermal average energy E_0. By ignoring states with these highly improbable energies we miss only a negligibly small extra contribution to S.

Thus, we can write $S = k_B \ln M \approx k_B \ln[g(E_0)\Delta E]$, where we have included only the states within the range $\pm\Delta E/2$ of E_0 (since $f(E)$ is sharply peaked).

Typically, as shown earlier, $\Delta E \sim 10^{-7}$ J and $g \sim 10^{10^{23}}$ (in whatever units of inverse energy!), and so, again,

$$S = k_B \ln g(E_0).$$

The above proof is equivalent to saying that we do not reduce the entropy of an open system by much if we suddenly isolate it from the surroundings with which it was in thermal contact and thermal equilibrium. We merely exclude the possibility of energy fluctuations that were in any case highly improbable.

7.4 Additivity of the entropy

We now show that the entropies of subsystems are additive.

Suppose we regard two subsystems, A and B, as subsystems of a composite system AB.

If $\tilde{p}_i^{\mathrm{A}}$ is the probability that subsystem A is in state i, and $\tilde{p}_j^{\mathrm{B}}$ is the probability that subsystem B is in state j, the **joint probability** $\tilde{p}_{ij}^{\mathrm{AB}}$ that AB is in the joint state ij in which A is in i and B is in j is

$$\tilde{p}_{ij}^{\mathrm{AB}} = \tilde{p}_i^{\mathrm{A}} \tilde{p}_j^{\mathrm{B}}.$$

Applying the definition (7.1) of the general entropy to the system AB, we have

$$\tilde{S}^{\mathrm{AB}} = -k_{\mathrm{B}} \sum_i \sum_j \tilde{p}_{ij}^{\mathrm{AB}} \ln \tilde{p}_{ij}^{\mathrm{AB}} = -k_{\mathrm{B}} \sum_i \sum_j \tilde{p}_i^{\mathrm{A}} \tilde{p}_j^{\mathrm{B}} \left(\ln \tilde{p}_i^{\mathrm{A}} + \ln \tilde{p}_j^{\mathrm{B}} \right).$$

But

$$\sum_i \tilde{p}_i^{\mathrm{A}} = 1 \quad \text{and} \quad \sum_j \tilde{p}_j^{\mathrm{B}} = 1.$$

Therefore,

$$\tilde{S}^{\mathrm{AB}} = -k_{\mathrm{B}} \sum_i \tilde{p}_i^{\mathrm{A}} \ln \tilde{p}_i^{\mathrm{A}} - k_{\mathrm{B}} \sum_j \tilde{p}_j^{\mathrm{B}} \ln \tilde{p}_j^{\mathrm{B}},$$

so that

$$\tilde{S}^{\mathrm{AB}} = \tilde{S}^{\mathrm{A}} + \tilde{S}^{\mathrm{B}}. \tag{7.11}$$

Thus, to get the entropy of the composite system we merely add the entropies of the subsystems.

The result (7.11) is true even for nonequilibrium entropies. Thus, even if $\tilde{S}^{\mathrm{A}}$, $\tilde{S}^{\mathrm{B}}$ and $\tilde{S}^{\mathrm{AB}}$ are changing with time, equation (7.11) is true at all times.

Examples

(i) The subsystems are **physically separate**:

$$\tilde{S}^{\mathrm{AB}} = \tilde{S}^{\mathrm{A}} + \tilde{S}^{\mathrm{B}}.$$

(ii) The subsystems **may occupy the same volume**. For example, in an N_2/O_2 mixture equation (7.11) applies if we treat the N_2 molecules as subsystem A and the O_2 molecules as subsystem. Thus,

$$\tilde{S}^{\mathrm{mixture}} = \tilde{S}^{\mathrm{N_2}} + \tilde{S}^{\mathrm{O_2}}.$$

It follows from examples (i) and (ii) that the increase of total entropy when two initially separated gases N_2 and O_2 are mixed is due entirely to the changes of $\tilde{S}^{\mathrm{N_2}}$ and $\tilde{S}^{\mathrm{O_2}}$ separately (e.g., doubling the volume accessible to either gas will give an increase of the entropy of that gas).

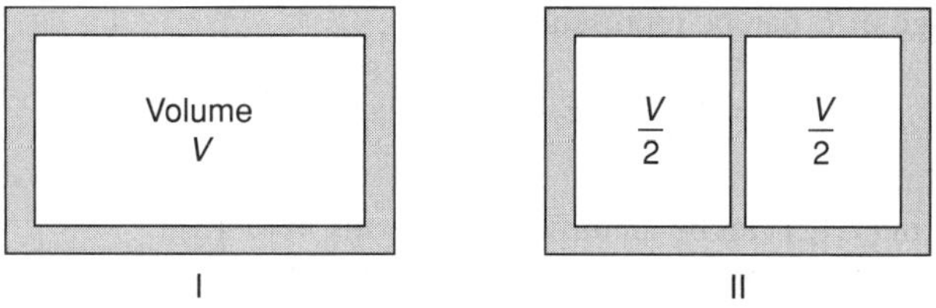

Figure 7.2 *Division of a volume into two equal subvolumes.*

Question: Can we regard a **subvolume** of a system as a "subsystem"?
Answer: Strictly, we cannot, since a volume, unlike a set of particles, cannot be said to be in a "state".

For systems in **thermodynamic equilibrium**, however, we show below that we can regard a subvolume of "reasonable" size as a subsystem, **with utterly negligible error**.

Suppose we have a gas in thermodynamic equilibrium in a closed container of volume V. We now place an (infinitesimally thin) adiabatic partition (wall) inside the container (see figure 7.2).

Although it is true that some of the states accessible in case I are not accessible in case II (for example, states in which all the particles are in one half of the box), these states form an utterly negligible fraction of the accessible states of I, in which states with essentially equal sharings of particles between the halves dominate. Thus, **to an almost perfect approximation**, we have

$$\mathrm{S}_{\mathrm{II}} = S_{\mathrm{I}}.$$

This means that we can **conceptually** divide a system in thermodynamic equilibrium into subvolumes (for example, unit volumes) and treat them as subsystems, and hence define entropy per **unit** volume (s):

$$s = \frac{S}{V}.$$

7.5 Statistical–mechanical description of the three types of energy transfer

We classified energy transfers to a thermodynamic system as being of **three** types:

(i) **Nondissipative work** (reversible work).
(ii) **Dissipative work** (irreversible work).
(iii) **Heat transfer** (reversible or irreversible).

Thus, if all three types are present,

$$\mathrm{d}U = đW_{\mathrm{rev}} + đW_{\mathrm{irrev}} + đQ. \tag{7.12}$$

In the statistical description, the internal energy U is

$$U = \langle E \rangle = \sum_i \tilde{p}_i E_i \left(= \int E\tilde{p}(E)g(E)\mathrm{d}E \equiv \int E\tilde{f}(E)\mathrm{d}E \right),$$

so **any** infinitesimal change of U can be represented as

$$\mathrm{d}U = \sum_i \mathrm{d}(\tilde{p}_i E_i) = \sum_i \tilde{p}_i \mathrm{d}E_i + \sum_i E_i \mathrm{d}\tilde{p}_i. \tag{7.13}$$

We now show how terms in (7.12) can be identified with terms in (7.13).

Firstly, note that the **quantum energy levels** E_i of the N-particle system are determined entirely by the form of the **potential** in which the particles move.

For example, suppose we take the case of a gas in a box and **fix** this potential by keeping the size of the box fixed.

We then do **dissipative work** (for example, by a paddle wheel) **and** allow heat transfer to or from the box, so that (7.12) becomes

$$\mathrm{d}U = đW_{\text{irrev}} + đQ.$$

Since the size (and hence potential) is fixed there is no change in the energy levels E_i, and so (7.13) becomes

$$\mathrm{d}U = \sum_i E_i \mathrm{d}\tilde{p}_i.$$

Thus, we identify

$$đW_{\text{irrev}} + đQ = \sum_i E_i \mathrm{d}\tilde{p}_i,$$

which states that **dissipative work and heat transfer change the probabilities of occupation of states, but do not change the energy levels**.

Suppose now, instead, that the system is thermally isolated, that there is no stirring, and that we subject the system to slow, frictionless compression by a piston.

Since this changes the size of the system, it changes the potential in which the particles move, and so produces changes $\mathrm{d}E_i$ in the energy levels E_i.

If the compression is **sufficiently slow** ("quasistatic"), no quantum jumps between states will be caused. Thus, there will be no change in the state probabilities $\tilde{p}_i$, that is, $\mathrm{d}\tilde{p}_i = 0$ (the **principle of invariance of microstate probabilities in quasistatic processes**). Thus, we can redefine **quasistatic work** as work in which $\mathrm{d}\tilde{p}_i = 0$ for all i.

Thus,

$$\mathrm{d}U = \sum_i \tilde{p}_i \mathrm{d}E_i$$

for this type of energy transfer.

But for quasistatic, nondissipative, reversible work, we have

$$\mathrm{d}U = đW_{\text{rev}}\ [= -p\mathrm{d}V].$$

Therefore, comparing, we have

$$đW_{\text{rev}} = \sum_i \tilde{p}_i \mathrm{d}E_i.$$

Thus, we have the identification

$$
\begin{aligned}
\mathrm{d}U &= \underbrace{đW_{\text{rev}}}_{\Updownarrow} + \underbrace{đW_{\text{irrev}} + đQ}_{\Updownarrow} \\
&= \sum_i \tilde{p}_i \mathrm{d}E_i + \sum_i E_i \mathrm{d}\tilde{p}_i .
\end{aligned}
\tag{7.14}
$$

We now apply this to the case of frictionless quasistatic compression of a gas in a cylinder in thermal contact with a heat reservoir at temperature T, as illustrated in figure 7.3.

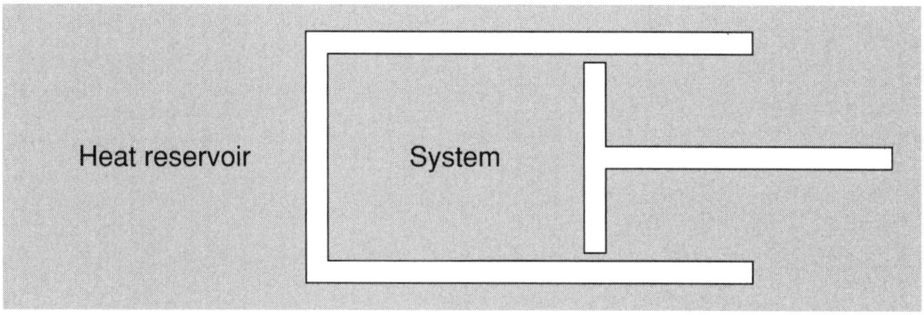

Figure 7.3 *Compression of a gas in thermal contact with a heat reservoir.*

Since the compression is frictionless, no dissipative work is done [$đW_{\text{irrev}} = 0$]. Also, for compression (with volume change $\mathrm{d}V < 0$), we have $đW_{\text{rev}} = -p\mathrm{d}V$.

For this case, (7.14) becomes

$$
\begin{aligned}
\mathrm{d}U &= -p\mathrm{d}V + đQ \\
&= \sum_i p_i \mathrm{d}E_i + \sum_i E_i \mathrm{d}p_i ,
\end{aligned}
\tag{7.15}
$$

where, since the system is open, in thermal contact with a reservoir at temperature T, p_i in this case is the Boltzmann distribution

$$
p_i = \frac{\mathrm{e}^{-E_i/k_{\mathrm{B}}T}}{Z} .
$$

With this p_i the change in entropy of the system is

$$
\begin{aligned}
\mathrm{d}S &= \mathrm{d}\left(-k_{\mathrm{B}} \sum_i p_i \ln p_i\right) = -k_{\mathrm{B}} \sum_i \mathrm{d}p_i \ln p_i - k_{\mathrm{B}} \sum_i \mathrm{d}p_i \\
&= -k_{\mathrm{B}} \sum_i \mathrm{d}p_i \left(-\frac{E_i}{k_{\mathrm{B}}T} - \ln Z\right) = \frac{1}{T} \sum_i \mathrm{d}p_i E_i \overset{(7.15)}{=} \frac{đQ}{T} ,
\end{aligned}
$$

where we have twice used that fact that $\sum_i \mathrm{d}p_i = 0$ (which follows from the fact that $\sum_i p_i = 1$). Thus, **in a process involving no dissipative work** (that is, involving only heat transfer and, possibly, nondissipative work), heat transfer $đQ$ to a system at temperature T increases the entropy of the system by $\mathrm{d}S = đQ/T$.

8
The Ideal Monatomic Gas

8.1 Quantum states of a particle in a three-dimensional box

Consider N free atoms, each of mass m, moving at random in a container (an oblong of sides L_x, L_y, L_z, that is, of volume $V = L_x L_y L_z$) and treat each atom as a structureless point particle.

Let the density be so low that atom–atom collisions are rare. (The only role of collisions then is to maintain thermal equilibrium by causing quantum jumps between states.)

Because the collisions (interactions) are rare, we can treat the N atoms as independent, and so we first consider the possible quantum states of one atom.

We represent the box by a potential well in which the potential energy $\mathcal{V}$ is zero inside the box, rising vertically to infinity at the walls (a square-well potential). For example, plotted along x, $\mathcal{V}$ (for values of y and z inside the well) has the form shown in figure 8.1.

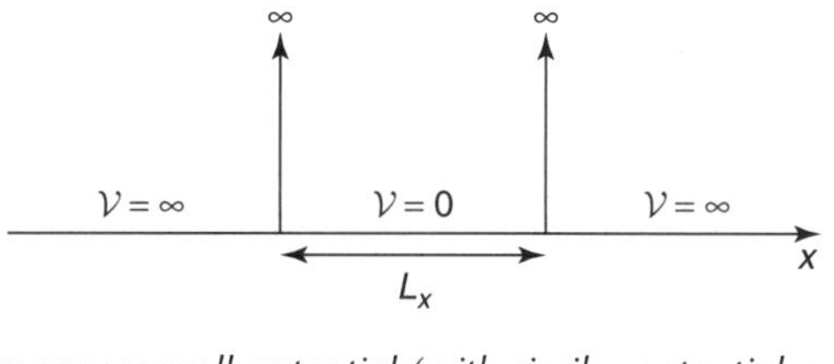

Figure 8.1 Rigid-box square-well potential (with similar potential profiles along y and z).

As already explained, the energy levels ε_i of a particle in this well are entirely determined by the potential. The space-dependent part of the wave function of a particle in this well has the form

$$\psi(x, y, z) = A \sin\left(\frac{2\pi x}{\lambda_x}\right) \sin\left(\frac{2\pi y}{\lambda_y}\right) \sin\left(\frac{2\pi z}{\lambda_z}\right), \tag{8.1}$$

where λ_x is the wavelength of the wave in the x direction (see figure 8.2).

A Course in Theoretical Physics, First Edition. P. J. Shepherd.

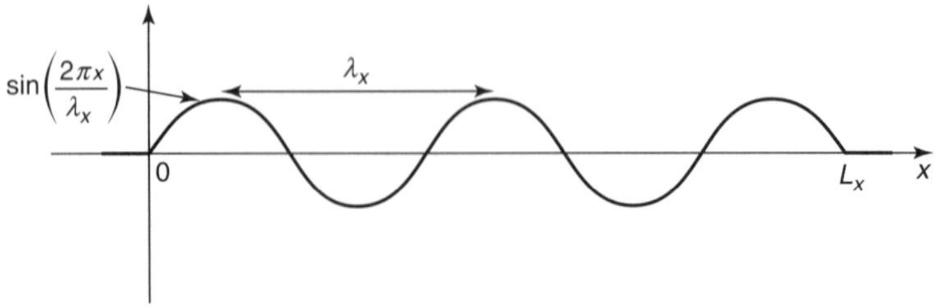

Figure 8.2 *The x-dependent factor of the wave function of a particle in a rigid box square-well potential.*

That is, ψ is constructed so as to vanish at the boundaries of the box (since ψ is continuous and must be zero outside the boundaries, where $\mathcal{V} = \infty$).

Thus, the length L_x has to be an integer number of half wavelengths:

$$L_x = n_x \frac{\lambda_x}{2} \quad (n_x \text{ integer}),$$

that is,

$$\lambda_x = \frac{2L_x}{n_x}. \tag{8.2}$$

Similarly, λ_y and λ_z are restricted to

$$\lambda_y = \frac{2L_y}{n_y} \quad \text{and} \quad \lambda_z = \frac{2L_z}{n_z}.$$

The one-dimensional **standing** wave $\sin(2\pi x/\lambda_x)$ is a superposition of two **running** waves:

$$\sin\left(\frac{2\pi x}{\lambda_x}\right) = \frac{e^{i2\pi x/\lambda_x} - e^{-i2\pi x/\lambda_x}}{2i}.$$

But the running waves $e^{i2\pi x/\lambda_x}$ and $e^{-i2\pi x/\lambda_x}$ are both in the form of wave functions $e^{ip_x x/\hbar}$ corresponding to definite values of the x-component p_x of momentum, namely, the values $p_x = h/\lambda_x$ and $p_x = -h/\lambda_x$, respectively $[\hbar \equiv h/2\pi]$.

Thus, the state (8.1) corresponds to a superposition of terms in which the component p_x takes the values

$$p_x = \pm\frac{h}{\lambda_x} = \pm\frac{n_x h}{2L_x} = \pm\frac{n_x \pi \hbar}{L_x}$$

and the allowed values of p_y and p_z are given by corresponding expressions.

Thus, the allowed energies of our one particle in the box are the kinetic energies (since $\mathcal{V} = 0$)

$$\varepsilon = \frac{p^2}{2m} = \frac{p_x^2 + p_y^2 + p_z^2}{2m} = \frac{\pi^2\hbar^2}{2m}\left(\frac{n_x^2}{L_x^2} + \frac{n_y^2}{L_y^2} + \frac{n_z^2}{L_z^2}\right) \equiv \varepsilon_{n_x n_y n_z}. \tag{8.3}$$

8.2 The velocity-component distribution and internal energy

Since p_x is quantized, that is, is restricted to the values

$$p_x = \pm n_x \frac{\pi \hbar}{L_x},$$

the x-component of velocity ($u_x = p_x/m$) is also quantized:

$$u_x = \pm n_x \frac{\pi \hbar}{m L_x}. \tag{8.4}$$

So the allowed values of u_x are separated by equal intervals, that is, are **uniformly spaced** (and **closely spaced**, if L_x is large). Therefore, the number of quantum states (for fixed values of u_y and u_z) for which the x-component of velocity takes a value in the range $(u_x, u_x + \mathrm{d}u_x)$ will be $A\mathrm{d}u_x$ with A a constant.

But if our one-particle system is **open**, that is, in thermal contact with a heat reservoir at temperature T, the probability p_i of finding the atom in a **particular** quantum state i is

$$p_i = \frac{1}{Z} \mathrm{e}^{-\frac{1}{2}m\left(u_x^2+u_y^2+u_z^2\right)/k_\mathrm{B}T},$$

and so the probability $f_{u_\alpha}(u_x)\mathrm{d}u_x$ of finding the x-component of velocity to be in the range $(u_x, u_x + \mathrm{d}u_x)$ is

$$f_{u_\alpha}(u_x)\mathrm{d}u_x = \frac{1}{Z} \mathrm{e}^{-\frac{1}{2}m\left(u_x^2+u_y^2+u_z^2\right)/k_\mathrm{B}T} \times A\mathrm{d}u_x$$

$$= B\mathrm{e}^{-\frac{1}{2}mu_x^2/k_\mathrm{B}T}\mathrm{d}u_x,$$

where

$$B = \frac{A\mathrm{e}^{-\frac{1}{2}m\left(u_y^2+u_z^2\right)/k_\mathrm{B}T}}{Z}.$$

But the state must have a value of u_x **somewhere** in the range $-\infty < u_x < \infty$; that is, we require

$$\int_{-\infty}^{\infty} f_{u_\alpha}(u_x)\mathrm{d}u_x = 1.$$

This gives

$$B \int_{-\infty}^{+\infty} \mathrm{e}^{-\frac{1}{2}mu_x^2/k_\mathrm{B}T}\mathrm{d}u_x = 1.$$

With the change of variables

$$w \equiv \sqrt{\frac{m}{2k_B T}} u_x, \text{ so that } du_x = \sqrt{\frac{2k_B T}{m}} dw,$$

this becomes

$$B\sqrt{\frac{2k_B T}{m}} \int_{-\infty}^{+\infty} dw e^{-w^2} = 1.$$

Using the standard integral

$$\int_{-\infty}^{+\infty} dw e^{-w^2} = \sqrt{\pi},$$

we get

$$B = \sqrt{\frac{m}{2\pi k_B T}}.$$

Thus, B has turned out to be independent of the values chosen for u_y and u_z, that is,

$$f_{u_\alpha}(u_x) du_x = \sqrt{\frac{m}{2\pi k_B T}} e^{-\frac{1}{2}mu_x^2/k_B T} du_x. \tag{8.5}$$

Here $f_{u_\alpha}(u_x)$ is the **Maxwell velocity-component distribution**.

[Similarly,

$$f_{u_\alpha}(u_y) du_y = \sqrt{\frac{m}{2\pi k_B T}} e^{-\frac{1}{2}mu_y^2/k_B T} du_y$$

and

$$f_{u_\alpha}(u_z) du_z = \sqrt{\frac{m}{2\pi k_B T}} e^{-\frac{1}{2}mu_z^2/k_B T} du_z.]$$

The distribution function $f_{u_\alpha}(u_x)$ is symmetric (see figure 8.3).

The average value of u_x is zero:

$$\overline{u_x} \equiv \int_{-\infty}^{+\infty} u_x f_{u_\alpha}(u_x) du_x = 0.$$

[Interpret this as u_x weighted with the probability $f_{u_\alpha}(u_x)du_x$ of finding the x-component of velocity to lie in the range $(u_x, u_x + du_x)$, "summed" over all intervals du_x.]

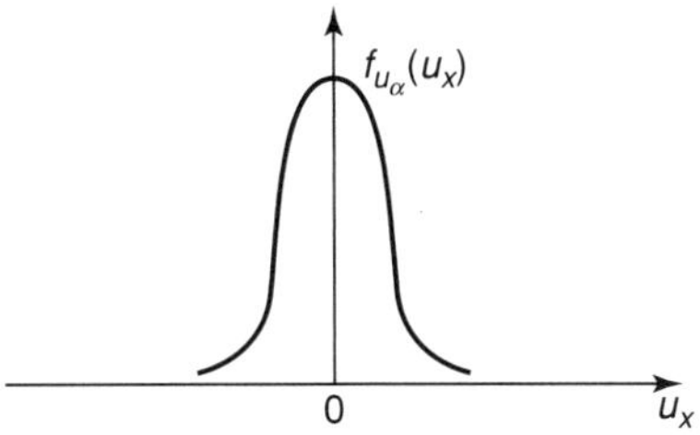

Figure 8.3 *Velocity-component distribution function $f_{u_\alpha}(u_x)$.*

Clearly, $\overline{u_x} = 0$ by inspection [since $f_{u_\alpha}(u_x)$ is symmetric (even)]. Equivalently, because the **integrand** $u_x f_{u_\alpha}(u_x)$ is an **odd** function of u_x (that is, changes sign when we replace u_x by $-u_x$), its integral between symmetric limits vanishes.

From (8.5), $f_{u_\alpha}(u_x)$ has the value $\mathrm{e}^{-1}[f_{u_\alpha}(u_x)]_{\max}$ at the two values of u_x for which

$$\frac{1}{2}mu_x^2/k_{\mathrm{B}}T = 1,$$

that is, at

$$u_x = \pm\sqrt{\frac{2k_{\mathrm{B}}T}{m}}.$$

Thus, the **width** of $f_{u_\alpha}(u_x)$ at e^{-1} of maximum is $2\sqrt{2k_{\mathrm{B}}T/m}$, that is, increases as $\sqrt{T}$.

The mean square of u_x is similarly defined as

$$\overline{u_x^2} = \int_{-\infty}^{\infty} u_x^2 f_{u_\alpha}(u_x)\mathrm{d}u_x = \sqrt{\frac{m}{2\pi k_{\mathrm{B}}T}} \int_{-\infty}^{\infty} u_x^2 \mathrm{e}^{-mu_x^2/2k_{\mathrm{B}}T}\mathrm{d}u_x.$$

The same change of variable

$$w \equiv \sqrt{\frac{m}{2k_{\mathrm{B}}T}}u_x \quad \left[\text{so that } u_x^2\mathrm{d}u_x = \left(\frac{2k_{\mathrm{B}}T}{m}\right)^{3/2} w^2\mathrm{d}w\right]$$

gives

$$\overline{u_x^2} = \frac{1}{\sqrt{\pi}}\left(\frac{2k_{\mathrm{B}}T}{m}\right)\int_{-\infty}^{\infty} w^2\mathrm{e}^{-w^2}\mathrm{d}x.$$

But the standard integral

$$\int_{-\infty}^{\infty} w^2\mathrm{e}^{-w^2}\mathrm{d}w = \frac{\sqrt{\pi}}{2}$$

(check by integration by parts), and so

$$\overline{u_x^2} = \frac{k_B T}{m}.$$

Similarly,

$$\overline{u_y^2} = \overline{u_z^2} = \frac{k_B T}{m}.$$

The **speed** u of the atom is given by

$$u^2 = u_x^2 + u_y^2 + u_z^2.$$

Therefore,

$$\overline{u^2} = \overline{u_x^2} + \overline{u_y^2} + \overline{u_z^2} = \frac{3k_B T}{m}. \tag{8.6}$$

Since the potential energy $\mathcal{V} = 0$ inside the box, the thermal average energy $\overline{\varepsilon}$ of the atom in this case is its thermal average **kinetic** energy, that is,

$$\overline{\varepsilon} = \frac{1}{2} m\overline{u^2}.$$

Therefore, from (8.6),

$$\overline{\varepsilon} = \frac{3}{2} k_B T.$$

For N independent atoms we therefore find for the internal energy $U = N\overline{\varepsilon}$ the result

$$U = \frac{3}{2} N k_B T.$$

8.3 The speed distribution

The velocity of an atom has components u_x, u_y, u_z, and can be represented by a point P in velocity space (see figure 8.4).

The point P is at a "distance" u from the origin, where u is the speed:

$$u^2 = u_x^2 + u_y^2 + u_z^2.$$

We want the probability

$$f_u(u)\mathrm{d}u$$

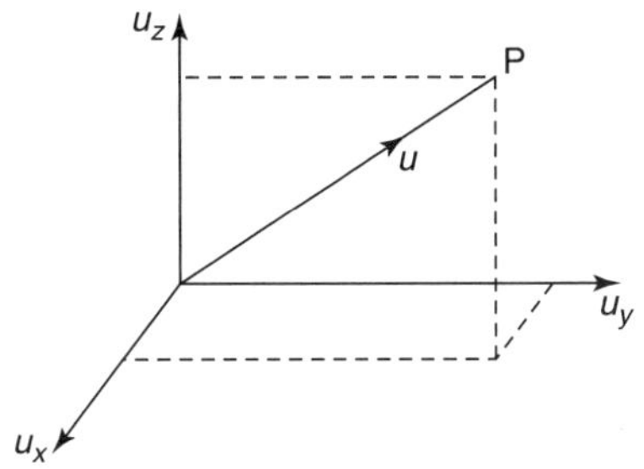

Figure 8.4 *Representative point in the velocity space of one atom.*

that the speed lies between u and $u + du$, that is, that the point representing the velocity lies in a spherical shell of radius u and thickness du.

Since u_x, u_y and u_z are have mutually independent probability distributions, the probability dp that the x-component of velocity lies in the interval $(u_x, u_x + du_x)$ and the y-component of velocity lies in the interval $(u_y, u_y + du_y)$ and the z-component of velocity lies in the interval $(u_z, u_z + du_z)$ is

$$\begin{aligned} dp &= [f_{u_\alpha}(u_x)du_x][f_{u_\alpha}(u_y)du_y][f_{u_\alpha}(u_z)du_z] \\ &= \left(\frac{m}{2\pi k_B T}\right)^{3/2} e^{-\frac{1}{2}m(u_x^2+u_y^2+u_z^2)/k_B T} du_x du_y du_z \\ &= \left(\frac{m}{2\pi k_B T}\right)^{3/2} e^{-\frac{1}{2}mu^2/k_B T} dV_u, \end{aligned}$$

where dV_u is an element of "volume" $du_x du_y du_z$ (with dimensions $[\text{velocity}]^3$) in velocity space, at the representative point specified by (u_x, u_y, u_z).

The corresponding probability per unit "volume" of velocity space is

$$\frac{dp}{dV_u} = \left(\frac{m}{2\pi k_B T}\right)^{3/2} e^{-\frac{1}{2}mu^2/k_B T}.$$

But the "volume" of the spherical shell corresponding to speeds between u and $u + du$ is $4\pi u^2 du$. Therefore, the probability that the atom is found in a state with speed in this interval $(u, u + du)$ is

$$f_u(u)du = \left(\frac{m}{2\pi k_B T}\right)^{3/2} e^{-\frac{1}{2}mu^2/k_B T} \times 4\pi u^2 du. \tag{8.7}$$

Here, $f_u(u)$ is the **Maxwell speed distribution**.

Because of the factor u^2, $f_u(u) \to 0$ as $u \to 0$. Also, the factor $\exp(-mu^2/2k_B T)$ ensures that $f_u(u) \to 0$ as $u \to \infty$. In figure 8.5 the curves a and b represent $f_u(u)$ for different temperatures ($T_b > T_a$). (Note that $u > 0$ always!)

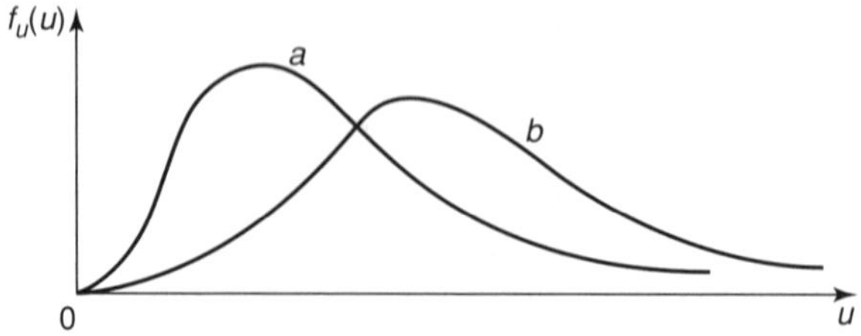

Figure 8.5 Maxwell speed distributions for two different temperatures ($T_b > T_a$).

As an exercise, you can use $f_u(u)$ to find the following results:

$$\overline{u} = \int_0^\infty u f_u(u)\mathrm{d}c = \left(\frac{8k_\mathrm{B}T}{\pi m}\right)^{1/2},$$

$$u_\mathrm{rms} = \sqrt{\overline{u^2}} = \left[\int_0^\infty u^2 f_u(u)\mathrm{d}u\right]^{1/2} = \left(\frac{3k_\mathrm{B}T}{m}\right)^{1/2}$$

(that is, $u_\mathrm{rms} \neq \overline{u}$).

You can also find the value of u at the maximum $u = u_\mathrm{max}$ of $f_u(u)$, by differentiating $u^2 \exp\left(-mu^2/2k_\mathrm{B}T\right)$ with respect to u and equating the result to zero. This gives

$$u_\mathrm{max} = \left(\frac{2k_\mathrm{B}T}{m}\right)^{1/2}$$

(the "most probable value" of u).

8.4 The equation of state

Suppose we do work on a one-particle system in a box by compressing the box quasistatically in the x-direction (see figure 8.6).

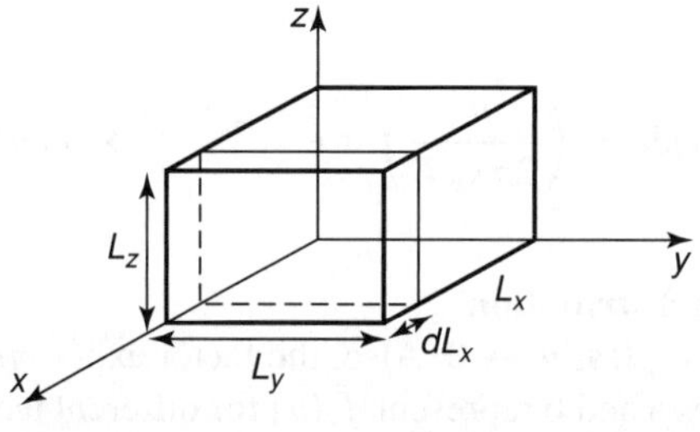

Figure 8.6 Infinitesimal quasistatic compression of a particle in a box.

Since this work is nondissipative, it goes entirely into changing the energy of the given state of the particle (and not into causing quantum jumps to a new state). If the particle is in the state $\psi_{n_x n_y n_z}$, with energy given by equation (8.3):

$$\varepsilon = \frac{\pi^2\hbar^2}{2m}\left(\frac{n_x^2}{L_x^2} + \frac{n_y^2}{L_y^2} + \frac{n_z^2}{L_z^2}\right),$$

its energy is changed by

$$\mathrm{d}\varepsilon = \frac{\partial\varepsilon}{\partial L_x}\mathrm{d}L_x = -\frac{\pi^2\hbar^2}{m}\frac{n_x^2}{L_x^3}\mathrm{d}L_x = -mu_x^2\frac{\mathrm{d}L_x}{L_x}$$

[since $u_x = \pm n_x \pi\hbar/mL_x$; see equation (8.4)].

Thus, the thermal average change of energy is

$$\overline{\mathrm{d}\varepsilon} = -m\overline{u_x^2}\frac{\mathrm{d}L_x}{L_x}. \tag{8.8}$$

But this averaged nondissipative work can also be written as

$$-p\mathrm{d}V = -pL_yL_z\mathrm{d}L_x. \tag{8.9}$$

Equating (8.8) and (8.9), we get

$$-pL_yL_z\mathrm{d}L_x = -m\overline{u_x^2}\frac{\mathrm{d}L_x}{L_x},$$

that is,

$$pV = m\overline{u_x^2}.$$

But

$$\overline{u_x^2} = \frac{\overline{u^2}}{3}.$$

Therefore,

$$pV = \frac{1}{3}m\overline{u^2}.$$

For a gas containing N atoms,

$$pV = \frac{1}{3}Nm\overline{u^2}, \quad \text{i.e.,} \quad p = \frac{1}{3}nm\overline{u^2},$$

where n is the number density $n = N/V$. But we found earlier that

$$\frac{1}{3}\overline{u^2} = \overline{u_x^2} = \frac{k_B T}{m}.$$

Therefore,

$$pV = Nk_BT,$$

which is the **equation of state** of an ideal monatomic gas.

8.5 Mean free path and thermal conductivity

Pick many molecules at random. The average of the distances travelled since the last collision of each molecule is called the **mean free path** l.

If we have a nonuniform macroscopic situation (for example, a temperature gradient), the velocity distribution appropriate for any given particle is assumed to be the Maxwell distribution characteristic of the temperature and number density **at the point of the previous collision**.

We now use this idea to calculate the thermal conductivity κ of a gas.

Assume a uniform temperature gradient $\mathrm{d}T/\mathrm{d}z$ in the z-direction.

Using the Maxwell speed distribution, we now show that the heat flux j_Q in the z direction (heat flow per unit area of the xy plane per second) takes the form

$$j_Q = -\kappa\frac{\mathrm{d}T}{\mathrm{d}z},$$

(that is, j_Q is in the positive z direction when $\mathrm{d}T/\mathrm{d}z < 0$), and thereby find an expression for the **thermal conductivity** κ.

In figure 8.7, S is the point of last collision for an atom at O that has travelled from S to O with speed u and at angle θ to the z-axis.

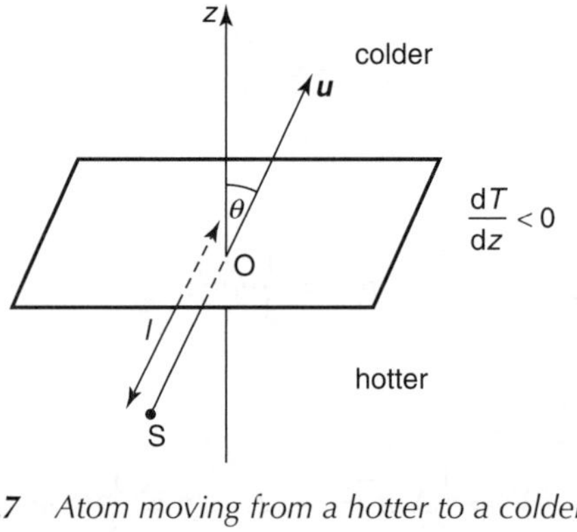

Figure 8.7 *Atom moving from a hotter to a colder region.*

(i) The number of molecules per unit volume with speeds in the range (u, $u + \mathrm{d}u$) and polar angles in the range $(\theta, \theta + \mathrm{d}\theta)$ is

$$n \times f_u(u)\mathrm{d}u \times \frac{1}{2}\sin\theta\mathrm{d}\theta,$$

where the third factor is the fraction of molecules whose velocity vectors have polar angles in the range $(\theta, \theta + \mathrm{d}\theta)$ [obtained by considering a sphere of radius R and dividing the area of that part of the surface lying between latitudes θ and $\theta + \mathrm{d}\theta$ (namely, $2\pi R \sin\theta \times R\mathrm{d}\theta$) by the total area $4\pi R^2$of the sphere].

(ii) The normal component of the velocity is $u\cos\theta$.

(iii) The excess energy per molecule by virtue of the temperature difference between S and the plane is

$$-c_{\mathrm{m}} \times l\cos\theta \times \frac{\mathrm{d}T}{\mathrm{d}z}\left(> 0 \quad \text{if} \, \frac{\mathrm{d}T}{\mathrm{d}z} < 0\right),$$

where c_{m} is the specific heat per molecule.

Taking the product of (i), (ii) and (iii), and integrating over all speeds u and over all polar angles from 0 to π (the latter allows for the molecules arriving from the cold side as well as those arriving from the hot side), we have

$$j_Q = \int_0^{\pi}\int_0^{\infty}\left(nf_u(u)\mathrm{d}u\tfrac{1}{2}\sin\theta\mathrm{d}\theta\right)(u\cos\theta)\left(-c_{\mathrm{m}}l\cos\theta\frac{\mathrm{d}T}{\mathrm{d}z}\right)$$

$$= -\frac{nc_{\mathrm{m}}l}{2}\frac{\mathrm{d}T}{\mathrm{d}z}\int_0^{\pi}\mathrm{d}\theta\sin\theta\cos^2\theta\int_0^{\infty}uf_u(u)\mathrm{d}u.$$

But (with $w = \cos\theta$, so that $\mathrm{d}w = -\sin\theta\mathrm{d}\theta$)

$$\int_0^{\pi}\mathrm{d}\theta\sin\theta\cos^2\mathrm{d}\theta = -\int_1^{-1}w^2\mathrm{d}w = \frac{2}{3}$$

and

$$\int_0^{\infty}uf_u(u)\mathrm{d}u = \overline{u}.$$

Therefore,

$$j_Q = -\frac{nc_{\mathrm{m}}l\overline{u}}{3}\frac{\mathrm{d}T}{\mathrm{d}z} \equiv -\kappa\frac{\mathrm{d}T}{\mathrm{d}z},$$

and so the thermal conductivity κ is given by

$$\kappa = \frac{1}{3} n c_{\mathrm{m}} l \bar{u} \quad \left(n = \frac{N}{V} \right).$$

We now make a crude estimate of the mean free path l in this expression for κ. Take the molecules to be hard spheres of **diameter** σ (see figure 8.8).

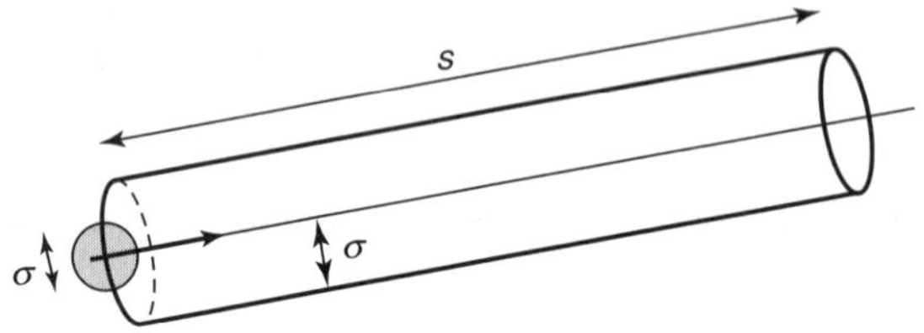

Figure 8.8 *Motion of a molecule represented by a hard sphere of diameter* σ.

A collision occurs if the centres of two molecules approach within a distance σ. Thus, in its motion through a distance s one molecule collides with as many other molecules as have their centres in a cylinder of volume $\pi\sigma^2 s$. There are $n\pi\sigma^2 s$ such molecules. Therefore, the number of collisions **per unit path** is $n\pi\sigma^2$, and so the average path per collision (mean free path) is

$$l \approx \frac{1}{n\pi\sigma^2}. \tag{8.10}$$

Therefore,

$$\kappa \approx \frac{c_{\mathrm{m}}\bar{u}}{3\pi\sigma^2}, \tag{8.11}$$

which is **independent of the number density**.

The relation (8.10) describes the experimental results well at ordinary pressures, but fails at very low pressures [when l is no longer proportional to $1/n$, as in (8.10), but becomes equal to the size of the box – that is, collisions occur only with the walls], and also at very high pressures, when the molecules are so close together that the concept of a mean free path breaks down.

9

Applications of Classical Thermodynamics

The **first law of thermodynamics** (for an infinitesimal processes) is

$$dU = đW + đQ,$$

where dU is the infinitesimal change of the internal energy U of the system (U is a state variable) that results from infinitesimal energy transfers (work $đW$ and heat transfer $đQ$) to the system. (The quantities $đW$ and $đQ$ are **not** infinitesimal changes of any state variable, but simply infinitesimal energies.)

The **second law of thermodynamics** can be stated in several ways. In the following sections we give the **entropy statement**, the **temperature statement**, and the **heat-engine statement**.

9.1 Entropy statement of the second law of thermodynamics

The **entropy statement** of the second law is: **There exists an additive function of state (state variable), known as the equilibrium entropy *S*, which can never decrease in a thermally isolated system.**

Treat the "system" + "surroundings" (that is, the "total system") as thermally isolated. Then the second law states:

$$\Delta S_{\text{tot}} \geq 0 \quad \text{in any process}$$

(Clausius' inequality).

For a reversible change we cannot have $\Delta S_{\text{tot}} > 0$, since the reversed process would have $\Delta S_{\text{tot}} < 0$, which would violate the second law. Therefore,

$$\Delta S_{\text{tot}} = 0 \quad \text{for reversible changes,}$$

that is,

$$\Delta S_{\text{system}} + \Delta S_{\text{surroundings}} = 0 \quad \text{for reversible changes,}$$

A Course in Theoretical Physics, First Edition. P. J. Shepherd.

or

$$\Delta S_{\text{system}} = -\Delta S_{\text{surroundings}} \quad \text{for reversible changes.}$$

Since, as we have shown, the probabilities of states (and hence the entropy) can be changed only by heat transfer and/or dissipative work, and the latter are both absent in a **reversible adiabatic** change (such as in the processes considered in mechanics), we have

$$\Delta S_{\text{system}} = \Delta S_{\text{surroundings}} = 0 \quad \text{for all reversible adiabatic changes.}$$

9.2 Temperature statement of the second law of thermodynamics

Consider an infinitesimal heat transfer between two bodies A and B with the constraints $(V_\text{A}, N_\text{A}, V_\text{B}, N_\text{B})$ held fixed. Then

$$đQ_\text{A} = \text{d}U_\text{A} = -\text{d}U_\text{B}.$$

If the system composed of A and B is thermally isolated, the entropy statement of the second law gives

$$\text{d}S_{\text{tot}} = \text{d}S_\text{A} + \text{d}S_\text{B} \geq 0.$$

But

$$\begin{aligned} \text{d}S_{\text{tot}} &= \text{d}S_\text{A} + \text{d}S_\text{B} \\ &= \left(\frac{\partial S_\text{A}}{\partial U_\text{A}}\right)_{V_\text{A},N_\text{A}} \text{d}U_\text{A} + \left(\frac{\partial S_\text{B}}{\partial U_\text{B}}\right)_{V_\text{B},N_\text{B}} \text{d}U_\text{B} \\ &= \left[\frac{\partial S_\text{A}}{\partial U_\text{A}} - \frac{\partial S_\text{B}}{\partial U_\text{B}}\right] đQ_\text{A} \geq 0. \end{aligned} \tag{9.1}$$

Now **define** the **thermodynamic temperature** by

$$\frac{1}{T} = \left(\frac{\partial S}{\partial U}\right)_{V,N}. \tag{9.2}$$

Then equation (9.1) becomes

$$\left(\frac{1}{T_\text{A}} - \frac{1}{T_\text{B}}\right) đQ_\text{A} \geq 0. \tag{9.3}$$

Call $1/T$ the "coldness"; that is, if A is colder than B, then

$$\frac{1}{T_\text{A}} > \frac{1}{T_\text{B}}.$$

So, from equation (9.3),

$$đQ_A > 0;$$

that is, A gains energy from B.

This leads to the **temperature statement** of the second law: **No process exists in which heat is transferred to a body from a colder body while the constraints on the bodies and the state of the rest of the world are left unchanged**.

[Recall that we defined the **statistical temperature** by

$$\frac{1}{k_B T} = \left(\frac{\partial \ln g}{\partial E}\right)_{V,N}.$$

The definition (9.2) is equivalent to this, since we showed that, for large systems, both open and closed, $S = k_B \ln g$.]

In a "simple" system (that is, a system with one chemical species), with fixed N, the only constraint is V, and so the equilibrium entropy $S = S(U, V)$. Then

$$dS = \left(\frac{\partial S}{\partial U}\right)_V dU + \left(\frac{\partial S}{\partial V}\right)_U dV$$

in **any** infinitesimal process between equilibrium states.

To find a general expression for $(\partial S/\partial V)_U$ consider a particular process for which it is easily calculated.

For example, consider a reversible adiabatic expansion or compression. For this process,

$$dU = -p dV,$$

and, since no heat enters the system and no dissipative work is done, we have $dS = 0$, that is,

$$0 = \left(\frac{\partial S}{\partial U}\right)_V dU + \left(\frac{\partial S}{\partial V}\right)_U dV = \frac{1}{T}(-p dV) + \left(\frac{\partial S}{\partial V}\right)_U dV.$$

Therefore,

$$\left(\frac{\partial S}{\partial V}\right)_U = \frac{p}{T}.$$

Thus,

$$dS = \left(\frac{\partial S}{\partial U}\right)_V dU + \left(\frac{\partial S}{\partial V}\right)_U dV = \frac{1}{T} dU + \frac{p}{T} dV,$$

that is,

$$dU = T dS - p dV, \tag{9.4}$$

which is a **general** expression for $\mathrm{d}U$ in terms of differential changes of state variables, that is, it is true for **any** infinitesimal process, whether reversible or irreversible, taking a system from an equilibrium initial state to an equilibrium final state.

9.3 Summary of the basic relations

For **any** infinitesimal process (whether reversible or irreversible)

$$\mathrm{d}U = đW + đQ. \tag{9.5}$$

For **any** infinitesimal process (whether reversible or irreversible) **between equilibrium states**,

$$\mathrm{d}U = T\mathrm{d}S - p\mathrm{d}V \tag{9.6}$$

($\mathrm{d}S$ is the change in **equilibrium** entropy, and T and p are the **equilibrium** temperature and pressure).

Consider **three** cases:

(I) In the case of **pure heat transfer** (whether reversible [i.e., isothermal] or irreversible), that is, when no work is done ($đW = 0$) and there is no change of the constraint V ($\mathrm{d}V = 0$), (9.5) and (9.6) give

$$\mathrm{d}U = đQ = T\mathrm{d}S,$$

that is, the change of entropy is

$$\mathrm{d}S = \frac{đQ}{T}.$$

(II) For the case of heat transfer plus **nondissipative** (reversible) work, that is, $đW = -p\mathrm{d}V$, comparison of equations (9.5) and (9.6) again gives

$$\mathrm{d}S = \frac{đQ}{T}.$$

(III) If **dissipative work** is involved in the given process with energy change $\mathrm{d}U$, then

$$đW \neq -p\mathrm{d}V,$$

and so, comparing equations (9.5) and (9.6), we have

$$đQ \neq T\mathrm{d}S.$$

Thus, in this case,

$$\mathrm{d}S \neq \frac{đQ}{T}.$$

For example, if the work involves fast compression, then $đW \neq -p\mathrm{d}V$ (since the local pressure behind the piston on the gas side is momentarily higher than the equilibrium pressure p), and so

$$đQ < T\mathrm{d}S,$$

that is,

$$\mathrm{d}S > \frac{đQ}{T} \quad \text{if dissipative work is done.}$$

Summarizing, we have shown that

$$\mathrm{d}S = \frac{đQ}{T} \quad \text{when} \quad đW_{\mathrm{diss}} = 0, \tag{9.7}$$

$$\mathrm{d}S > \frac{đQ}{T} \quad \text{when} \quad đW_{\mathrm{diss}} \neq 0. \tag{9.8}$$

9.4 Heat engines and the heat-engine statement of the second law of thermodynamics

Define a **heat source** and **heat sink** as bodies that provide and receive energy purely by heat transfer, that is, $\mathrm{d}S = đQ/T$ for heat sources and heat sinks ($đQ < 0$ for a heat source; $đQ > 0$ for a heat sink).

Define a **heat engine** is a device that turns heat abstracted from a heat source into work.

The removal of heat from a heat source lowers the entropy of the source.

Suppose that **all** the heat extracted from the source is converted into nondissipative work on a system. Since such work does not change the entropy of the system (this follows from case II above), this would imply that we could decrease the total entropy (the entropy of the heat source plus the entropy of the system on which the work is done), contrary to the entropy statement of the second law.

Hence we have the **heat engine statement** (Kelvin statement) of the second law: **No process exists in which heat is extracted from a source and converted entirely into useful work, leaving the rest of the world unchanged**.

To satisfy the rule $\Delta S_{\mathrm{tot}} \geq 0$, it is clear that we must sacrifice as much heat to a heat sink as will increase the entropy of the heat sink by at least enough to cancel the decrease of entropy of the heat source.

Consider a heat engine that works in a **cycle**, in which the device returns to its original thermodynamic state at the end of the cycle.

In one cycle the working substance of the engine (for example, a gas) takes heat Q_{h} from a very large <u>h</u>ot heat source at a high temperature T_{h}, does useful work W, and rejects waste heat Q_{c} to a very large <u>c</u>old heat sink at a lower temperature T_{c}. Such a device is a **two-port engine**.

At the end of the cycle the internal energy and entropy of the <u>w</u>orking <u>s</u>ubstance are unchanged ($\Delta U_{\mathrm{ws}} = 0$, $\Delta S_{\mathrm{ws}} = 0$). Therefore, by energy conservation,

$$W = Q_{\mathrm{h}} - Q_{\mathrm{c}}.$$

Also,

$$\begin{aligned}\Delta S_{tot} &= \Delta S_{ws} + \Delta S_{heat\ source} + \Delta S_{heat\ sink} \\ &= 0 - \frac{Q_h}{T_h} + \frac{Q_c}{T_c} \\ &\geq 0 \text{ by the entropy statement of the second law.}\end{aligned}$$

(The = sign in the third line applies for **reversible** processes, that is, when $đW_{diss} = 0$ **and** the heat transfer is isothermal.)

Thus,

$$Q_c \geq \left(\frac{T_c}{T_h}\right) Q_h,$$

that is, we must sacrifice heat Q_c at least equal to $(T_c/T_h)\,Q_h$ to the cold heat sink.

Define the **thermodynamic-engine efficiency** η_e as

$$\begin{aligned}\eta_e &= \frac{\text{work extracted per cycle}}{\text{heat put in at } T_h \text{ per cycle}} \\ &= \frac{W}{Q_h} = \frac{Q_h - Q_c}{Q_h} = 1 - \frac{Q_c}{Q_h}.\end{aligned}$$

Naturally, the efficiency is a maximum (for the given T_c and T_h) when the waste heat Q_c is a minimum, that is, when $Q_c = (T_c/T_h)\,Q_h$, which applies for a **reversible** cycle (one with $\Delta S_{tot} = 0$, that is, with **no** dissipative work and with isothermal heat transfers between the heat source/sink and the working substance).

Thus,

$$\eta_{e,max} = 1 - \frac{T_c}{T_h}.$$

The quantity $\eta_{e,max}$ is the **ideal thermodynamic-engine efficiency**.

A reversible two-port engine is known as a **Carnot engine**. The above results can be stated as **Carnot's principle**:

All reversible two-port engines working between the same pair of temperatures have the same efficiency, and no irreversible engine working between the same pair of temperatures can have a greater efficiency.

(In fact, Kelvin defined the thermodynamic temperature in terms of the efficiency $\eta_{e,max} = 1 - T_c/T_h$ for a heat sink and heat source one of which was water at its triple point T_c (or T_h) = 273.16 K.)

We can increase $\eta_{e,max} = 1 - T_c/T_h$ by using a lower T_c and/or a higher T_h. Since the cold sink is often the atmosphere (for example, in the case of an internal-combustion engine) or a river or the sea (in the case of a power-plant steam turbine), it is not usually possible to control T_c. So we must use heat **sources** at the highest practicable temperatures T_h.

For example, for a power-plant turbine, typically, $T_h = 810\text{K}$ and $T_c = 300\text{K}$, so that $\eta_{e,max} = 1 - 300/810 = 0.63$. In practice, because of irreversibility (dissipative work and nonisothermal heat transfers) the best efficiencies achieved are ~ 0.47.

9.5 Refrigerators and heat pumps

Suppose that our heat engine is run backwards. In this case the cold heat sink becomes a cold heat **source**, supplying heat to the working substance at T_c, work W is done on the "working substance", and heat Q_h ($= Q_c + W$ by the first law) is transferred to the hot heat **sink** at T_h.

As before, in one whole cycle,

$$\Delta S_{\text{total}} = \Delta S_{\text{ws}} + \Delta S_{\text{source}} + \Delta S_{\text{sink}} = 0 - \frac{Q_c}{T_c} + \frac{Q_h}{T_h} \geq 0.$$

Therefore,

$$Q_c \leq \left(\frac{T_c}{T_h}\right) Q_h.$$

This engine can be viewed **either** as a **heat pump** (extracting heat Q_c from a cold environment and pumping it, by expenditure of work W, into a hotter system, for example, a house), **or** as a **refrigerator**.

For a heat pump the aim is to maximize Q_h for a given W. For a refrigerator the aim is to maximize Q_c (the heat taken from the cold source) for a given W.

Thus, the "coefficient of performance" for a heat pump is

$$\eta_{\text{hp}} = \frac{Q_h}{W} = \frac{Q_h}{Q_h - Q_c} = \frac{1}{1 - \dfrac{Q_c}{Q_h}}.$$

Therefore,

$$\eta_{\text{hp,max}} = \left(\frac{Q_h}{W}\right)_{\max} = \frac{1}{1 - \left(\dfrac{Q_c}{Q_h}\right)_{\max}}$$
$$= \frac{1}{1 - \dfrac{T_c}{T_h}} = \frac{T_h}{T_h - T_c} \left(= \eta_{\text{e,max}}^{-1}\right).$$

The refrigerator coefficient of performance is, naturally,

$$\eta_r = \frac{Q_c}{W} = \frac{Q_c}{Q_h - Q_c} = \frac{Q_c/Q_h}{1 - \dfrac{Q_c}{Q_h}},$$

and so

$$\eta_{\text{r,max}} = \left(\frac{Q_c}{W}\right)_{\max} = \frac{(Q_c/Q_h)_{\max}}{1 - (Q_c/Q_h)_{\max}}$$
$$= \frac{T_c}{T_h - T_c} = \eta_{\text{hp,max}} - 1.$$

Thus, heat pumps and refrigerators are at their most efficient when $T_h \approx T_c$.

9.6 Example of a Carnot cycle

Consider gas in a cylinder, undergoing the four-stage cycle shown in figure 9.1.

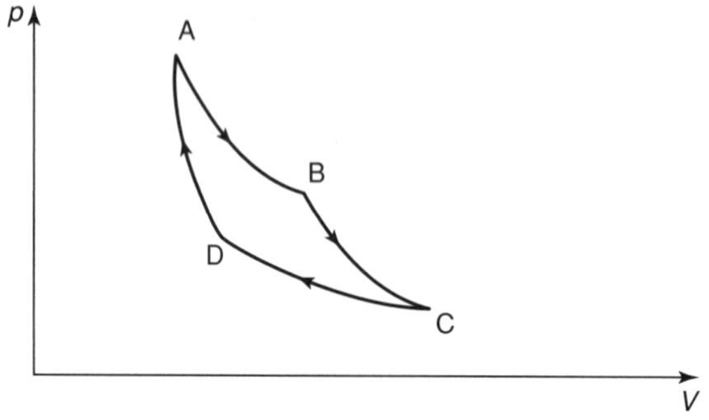

Figure 9.1 *Carnot cycle (see the text).*

Start from A:

A → B: The gas is in thermal contact with a heat source (reservoir) at thermodynamic temperature T_1, absorbs heat Q_1 from this source, and expands **isothermally**, doing nondissipative work on the piston. In this process, $\Delta S_{\text{gas}} = Q_1/T_1$.

B → C: The heat source is removed. The gas continues to expand (**adiabatically**), doing nondissipative work on the piston and cooling in the process to T_2. In this process, $\Delta S_{\text{gas}} = 0$.

C → D: The gas is compressed **isothermally** (the piston does nondissipative work on the gas) and heat Q_2 is transferred to a cold heat source at thermodynamic temperature T_2. In this process, $\Delta S_{\text{gas}} = -Q_2/T_2$.

D → A: The gas is compressed **adiabatically** by the piston (as in BC, there is no contact with a heat source) until the thermodynamic temperature rises to T_1 at A. In this process, $\Delta S_{\text{gas}} = 0$.

The net work W done on the gas in the cycle is

$$W = -\oint p\mathrm{d}V = \text{minus the area enclosed by ABCDA}$$

[$\oint$ means "integral round the cycle"].
By the first law,

$$W = Q_1 - Q_2.$$

Also, since the change in the entropy of the gas around the cycle is zero,

$$\frac{Q_1}{T_1} - \frac{Q_2}{T_2} = 0. \tag{9.9}$$

Now assume that the gas is **ideal**. Define the **ideal-gas temperature** θ to be such that

$$pV = nR\theta \tag{9.10}$$

(that is, $\theta \propto p$ for a constant-volume gas thermometer, in accordance with our earlier definition of the ideal-gas temperature and as demonstrated experimentally by measurements in the low-density limit).

The other key property of an **ideal** gas is that, for a given N, the internal energy U depends only on θ, and not on V.

We now show that $\theta \propto T$.

For each infinitesimal part of the cycle ABCDA we have

$$\frac{đQ}{\theta} = \frac{\mathrm{d}U - đW}{\theta}.$$

Since the work is always nondissipative, $đW = -p\mathrm{d}V$, that is,

$$\frac{đQ}{\theta} = \frac{\mathrm{d}U}{\theta} + \frac{p\mathrm{d}V}{\theta}.$$

Thus, from equation (9.10),

$$\frac{đQ}{\theta} = \frac{\mathrm{d}U}{\theta} + nR\frac{\mathrm{d}V}{V}.$$

Therefore,

$$\oint \frac{đQ}{\theta} = \oint \frac{\mathrm{d}U}{\theta} + nR\oint \frac{\mathrm{d}V}{V}.$$

Since U depends only on θ the quantity $\mathrm{d}U/\theta$ is an exact differential of a quantity that depends only on θ, and since in the complete cycle the gas returns to the same state A with unchanged θ, we have

$$\oint \frac{\mathrm{d}U}{\theta} = 0.$$

Similarly, since $\mathrm{d}V/V$ is the exact differential $\mathrm{d}(\ln V)$, and since in the complete cycle $\ln V$ returns to its starting value,

$$\oint \frac{\mathrm{d}V}{V} = 0.$$

Therefore,

$$\oint \frac{đQ}{\theta} = 0,$$

that is,

$$\frac{Q_1}{\theta_1} - \frac{Q_2}{\theta_2} = 0, \tag{9.11}$$

where θ_1 and θ_2 are the **ideal-gas temperatures** of the hot and cold heat sources.

Comparing (9.9) and (9.11), we find

$$\frac{T_1}{T_2} = \frac{\theta_1}{\theta_2},$$

that is, the thermodynamic temperature T is proportional to the ideal-gas temperature θ.

Since, by definition, $T = (\partial U/\partial S)_V$, by choosing suitable units for S [i.e., a suitable value for k_B in $S = k_B \ln g(U)$], we can make T **equal to** θ.

The value of k_B that ensures that $T_{tr}(H_2O) = 273.16$ K is

$$k_B = 1.38[0658] \times 10^{-23}\,\mathrm{J\,K^{-1}}$$

(Boltzmann's constant).

9.7 The third law of thermodynamics

Entropy changes can be measured in a process A → B as

$$\Delta S(\mathrm{A} \to \mathrm{B}) = \int_{\mathrm{A}}^{\mathrm{B}} \frac{đQ}{T}$$

(in the absence of dissipative work).

How can we measure **absolute entropies**?

Experimentally, entropy **changes** tend to be **very small** in processes in which the system goes from one **low-temperature** equilibrium state to another at the same temperature (for example, isothermal processes with nondissipative work). Moreover, for the given system this entropy change get smaller and smaller when the same process is implemented at ever lower temperatures, suggesting that in the low-temperature limit the entropy comes to depend only on T. We postulate, therefore, that **the absolute entropies of all substances approach the same value as $T \to 0$**.

Consider, for example, a closed system of N atomic magnetic moments ("spins") that can align themselves in just two ways in a magnetic field of magnetic induction $\boldsymbol{B}$.

Since the magnets like to align themselves parallel to the field, the lowest-energy state corresponds to

$$\uparrow\uparrow\uparrow\uparrow\uparrow\uparrow\uparrow \cdots\cdots \uparrow\uparrow\uparrow\uparrow\uparrow \quad \Uparrow \boldsymbol{B} \quad (N_{\mathrm{up}} = N, N_{\mathrm{down}} = 0)$$

Thus,

$$W(N_{\mathrm{up}}, N_{\mathrm{down}}) = W(N, 0) = 1.$$

It requires extra energy to "flip a spin" so that it points opposite to the field:

$$\uparrow\uparrow\downarrow\uparrow\uparrow\uparrow\uparrow \cdots\cdots \uparrow\uparrow\uparrow\uparrow\uparrow \quad \Uparrow \boldsymbol{B} \quad (N_{\mathrm{up}} = N-1, N_{\mathrm{down}} = 1)$$

This situation (one reversed spin) can be realized in N ways, that is,

$$W(N_{\mathrm{up}}, N_{\mathrm{down}}) = W(N-1, 1) = N.$$

The situation with two spins reversed can be realized in

$$W(N_{\text{up}}, N_{\text{down}}) = W(N-2, 2) = \frac{N(N-1)}{2!} \text{ways}$$

(N ways of choosing the first and $N - 1$ ways of choosing the second, the order of the two being irrelevant). This is, of course, a specific instance of the formula

$$W(N_{\text{up}}, N_{\text{down}}) = \frac{N!}{N_{\text{up}}!N_{\text{down}}!}.$$

But the entropy S for a closed system is

$$S = k_{\text{B}} \ln W.$$

So, for the lowest-energy configuration,

$$S = k_{\text{B}} \ln 1 = 0.$$

We therefore postulate the **third law of thermodynamics** (**Nernst's theorem**):

The equilibrium entropies of all systems and the entropy changes in all reversible isothermal processes tend to zero as $T \to 0$.

In some cases it is necessary to get down to temperatures $< 10^{-5}$ K before the equilibrium entropies approach zero.

For example, for copper the components of the nuclear-spin angular momentum can take four possible values

$$m_{\text{n}}\hbar = \frac{3}{2}\hbar, \frac{1}{2}\hbar, -\frac{1}{2}\hbar, -\frac{3}{2}\hbar.$$

(The total nuclear spin quantum number I, defined as $I = m_{\text{n,max}}$, is $3/2$ for the copper nucleus; clearly, the number of possible values of m_{n} is $2I + 1$, equal to 4 for copper.)

Even at $T = 10^{-4}$ K, states in which all $2I + 1$ ($= 4$) spin orientations occur for each nucleus are present in zero field.

Since each spin has $2I + 1$ possible orientations, the N spins can be arranged in $(2I + 1)^N$ ways. Thus,

$$\begin{aligned} S &= k_{\text{B}} \ln(2I + 1)^N = N k_{\text{B}} \ln(2I+1) \\ &= N k_{\text{B}} \ln 4 \text{ in this case} \\ &\neq 0. \end{aligned}$$

This is because the interactions between the nuclear spins are so weak that, in zero external magnetic field, the energy spacings between N-spin states with different configurations of spin orientations over the sites are smaller than $k_{\text{B}}T$ even when $T = 10^{-4}$ K, and so even this low temperature is high enough to ensure that **all** possible nuclear N-spin states are occupied with essentially equal probability.

When $T < 10^{-5}$ K, the higher-energy spin states become occupied with lower probability and, therefore, fewer states contribute to the entropy $S = k_B \ln W$. We expect that there will be **just one** state with an energy lower than all the other state energies (that is, a **nondegenerate ground state**), and this state will be the only one to contribute to S in the limit $T \to 0$. Thus, we can expect that $S \to 0$ for $T \to 0$ even in the nuclear-spin case.

Note that the third law applies only to **equilibrium entropies**. If a system becomes frozen into **nonequilibrium states**, so that (for example, because of energy barriers) it cannot explore all the microstates consistent with the given U, V and N, we cannot even measure the equilibrium entropy in this case.

9.8 Entropy-change calculations

Infinitesimal heat transfer $đQ$ to a body at constant volume in the absence of dissipative work produces a temperature change $\mathrm{d}T$ such that, by definition of C_V (see section 5.16.1),

$$đQ = C_V \mathrm{d}T.$$

If in a succession of such infinitesimal processes the temperature of the body is increased from T_1 to T_2, the total heat transfer to the body is

$$Q = \int_{T=T_1}^{T=T_2} đQ = \int_{T_1}^{T_2} C_V \mathrm{d}T,$$

where, in general, C_V itself depends on T.

The corresponding entropy change is

$$\Delta S = \int_{T=T_1}^{T=T_2} \frac{đQ}{T} = \int_{T_1}^{T_2} \frac{C_V \mathrm{d}T}{T}.$$

If, instead, the process is carried out at **constant pressure**, the corresponding relations are

$$Q = \int_{T=T_1}^{T=T_2} đQ = \int_{T_1}^{T_2} C_p \mathrm{d}T,$$

$$\Delta S = \int_{T=T_1}^{T=T_2} \frac{đQ}{T} = \int_{T_1}^{T_2} \frac{C_p \mathrm{d}T}{T}.$$

Example Each of two bodies has

$$C_V = AT^2, \quad \text{with} \quad A = 0.01\,\mathrm{J\,K^{-3}}$$

(a) If they form a closed system and are initially at temperatures 200 K and 400 K, respectively, and are then brought into thermal contact until thermal equilibrium is reached:
   (i) What is their final common temperature?
   (ii) What is the increase of entropy?

Solution

(b) Body 1 is heated from 200 K to final temperature T_f, and increases its energy by

$$Q^{(\text{to }1)} = \int_{T=200\,\text{K}}^{T=T_f} đQ^{(\text{to }1)} = \int_{200}^{T_f} C_V^{(1)}\text{d}T$$
$$= \int_{200}^{T_f} AT^2\text{d}T = \left.\frac{AT^3}{3}\right|_{200}^{T_f} = \frac{A}{3}\left(T_f^3 - 200^3\right).$$

Body 2 cools from 400 K to T_f, and "gains" energy equal to

$$Q^{(\text{to }2)} = \int_{T=400\,\text{K}}^{T=T_f} đQ^{(\text{to }2)} = \int_{400}^{T_f} C_V^{(2)}\text{d}T$$
$$= \int_{400}^{T_f} AT^2\text{d}T = \left.\frac{AT^3}{3}\right|_{400}^{T_f} = \frac{A}{3}\left(T_f^3 - 400^3\right).$$

Since the system is closed (and no work is done), the first law gives

$$Q^{(\text{to }1)} + Q^{(\text{to }2)} = 0 \left(\text{i.e.,}\ \ Q^{(\text{to }2)} = -Q^{(\text{to }1)}\right)$$

that is,

$$\frac{A}{3}\left(T_f^3 - 200^3 + T_f^3 - 400^3\right) = 0,$$

so that $T_f = 330\,\text{K}$.
(ii) The increase in entropy is

$$\Delta S = \Delta S^{(1)} + \Delta S^{(2)}$$
$$= \int_{T=200\,\text{K}}^{T=T_f} \frac{đQ^{(\text{to }1)}}{T} + \int_{T=400\,\text{K}}^{T=T_f} \frac{đQ^{(\text{to }2)}}{T} = \int_{200}^{T_f} \frac{C_V^{(1)}\text{d}T}{T} + \int_{400}^{T_f} \frac{C_V^{(2)}\text{d}T}{T}$$
$$= A\int_{200}^{T_f} \frac{T^2\text{d}T}{T} + A\int_{400}^{T_f} \frac{T^2\text{d}T}{T} = A\left(\left.\frac{T^2}{2}\right|_{200}^{T_f} + \left.\frac{T^2}{2}\right|_{400}^{T_f}\right)$$
$$= \frac{A}{2}\left(T_f^2 - 200^2 + T_f^2 - 400^2\right) = \frac{0.01}{2}(2 \times 330^2 - 200^2 - 400^2) = 89\,\text{J}\,\text{K}^{-1},$$

that is, there is an increase of the total entropy, as expected.

(b) Now let the cooling of the hotter body and the heating of the cooler body occur in the presence of a working substance.

What is the maximum work we can extract in the process, and what is the final common temperature of the two bodies in the maximum-work process?

Solution

We extract the maximum work when the heat transfer to the cooler body is only just enough to ensure that $\Delta S_{\text{tot}} = 0$, that is, T_f in this case is such that

$$\Delta S^{(1)} + \Delta S^{(2)} = 0,$$

that is,

$$\int_{T=200\,\text{K}}^{T=T_\text{f}} \frac{đQ^{(\text{to }1)}}{T} + \int_{T=400\,\text{K}}^{T=T_\text{f}} \frac{đQ^{(\text{to }2)}}{T} = 0.$$

Therefore,

$$\frac{A}{2}\left(T_\text{f}^2 - 200^2 + T_\text{f}^2 - 400^2\right) = 0,$$

so that $T_\text{f} = 316.22\,\text{K}$.

Thus, the heat "gained" by the hotter body is

$$\begin{aligned} Q^{(\text{to }2)} &= \int_{T=400\,\text{K}}^{T=316.22\,\text{K}} đQ^{(\text{to }2)} = \int_{400}^{316.22} C_V^{(2)}\text{d}T \\ &= A\int_{400}^{316.22} T^2\text{d}T = \frac{0.01}{3}(316.22^3 - 400^3) \\ &= -1.08 \times 10^5\,\text{J} \end{aligned}$$

(which is a heat **loss**, as expected).

The heat gained by the cooler body is

$$Q^{(\text{to }1)} = \int_{T=200\,\text{K}}^{T=316.22\,\text{K}} đQ^{(1)} = \frac{0.01}{3}(316.22^3 - 200^3) = 0.78 \times 10^5\,\text{J}.$$

Therefore, the total heat "gained" by the two bodies is

$$(-1.08 + 0.78) \times 10^5\,\text{J} = -3 \times 10^4\,\text{J}.$$

Since the system is closed, this net heat loss of 3×10^4 J must have gone into work:

$$W_{\text{max}} = 3 \times 10^4\,\text{J}.$$

10

Thermodynamic Potentials and Derivatives

10.1 Thermodynamic potentials

We now introduce three other important state variables, that is, properties of a system that take a well-defined value in a thermodynamic equilibrium state of the system.

They all have the dimensions of energy and, together with the internal energy U, are called **thermodynamic potentials**.

They are the **Helmholtz free energy** F, defined as

$$F = U - TS,$$

the **enthalpy** H, defined as

$$H = U + pV,$$

and the **Gibbs free energy** (or Gibbs potential) G, defined as

$$G = U - TS + pV.$$

In an infinitesimal process taking a system (whether reversibly or irreversibly) between two thermodynamic equilibrium states, the differential change in U is given by the previously proved **thermodynamic identity**

$$dU = T\,dS - p\,dV.$$

The accompanying change in the Helmholtz free energy F of the system will be

$$
\begin{aligned}
dF = d(U - TS) &= dU - d(TS) \\
&= dU - T\,dS - S\,dT = T\,dS - p\,dV - T\,dS - S\,dT \\
&= -S\,dT - p\,dV.
\end{aligned}
$$

A Course in Theoretical Physics, First Edition. P. J. Shepherd.
© 2013 John Wiley & Sons, Ltd. Published 2013 by John Wiley & Sons, Ltd.

Similarly, the accompanying change of the enthalpy H of the system will be

$$\begin{aligned} \mathrm{d}H &= \mathrm{d}(U + pV) = \mathrm{d}U + \mathrm{d}(pV) \\ &= \mathrm{d}U + p\mathrm{d}V + V\mathrm{d}p = T\mathrm{d}S - p\mathrm{d}V + p\mathrm{d}V + V\mathrm{d}p \\ &= T\mathrm{d}S + V\mathrm{d}p. \end{aligned}$$

The accompanying change of the Gibbs free energy G of the system will be

$$\begin{aligned} \mathrm{d}G &= \mathrm{d}(U - TS + pV) = \mathrm{d}U - \mathrm{d}(TS) + \mathrm{d}(pV) \\ &= \mathrm{d}U - T\mathrm{d}S - S\mathrm{d}T + p\mathrm{d}V + V\mathrm{d}p \\ &= T\mathrm{d}S - p\mathrm{d}V - T\mathrm{d}S - S\mathrm{d}T + p\mathrm{d}V + V\mathrm{d}p \\ &= -S\mathrm{d}T + V\mathrm{d}p. \end{aligned}$$

Note that the relations

$$\begin{aligned} \mathrm{d}U &= T\mathrm{d}S - p\mathrm{d}V \\ \mathrm{d}F &= -S\mathrm{d}T - p\mathrm{d}V \\ \mathrm{d}H &= T\mathrm{d}S + V\mathrm{d}p \\ \mathrm{d}G &= -S\mathrm{d}T + V\mathrm{d}p \end{aligned} \tag{10.1}$$

are of the form

$$\begin{aligned} \mathrm{d}U &= \left(\frac{\partial U}{\partial S}\right)_V \mathrm{d}S + \left(\frac{\partial U}{\partial V}\right)_S \mathrm{d}V \\ \mathrm{d}F &= \left(\frac{\partial F}{\partial T}\right)_V \mathrm{d}T + \left(\frac{\partial F}{\partial V}\right)_T \mathrm{d}V \\ \mathrm{d}H &= \left(\frac{\partial H}{\partial S}\right)_p \mathrm{d}S + \left(\frac{\partial H}{\partial p}\right)_S \mathrm{d}p \\ \mathrm{d}G &= \left(\frac{\partial G}{\partial T}\right)_p \mathrm{d}T + \left(\frac{\partial G}{\partial p}\right)_T \mathrm{d}p \end{aligned} \tag{10.2}$$

Comparison of the first equation (10.1) with the first equation (10.2) shows that

$$\left(\frac{\partial U}{\partial S}\right)_V = T \quad \text{and} \quad \left(\frac{\partial U}{\partial V}\right)_S = -p, \tag{10.3}$$

that is, knowledge of U as a function of S and V will enable us to find T and p as functions of S and V via (10.3), that is, to find

$$T = T(S, V) \quad \text{and} \quad p = p(S, V).$$

These are two equations involving four variables, so that from them we can eliminate any one variable and find the equation relating any chosen remaining variable to the other two.

Thus, knowledge of $U(S, V)$) for a given system (that is, knowledge of U as a function of S and V) would enable us to find all the relations connecting the state variables for that system.

We say that the relation

$$U = U(S, V)$$

is a **fundamental relation** (the U relation) for the system.

If, instead, we were given U as a function of any other pair of variables, for example T and V, the corresponding partial derivatives of U do not equate to state variables as in the above comparison of (10.1) and (10.2), and so it is not possible to deduce all thermodynamic relations from the relation $U = U(T, V)$.

We say that S and V are the **natural variables** of the thermodynamic potential U.

(Recall that the thermodynamic potentials are considered as functions of just two state variables, since with N fixed, as here, a one-component system has only two thermodynamic degrees of freedom, that is, fixing any two state variables determines the thermodynamic equilibrium state and hence the corresponding thermodynamic potential.)

Similarly, comparing (10.1) and (10.2) for dF, for dH, and for dG, we see that:

- the natural variables for F are T and V,
- the natural variables for H are S and p
- the natural variables for G are T and p,

so that the F, H and G fundamental relations are of the forms

$$F = F(T, V), \quad H = H(S, p), \quad G = G(T, p).$$

From these, by taking partial derivatives and comparing (10.1) and (10.2), we can find:

- S and p as functions of T and V
- T and V as functions of S and p
- S and V as functions of T and p.

10.2 The Maxwell relations

The result of successive partial differentiation of a function $f(x, y)$ with respect to x and y is independent of the order in which the derivatives are taken:

$$\left(\frac{\partial}{\partial y}\left(\frac{\partial f}{\partial x}\right)_y\right)_x = \left(\frac{\partial}{\partial x}\left(\frac{\partial f}{\partial y}\right)_x\right)_y.$$

Applying this to the function $U(S, V)$, we have

$$\left(\frac{\partial}{\partial V}\left(\frac{\partial U}{\partial S}\right)_V\right)_S = \left(\frac{\partial}{\partial S}\left(\frac{\partial U}{\partial V}\right)_S\right)_V.$$

Thus, from (10.1) and (10.2) we have

$$\left(\frac{\partial T}{\partial V}\right)_S = -\left(\frac{\partial p}{\partial S}\right)_V.$$

This is "S, V" **Maxwell relation**, derived by inspection of the relation for dU in (10.1).

Those derived from inspection of the relations for dF, dH and dG in (10.1) are, clearly,

$$\left(\frac{\partial S}{\partial V}\right)_T = \left(\frac{\partial p}{\partial T}\right)_V, \quad \left(\frac{\partial T}{\partial p}\right)_S = \left(\frac{\partial V}{\partial S}\right)_p, \quad -\left(\frac{\partial S}{\partial p}\right)_T = \left(\frac{\partial V}{\partial T}\right)_p$$

The Maxwell relations are very useful in the calculation of thermodynamic derivatives.

10.3 Calculation of thermodynamic derivatives

In this section we derive some useful general rules for calculating thermodynamic derivatives.

Suppose we have a function $z(x, y)$. Then, since

$$\mathrm{d}z = \left(\frac{\partial z}{\partial x}\right)_y \mathrm{d}x + \left(\frac{\partial z}{\partial y}\right)_x \mathrm{d}y,$$

we have, for a change in which $\mathrm{d}z = 0$,

$$0 = \left(\frac{\partial z}{\partial x}\right)_y \mathrm{d}x + \left(\frac{\partial z}{\partial y}\right)_x \mathrm{d}y,$$

that is,

$$\left.\frac{\mathrm{d}y}{\mathrm{d}x}\right|_{\mathrm{d}z=0} = -\frac{(\partial z/\partial x)_y}{(\partial z/\partial y)_x}.$$

But the left-hand side can be identified with $(\partial y/\partial x)_z$, and so

$$\left(\frac{\partial y}{\partial x}\right)_z = -\frac{(\partial z/\partial x)_y}{(\partial z/\partial y)_x}. \qquad \text{(Rule 1)}$$

We also have

$$\left(\frac{\partial x}{\partial y}\right)_z = \frac{1}{(\partial y/\partial x)_z} \qquad \text{(Rule 2)}$$

and

$$\left(\frac{\partial y}{\partial x}\right)_z = \frac{(\partial y/\partial w)_z}{(\partial x/\partial w)_z} \qquad \text{(Rule 3)}$$

where w is some other variable, expressible as a function of x and y (and hence of x and z or of y and z).

[Rule 3 is just a version of the obvious rule that, when w is treated as a function of x and z, we have

$$\left(\frac{\partial y}{\partial w}\right)_z = \left(\frac{\partial y}{\partial x}\right)_z \left(\frac{\partial x}{\partial w}\right)_z .]$$

Our aim is to give a general procedure for calculating an arbitrary thermodynamic derivative

$$\left(\frac{\partial X}{\partial Y}\right)_Z$$

in terms of the state variables and the following experimentally determinable quantities:
the thermal expansivity

$$\alpha = \frac{1}{V}\left(\frac{\partial V}{\partial T}\right)_p,$$

the isothermal compressibility

$$\kappa_T = -\frac{1}{V}\left(\frac{\partial V}{\partial p}\right)_T,$$

the constant-volume heat capacity

$$C_V = \left(\frac{đQ}{\mathrm{d}T}\right)_V = T\left(\frac{\partial S}{\partial T}\right)_V,$$

and the constant-pressure heat capacity

$$C_p = \left(\frac{đQ}{\mathrm{d}T}\right)_p = T\left(\frac{\partial S}{\partial T}\right)_p.$$

Step 1 If in $(\partial Y/\partial X)_Z$ the quantity Z is one of the thermodynamic potentials, bring it into the numerator position by Rule 1 and then use the appropriate expression (10.1) for dZ.

Example:

$$\left(\frac{\partial p}{\partial T}\right)_F = -\frac{(\partial F/\partial T)_p}{(\partial F/\partial p)_T} = -\left[\frac{-S - p(\partial V/\partial T)_p}{-p(\partial V/\partial p)_T}\right] = \frac{S + pV\alpha}{pV\kappa_T}.$$

Step 2 Derivatives containing S are then calculated as follows:

(i) Re-express partial derivatives at constant S in terms of derivatives of S by Rule 1, for example,

$$\left(\frac{\partial T}{\partial V}\right)_S = -\frac{(\partial S/\partial V)_T}{(\partial S/\partial T)_V}$$

(ii) Partial derivatives of the form $(\partial S/\partial T)_X$ ($X = V$ or p) can then be expressed in terms of C_p or C_V.

(iii) Partial derivatives of the form $(\partial S/\partial X)_T$ with $X = p$ or V can be re-expressed using the appropriate Maxwell relation, for example,

$$\left(\frac{\partial S}{\partial V}\right)_T = \left(\frac{\partial p}{\partial T}\right)_V,$$

which is the "T, V" Maxwell relation, following from the expression for dF:

$$\mathrm{d}F = -S\mathrm{d}T - p\mathrm{d}V.$$

(iv) Partial derivatives of the form $(\partial S/\partial X)_Y$ where neither X nor Y is T are re-expressed as a ratio of temperature derivatives by Rule 3. For example,

$$\left(\frac{\partial S}{\partial p}\right)_V = \frac{(\partial S/\partial T)_V}{(\partial p/\partial T)_V} = \frac{C_V}{T\,(\partial p/\partial T)_V}.$$

Step 3 If your result contains $(\partial p/\partial T)_V$ (or its reciprocal $(\partial T/\partial p)_V$), use Rule 1:

$$\left(\frac{\partial p}{\partial T}\right)_V = -\frac{(\partial V/\partial T)_p}{(\partial V/\partial p)_T} = \frac{\alpha}{\kappa_T}.$$

Finally, check that the physical dimensions of your answer are correct!

11

Matter Transfer and Phase Diagrams

11.1 The chemical potential

So far we have considered one-component systems with fixed N. If we allow N to vary, the differential change of U when $S \to S + dS$, $V \to V + dV$, and $N \to N + dN$ is

$$dU = T dS - p dV + \mu dN, \tag{11.1}$$

which defines the chemical potential μ as

$$\mu = \left(\frac{\partial U}{\partial N} \right)_{S,V}.$$

Equation (11.1) can be written as

$$dS = \frac{dU}{T} + \frac{p}{T} dV - \frac{\mu}{T} dN. \tag{11.2}$$

Suppose we have two subsystems, 1 and 2, separated by a diathermal, movable, **permeable** wall.

If the two systems form a composite system in thermodynamic equilibrium, the entropy will be a maximum, and infinitesimal transfers of energy, volume and particles will give zero first-order differential change of

$$S_{\text{tot}} = S_1 + S_2,$$

that is,

$$
\begin{aligned}
dS_{\text{tot}} &= dS_1 + dS_2 \\
&= \frac{1}{T_1} dU_1 + \frac{p_1}{T_1} dV_1 - \frac{\mu_1}{T_1} dN_1 + \frac{1}{T_2} dU_2 + \frac{p_2}{T_2} dV_2 - \frac{\mu_2}{T_2} dN_2 \\
&= \left(\frac{1}{T_1} - \frac{1}{T_2} \right) dU_1 + \left(\frac{p_1}{T_1} - \frac{p_2}{T_2} \right) dV_1 - \left(\frac{\mu_1}{T_1} - \frac{\mu_2}{T_2} \right) dN_1 \\
&= 0.
\end{aligned}
$$

A Course in Theoretical Physics, First Edition. P. J. Shepherd.
© 2013 John Wiley & Sons, Ltd. Published 2013 by John Wiley & Sons, Ltd.

Here, we have used equation (11.2) and the facts that $\mathrm{d}U_2 = -\mathrm{d}U_1$, $\mathrm{d}V_2 = -\mathrm{d}V_1$ and $\mathrm{d}N_2 = -\mathrm{d}N_1$.

Since this result must be true for arbitrary infinitesimal transfers, we must have

$$T_1 = T_2 \quad p_1 = p_2, \quad \mu_1 = \mu_2$$

at equilibrium.

11.2 Direction of matter flow

Imagine now that the two systems 1 and 2 are in thermal and mechanical equilibrium (so that $T_1 = T_2$ and $p_1 = p_2$) but, because of an impermeable wall between them, are not in material equilibrium (i.e., $\mu_1 \neq \mu_2$).

Briefly remove the impermeability constraint, and then replace it. That is, punch holes in the wall and then quickly block them!

An irreversible infinitesimal transfer of particles $\mathrm{d}N_1(= -\mathrm{d}N_2)$ occurs.

The corresponding change in the total entropy must be positive:

$$\begin{aligned} \mathrm{d}S_{\mathrm{tot}} &= \mathrm{d}S_1 + \mathrm{d}S_2 \\ &= -\left(\frac{\mu_1}{T_1} - \frac{\mu_2}{T_2}\right)\mathrm{d}N_1 = -\left(\frac{\mu_1}{T_1} - \frac{\mu_2}{T_1}\right)\mathrm{d}N_1 \\ &> 0, \end{aligned}$$

so that, for $\mathrm{d}N_1$ to be positive, we must have $\mu_2 > \mu_1$. Thus, **matter flows from regions of high chemical potential to regions of low chemical potential**.

11.3 Isotherms and phase diagrams

In the case of an ideal gas we found the equation of state

$$p = \frac{Nk_{\mathrm{B}}T}{V}.$$

Thus, the p,V isotherms for various temperatures are curves of the form shown in figure 11.1, in which it can be seen that, on each isotherm, for every value of p there is just one value of V.

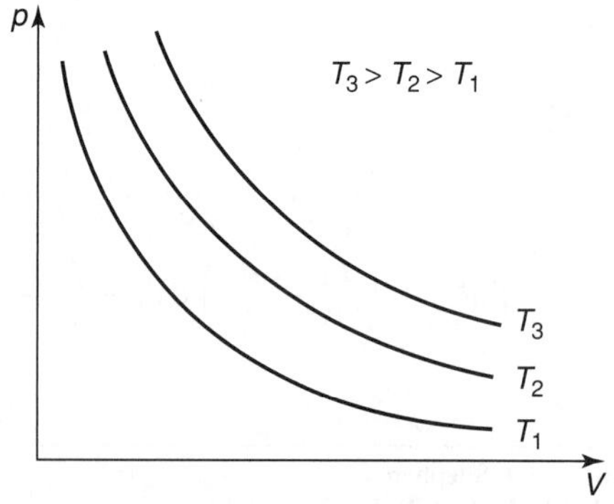

Figure 11.1 *p, V isotherms.*

For a system of interacting particles the equation of state is more complicated (see figure 11.2). On each isotherm above a certain temperature T_c one value of the volume V corresponds to each value of the pressure p. On each isotherm below T_c there is a particular value of p for which there are two values of V, corresponding to the coexistence of two phases. The exception to this is that at one particular temperature T_{tr} one pressure value p_{tr} corresponds to three distinct values of V, corresponding to the coexistence of three phases (the triple point).

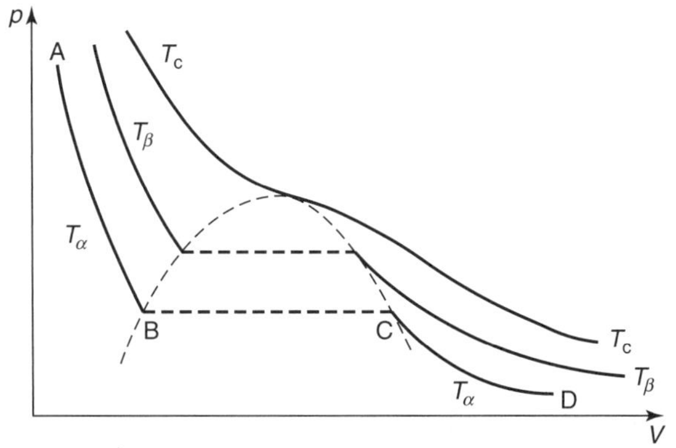

Figure 11.2 *Set of isotherms for a system of interacting particles.*

For example, on the isotherm marked T_α there are two values of V corresponding to the same pressure p. A system with all N particles in the thermodynamic equilibrium state marked B clearly has a higher density than a system with all N particles in the thermodynamic equilibrium state marked C.

A system in state B is said to be in the **liquid phase**, and a system in state C is said to be in the **vapour phase**.

When the two phases coexist in equilibrium there is a clear distinction between them, arising from their different volumes per particle (and hence different densities). The two phases also have different entropies per particle.

If we compress the gas from the point D along the isotherm T_α, the pressure increases until we reach the point C, after which compression causes no further pressure increase. Instead, droplets of the liquid phase (corresponding to point B) form. Further compression converts all the vapour to liquid in the state B, and then causes the pressure to increase along BA.

If, instead, we had followed a higher-temperature isotherm, for example, the T_β one, the density difference of the liquid and vapour phases is clearly smaller.

At temperatures at and above the critical temperature T_c the density difference (and entropy difference) between the two phases vanishes, and consequently so does the distinction between the phases. There can be no coexistence of different phases above T_c.

Each pair of points like B and C corresponds to states with the same p and T.

Thus, each such pair of coexisting phases corresponds to a single point in the p, T plane, and all the pairs (that is, with differing values of the density difference) form a line in the p, T plane.

Similarly, the points corresponding to coexisting liquid and solid phases lie on another line in the p, T plane, and so, too, do the points corresponding to coexisting solid and vapour phases.

The plot of these **coexistence lines** in the p, T plane is called the **phase diagram**.

A typical phase diagram is shown in figure 11.3.

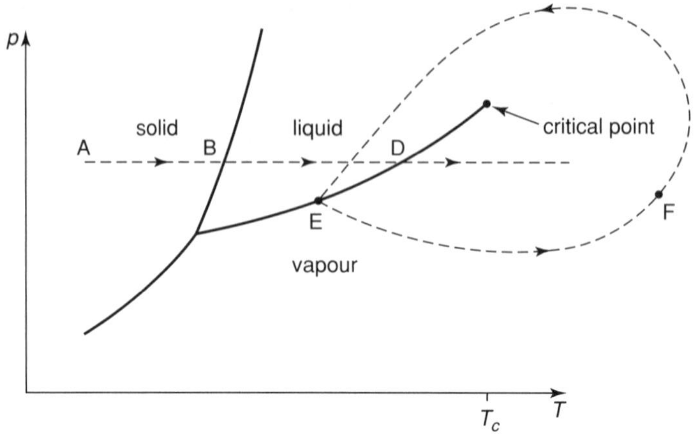

Figure 11.3 Typical phase diagram.

If, at some point in the p, T plane, solid and liquid can coexist in equilibrium **and** liquid and vapour can coexist in equilibrium, the zeroth law of thermodynamics tells us that at this point the solid and vapour can also coexist in equilibrium. Thus, **the three coexistence lines intersect at one point** (the **triple point**).

Imagine that, starting at state A, the system (in the solid phase) is heated at constant pressure. When the state B is reached, further heating does not increase the temperature but converts solid to liquid (a phase with a different density and a higher entropy per particle).

When all the solid has been converted to liquid by the heating, there will have been an entropy change $S_{\text{liquid}} - S_{\text{solid}}$ at the constant temperature T_{B} (the melting temperature at the pressure used). We must have

$$S_{\text{liquid}} - S_{\text{solid}} = \frac{L_{\text{f}}}{T_{\text{B}}},$$

where L_{f} is the heat supplied to melt the solid completely, called **latent heat of fusion**. [Dividing L_{f} by the number of moles (n_{m}) of the substance, we get l_{f}, the **molar latent heat of fusion**:

$$l_{\text{f}} = \frac{L_{\text{f}}}{n_{\text{m}}} = \frac{1}{n_{\text{m}}}(S_{\text{liquid}} - S_{\text{solid}})T_{\text{B}}.$$

It is clear from this that l_{f} varies along the liquid–solid coexistence line.]

Further heating takes the system through a single-phase region along the path BD. At D vapour begins to form, and the system remains at point D on the liquid–vapour coexistence line until all the liquid has been converted to vapour.

When all the liquid has been converted to vapour by the heating, there will have been an entropy change $S_{\text{vapour}} - S_{\text{liquid}}$ at the constant temperature T_{D} (the boiling temperature at the pressure used). We must have

$$S_{\text{vapour}} - S_{\text{liquid}} = \frac{L_{\text{v}}}{T_{\text{D}}},$$

where L_v is the heat supplied to vaporize the liquid completely, called the **latent heat of vaporization**. [Dividing L_v by n_m (the number of moles of the substance), we get l_v, the **molar latent heat of vaporization**:

$$l_v = \frac{L_v}{n_m} = \frac{1}{n_m}(S_{\text{vapour}} - S_{\text{liquid}})T_D.$$

It is clear from this that l_v varies along the vapour–liquid coexistence line.]

Suppose now that our system is entirely in the vapour phase, in the **vapour** thermodynamic equilibrium state specified by point E on the phase diagram.

By withdrawing heat from the vapour (or compressing it) we can convert it (via a two-phase situation) entirely into the **liquid** thermodynamic equilibrium state specified by the same point E on the phase diagram.

In the process we can see the **boundary** between the two phases.

Alternatively, we could take the system from the vapour phase (no liquid) at E to the liquid phase (no vapour) at E by following a path such as EFE. In this case, **no phase boundary ever appears**! There is no point at which we can say that "the vapour becomes liquid"!

Thus, **there is no distinction in principle between the liquid and vapour phases**. (We can use the term "fluid phase" for both.)

We can distinguish them only when they coexist, and then only by a **quantitative** measure such as their density difference.

The solid–liquid coexistence line does not terminate at a critical point, and so there **is** a distinction in principle between solid and liquid phases (they have **different symmetry**).

11.4 The Euler relation

The internal energy can be expressed in terms of its natural variables as

$$U = U(S, V, N).$$

Consider now a scaled system, λ times as big (or a composite system consisting of λ replicas of the original system).

The values of S, V and N are all multiplied by λ when we go to the scaled system. (Such quantities are said to be **extensive**.) Similarly, since U is also extensive, U becomes λU. Thus, we must have

$$\lambda U = U(\lambda S, \lambda V, \lambda N).$$

Taking the ordinary derivative of both sides with respect to λ, we have

$$\begin{aligned} U &= \left(\frac{\partial U}{\partial(\lambda S)}\right)_{\lambda V,\lambda N}\frac{d(\lambda S)}{d\lambda} + \left(\frac{\partial U}{\partial(\lambda V)}\right)_{\lambda S,\lambda N}\frac{d(\lambda V)}{d\lambda} + \left(\frac{\partial U}{\partial(\lambda N)}\right)_{\lambda S,\lambda V}\frac{d(\lambda N)}{d\lambda} \\ &= \left(\frac{\partial U}{\partial(\lambda S)}\right)_{\lambda V,\lambda N} S + \left(\frac{\partial U}{\partial(\lambda V)}\right)_{\lambda S,\lambda N} V + \left(\frac{\partial U}{\partial(\lambda N)}\right)_{\lambda S,\lambda V} N. \end{aligned}$$

Since this is true for arbitrary λ, it must be true, in particular, for $\lambda = 1$. Hence,

$$U = \left(\frac{\partial U}{\partial S}\right)_{V,N} S + \left(\frac{\partial U}{\partial V}\right)_{S,N} V + \left(\frac{\partial U}{\partial N}\right)_{S,V} N.$$

Therefore,

$$U = TS - pV + \mu N,$$

which is called the **Euler relation** (because it is an example of the general relation of Euler for **homogeneous** forms).

(The proof would not have worked if we had fixed (and ignored) N, as earlier in the this book, since we could not scale the system without changing N. So, even if we have fixed and ignored N, it is **not true** that $U = TS - pV$.)

11.5 The Gibbs–Duhem relation

Take an infinitesimal variation of the Euler relation:

$$\mathrm{d}U = T\mathrm{d}S + S\mathrm{d}T - p\mathrm{d}V - V\mathrm{d}p + \mu\mathrm{d}N + N\mathrm{d}\mu,$$

and subtract from it the thermodynamic identity with the same variations of S, V and N:

$$\mathrm{d}U = T\mathrm{d}S - p\mathrm{d}V + \mu\mathrm{d}N.$$

The result is

$$0 = S\mathrm{d}T - V\mathrm{d}p + N\mathrm{d}\mu,$$

which relates the variations of the three **intensive** (that is, invariant under the scaling) parameters in any infinitesimal process between thermodynamic equilibrium states.

For example, if the temperature changes by $\mathrm{d}T$ and the pressure by $\mathrm{d}p$ the corresponding infinitesimal change in μ is given by the **Gibbs–Duhem relation**

$$\begin{aligned}\mathrm{d}\mu &= -\frac{S}{N}\mathrm{d}T + \frac{V}{N}\mathrm{d}p \\ &= -s\mathrm{d}T + v\mathrm{d}p,\end{aligned}$$

where s is the entropy per particle and v is the volume per particle.

11.6 Slopes of coexistence lines in phase diagrams

Coexisting phases are in material (as well as thermal and mechanical) equilibrium, and so must have equal values of μ.

Let us start at a point on the liquid–vapour coexistence line in the phase diagram, and move an infinitesimal distance along it in the p, T plane, so that the temperature and pressure changes are $\mathrm{d}T$ and $\mathrm{d}p$.

To maintain $\mu_{\text{liq}} = \mu_{\text{vap}}$, the corresponding **changes** in μ_{liq} and μ_{vap} must also be equal.

But the Gibbs–Duhem relations for the liquid and vapour are

$$\begin{aligned}\mathrm{d}\mu_{\text{liq}} &= -s_{\text{liq}}\mathrm{d}T + v_{\text{liq}}\mathrm{d}p, \\ \mathrm{d}\mu_{\text{vap}} &= -s_{\text{vap}}\mathrm{d}T + v_{\text{vap}}\mathrm{d}p.\end{aligned}$$

Equating these variations, we have

$$-s_{\text{liq}}\mathrm{d}T + v_{\text{liq}}\mathrm{d}p = -s_{\text{vap}}\mathrm{d}T + v_{\text{vap}}\mathrm{d}p,$$

that is,

$$(s_{\text{vap}} - s_{\text{liq}})\mathrm{d}T = (v_{\text{vap}} - v_{\text{liq}})\mathrm{d}p$$

Thus, the slope of the coexistence line is

$$\frac{\mathrm{d}p}{\mathrm{d}T} = \frac{s_{\text{vap}} - s_{\text{liq}}}{v_{\text{vap}} - v_{\text{liq}}} = \frac{S_{\text{vap}} - S_{\text{liq}}}{V_{\text{vap}} - V_{\text{liq}}},$$

where the quantities S and V here are the total entropies and volumes when all N particles are in the indicated phases.

This gives us the **Clapeyron–Clausius relation**

$$\frac{\mathrm{d}p}{\mathrm{d}T} = \frac{L_{\text{v}}}{T(V_{\text{vap}} - V_{\text{liq}})} = \frac{L_{\text{v}}}{T\Delta V},$$

where L_{v} and ΔV are the total latent heat of vaporization and volume difference for a given amount of the substance at the point where the slope is calculated, and T is the temperature at this point.

12

Fermi–Dirac and Bose–Einstein Statistics

12.1 The Gibbs grand canonical probability distribution

In section 6.4 we derived the probability distribution for the microstates i of a system A of constant volume V and fixed particle number N in thermal contact with a large heat reservoir R of temperature T, with A and R together forming a closed system. (The set of possible states of A in this case is called **the canonical ensemble**.) The result was the **Boltzmann canonical probability distribution**

$$p_i = \frac{1}{Z} e^{-E_i/k_B T}, \tag{6.9}$$

with the **partition function** Z given by

$$Z = \sum_i e^{-E_i/k_B T}, \tag{6.10}$$

and was derived by applying the principle of equal equilibrium probability to all the states of the reservoir R when the system A was constrained to be in the microstate i.

We now allow the wall between the system A and the reservoir R to be permeable to particles, so that the number N of particles in system A is no longer fixed, and ask "what is the probability that system A is in the N-particle microstate $i(N)$ [with corresponding energy $E_{i(N)}$]?". Because we are now considering the states for all possible values of N, the number of states under consideration is vastly greater than the number in the canonical ensemble (which has fixed N), and the corresponding statistical ensemble of states of system A is called the **grand canonical ensemble**.

The probability $p_{N,i(N)}$ that system A has N particles and is in the N-particle microstate $i(N)$ is then, in analogy with the derivation of (6.8), proportional to the number of accessible states of the reservoir R when this is the case:

$$p_{N,i(N)} \propto g_R(E_R, N_R), \tag{12.1}$$

A Course in Theoretical Physics, First Edition. P. J. Shepherd.

which differs from (6.8) only in the fact that we must now take into account the dependence of the density of states $g_{\rm R}$ of the reservoir on the number $N_{\rm R}$ of particles in it. We now use the fact that the entropy $S_{\rm R}$ of the reservoir is given by $S_{\rm R} = k_{\rm B} \ln g_{\rm R}$, and use the definitions of temperature and chemical potential [with the latter definition in the form that follows from (12.1)]:

$$\frac{1}{k_{\rm B}T_{\rm R}} = \left(\frac{\partial \ln g_{\rm R}(E_{\rm R}, N_{\rm R})}{\partial E_{\rm R}}\right)_{N_{\rm R}, V_{\rm R}}, \qquad -\frac{\mu_{\rm R}}{k_{\rm B}T_{\rm R}} = \left(\frac{\partial \ln g_{\rm R}(E_{\rm R}, N_{\rm R})}{\partial N_{\rm R}}\right)_{E_{\rm R}, V_{\rm R}}.$$

Because the reservoir is very large, $T_{\rm R}$ and $\mu_{\rm R}$ in these expressions are essentially constants, equal to the temperature T and chemical potential μ of our system A because of the assumed thermal and material equilibrium between this system and the reservoir. From this we deduce that (12.1) becomes

$$p_{N,i(N)} \propto g_{\rm R}(E_{\rm R}, N_{\rm R}) \propto {\rm e}^{(E_{\rm R}-\mu N_{\rm R})/k_B T}. \tag{12.2}$$

But when our system A has N particles and is in microstate $i(N)$, with energy $E_{i(N)}$, we have

$$E_{\rm R} = E_{\rm tot} - E_{i(N)} \quad \text{and} \quad N_{\rm R} = N_{\rm tot} - N,$$

where $E_{\rm tot}$ and $N_{\rm tot}$ are the fixed total energy and total number of particles in the closed combined system "A plus R". Therefore, (12.2) becomes

$$p_{N,i(N)} \propto {\rm e}^{-(E_{i(N)}-\mu N)/k_B T}.$$

To ensure that $\sum\limits_{N}\sum\limits_{i(N)} p_{N,i(N)} = 1$, we must have

$$p_{N,i(N)} = \frac{1}{\mathcal{Z}}{\rm e}^{-(E_{i(N)}-\mu N)/k_B T}, \tag{12.3}$$

where

$$\mathcal{Z} = \sum_{N}\sum_{i(N)} {\rm e}^{-(E_{i(N)}-\mu N)/k_B T} \tag{12.4}$$

is the **grand partition function**, which is clearly a function of T, V (through the energy eigenvalues $E_{i(N)}$) and μ.

In analogy with the definition of the general equilibrium entropy S for the canonical ensemble [see (7.1)], the general definition of the equilibrium entropy S of system A for the grand canonical ensemble is

$$S = -k_B \sum_{N}\sum_{i(N)} p_{N,i(N)} \ln p_{N,i(N)},$$

which, with (12.3), gives

$$S = -k_B \sum_{N}\sum_{i(N)} p_{N,i(N)}\left(-\ln \mathcal{Z} + \frac{E_{i(N)}}{k_B T} - \frac{\mu N}{k_B T}\right) = k_B \ln \mathcal{Z} - \frac{U}{T} + \frac{\mu N}{T}, \tag{12.5}$$

where, as before, U is the thermal average energy $\overline{E}$ of system A, and $N \equiv \overline{N}$ is the thermal average number of particles in system A. The expression (12.5) defines a new thermodynamic potential Ω (the **grand potential**), which, like $\mathcal{Z}$, is fully determined by T, V, and μ:

$$\Omega = \Omega(T, V, \mu) = -k_B T \ln \mathcal{Z}(T, V, \mu) = U - ST - \mu N = F - \mu N. \tag{12.6}$$

In terms of Ω, the microstate probability distribution (12.3) for the grand canonical ensemble can clearly be written as

$$p_{N,i(N)} = \mathrm{e}^{(\Omega + \mu N - E_{i(N)})/k_B T}. \tag{12.7}$$

12.2 Systems of noninteracting particles

As noted in chapter 8 on the ideal monatomic gas, when the particles in a gas are essentially noninteracting (a situation that arises, for example, when the gas is sufficiently rarefied), the many-particle states i can be constructed by assigning particles to the single-particle eigenstates, with single-particle energies ε_k, of the system, where k represents the set of quantum numbers that define the state of the particle. Let N_k represent the number of particles that (in a given microstate of the whole system) are in the k-th single-particle state (N_k is called the **occupation number** of this state), and let $\overline{N_k}$ denote the thermal average of N_k.

We now consider a single particle in the gas and (as we are entitled to do if it is essentially not interacting with the other particles) treat it as a system itself, in thermal equilibrium with the reservoir of temperature T. In the situation with fixed N, the probability that this particle is in the k-th single-particle quantum state is given by the Boltzmann probability distribution

$$p_k \propto \mathrm{e}^{-\varepsilon_k/k_\mathrm{B} T}. \tag{6.9}$$

If conditions are such that the thermal average occupation number $\overline{N_k} \ll 1$ for all single-particle states k (conditions that obtain in practice for all ordinary molecular or atomic gases), then $\overline{N_k}$ is proportional to p_k, and so

$$\overline{N_k} = a\mathrm{e}^{-\varepsilon_k/k_\mathrm{B} T}, \tag{12.8}$$

where the constant a is determined by the normalization condition

$$\sum_k \overline{N_k} = N$$

(N is the total number of particles in the gas).

We now derive an expression for the constant a in (12.8) in terms of the chemical potential μ of the gas. To do this, we shall return to the grand canonical ensemble, so that the probability that our system has N particles and is in state $i(N)$ is given by (12.7). This time, however, we shall take our system to be all particles that are in the single-particle state k. We can do this, since we are dealing with an ideal gas, and so there is no interaction between these particles and the others in the system [and no quantum-exchange effects with particles in other states either, since such effects (see the next section) come into play only between particles in the same single-particle state]. Then, for this system, we have $N = N_k$ and $E = N_k\varepsilon_k$, and (12.7) becomes

$$p_{N_k} = \mathrm{e}^{[\Omega_k + N_k(\mu-\varepsilon_k)]/k_BT}. \tag{12.9}$$

From this, the probability that there are no particles at all in the single-particle state k is

$$p_{0_k} = \mathrm{e}^{\Omega_k/k_BT}.$$

In the case of interest here, in which $\overline{N_k} \ll 1$ for all k, the probability p_{0_k} is clearly almost unity, and so (12.9) becomes

$$p_{N_k} = \mathrm{e}^{[N_k(\mu-\varepsilon_k)]/k_BT}. \tag{12.10}$$

As above, in the case when $\overline{N_k} << 1$, the thermal average occupation number $\overline{N_k}$ can be identified with the probability p_{1_k} that there is one particle in single-particle state k. Thus, from (12.10), we have

$$\overline{N_k} = p_{1_k} = \mathrm{e}^{(\mu-\varepsilon_k)/k_BT}. \tag{12.11}$$

This expresses the coefficient a in the Boltzmann single-particle state probability distribution (12.8) in terms of the chemical potential μ of the gas.

12.3 Indistinguishability of identical particles

Suppose we have a system of two identical noninteracting particles in a two-particle state ψ_{jk}, in which one particle is in single-particle state ϕ_j and the other is in single-particle state ϕ_k. Then, because of the indistinguishability of identical particles, the probability of finding one particle in infinitesimal volume $\mathrm{d}^3\boldsymbol{r}_1$ at $\boldsymbol{r}_1$ and the other in $\mathrm{d}^3\boldsymbol{r}_2$ at $\boldsymbol{r}_2$ remains the same under exchange of the particles, that is,

$$|\psi_{jk}(\boldsymbol{r}_1, \boldsymbol{r}_2)|^2\mathrm{d}^3\boldsymbol{r}_1\mathrm{d}^3\boldsymbol{r}_2 = |\psi_{jk}(\boldsymbol{r}_2, \boldsymbol{r}_1)|^2\mathrm{d}^3\boldsymbol{r}_1\mathrm{d}^3\boldsymbol{r}_2.$$

From this, it follows that the two-particle wave function must be either symmetric under exchange of particles (such particles are called **bosons**), or antisymmetric (such particles are called **fermions**):

$$\psi_{jk}(\boldsymbol{r}_2, \boldsymbol{r}_1) = \pm\psi_{jk}(\boldsymbol{r}_1, \boldsymbol{r}_2).$$

Thus, for two noninteracting identical bosons the wave function must be of the form

$$\psi_{jk}(\boldsymbol{r}_1, \boldsymbol{r}_2) = \frac{1}{\sqrt{2}}\left(\phi_j(\boldsymbol{r}_1)\phi_k(\boldsymbol{r}_2) + \phi_k(\boldsymbol{r}_1)\phi_j(\boldsymbol{r}_2)\right). \tag{12.12}$$

For two noninteracting identical fermions the wave function must be of the form

$$\psi_{jk}(\boldsymbol{r}_1, \boldsymbol{r}_2) = \frac{1}{\sqrt{2}}\left(\phi_j(\boldsymbol{r}_1)\phi_k(\boldsymbol{r}_2) - \phi_k(\boldsymbol{r}_1)\phi_j(\boldsymbol{r}_2)\right). \tag{12.13}$$

Note that if we set $k = j$, that is, assign the two fermions to the same single-particle state, the two-particle wave function $\psi_{jk}(\boldsymbol{r}_1, \boldsymbol{r}_2)$ becomes identically zero – it is not possible for two identical fermions to occupy the same single-particle state. This is the statement of the well-known **Pauli principle**.

12.4 The Fermi–Dirac and Bose–Einstein distributions

In deriving the expression (12.11) for the thermal average occupation number of single-particle state k we made the assumption that conditions are such that $\overline{N_k} << 1$ for all k. If the temperature is sufficiently low, this will no longer be the case for the lowest-energy single-particle states k. In this case we proceed as follows. First note that, from (12.6), we have

$$d\Omega = dF - d(\mu N) = -SdT - pdV + \mu dN - \mu dN - Nd\mu = -SdT - pdV - Nd\mu, \quad (12.14)$$

and so the thermal average number of particles in the system is given by

$$\overline{N} = N = -\left(\frac{\partial \Omega}{\partial \mu}\right)_{T,V}. \quad (12.15)$$

As in the derivation of the Boltzmann distribution, we take our system to consist of all particles that are in the k-th single-particle state. For this system, (12.15) reads:

$$\overline{N_k} = -\left(\frac{\partial \Omega_k}{\partial \mu}\right)_{T,V}, \quad (12.16)$$

where the grand potential for this system is written as [see (12.6)]

$$\Omega_k = \Omega_k(T, V, \mu) \equiv -k_B T \ln \mathcal{Z}_k(T, V, \mu), \quad (12.17)$$

in which $\mathcal{Z}_k$ is the appropriate grand partition function

$$\mathcal{Z}_k = \sum_{N_k} \left(e^{(\mu-\varepsilon_k)/k_B T}\right)^{N_k}. \quad (12.18)$$

Firstly we consider fermions, for which N_k can take only the values 0 and 1. In this case, (12.17) becomes

$$\Omega_k = -k_B T \ln\left(1 + e^{(\mu-\varepsilon_k)/k_B T}\right) \quad (12.19)$$

and (12.16) gives

$$\overline{N_k} = \frac{e^{(\mu-\varepsilon_k)/k_B T}}{1 + e^{(\mu-\varepsilon_k)/k_B T}} = \frac{1}{e^{(\varepsilon_k-\mu)/k_B T} + 1}, \quad (12.20)$$

which is the **Fermi–Dirac distribution function** for $\overline{N_k}$.

It can be seen from (12.20) that, for fermions at all temperatures, $\overline{N_k} = 1/2$ for single-particle states k with energies ε_k equal to the chemical potential μ for the given temperature. (Since the single-particle energies $\varepsilon_k = \hbar^2 k^2/2m$ for noninteracting particles of momentum $\hbar k$ are all positive, this tells us the chemical potential of a system of free fermions at $T = 0$ is positive.) In the limit of zero temperature $T = 0$, (12.19) shows that for states with $\varepsilon_k < \varepsilon_F = \mu(T = 0)$ we have $\overline{N_k} = 1$, while for states with $\varepsilon_k > \varepsilon_F = \mu(T = 0)$ we have $\overline{N_k} = 0$. The energy ε_F thus defined is called the **Fermi energy**. As we increase the temperature T, the occupation of single-particle states with energies greater than ε_F becomes greater than zero and that of single-particle states with energies less than ε_F becomes less than one. What was a step distribution at $T = 0$ becomes "flattened out" as T increases, and the chemical potential μ (which is the energy at which $\overline{N_k} = 1/2$) becomes less than the value (ε_F) that it takes at $T = 0$. At a certain temperature T_F (the **Fermi temperature**) the chemical potential μ becomes zero, and at temperatures $T > T_F$ the chemical potential μ becomes negative. In this high-temperature limit, we have $e^{(\mu-\varepsilon_k)/k_B T} << 1$, and the Fermi–Dirac distribution (12.20) becomes

$$\overline{N_k} = e^{(\mu-\varepsilon_k)/k_B T}, \tag{12.21}$$

which decreases exponentially with the single-particle energy ε_k (recall that $\varepsilon_k \geq 0$ for all k, since the particles we are considering are free and noninteracting), and takes the value $e^{\mu/k_B T} << 1$ at $\varepsilon_k = 0$ (recall that, at the temperatures considered, μ is negative and greater in modulus than $k_B T$). Thus, in this limit, the Fermi–Dirac distribution becomes identical with the Boltzmann distribution (12.11).

For **bosons** there is no quantum-mechanical restriction on the number of particles occupying the k-th single-particle state, and so the grand partition function takes the form

$$\mathcal{Z}_k = \sum_{N_k=0}^{\infty} \left(e^{(\mu-\varepsilon_k)/k_B T}\right)^{N_k}. \tag{12.22}$$

Provided that $x \equiv e^{(\mu-\varepsilon_k)/k_B T} < 1$ (which is ensured for all single-particle states k if, as is the case, μ is always negative for bosons), we can use the standard geometric sum

$$\sum_{n=0}^{\infty} x^n = \frac{1}{1-x}$$

to find, using (12.16) and (12.17),

$$\begin{aligned}
\overline{N_k} &= -\left(\frac{\partial \Omega_k}{\partial \mu}\right)_{T,V} = k_B T \left(\frac{\partial \ln \mathcal{Z}_k}{\partial \mu}\right)_{T,V} \\
&= k_B T \left(\frac{\partial}{\partial \mu} \ln \frac{1}{1 - e^{(\mu-\varepsilon_k)/k_B T}}\right)_{T,V} = -k_B T \left(\frac{\partial \ln\left(1 - e^{(\mu-\varepsilon_k)/k_B T}\right)}{\partial \mu}\right)_{T,V} \\
&= \frac{e^{(\mu-\varepsilon_k)/k_B T}}{1 - e^{(\mu-\varepsilon_k)/k_B T}} = \frac{1}{e^{(\varepsilon_k-\mu)/k_B T} - 1}.
\end{aligned} \tag{12.23}$$

This is the **Bose–Einstein distribution function**. In the high-temperature limit when $e^{(\mu-\varepsilon_k)/k_B T} << 1$ (recall that $\varepsilon_k \geq 0$ for all k, and that, for bosons, $\mu < 0$ always), it is clear that the Bose–Einstein distribution function (12.23), like the Fermi–Dirac distribution (12.20), tends to the Boltzmann distribution (12.21). The latter may be regarded as the classical limit of the Fermi–Dirac and Bose–Einstein distributions.

12.5 The entropies of noninteracting fermions and bosons

Using (12.14) we can write the equilibrium entropy S of a system as

$$S = -\left(\frac{\partial \Omega}{\partial T}\right)_{V,\mu}.$$

In the case of noninteracting particles the contribution S_k of the single-particle state k to this is

$$S_k = -\left(\frac{\partial \Omega_k}{\partial T}\right)_{V,\mu},$$

with Ω_k given by (12.17) and (12.18) for both fermions and bosons.

For **fermions** there are just two terms in $\mathcal{Z}_k$, and so Ω_k takes the form (12.19). Then

$$\begin{aligned}
S_k &= -\left(\frac{\partial}{\partial T}(-k_B T \ln\left(1 + e^{(\mu-\varepsilon_k)/k_B T}\right)\right)_{V,\mu} \\
&= k_B \ln\left(1 + e^{(\mu-\varepsilon_k)/k_B T}\right) + k_B T\left(\frac{e^{(\mu-\varepsilon_k)/k_B T}}{1 + e^{(\mu-\varepsilon_k)/k_B T}}\right)\left(-\frac{\mu-\varepsilon_k}{k_B T^2}\right) \\
&= k_B \ln\left(1 + e^{(\mu-\varepsilon_k)/k_B T}\right) + k_B \overline{N_k}\left(-\frac{\mu-\varepsilon_k}{k_B T}\right).
\end{aligned}$$

Here, we have used the Fermi–Dirac distribution function $\overline{N_k}$ (12.20). We use (12.20) again, in the forms

$$1 - \overline{N_k} = \frac{1}{1 + e^{(\mu-\varepsilon_k)/k_B T}} \quad \text{and} \quad \frac{\mu-\varepsilon_k}{k_B T} = \ln \overline{N_k} - \ln(1 - \overline{N_k}).$$

Then S_k becomes

$$\begin{aligned}
S_k &= -k_B \ln(1 - \overline{N_k}) - k_B \overline{N_k}\left(\ln \overline{N_k} - \ln(1 - \overline{N_k})\right) \\
&= -k_B\left(\overline{N_k} \ln \overline{N_k} + (1 - \overline{N_k}) \ln(1 - \overline{N_k})\right).
\end{aligned}$$

Thus, the entropy S of a system of noninteracting **fermions** is

$$S_k = -k_B \sum_k \left(\overline{N_k} \ln \overline{N_k} + (1 - \overline{N_k}) \ln(1 - \overline{N_k})\right). \tag{12.24}$$

We now find the entropy of a system of interacting **bosons**, using the same method used above for fermions, except that now the grand partition function is given by (12.22) and $\overline{N_k}$ is given by the Bose–Einstein distribution function (12.23). In this case the contribution S_k to the entropy S is given by

$$\begin{aligned}
S_k = -\left(\frac{\partial \Omega_k}{\partial T}\right)_{V,\mu} &= -k_B \ln\left(1 - e^{(\mu-\varepsilon_k)/k_B T}\right) - k_B T\left(\frac{\partial}{\partial T}\ln\left(1 - e^{(\mu-\varepsilon_k)/k_B T}\right)\right)_{V,\mu} \\
&= -k_B \ln\left(1 - e^{(\mu-\varepsilon_k)/k_B T}\right) - k_B T\frac{\left(-e^{(\mu-\varepsilon_k)/k_B T}\right)}{1 - e^{(\mu-\varepsilon_k)/k_B T}}\left(-\frac{\mu-\varepsilon_k}{k_B T^2}\right) \\
&= -k_B \ln\left(1 - e^{(\mu-\varepsilon_k)/k_B T}\right) + k_B \overline{N_k}\left(\frac{\varepsilon_k - \mu}{k_B T}\right).
\end{aligned}$$

But, from (12.23), we have

$$\frac{\varepsilon_k - \mu}{k_\mathrm{B} T} = \ln(1 + \overline{N_k}) - \ln \overline{N_k}$$

and

$$\ln\left(1 - \mathrm{e}^{(\mu - \varepsilon_k)/k_\mathrm{B} T}\right) = (\mu - \varepsilon_k)/k_\mathrm{B} T - \ln \overline{N_k} = -\ln(1 + \overline{N_k})$$

so that

$$\begin{aligned} S_k &= k_B \ln(1 + \overline{N_k}) + k_B \overline{N_k} \left(\ln(1 + \overline{N_k}) - \ln \overline{N_k}\right) \\ &= k_B \left((1 + \overline{N_k}) \ln(1 + \overline{N_k}) - \overline{N_k} \ln \overline{N_k}\right) \end{aligned}$$

Thus, the entropy of a system of noninteracting **bosons** is

$$S = k_B \sum_k \left((1 + \overline{N_k}) \ln(1 + \overline{N_k}) - \overline{N_k} \ln \overline{N_k}\right). \tag{12.25}$$

Note the difference between equations (12.24) and (12.25). Note also that the expression (12.24) has a formal resemblance to (7.1) for the general entropy S of a system. Indeed, it is often derived formally from (7.1) by treating the single-particle state k as a "system" with two states (unoccupied and singly occupied, for fermions – recall that the number of particles in a system is not fixed in the grand canonical ensemble), and identifying the thermal average occupation number $\overline{N_k}$ for fermions with the thermal-average probability p_k of occupation of that single-particle state k.

Module III

Many-Body Theory

13

Quantum Mechanics and Low-Temperature Thermodynamics of Many-Particle Systems

13.1 Introduction

By a "many-particle system" we shall mean a system of macroscopic size (for example, one for which the number of particles is of the order of 10^{23}).

By "low-temperature" we shall mean that the temperature T is such that (a) the system is only weakly excited relative to the ground state and (b) quantum statistics is necessary (Fermi–Dirac for fermions, Bose–Einstein for bosons), that is, the quantum-mechanical principle of indistinguishability of identical particles has macroscopic consequences. (An example of the latter is the dramatic difference in the properties of liquid helium-3 and liquid helium-4.)

The properties to be studied in this chapter include:

(a) the wave function and energy of the ground state of the system;
(b) the wave functions, energies and widths (inverse lifetimes) of weakly excited states;
(c) the thermodynamic state variables [found from the energies from (b) by the methods of statistical mechanics].

13.2 Systems of noninteracting particles

The hamiltonian $\hat{H}$ of a system of noninteracting particles can be written as:

$$\hat{H} = \sum_{n=1}^{N} \hat{H}_n, \tag{13.1}$$

where $\hat{H}_n$ is the hamiltonian of particle n, and $\hat{H}$ clearly contains no potential energies of interaction of the type $\hat{V}_{nm}$.

A Course in Theoretical Physics, First Edition. P. J. Shepherd.

The single-particle energy eigenstates (eigenstates of $\hat{H}_n$) may be written as

$$\psi_{\alpha_i}(\boldsymbol{r}_n, \tau_n) \equiv \psi_{\alpha_i}(\xi_n), \tag{13.2}$$

with corresponding energy eigenvalues ε_{α_i}:

$$\hat{H}_n \psi_{\alpha_i}(\boldsymbol{r}_n, \tau_n) = \varepsilon_{\alpha_i} \psi_{\alpha_i}(\boldsymbol{r}_n, \tau_n). \tag{13.3}$$

The symbol α denotes the set of quantum numbers corresponding to a complete set of single-particle observables. For example, for an electron moving in the Coulomb potential of a nucleus the set α could be the set $\{n, l, m, m_s\}$ (see section 1.17 on hydrogenic states), while for a free particle with spin the set α could be the set $\{k_x, k_y, k_z, m_s\}$. The different sets of possible numerical values of the quantum numbers in a given set α are labelled by the index i, which may be considered to take integer values from 1 to ∞, using any chosen ordering convention. For example, we could choose $i = 1$ to correspond to the set of quantum-number values for the ground state of the particle (the eigenstate ψ_{α_1}, with energy ε_{α_1}).

We shall work with single-particle energy eigenstates ψ_{α_i} that are orthonormal, that is, normalized and constructed to be mutually orthogonal:

$$\int \psi^*_{\alpha_i}(\xi_n)\psi_{\alpha_j}(\xi_n)\mathrm{d}\xi_n = \delta_{\alpha_i,\alpha_j}, \tag{13.4}$$

where the Kronecker delta $\delta_{\alpha_i,\alpha_j}$ is equal to 1 when the sets α_i and α_j are numerically identical, and equal to 0 otherwise. The integral in (13.4) is to be interpreted as an integral over the volume of the space occupied by particle n (that is, a "continuous sum" over the allowed position coordinates $\boldsymbol{r}_n$ of the particle), combined with a discrete sum over the possible "spin coordinates" τ_n of particle n. If, as is customary (see chapter 2), we use the **proper representation** to describe the spin part of the wave function, the "spin coordinates" here may be chosen to be the allowed values of the quantum number m_s of the z-component of spin.

13.2.1 Bose systems

The wave function of any system of bosons (whether the latter are interacting or not) must be symmetric under the interchange of the coordinates (both position and spin) of any two bosons. In particular, every N-particle energy eigenstate [eigenstate of the hamiltonian (13.1)] of a system of noninteracting bosons has the general form

$$\Phi_{N_1,N_2,\ldots,N_\infty}(\xi_1, \ldots, \xi_N) = C \sum_{\mathcal{P}} \mathcal{P}\, \{\psi_{\alpha_i}(\xi_1)\psi_{\alpha_j}(\xi_2) \ldots \psi_{\alpha_l}(\xi_N)\}, \tag{13.5}$$

where N_i is the number of particles that occupy (in the given N-particle state) the single-particle state specified by the quantum-number set α_i, and the sum over $\mathcal{P}$ is over all **distinct** permutations of the N particles amongst the occupied single-particle states $\psi_{\alpha_i}, \psi_{\alpha_j}, \ldots, \psi_{\alpha_l}$. The wave function (13.5) is clearly an eigenfunction of the hamiltonian operator $\hat{H}$ (13.1) with eigenvalue

$$E_{N_1,N_2,\ldots} = \sum_{i=1}^{\infty} N_i \varepsilon_{\alpha_i}. \tag{13.6}$$

[This is because every permutation term in the wave function (13.5) is clearly an eigenfunction of the operator $\hat{H}$ (13.1) with the same eigenvalue (13.6).] Thus, for noninteracting particles the single-particle energies are additive, each being multiplied by the corresponding occupation number as in equation (13.6).

As an example, a particular six-boson energy eigenstate is described by the wave function

$$\Phi_{3,0,1,1,0,0,1,\dots}(\xi_1,\dots,\xi_6) = C\sum_{\mathcal{P}} \mathcal{P}\,\{\psi_{\alpha_1}(\xi_1)\psi_{\alpha_1}(\xi_2)\psi_{\alpha_1}(\xi_3)\psi_{\alpha_3}(\xi_4)\psi_{\alpha_4}(\xi_5)\psi_{\alpha_7}(\xi_6)\} \tag{13.7}$$

and has energy $E_{3,0,1,1,0,0,1,\dots} = 3\varepsilon_{\alpha_1} + \varepsilon_{\alpha_3} + \varepsilon_{\alpha_4} + \varepsilon_{\alpha_7}$.

The number of terms in the sum (13.5) is $N!/N_1!N_2!\dots N_\infty!$ (recall that $0! = 1$). The condition that the N-boson function Φ (13.5) be normalized to unity is written as

$$\begin{aligned} 1 &= \int \Phi^*_{N_1,N_2,\dots,N_\infty}(\xi_1,\dots,\xi_N)\Phi_{N_1,N_2,\dots,N_\infty}(\xi_1,\dots,\xi_N)\mathrm{d}\xi_1\dots\mathrm{d}\xi_N \\ &= |C|^2\frac{N!}{N_1!N_2!\dots N_\infty!}, \end{aligned} \tag{13.8}$$

since all cross terms vanish by virtue of the orthonormality condition (13.4) (the cross terms involve distinct permutations) and the other $N!/N_1!N_2!\dots N_\infty!$ terms each give unity. Therefore, apart from an arbitrary phase factor $\mathrm{e}^{i\theta}$ that we choose to set equal to unity, we have:

$$C = \left(\frac{N_1!N_2!\dots N_\infty}{N!}\right)^{1/2} \tag{13.9}$$

The wave functions $\Phi_{N_1,N_2,\dots,N_\infty}(\xi_1,\dots,\xi_N)$ (13.5), with all possible integer (including zero) values of the occupation numbers $N_i (i = 1, 2, \dots, \infty)$ subject to the condition

$$\sum_{i=1}^{\infty} N_i = N, \tag{13.10}$$

describe all the possible energy eigenstates of a system of N noninteracting bosons.

For particles that are not only noninteracting but also free, the hamiltonian for particle n is

$$\hat{H}_n = -\frac{\hbar^2}{2m}\nabla_n^2 \equiv -\frac{\hbar^2}{2m}\left(\frac{\partial^2}{\partial x_n^2} + \frac{\partial^2}{\partial y_n^2} + \frac{\partial^2}{\partial z_n^2}\right), \tag{13.11}$$

where x_n, y_n, z_n are the components of the position vector $\boldsymbol{r}_n$ of particle n, and m is the particle mass. The corresponding single-particle energy eigenfunctions take the form of plane waves

$$\psi_{\alpha_i}(\boldsymbol{r}_n) = \frac{1}{\sqrt{V}}\mathrm{e}^{i\boldsymbol{k}_i\cdot\boldsymbol{r}_n} \tag{13.12}$$

(we have used box normalization, with a box of volume V), with corresponding energy eigenvalue

$$\varepsilon_{\alpha_i} = \frac{\hbar^2 k_i^2}{2m}. \tag{13.13}$$

In this case, the eigenvalue (13.6) of $\hat{H}$ (13.1) is of the form

$$E_{N_1,N_2,\ldots} = \sum_{i=1}^{\infty} N_i \frac{\hbar^2 k_i^2}{2m}. \tag{13.14}$$

The ground state is that for which all N particles are in the single-particle state with zero wave vector $\boldsymbol{k} = \boldsymbol{k}_1 = 0$ (we label this state $i = 1$, by convention), and its energy (the ground-state energy E_0) is

$$E_0 = E_{N,0,0,0,\ldots} = 0. \tag{13.15}$$

A "singly excited state" of the N-particle system is a state in which only one particle is not in the single-particle state $i = 1$ (with zero wave vector $\boldsymbol{k}$). For example, the singly excited state in which all the N_i are zero except for $N_1(= N - 1)$ and $N_4(= 1)$ clearly has energy

$$E_{N-1,0,0,1,0,\ldots} = \frac{\hbar^2 k_4^2}{2m}. \tag{13.16}$$

The corresponding **excitation energy**, defined as the difference between this energy and the ground-state energy (13.15), is

$$\varepsilon_{\mathrm{ex}}(\boldsymbol{k}_4) = E_{N-1,0,0,1,0,\ldots} - E_{N,0,0,0,\ldots} = \frac{\hbar^2 k_4^2}{2m}. \tag{13.17}$$

Since the states (13.5) are exact energy eigenstates (stationary states), the time-dependent part of the wave function is simply $\mathrm{e}^{-iEt/\hbar}$, with the energy E given by (13.14).

Our knowledge of all the energy eigenstates and of the corresponding energies of a system of N free bosons makes it possible to apply Bose–Einstein statistics (see chapter 12) to calculate all the thermodynamic properties (state variables) of the system.

13.2.2 Fermi systems

The wave function of a system of fermions (whether interacting or noninteracting) must be antisymmetric under the interchange of the coordinates (both position and spin) of any two fermions. For N noninteracting fermions any particular stationary state (energy eigenstate) is one in which a particular set of N single-particle eigenstates is occupied (with one fermion per state) and all other single-particle states are unoccupied. The antisymmetrization is ensured by the determinantal form

$$\begin{aligned}\Phi_{N_1,N_2,\ldots,N_\infty}(\xi_1,\ldots,\xi_N) &= D\sum_{\mathcal{P}}(-1)^{\mathcal{P}}\mathcal{P}\{\psi_{\alpha_i}(\xi_1)\psi_{\alpha_j}(\xi_2)\ldots\psi_{\alpha_l}(\xi_N)\}\\ &= D\begin{vmatrix}\psi_{\alpha_i}(\xi_1) & \psi_{\alpha_i}(\xi_2) & \cdot & \cdot & \cdot & \psi_{\alpha_i}(\xi_N)\\ \psi_{\alpha_j}(\xi_1) & \psi_{\alpha_j}(\xi_2) & \cdot & \cdot & \cdot & \psi_{\alpha_j}(\xi_N)\\ \cdot & \cdot & \cdot & \cdot & \cdot & \cdot\\ \cdot & \cdot & \cdot & \cdot & \cdot & \cdot\\ \cdot & \cdot & \cdot & \cdot & \cdot & \cdot\\ \psi_{\alpha_l}(\xi_1) & \psi_{\alpha_l}(\xi_2) & \cdot & \cdot & \cdot & \psi_{\alpha_l}(\xi_N)\end{vmatrix}.\end{aligned} \tag{13.18}$$

Here, the factor $(-1)^{\mathcal{P}}$ symbolizes +1 for even permutations $\mathcal{P}$ of $\xi_1, \ldots, \xi_N$ (that is, permutations obtainable from $\xi_1, \ldots, \xi_N$ by an even number of pairwise interchanges), and symbolizes –1 for odd permutations $\mathcal{P}$ of $\xi_1, \ldots, \xi_N$.

If any single-particle state were to be occupied by more than one particle, at least two rows of the above determinant would be identical and the wave function (13.18) would vanish. This is the mathematical statement of the **Pauli principle**, namely, that no two fermions can occupy the same single-particle state. In other words, in (13.18), each occupation number N_i can only take either the value 0 or the value 1. The number of terms in the above antisymmetrized sum (13.18) over permutations is thus $N!/N_1!N_2!\ldots N_\infty! = N!$, since every $N_i! = 1$. The condition that (13.18) be normalized to unity is then

$$|D|^2\, N! = 1,$$

so that, apart from an arbitrary phase factor $e^{i\theta}$ that we again choose to set equal to unity,

$$D = \frac{1}{\sqrt{N!}}. \tag{13.19}$$

The energy of an N-fermion system in the state $\Phi_{N_1,N_2,\ldots,N_\infty}(\xi_1, \ldots, \xi_N)$ is then

$$\begin{aligned}
E_{N_1,N_2,\ldots} &= \int \Phi^*_{N_1,N_2,\ldots,N_\infty}(\xi_1, \ldots, \xi_N)\left(\sum_{n=1}^{N} \hat{H}_n\right)\Phi_{N_1,N_2,\ldots,N_\infty}(\xi_1, \ldots, \xi_N)\mathrm{d}\xi_1\ldots\mathrm{d}\xi_N \\
&= \frac{1}{N!}\int\left(\sum_{\mathcal{P}}(-1)^{\mathcal{P}}\mathcal{P}\left[\psi_{\alpha_i}(\xi_1)\ldots\psi_{\alpha_l}(\xi_N)\right]\right)^*\left(\sum_{n=1}^{N}\hat{H}_n\right)\left(\sum_{\mathcal{P}'}(-1)^{\mathcal{P}'}\mathcal{P}'\left[\psi_{\alpha_i}(\xi_1)\ldots\psi_{\alpha_l}(\xi_N)\right]\right)\mathrm{d}\xi_1\ldots\mathrm{d}\xi_N \\
&= \left(\sum_{i=1}^{\infty}N_i\varepsilon_{\alpha_i}\right)\frac{1}{N!}\int\left(\sum_{\mathcal{P}}(-1)^{\mathcal{P}}\mathcal{P}\left[\psi_{\alpha_i}(\xi_1)\ldots\psi_{\alpha_l}(\xi_N)\right]\right)^*\left(\sum_{\mathcal{P}'}(-1)^{\mathcal{P}'}\mathcal{P}'\left[\psi_{\alpha_i}(\xi_1)\ldots\psi_{\alpha_l}(\xi_N)\right]\right)\mathrm{d}\xi_1\ldots\mathrm{d}\xi_N.
\end{aligned} \tag{13.20}$$

In (13.20) we have used the fact that every term in the sum over $\mathcal{P}'$ here is an eigenfunction of the operator $\hat{H} = \sum_{n=1}^{N}\hat{H}_n$ with the same eigenvalue $\sum_{i=1}^{\infty}N_i\varepsilon_{\alpha_i}$ [see (13.3)]. (Here, every N_i is equal to 0 or 1.) There are $N!$ such terms (permutations), and the integral over $\mathrm{d}\xi_1\ldots\mathrm{d}\xi_N$ of each one multiplied by the term with the identical permutation in $\Phi^*_{N_1,N_2,\ldots,N_\infty}(\xi_1, \ldots, \xi_N)$ gives unity (a product of N integrals, each of which is unity by virtue of the normalization (13.4) of the single-particle wave functions). The other integrals (those involving distinct permutations $\mathcal{P}' \neq \mathcal{P}$ in $\Phi_{N_1,N_2,\ldots,N_\infty}(\xi_1, \ldots, \xi_N)$ and $\Phi^*_{N_1,N_2,\ldots,N_\infty}(\xi_1, \ldots, \xi_N)$) vanish by virtue of the orthogonality (13.4) of the single-particle wave functions, and so the value of the integral in the last line of (13.20) is $N!$. Thus, (13.20) becomes

$$E_{N_1,N_2,\ldots} = \sum_{i=1}^{\infty}N_i\varepsilon_{\alpha_i}.$$

This has the same form as that found for bosons [see (13.6)]. However, whereas in the boson case every N_i can take any zero or integer value subject only to the restriction $\sum_{i=1}^{\infty} N_i = N$, for fermions every N_i can take only the value 0 or 1 (subject, of course, to the same restriction), so that

$$E_{N_1,N_2,\ldots} = \sum_{\text{occupied } i} \varepsilon_{\alpha_i} \text{ for fermions.} \tag{13.21}$$

If the fermions are not only noninteracting but also free, the appropriate single-particle energy eigenfunctions are the plane waves (13.12), labelled by the wave vectors $\boldsymbol{k}_i$, with corresponding energy eigenvalues (13.13). Because of the Pauli principle, no more than two fermions [for example, one with $m_s = +1/2$ ("spin up"), and one with $m_s = -1/2$ ("spin down")] can have the same value of $\boldsymbol{k}_i$.

13.2.2.1 The ground state of a system of free fermions

The ground state of a system of N free fermions may be viewed as that in which the fermions have been fed, two at a time, into the $N/2$ lowest-energy plane-wave states. Therefore, the ground-state energy is given by (13.21) in which the occupied states i are those with wave vectors $\boldsymbol{k}_i$ lying within a sphere (the **Fermi sphere**) in $\boldsymbol{k}$-space, the surface of which (the **Fermi surface**) corresponds to the wave vectors $\boldsymbol{k}$ of the occupied single-particle states of greatest energy. This single-particle energy (the **Fermi energy** ε_F) can be written as

$$\varepsilon_F = \frac{\hbar^2 k_F^2}{2m}, \tag{13.22}$$

where k_F (the **Fermi wave number**, corresponding to the **Fermi momentum** $p_F = \hbar k_F$) is the radius of the Fermi sphere. Thus, for the ground-state energy E_0 equation (13.21) gives

$$E_0 = 2 \sum_{\substack{\boldsymbol{k} \\ (|\boldsymbol{k}| \le k_F)}} \frac{\hbar^2 k^2}{2m}. \tag{13.23}$$

Suppose, for example, that the fermions are in a cubic box of side L, with rigid walls. Then the allowed values of the component k_x of $\boldsymbol{k}$ are

$$k_x = \frac{2\pi n_x}{L} \quad (n_x = 0, \pm 1, \pm 2, \ldots), \tag{13.24}$$

with analogous expressions for the allowed values of k_y and k_z, so that

$$\boldsymbol{k} = \frac{2\pi \boldsymbol{n}}{L}. \tag{13.25}$$

But any sum of terms $f(\boldsymbol{k})$ over a set of the allowed values of the vector $\boldsymbol{k}$ [such as the sum (13.23)] can be written as the corresponding sum over the corresponding allowed values of the dimensionless vector $\boldsymbol{n}$, with subsequent conversion of the latter to an integral, as follows:

$$\sum_{\boldsymbol{k}} f(\boldsymbol{k}) = \sum_{\boldsymbol{n}} f\left(\frac{2\pi \boldsymbol{n}}{L}\right).$$

When, as in the case of (13.23), the components of $\boldsymbol{n}$ take all integer values (including zero) within certain ranges and when it is also the case that the function f vanishes for $\boldsymbol{n} = 0$, the sum over the dimensionless vector $\boldsymbol{n}$ can be replaced by an integral over $\boldsymbol{n}$, as follows:

$$\sum_{\boldsymbol{k}} f(\boldsymbol{k}) = \sum_{\boldsymbol{n}} f\left(\frac{2\pi \boldsymbol{n}}{L}\right) \to \int \mathrm{d}^3\boldsymbol{n}\, f\left(\frac{2\pi \boldsymbol{n}}{L}\right) = \left(\frac{L}{2\pi}\right)^3 \int \mathrm{d}^3\boldsymbol{k}\, f(\boldsymbol{k}) = \frac{V}{(2\pi)^3} \int \mathrm{d}^3\boldsymbol{k}\, f(\boldsymbol{k}),$$

where $V = L^3$ is the volume of the system. The ground-state energy (13.23) can then be written as

$$\begin{aligned} E_0 &= 2\frac{V}{(2\pi)^3} \int\limits_{|\boldsymbol{k}|\le k_\mathrm{F}} \mathrm{d}^3\boldsymbol{k}\, \frac{\hbar^2 k^2}{2m} = 2\frac{V}{(2\pi)^3} \int_0^{2\pi} \mathrm{d}\varphi \int_0^{\pi} \sin\theta\, \mathrm{d}\theta \int_0^{k_\mathrm{F}} k^2 \mathrm{d}k\, \frac{\hbar^2 k^2}{2m} \\ &= \frac{8\pi V}{(2\pi)^3} \frac{\hbar^2}{2m} \frac{k_\mathrm{F}^5}{5} = \frac{\hbar^2 k_\mathrm{F}^5 V}{10\pi^2 m}. \end{aligned} \tag{13.26}$$

The Fermi wave number k_F can be found from the fact that in the ground state the N free fermions occupy the plane-wave states inside the Fermi sphere (two fermions to each state):

$$N = 2 \sum_{\substack{\boldsymbol{k} \\ (|\boldsymbol{k}|\le k_\mathrm{F})}} 1 = 2\frac{V}{(2\pi)^3} \int_0^{k_\mathrm{F}} \mathrm{d}^3\boldsymbol{k} = \frac{2V}{(2\pi)^3} \frac{4}{3}\pi k_\mathrm{F}^3 = \frac{V k_\mathrm{F}^3}{3\pi^2}, \tag{13.27}$$

so that

$$k_\mathrm{F} = \left(\frac{3\pi^2 N}{V}\right)^{1/3} = (3\pi^2 \rho)^{1/3}, \tag{13.28}$$

where $\rho = N/V$ is the particle-number density. From (13.26) and (13.27) we have

$$E_0 = \frac{3N}{5} \frac{\hbar^2 k_\mathrm{F}^2}{2m} = \frac{3}{5} N \varepsilon_\mathrm{F}, \tag{13.29}$$

that is, the average energy per particle in the ground state is three-fifths of the Fermi energy ε_F (13.22).

13.2.2.2 Singly excited states of a system of free fermions

Suppose we take N free fermions in the ground state and excite one of the particles by some means that results in transfer of momentum $\hbar\boldsymbol{k}$ to the particle. It is clear that only particles with a momentum $\hbar\boldsymbol{q}$ (inside the Fermi sphere) such that $\hbar\boldsymbol{q} + \hbar\boldsymbol{k}$ lies outside the Fermi sphere can be excited, since if $\hbar(\boldsymbol{q} + \boldsymbol{k})$ lies inside the sphere the Pauli principle forbids a scattering into that state as the latter is already occupied. For a given momentum transfer $\hbar\boldsymbol{k}$, for what initial momentum $\hbar\boldsymbol{q}$ is the excitation energy smallest? The general expression for the excitation energy is

$$\varepsilon_\mathrm{ex}(\boldsymbol{q} + \boldsymbol{k} \leftarrow \boldsymbol{q}) = \frac{\hbar^2(\boldsymbol{q} + \boldsymbol{k})^2}{2m} - \frac{\hbar^2 q^2}{2m} = \hbar^2 \left(\frac{\boldsymbol{k}\cdot\boldsymbol{q}}{m} + \frac{k^2}{2m}\right). \tag{13.30}$$

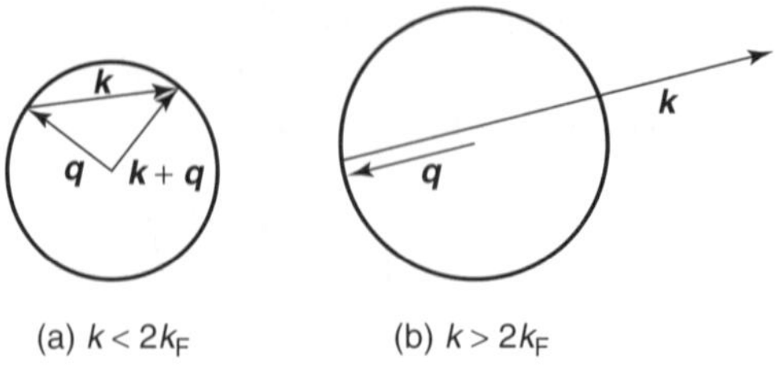

Figure 13.1 *Constructions to determine $\varepsilon_{\rm ex}^{\rm min}(\boldsymbol{k})$ for free fermions when (a) $k < 2k_{\rm F}$ and (b) $k \geq 2k_{\rm F}$.*

If $k < 2k_{\rm F}$ we can always find a $\boldsymbol{q}$ such that $\varepsilon_{\rm ex}(\boldsymbol{q}+\boldsymbol{k} \leftarrow \boldsymbol{q}) = 0$ (see figure 13.1a). In other words,

$$\varepsilon_{\rm ex}^{\rm min}(\boldsymbol{k}) = 0 \text{ when } k < 2k_{\rm F}. \tag{13.31}$$

If, on the other hand, $k > 2k_{\rm F}$, we get the smallest possible excitation energy when $\boldsymbol{q}$ is in the opposite direction to the given $\boldsymbol{k}$ and has magnitude $|\boldsymbol{q}| = k_{\rm F}$ (see figure 13.1b). In this case, (13.30) gives

$$\varepsilon_{\rm ex}^{\rm min}(\boldsymbol{k}) = \hbar^2\left(-\frac{kk_{\rm F}}{m} + \frac{k^2}{2m}\right) \text{ when } k \geq 2k_{\rm F}. \tag{13.32}$$

Thus, (13.31) and (13.32) give the plot of $\varepsilon_{\rm ex}^{\rm min}$ against k that is shown in figure 13.2.

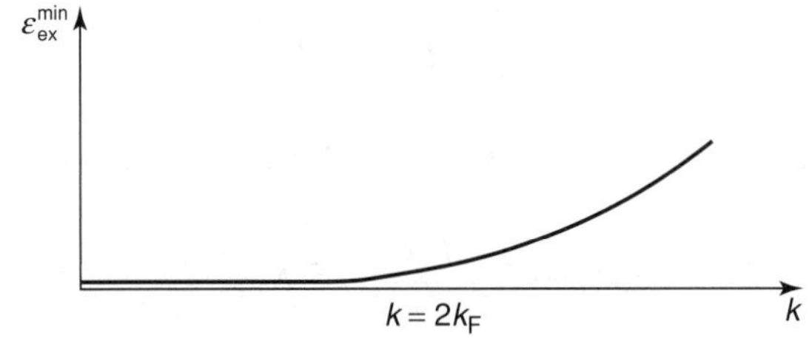

Figure 13.2 *Minimum single-particle excitation energy $\varepsilon_{\rm ex}^{\rm min}(\boldsymbol{k})$ for free fermions as a function of k.*

Whatever the magnitude of $\boldsymbol{k}$, we get the **greatest** possible excitation energy when $\boldsymbol{q}$ is in the same direction as the given $\boldsymbol{k}$ and has magnitude $|\boldsymbol{q}| = k_{\rm F}$ (see figure 13.3).
Thus,

$$\varepsilon_{\rm ex}^{\rm max}(\boldsymbol{k}) = \hbar^2\left(\frac{kk_{\rm F}}{m} + \frac{k^2}{2m}\right) \text{ for any } \boldsymbol{k}, \tag{13.33}$$

and so the excitation spectrum has the form shown in figure 13.4. The shaded area is known as the particle–hole continuum.

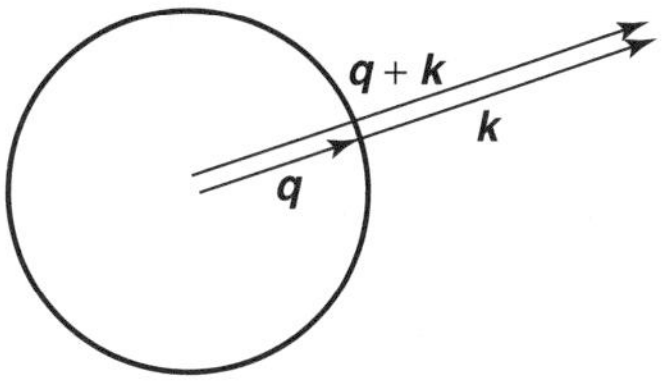

Figure 13.3 *Construction to determine $\varepsilon_{ex}^{max}(\boldsymbol{k})$ for free fermions.*

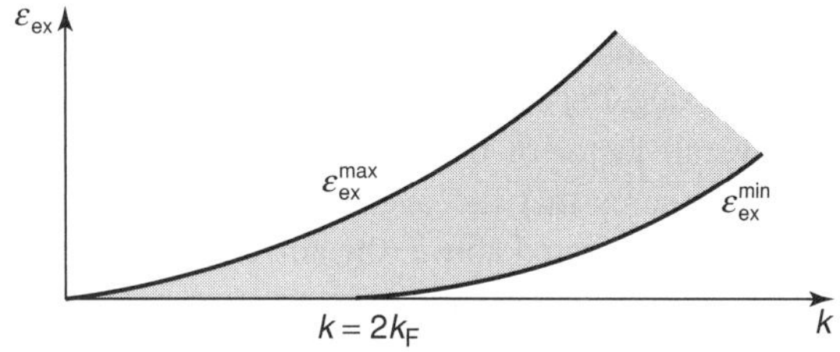

Figure 13.4 *Single-particle excitation spectrum $\varepsilon_{ex}(\boldsymbol{k})$ for free fermions.*

13.3 Systems of interacting particles

In the case of a system of **interacting** fermions or bosons we can no longer construct exact energy eigenstates by "feeding" the particles into single-particle energy eigenstates ("orbitals") as we did for noninteracting particles in section 13.2.

Because of the interaction terms in the hamiltonian of the N-particle system, the exact energy eigenfunctions with energy E will be **nonfactorizable** functions $\Psi_E^{(\alpha)}(\xi_1, \ldots, \xi_N)$ of all the coordinates. This is the space–spin part, the time-dependent part being, of course, the oscillatory factor $e^{-iEt/\hbar}$, and the index α labels the different degenerate N-particle states of energy E.

The space–spin parts $\Phi_{N_1,N_2,\ldots,N_\infty}(\xi_1, \ldots, \xi_N)$ (with $\sum_{i=1}^{\infty} N_i = N$) of the energy eigenfunctions of a system of N **noninteracting** particles form a **complete set**; that is, any normalizable N-particle space–spin wave function can be expressed as a linear combination of them. In particular, therefore, this must be true of the above exact energy eigenstates (stationary states) $\Psi_E^{(\alpha)}(\xi_1, \ldots, \xi_N)$ of the system of interacting particles:

$$\Psi_E^{(\alpha)}(\xi_1, \ldots, \xi_N) = \sum_{N_1=0} \sum_{\substack{N_2=0 \\ \left(\sum_{i=1}^{\infty} N_i = N\right)}} \cdots \sum_{N_\infty=0} C_E^{(\alpha)}(N_1, N_2, \ldots, N_\infty)\Phi_{N_1,N_2,\ldots N_\infty}(\xi_1, \ldots, \xi_N). \quad (13.34)$$

[For example, (13.34) could be a perturbation expansion of an exact energy eigenstate of the system of interacting particles in terms of the "unperturbed" states $\Phi_{N_1,N_2,\ldots,N_\infty}(\xi_1, \ldots, \xi_N)$.]

Like any linear transformation with nonzero determinant, (13.34) can be inverted:

$$\Phi_{N_1,N_2,\ldots N_\infty}(\xi_1, \ldots, \xi_N) = \sum_{E,\alpha(E)} D_{N_1,N_2,\ldots,N_\infty}(E, \alpha(E))\Psi_E^{(\alpha)}(\xi_1, \ldots, \xi_N). \quad (13.35)$$

In certain circumstances, as we shall see below, the factorizable wave functions $\Phi_{N_1,N_2,\ldots,N_\infty}(\xi_1,\ldots,\xi_N)$ that form the exact energy eigenstates for a system of noninteracting particles can be used as an approximate description of states of the corresponding system of interacting particles. These functions, according to (13.35), are wave packets of the exact energy eigenstates $\Psi_E^{(\alpha)}(\xi_1,\ldots,\xi_N)$ with different energies E, and so correspond to nonstationary states that do not correspond to a single well-defined energy. Indeed, the packet (13.35) may be thought of as having a **width** γ, defined as a quantity of the order of the range of energies E within which the coefficients $D_{N_1N_{2,\ldots,}N_\infty}(E,\alpha(E))$ are substantially nonzero. Building the time dependence into the packet (13.35), that is, multiplying each $\Psi_E^{(\alpha)}(\xi_1,\ldots,\xi_N)$ in the sum over E in (13.35) by the corresponding factor $e^{-iEt/\hbar}$, leads, in the general case, to a complicated overall time dependence for the sum. However, in the circumstances in which a given factorizable function $\Phi_{N_1,N_2,\ldots,N_\infty}(\xi_1,\ldots,\xi_N)$ provides a reasonably good approximate description of a state of the system of interacting particles (circumstances, usually, in which the corresponding packet (13.35) is narrow, that is, γ is small), this time dependence can usually be approximated by an overall factor of the form $e^{-i(E_0-i\gamma)t/\hbar}$, where E_0 is the central energy of the packet. The factor $e^{-\gamma t/\hbar}$ here ensures that the overall wave function (probability amplitude) decays exponentially in time to zero, so that, as asserted above, the corresponding state is **nonstationary**.

Every stationary state $\Psi_E^{(\alpha)}(\xi_1,\ldots,\xi_N;t)$ of a system of interacting particles (that is, the exact ground state and each of the exact stationary excited states) takes the form of a product of the simple time-dependent factor $e^{-iEt/\hbar}$ (with the appropriate energy E) and a space–spin part in the form [see (13.34)] of an enormously complicated, nonfactorizable N-particle wave function. We have absolutely no hope of solving the relevant time-dependent Schrödinger equation for the functions $\Psi_E^{(\alpha)}(\xi_1,\ldots,\xi_N;t)$.

But does it matter that we cannot find the exact stationary states? Do they have any direct relevance in physical experiments? Fortunately, the answer to both questions is an unequivocal no, for essentially two reasons. Firstly, in order to be sure that, in a given physical system, we were dealing with an exact energy eigenstate, we should have to measure the energy with an uncertainty smaller than the gap between adjacent energy levels. This is perfectly possible, for example, when we are considering the relative motion of an electron and a proton in a hydrogen atom, at least for the lowest-lying energy levels. In a many-body system, however, where the density of states per unit energy interval is of the order of e^N (in any conventional units of $[\text{energy}]^{-1}$!), the mean spacing between states is of the order of $e^{-N} \sim e^{-10^{23}}$, that is, the measurement uncertainty ΔE must be fantastically small. From the energy–time uncertainty relation $(\Delta E \Delta t)_{\min} \sim \hbar$ we should need an effectively infinite measurement time Δt to pinpoint the energy to the required degree of certainty. Such conditions can never obtain in any real physical measurement.

Secondly, any two given exact many-body states are "connected by an operator" (in the sense of yielding a nonzero matrix element of that operator) only if the operator is of unimaginable complexity – the operator would have to act differently on all the particle coordinates forming the arguments of one of the corresponding wave functions, and in precisely the way required to change it either into the other wave function or into a wave function with a large component of the other wave function (out of the infinity of possible components it could have) in its expansion in terms of the exact energy eigenfunctions of the system. The available physical probes, on the other hand, that is, the probes available in real physical experiments, are always of very much simpler types, representable by operators acting either on the coordinates of one or two particles or uniformly on the coordinates of all the particles. Thus, if, for example, a system of interacting particles could be prepared in its ground state, no operation ever devised by physicists could carry the system into an exact excited eigenstate. The methods used (for example, hitting the system with a neutron so that the latter transfers a certain momentum $\hbar\boldsymbol{k}$ and energy $\hbar\omega$ to the system) all knock the system into an excited state which is not an exact energy eigenstate (stationary state) but has a width (energy uncertainty) and hence a finite lifetime. The reason is obvious. The neutron will interact with, at most, one particle and the quantum-mechanical operator corresponding to this interaction will act on the coordinates of only one particle of the system, taking the system from the ground state (if this were attainable) to a state that is certainly not an exact energy eigenstate.

It should now be clear that the study of the exact energy eigenstates of a many-particle system has no relevance to physics. And it should be equally clear that what we must study are precisely the **nonstationary states** obtainable by the use of the available physical "probes". These inexact, physically producible excitations of the system are called **elementary excitations**.

As we have seen above [see (13.35)], the wave functions of these physically obtainable nonstationary excited states can be represented by wave packets (linear combinations) of exact energy eigenstates, the range of energy values that contribute being a measure of the energy width γ of the state.

13.4 Systems of interacting fermions (the Fermi liquid)

Imagine that we have a closed system of noninteracting fermions (an ideal Fermi gas) in a constant potential, and that the system is in its ground state, represented by a filled Fermi sphere. Now suppose that interactions between the particles are switched on adiabatically (meaning, in this context, infinitely slowly). If the result of this process is that we go over continuously from the ground state of the ideal gas to the ground state of the nonideal gas (the real system) by the time the interactions have reached their full physical value, the system is said to be a **normal Fermi system**. (If, for some reason, the Fermi surface of the ideal system is anisotropic, it is in general one of the **excited** states of the ideal system that goes over continuously into the ground state of the real system as the interaction is switched on adiabatically, but this point need not concern us here.) The reason that the imagined switching on of the physical interactions here must be adiabatic is that, if the interactions were switched on suddenly, with a rapid time development $\Delta V(t)$, the jolt would cause a transition from the ground state of the ideal system into many possible excited states of the real system, with excitation energies corresponding to the frequencies present in the Fourier decomposition of $\Delta V(t)$.

An example of a real Fermi system that is not normal in this sense is a superconducting Fermi system, whose ground state cannot be obtained continuously from the ground state of the system without interactions. A direct consequence of this is that perturbation theory cannot be applied to the problem of the ground state of a superconductor, as we shall see in more detail in chapter 16.

To consider the elementary excitations of the **real** system, we begin from a particular excited state of an **ideal** $(N + 1)$-particle system, namely, the state in which N particles completely occupy a Fermi sphere and one particle has momentum $\boldsymbol{p}$ outside the Fermi sphere. This state is an exact eigenstate of the $(N + 1)$-particle hamiltonian, with energy $E_0^{\text{ideal}}(N) + (p^2/2m)$. We now imagine that, as in the ground-state case considered above, the physical interactions are switched on adiabatically. With interactions present, the "extra" particle can, of course, scatter into states with other momenta. However, as we shall see, there are circumstances in which the possible available final states for scattering processes are very few in number, that is, "the phase space for scattering is small" (this is the case, in particular, when $\boldsymbol{p}$ is not too far above the Fermi surface); we shall confine ourselves to considering this situation. In this case, the extra particle, outside the Fermi sphere, is likely to remain unscattered until well after the imagined process of switching on of the interactions is complete. Because the phase space for scattering, though small, is nonzero, the resulting state is not an exact stationary state of the full hamiltonian, but may be termed **quasistationary**.

The extra particle, while still having momentum $\boldsymbol{p}$, is then nonetheless interacting with all the other particles, and these interactions cause virtual transitions that "renormalize" (change) the energy of the particle. (These are essentially the virtual, energy-nonconserving transitions that are found in the second- and higher-order terms in the standard time-independent perturbation-theory expansion for the energy [see section 3.1]). With this extra "self-energy", the particle may be regarded as a particle that is "dressed" by its interactions with the other particles, in contrast to the original "bare" particle of the ideal system. This dressed particle may be regarded as a completely independent particle and is known as a **quasiparticle** or, better, **Landau quasiparticle**. In view of the fact that the resulting excited state of the real $(N + 1)$-particle system is obtainable continuously from the corresponding excited state of the ideal $(N+1)$-particle system, which had energy $E_0^{\text{ideal}}(N)+(p^2/2m)$, we may expect that its energy will be of the same general form, viz., $E_0^{\text{real}}(N)+(p^2/2m^*)$, but with a "renormalized" N-particle ground-state energy $E_0^{\text{real}}(N)$ and with a "renormalized", **effective mass**. In other words, we should expect the excitation spectrum to remain quadratic in p, and this may be taken as an alternative definition of a normal Fermi system.

We may take matters a step further by putting all the particles on an equal footing. Then each particle carries with it a "self-energy cloud", such that the renormalization of the single-particle energy (and, correspondingly, the deviation of m^* from m) is greater for states near the Fermi surface than for those deep inside the Fermi sphere. This is because the deeper is the state, the greater is the difference in energy between it and lowest states to which transitions are kinematically allowed (that is, not forbidden by the Pauli principle), and the perturbation-theory correction to the energy of a single-particle state deep inside the Fermi sphere is then very small because of the large energy-difference denominators involved (see section 3.1). In this picture, the total number of quasiparticles is equal to the total number of particles.

We stress again, however, that the quasiparticle picture has validity only if the N-particle state is such that real scatterings are strongly kinematically suppressed. This, as we shall see, is the case when the system is in a weakly excited state.

In precisely the same way, we may define **quasiholes** as holes in the ground-state distribution for the real system that are obtained from the corresponding "holed" configuration of the ideal system by adiabatic switching on of the interactions. As in the case of a quasiparticle, a quasihole will be reasonably well defined only if real scatterings (corresponding to filling of the hole) are strongly suppressed by the Pauli principle. As with quasiparticles, this is the case if the hole is near the Fermi surface.

It is clear from our description of quasiparticles and quasiholes that they have the spin of the corresponding particles and holes of the ideal system, that is, half-odd-integer spins, since the particles are fermions. Since in any process in a closed quantum system the quantum number m_s of the z-component of spin can change only by an integer, quasiparticle and quasihole excitations cannot be produced singly, but only in quasiparticle–quasihole pairs. This is an essential characteristic of a Fermi-type excitation spectrum. Bose-type excitations have zero (or integer) spin, and so can appear or disappear singly without violating any quantum selection rules.

Thus, we may represent any weakly excited state of a closed many-fermion system with interactions as a state in which we have equal numbers of quasiholes and excited quasiparticles. Moreover, since the system is **normal** we may expect that the quasiparticle–quasihole excitation spectrum will have essentially the same form as the particle–hole excitation spectrum of the ideal system, illustrated in figure 13.4. Since the thermodynamics of the system depends on the excitation spectrum, via the partition function, we may thus expect the normal Fermi liquid and ideal Fermi gas to have qualitatively similar thermodynamic properties. For example, the specific heat of a Fermi liquid, like that of the corresponding ideal Fermi gas, will have a term that is linear in the temperature T, albeit with a different coefficient because of the change from the mass m to the effective mass m^*.

We now consider the **lifetime** of an excited quasiparticle, and hence of the corresponding many-particle state. Suppose, for example, that a fermion of momentum $\boldsymbol{k}$ is injected into an N-fermion system in its ground state. The momentum $\boldsymbol{k}$ lies outside the Fermi sphere (we assume that something like the latter is preserved in the presence of the interactions). The added fermion will scatter off other fermions. For example, it may excite a fermion (of initial momentum $\boldsymbol{k}'$, say) out of the Fermi sphere, losing some of its own momentum in the process. All that is necessary for this to be possible is that the final momenta of the two fermions ($\boldsymbol{k}''$and $\boldsymbol{k}'''$) should lie outside the Fermi sphere (all states with momenta inside are occupied) and that their sum should be equal to $\boldsymbol{k}+\boldsymbol{k}'$ (since the scattering must conserve momentum). Thus, the probability of finding a particle with momentum $\boldsymbol{k}$ well outside the Fermi sphere will decay in time, so that the single-particle state created by the corresponding particle injection cannot be an exact energy eigenstate of the hamiltonian of the system of interacting fermions.

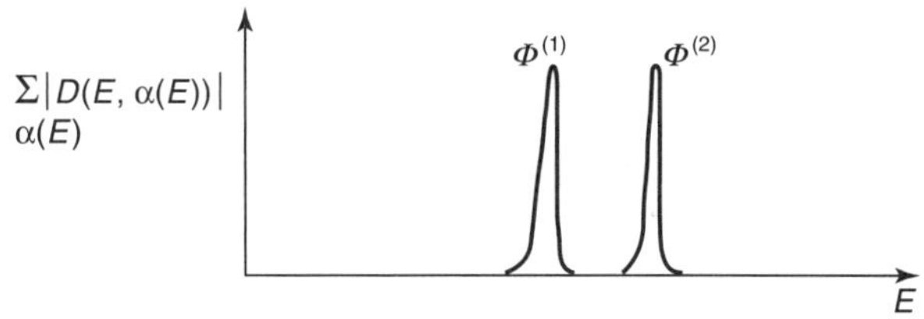

Figure 13.5 *Nonoverlapping wave packets, describing low-lying excited states.*

We can now ask the following question: Under what circumstances will an approximate energy eigenstate of this kind be an almost exact energy eigenstate, that is, a state of almost infinite lifetime? The answer is clear: it will be when the ways in which the extra particle can change its momentum by scattering are few in number, that is, when the scattering can lead to only a few possible final momenta of the two participating particles. Because (by virtue of the Pauli principle) there can be no scattering into single-particle states already occupied, that is, into single-particle states inside the Fermi sphere, we should expect the scattering probability to decrease as the momentum $\boldsymbol{k}$ of the injected particle gets closer to the Fermi surface. [In fact, by simple phase-space arguments, it can be shown that the scattering probability is proportional to $(k - k_F)^2$.]

The decrease in scattering probability with approach to the Fermi surface may be interpreted in terms of the exact stationary states of the real system. Suppose that a given excited state is described by a wave function of the form (13.35), which is not an exact eigenstate but a wave packet of exact energy eigenstates $\Psi_E^{(\alpha)}(\xi_1, \ldots, \xi_N; t)$ of the full hamiltonian $\hat{H}$. For example, suppose that the wave packets of two such excited states $\Phi^{(1)}$ and $\Phi^{(2)}$ are as represented in figure 13.5 while the wave packets of two other excited states $\Phi^{(3)}$ and $\Phi^{(4)}$ are as represented in figure 13.6.

Since $\hat{H}\Psi_E^{(\alpha)} = E\Psi_E^{(\alpha)}$, it is clear that $\hat{H}$ has no nonzero matrix elements between $\Phi^{(1)}$ and $\Phi^{(2)}$, but does between $\Phi^{(3)}$ and $\Phi^{(4)}$. In other words, wider wave packets (such as $\Phi^{(3)}$ and $\Phi^{(4)}$), describing more-highly excited states, can overlap, and scattering from one such state to another is then possible. Such states have a finite lifetime: instead of the time dependence $e^{-iEt/\hbar}$ the wave function describing such a state has the time dependence $e^{-i(E_0 - i\gamma)t/\hbar} = e^{-iE_0t/\hbar}e^{-\gamma t/\hbar}$, where E_0 is the central energy of the packet. Thus, the state may be thought of as having a complex energy in which the imaginary part is a measure of the energy width of the packet and is inversely proportional to the lifetime of the state.

Thus, provided that the quasiparticle concept is valid, so that the Φ-type states (that is, states constructed by feeding particles into single-particle orbitals) are narrow wave packets, that is, reasonably long lived, we may use a Φ-type (**independent particle**) description.

Further examples of such a description are the familiar shell model of the atom and the shell model of the nucleus. In both cases, the existence of the enormously strong interactions between the bare particles might lead one to think that approximating the wave function by feeding particles into single-particle orbitals

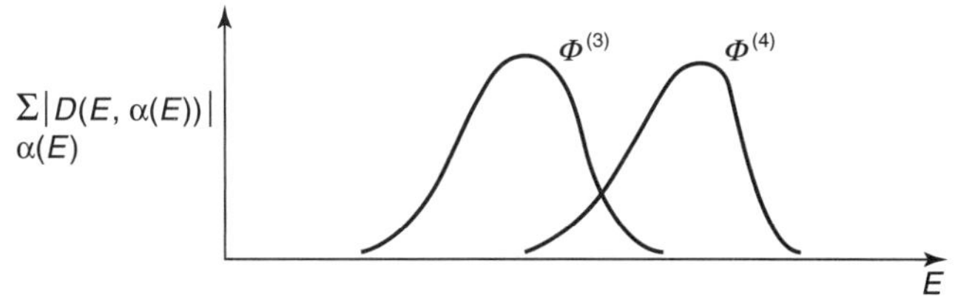

Figure 13.6 *Overlapping wave packets, describing high-lying excited states.*

would lead to a very unsatisfactory description. In fact, however, because of the Pauli-principle ("kinematic") restrictions on the possible scattering processes, the shell-model (single-particle) descriptions are remarkably good.

In the following section, we return to the degenerate ($T = 0$) Fermi liquid and consider Landau's phenomenological, independent-particle theory of this system.

13.5 The Landau theory of the normal Fermi liquid

Landau's theory, developed below, is intended to be applicable in precisely those circumstances in which the quasiparticle concept, as discussed above, is valid, that is, when the probability of scattering of a quasihole or excited quasiparticle by a particle inside the Fermi sphere is negligible by virtue of the Pauli principle. Note that this does not rule out the **mutual** scattering of excited quasiparticles and the scattering of excited quasiparticles by quasiholes (with consequent decay of the corresponding many-particle state), and the theory will take such processes into account.

As we have already seen, the energy of the system of $N + 1$ interacting fermions (Fermi liquid) obtained by adding a quasiparticle of momentum $\boldsymbol{p}_1$ (of magnitude close to the Fermi momentum p_F) to the N-fermion liquid in its ground state can be expressed as

$$E_0^{\text{liq}}(N) + \frac{p_1^2}{2m^*} \quad (p_1 > p_F), \tag{13.36}$$

where $E_0^{\text{liq}}(N)$ is the N-particle ground-state energy and m^* is the effective mass of the added quasiparticle. Addition of a further quasiparticle, of momentum $\boldsymbol{p}_2$ (again of magnitude close to the Fermi momentum p_F), gives an $(N + 2)$-fermion state with energy

$$E_0^{\text{liq}}(N) + \frac{p_1^2}{2m^*} + \frac{p_2^2}{2m^*} + f(\boldsymbol{p}_1, \boldsymbol{p}_2) \quad (p_1, p_2 > p_F), \tag{13.37}$$

where $f(\boldsymbol{p}_1, \boldsymbol{p}_2)$ (with $\boldsymbol{p}_1$ and $\boldsymbol{p}_2$ lying just above the Fermi surface) denotes the energy of the interaction of the two added quasiparticles in the presence of the original N fermions. (The energy of the interaction of the added quasiparticles with the previously present N fermions in their ground state is, of course, taken into account in the fact that the effective mass m^* differs from the bare-particle mass m.) Similarly, the $(N - 2)$-fermion state obtained from a system of N interacting fermions in its ground state by the removal of two quasiparticles of momenta $\boldsymbol{p}_1$ and $\boldsymbol{p}_2$ (both of magnitude close to the Fermi momentum p_F) has energy

$$E_0^{\text{liq}}(N) - \frac{p_1^2}{2m^*} - \frac{p_2^2}{2m^*} + f(\boldsymbol{p}_1, \boldsymbol{p}_2) \quad (p_1, p_2 < p_F), \tag{13.38}$$

where $f(\boldsymbol{p}_1, \boldsymbol{p}_2)$ (with $\boldsymbol{p}_1$ and $\boldsymbol{p}_2$ lying just below the Fermi surface) denotes the energy of the interaction of the resulting two quasiholes. Finally, the N-fermion state obtained from a system of N interacting fermions in its ground state by the removal of a quasiparticle of momentum $\boldsymbol{p}_1$ and the addition of a quasiparticle of momentum $\boldsymbol{p}_2$ (both of magnitude close to the Fermi momentum p_F) has energy

$$E_0^{\text{liq}}(N) - \frac{p_1^2}{2m^*} + \frac{p_2^2}{2m^*} - f(\boldsymbol{p}_1, \boldsymbol{p}_2) \quad (p_1 < p_F, p_2 > p_F), \tag{13.39}$$

where $-f(\boldsymbol{p}_1, \boldsymbol{p}_2)$ (with $\boldsymbol{p}_1$ lying just below the Fermi surface and $\boldsymbol{p}_2$ lying just above it) denotes the energy of the interaction of the quasihole and excited quasiparticle. The inclusion of the minus sign in the definition of this energy (that is, for the case when one momentum argument lies below the Fermi surface and the other lies above it) enables us to write a general expression for the energy of a weakly excited Fermi liquid in the form

$$E = E_0^{\text{liq}}(N) + \sum_{\boldsymbol{p}} \varepsilon(p)\delta n_{\boldsymbol{p}} + \frac{1}{2}\sum_{\boldsymbol{p}}\sum_{\boldsymbol{p}'} f(\boldsymbol{p}, \boldsymbol{p}')\delta n_{\boldsymbol{p}}\delta n_{\boldsymbol{p}'}, \tag{13.40}$$

where $\varepsilon(p) \equiv p^2/2m^*$ denotes the single-quasiparticle energy [see (13.36)], the sums are over all single-particle momenta allowed by the boundary conditions of the problem, and $\delta n_{\boldsymbol{p}}$ denotes the deviation (in the given many-particle state) of the occupation number of the single-particle state with momentum $\boldsymbol{p}$ from its value in the ground state of the N-particle system. For example, the expression (13.40) yields the expected result (13.37) for the case when all $\delta n_{\boldsymbol{p}}$ are zero except for $\delta n_{\boldsymbol{p}_1} = \delta n_{\boldsymbol{p}_2} = +1$. Similarly, as it should, it yields (13.38) for the case when all $\delta n_{\boldsymbol{p}}$ are zero except for $\delta n_{\boldsymbol{p}_1} = \delta n_{\boldsymbol{p}_2} = -1$, and it yields (13.39) for the case when all $\delta n_{\boldsymbol{p}}$ are zero except for $\delta n_{\boldsymbol{p}_1} = -1$ and $\delta n_{\boldsymbol{p}_2} = +1$.

The expression (13.40) shows that addition of an extra quasiparticle into a previously empty quasiparticle state with momentum $\boldsymbol{p}_1$ (so that $\delta n_{\boldsymbol{p}_1} = +1$) increases the energy of the system by the **quasiparticle energy** $\tilde{\varepsilon}(\boldsymbol{p}_1)$, given by

$$\tilde{\varepsilon}(\boldsymbol{p}_1) = \varepsilon(p_1) + \delta n_{\boldsymbol{p}_1}\sum_{\boldsymbol{p}} f(\boldsymbol{p}, \boldsymbol{p}_1)\delta n_{\boldsymbol{p}} = \varepsilon(p_1) + \sum_{\boldsymbol{p}} f(\boldsymbol{p}, \boldsymbol{p}_1)\delta n_{\boldsymbol{p}}. \tag{13.41}$$

The expressions (13.37)–(13.41) take only the **spin-independent** interactions of the quasiparticles and quasiholes into account. The spin angular-momentum vector $\boldsymbol{S} = \frac{1}{2}\hbar\boldsymbol{\sigma}$ has a nonzero expectation value $\langle \boldsymbol{S}\rangle_{\boldsymbol{p}} = \frac{1}{2}\hbar\langle\boldsymbol{\sigma}\rangle_{\boldsymbol{p}}$ for an **unpaired** quasiparticle or quasihole of momentum $\boldsymbol{p}$, that is, one associated with a momentum $\boldsymbol{p}$ for which $\delta n_{\boldsymbol{p}} = +1 \text{or} - 1$, respectively, while the expectation value of the total spin vector for two quasiparticles in the same single-particle momentum state is zero (so that, for example, a completely filled Fermi sphere has zero expectation value of the total spin vector). Thus, with inclusion of the **spin-dependent** interactions, equations (13.40) and (13.41) are replaced by

$$E = E_0^{\text{liq}}(N) + \sum_{\boldsymbol{p}} \varepsilon(p)\delta n_{\boldsymbol{p}} + \frac{1}{2}\sum_{\boldsymbol{p}}\sum_{\boldsymbol{p}'} f(\boldsymbol{p}, \boldsymbol{p}')\delta n_{\boldsymbol{p}}\delta n_{\boldsymbol{p}'} + \frac{1}{2}\sum_{\boldsymbol{p}}\sum_{\boldsymbol{p}'} \zeta(\boldsymbol{p}, \boldsymbol{p}')\langle\boldsymbol{\sigma}\rangle_{\boldsymbol{p}}\cdot\langle\boldsymbol{\sigma}\rangle_{\boldsymbol{p}'}, \tag{13.42}$$

and

$$\tilde{\varepsilon}(\boldsymbol{p}_1) = \varepsilon(p_1) + \sum_{\boldsymbol{p}} f(\boldsymbol{p}, \boldsymbol{p}_1)\delta n_{\boldsymbol{p}} + \langle\boldsymbol{\sigma}\rangle_{\boldsymbol{p}_1}\cdot\sum_{\boldsymbol{p}} \zeta(\boldsymbol{p}, \boldsymbol{p}_1)\langle\boldsymbol{\sigma}\rangle_{\boldsymbol{p}}, \tag{13.43}$$

respectively. The dot-product form of the spin–spin interaction energies in (13.42) and (13.43) ensures that these energies are scalars (invariants) under rotation of the coordinate axes, as they should be, and, because the factor $\boldsymbol{\sigma}$ in $\boldsymbol{S}$ is physically dimensionless, the quantities $\zeta(\boldsymbol{p}, \boldsymbol{p}')$, like $f(\boldsymbol{p}, \boldsymbol{p}')$, have the physical dimension of energy.

We now derive a formula for the effective mass m^* (the **Landau effective-mass formula**) of a quasiparticle just above the Fermi surface in terms of a particular part of the quasiparticle–quasiparticle interaction energy $f(\boldsymbol{p}, \boldsymbol{p}')$. Consider a system of $N + 1$ interacting fermions obtained by taking a system of N interacting fermions in its ground state and adding a quasiparticle of momentum $\boldsymbol{p}$ just outside the Fermi sphere. Suppose next that every particle in this system is given the same infinitesimal extra momentum $\delta\boldsymbol{q}$. We then calculate the resulting change in the total energy of the system in two ways (one in terms of the picture of interacting particles, and the other in terms of the picture of interacting quasiparticles) and equate the two results.

In the particle picture, changing the velocity of every particle by the same amount $\delta\boldsymbol{q}/m$ does not change the potential energy of interaction of the particles, and so the change in the total energy of the system is equal to the change in the total kinetic energy of the particles:

$$\delta E = \sum_{n=1}^{N+1}\left(\frac{(\boldsymbol{p}_n+\delta\boldsymbol{q})^2}{2m}-\frac{p_n^2}{2m}\right) = \sum_{n=1}^{N+1}\frac{\boldsymbol{p}_n\cdot\delta\boldsymbol{q}}{m}+O\left[(\delta\boldsymbol{q})^2\right] = \frac{\delta\boldsymbol{q}\cdot\boldsymbol{p}}{m}+O\left[(\delta\boldsymbol{q})^2\right], \tag{13.44}$$

where we have identified $\boldsymbol{p}_{N+1}$ as the momentum $\boldsymbol{p}$ of the quasiparticle outside the Fermi sphere and used the fact that then $\sum_{n=1}^{N}\boldsymbol{p}_n = 0$ for the momenta inside the Fermi sphere.

In the quasiparticle picture, the energy of the system **before** we make the uniform shift $\delta\boldsymbol{q}$ in the momenta is [see (13.36)]

$$E_{\text{before}}^{\text{liq}}(N+1) = E_0^{\text{liq}}(N)+\frac{p^2}{2m^*}, \tag{13.45}$$

and the occupation of the quasiparticle states inside the Fermi sphere is described by the ground-state distribution $n_{\boldsymbol{p}}^{(0)}$. **After** the uniform shift in the momenta of all the particles the whole quasiparticle-momentum distribution is shifted in momentum space by $\delta\boldsymbol{q}$, giving a new distribution $n_{\boldsymbol{p}} = n_{\boldsymbol{p}-\delta\boldsymbol{q}}^{(0)}$. (This may be viewed as creating excited quasiparticles just above the Fermi surface on one side of the Fermi sphere – the hemisphere with symmetry axis in the direction of $\delta\boldsymbol{q}$, while leaving quasiholes just below the Fermi surface on the other side.) The corresponding change in the quasiparticle-momentum distribution is

$$\delta n_{\boldsymbol{p}} = n_{\boldsymbol{p}} - n_{\boldsymbol{p}}^{(0)} = n_{\boldsymbol{p}-\delta\boldsymbol{q}}^{(0)} - n_{\boldsymbol{p}}^{(0)} = -\delta\boldsymbol{q}\cdot\nabla_{\boldsymbol{p}} n_{\boldsymbol{p}}^{(0)}, \tag{13.46}$$

where $\nabla_{\boldsymbol{p}} \equiv \hat{\boldsymbol{i}}\partial/\partial p_x+\hat{\boldsymbol{j}}\partial/\partial p_y+\hat{\boldsymbol{k}}\partial/\partial p_z$ and, in the last term, we are treating the dimensionless number $n_{\boldsymbol{p}}^{(0)}$ as a continuous function of the quasiparticle momentum $\boldsymbol{p}$. The total energy is given by equation (13.42), in which, however, the final, spin-dependent term is absent by virtue of the fact that only the unpaired quasiparticle with momentum $\boldsymbol{p}$ has a nonzero expectation value of the spin vector (the other, spin-paired quasiparticles, with momenta $\boldsymbol{p}'$, all have $\langle\boldsymbol{\sigma}\rangle_{\boldsymbol{p}'} = 0$):

$$\begin{aligned} E_{\text{after}}^{\text{liq}}(N+1) = E_0^{\text{liq}}(N) &+ \sum_{\substack{\boldsymbol{p}' \\ (\neq \boldsymbol{p}+\delta\boldsymbol{q})}} \varepsilon(p')\delta n_{\boldsymbol{p}'} + \frac{1}{2}\sum_{\substack{\boldsymbol{p}',\boldsymbol{p}'' \\ (\neq \boldsymbol{p}+\delta\boldsymbol{q})}} f(\boldsymbol{p}',\boldsymbol{p}'')\delta n_{\boldsymbol{p}'}\delta n_{\boldsymbol{p}''} \\ &+\frac{(\boldsymbol{p}+\delta\boldsymbol{q})^2}{2m^*} + \sum_{\substack{\boldsymbol{p}' \\ (\neq \boldsymbol{p}+\mathrm{d}\boldsymbol{q})}} f(\boldsymbol{p}+\delta\boldsymbol{q},\boldsymbol{p}')\delta n_{\boldsymbol{p}'}, \end{aligned} \tag{13.47}$$

where we have separated out the contributions associated with the unpaired quasiparticle with (shifted) momentum $\boldsymbol{p}+\delta\boldsymbol{q}$. In this expression the second term on the right-hand side vanishes because $\delta n_{-\boldsymbol{p}'} = -\delta n_{\boldsymbol{p}'}$

and $\varepsilon(p') \equiv p'^2/2m^*$ is even in $p' \equiv |\boldsymbol{p}'|$, while, after substitution of (13.46), the third term is clearly of order $(\delta\boldsymbol{q})^2$. Thus, to order $\delta\boldsymbol{q}$, equation (13.47) becomes

$$E^{\text{liq}}_{\text{after}}(N+1) = E^{\text{liq}}_0(N) + \frac{p^2}{2m} + \frac{\boldsymbol{p}\cdot\delta\boldsymbol{q}}{m^*} + \sum_{\boldsymbol{p}'} f(\boldsymbol{p},\boldsymbol{p}')\left(-\delta\boldsymbol{q}\cdot\nabla_{\boldsymbol{p}'}n^{(0)}_{\boldsymbol{p}'}\right). \tag{13.48}$$

Subtracting (13.45) from (13.48), we have, to order $\delta\boldsymbol{q}$,

$$\delta E = E^{\text{liq}}_{\text{after}}(N+1) - E^{\text{liq}}_{\text{before}}(N+1) = \frac{\boldsymbol{p}\cdot\delta\boldsymbol{q}}{m^*} - \delta\boldsymbol{q}\cdot\sum_{\boldsymbol{p}'} f(\boldsymbol{p},\boldsymbol{p}')\nabla_{\boldsymbol{p}'}n^{(0)}_{\boldsymbol{p}'}. \tag{13.49}$$

Comparing this with the expression (13.44) for δE calculated in the particle picture, we find

$$\frac{\boldsymbol{p}}{m} = \frac{\boldsymbol{p}}{m^*} - \sum_{\boldsymbol{p}'} f(\boldsymbol{p},\boldsymbol{p}')\nabla_{\boldsymbol{p}'}n^{(0)}_{\boldsymbol{p}'}, \tag{13.50}$$

which can be read as stating that the quasiparticle velocity $\boldsymbol{p}/m^*$ is equal to the particle velocity $\boldsymbol{p}/m$ plus a "**backflow velocity**" $\sum_{\boldsymbol{p}'} f(\boldsymbol{p},\boldsymbol{p}')\nabla_{\boldsymbol{p}'}n^{(0)}_{\boldsymbol{p}'}$ in the opposite direction.

Recall that we are considering a quasiparticle with momentum $\boldsymbol{p}$ near the Fermi surface (that is, with $|\boldsymbol{p}| \approx p_{\text{F}}$). Also, because $\nabla_{\boldsymbol{p}'}n^{(0)}_{\boldsymbol{p}'}$ is nonzero only for momenta $\boldsymbol{p}'$ near the Fermi surface, the only nonzero terms in the sum in (13.50) are those with $|\boldsymbol{p}'| \approx p_{\text{F}}$, and the quantities $f(\boldsymbol{p},\boldsymbol{p}')$ can be written as $f(p_{\text{F}},\theta)$, where θ is the angle between $\boldsymbol{p}$ and $\boldsymbol{p}'$. The sum over the momenta in (13.50) can be replaced by one-half times a sum over states (since two spin states are associated with each momentum state). If we then replace the sum over states by an integral, equation (13.50) becomes

$$\frac{\boldsymbol{p}}{m} = \frac{\boldsymbol{p}}{m^*} - \frac{1}{2}\int \mathrm{d}\varepsilon' D(\varepsilon') \int \frac{\mathrm{d}\Omega}{4\pi} f(p_{\text{F}},\theta)\nabla_{\boldsymbol{p}'}n^{(0)}_{\boldsymbol{p}'}, \tag{13.51}$$

where $D(\varepsilon')$ is the density of quantum states per unit energy interval ($\varepsilon' = p'^2/2m^*$), and $\int \mathrm{d}\Omega \ldots \equiv \int_0^{2\pi} \mathrm{d}\varphi \int_0^{\pi} \sin\theta \mathrm{d}\theta \ldots$. In (13.51) we write

$$\nabla_{\boldsymbol{p}'}n^{(0)}_{\boldsymbol{p}'} = (\nabla_{\boldsymbol{p}'}\varepsilon')\frac{\partial n^{(0)}_{\boldsymbol{p}'}}{\partial \varepsilon'} = \frac{\boldsymbol{p}'}{m^*}\frac{\partial n^{(0)}_{\boldsymbol{p}'}}{\partial \varepsilon'} = -2\delta(\varepsilon' - \varepsilon_{\text{F}})\frac{\boldsymbol{p}'}{m^*}, \tag{13.52}$$

where we have used the fact that $n^{(0)}_{\boldsymbol{p}'}$ is a step function of ε', equal to 2 for $\varepsilon' < \varepsilon_{\text{F}}$ and equal to zero for $\varepsilon' > \varepsilon_{\text{F}}$, so that $\partial n^{(0)}_{\boldsymbol{p}'}/\partial\varepsilon' = -2\delta(\varepsilon' - \varepsilon_{\text{F}})$. Putting (13.52) into (13.51) and carrying out the energy integration, we find

$$\frac{\boldsymbol{p}}{m} = \frac{\boldsymbol{p}}{m^*} + \frac{D(\varepsilon_{\text{F}})}{m^*}\int \frac{\mathrm{d}\Omega}{4\pi} f(p_{\text{F}},\theta)\boldsymbol{p}'. \tag{13.53}$$

Taking the dot-product of (13.53) with the vector $\boldsymbol{p}$, and setting $\boldsymbol{p}\cdot\boldsymbol{p}\approx p_{\mathrm{F}}^2$ and $\boldsymbol{p}\cdot\boldsymbol{p}'\approx p_{\mathrm{F}}^2\cos\theta$, we find

$$\frac{1}{m}=\frac{1}{m^*}+\frac{D(\varepsilon_{\mathrm{F}})}{m^*}\int\frac{\mathrm{d}\Omega}{4\pi}f(p_{\mathrm{F}},\theta)\cos\theta. \tag{13.54}$$

We now write the physically dimensionless quantity $D(\varepsilon_{\mathrm{F}})f(p_{\mathrm{F}},\theta)$ as an expansion in Legendre polynomials of $\cos\theta$:

$$D(\varepsilon_{\mathrm{F}})f(p_{\mathrm{F}},\theta)=\sum_l F_l P_l(\cos\theta). \tag{13.55}$$

Noting that the factor $\cos\theta$ in (13.54) is just $P_1(\cos\theta)$, and using the orthogonality relationship for Legendre polynomials [see (4.173)]

$$\int\mathrm{d}\Omega\, P_l(\cos\theta)P_{l'}(\cos\theta)=\frac{4\pi\,\delta_{ll'}}{2l+1}, \tag{13.56}$$

we can perform the angular integration in (13.54) and obtain

$$\frac{1}{m}=\frac{1}{m^*}+\frac{F_1}{3m^*}, \tag{13.57}$$

so that the effective mass m^* is given by the expression

$$\frac{m^*}{m}=1+\frac{F_1}{3}, \tag{13.58}$$

which is the **Landau effective-mass formula**.

We now use Landau Fermi-liquid theory to find an expression for the **speed of sound** u in a Fermi liquid at zero temperature T. We shall do this by expressing u in terms of a derivative of the chemical potential μ of the system and using the fact that the latter (at $T=0$) can be identified with the increase in energy of the system at $T=0$ when one quasiparticle, with momentum of magnitude p_{F}, is added to it.

The square of the speed of sound is given by

$$u^2=\left(\frac{\partial P}{\partial\rho}\right)_{N,S}=\left(\frac{\partial P}{\partial V}\right)_{N,S}\Big/\left(\frac{\partial\rho}{\partial V}\right)_{N,S}=-\frac{V^2}{mN}\left(\frac{\partial P}{\partial V}\right)_{N,S}=-\frac{V^2}{mN}\left(\frac{\partial P}{\partial V}\right)_{N,T}, \tag{13.59}$$

where P is the pressure, $\rho=mN/V$ is the mass density, S is the entropy, and in the last step we have used the fact that $\kappa_T=-(1/V)(\partial V/\partial P)_{N,T}$ and $\kappa_S=-(1/V)(\partial V/\partial P)_{N,S}$ (the isothermal and adiabatic compressibilities) are equal at $T=0$. But

$$\left(\frac{\partial P}{\partial V}\right)_{N,T}=\frac{N}{V}\left(\frac{\partial\mu}{\partial V}\right)_{N,T}=\frac{N}{V^2}\left(\frac{\partial\mu}{\partial\ln V}\right)_{N,T}=-\frac{N}{V^2}\left(\frac{\partial\mu}{\partial\ln\rho}\right)_{N,T}=-\frac{N}{V^2}\left(\frac{\partial\mu}{\partial\ln N}\right)_{V,T} \tag{13.60}$$

where the first equality follows from the Gibbs–Duhem relation (see section 11.5)

$$\mathrm{d}\mu=-\frac{S}{N}\mathrm{d}T+\frac{V}{N}\mathrm{d}P \tag{13.61}$$

and in the third and fourth equalities we have used two ways of writing a partial derivative of μ with respect to the variable $\ln\rho = \ln(mN/V) = \ln m + \ln N - \ln V$. Putting (13.60) into (13.59) we have

$$u^2 = -\frac{V^2}{mN}\left(\frac{\partial P}{\partial V}\right)_{N,T} = \frac{1}{m}\left(\frac{\partial\mu}{\partial\ln N}\right)_{V,T} = \frac{N}{m}\left(\frac{\partial\mu}{\partial N}\right)_{V,T}. \tag{13.62}$$

To calculate the derivative in (13.62) we must find the change $\delta\mu$ in μ (at $T=0$) brought about by increasing the number of particles from N to $N+\delta N$. The chemical potential $\mu(T=0)$ before this change is simply the increase in energy of the system when one quasiparticle, with momentum $\boldsymbol{p}_1$ of magnitude $p_{\mathrm{F}}(N)$, is added to the N-fermion system in its ground state, that is,

$$\mu(N, T=0) = \varepsilon(p_{\mathrm{F}}(N)) = \frac{p_{\mathrm{F}}^2(N)}{2m^*}, \tag{13.63}$$

where $p_{\mathrm{F}}(N)$ is the Fermi momentum for the N-fermion system, related to N by [see (13.27)]:

$$N = \frac{Vp_{\mathrm{F}}^3}{3\pi^2\hbar^3}. \tag{13.64}$$

Similarly, the chemical potential $\mu(N+\delta N, T=0)$ that obtains after the replacement of N by $N+\delta N$ is simply the increase in energy of the system when one quasiparticle, with momentum $\boldsymbol{p}_1$ of magnitude $p_{\mathrm{F}}(N+\delta N)$, is added to the $(N+\delta N)$-fermion system in its ground state. Because the latter state can be thought of as a state obtained from the ground state of the N-fermion system by the addition of quasiparticles with momenta $\boldsymbol{p}$ distributed uniformly just above the N-fermion Fermi surface, this addition being represented by nonzero quantities $\delta n_{\boldsymbol{p}}$ (all equal to 2) that sum to δN, and because the subsequently added quasiparticle with momentum $\boldsymbol{p}_1$ interacts with these quasiparticles, we have

$$\mu(N+\delta N, T=0) = \tilde{\varepsilon}(\boldsymbol{p}_1) = \varepsilon(p_{\mathrm{F}}+\delta p_{\mathrm{F}}) + \sum_{\boldsymbol{p}} f(\boldsymbol{p}, \boldsymbol{p}_1)\delta n_{\boldsymbol{p}}, \tag{13.65}$$

where we have used (13.43) with the spin–spin interaction term set to zero (since there are no unpaired spins with which the unpaired added quasiparticle of momentum $\boldsymbol{p}_1$ can interact). The sum over $\boldsymbol{p}$ in (13.65) is independent of the direction of $\boldsymbol{p}_1$ (because of the spherical symmetry of the quantities $\delta n_{\boldsymbol{p}}$ representing the change δN in the number of particles). Because we are considering a quasiparticle with momentum $\boldsymbol{p}_1$ at the $(N+\delta N)$-particle Fermi surface (i.e., with $|\boldsymbol{p}_1| \approx p_{\mathrm{F}}+\delta p_{\mathrm{F}}$), and because $\delta n_{\boldsymbol{p}}$ is nonzero (in fact, 2) only for momenta $\boldsymbol{p}$ with magnitudes infinitesimally greater than p_{F}, neglecting quantities of second order of smallness we can write the quantities $f(\boldsymbol{p}, \boldsymbol{p}')$ in (13.65) as $f(p_{\mathrm{F}}, \theta)$, where θ is the angle between $\boldsymbol{p}$ and $\boldsymbol{p}_1$. Then equation (13.65) becomes

$$\begin{aligned}\mu(N+\delta N, T=0) = \tilde{\varepsilon}(\boldsymbol{p}_1) &= \varepsilon(p_{\mathrm{F}}+\delta p_{\mathrm{F}}) + \int\frac{\mathrm{d}\Omega}{4\pi}f(p_{\mathrm{F}},\theta)\sum_{\boldsymbol{p}}\delta n_{\boldsymbol{p}} \\ &= \varepsilon(p_{\mathrm{F}}) + \left.\frac{\mathrm{d}\varepsilon(p)}{\mathrm{d}p}\right|_{p=p_{\mathrm{F}}}\delta p_{\mathrm{F}} + \delta N\int\frac{\mathrm{d}\Omega}{4\pi}f(p_{\mathrm{F}},\theta).\end{aligned} \tag{13.66}$$

But $d\varepsilon/dp|_{p=p_F} = p_F/m^*$. Also, from (13.64),

$$\delta N = \frac{V p_F^2 \delta p_F}{\pi^2 \hbar^3}, \text{ so that } \delta p_F = \frac{\pi^2 \hbar^3}{V p_F^2} \delta N \tag{13.67}$$

and (13.66) becomes

$$\mu(N + \delta N, T = 0) = \varepsilon(p_F) + \frac{\pi^2 \hbar^3}{V m^* p_F} \delta N + \delta N \int \frac{d\Omega}{4\pi} f(p_F, \theta). \tag{13.68}$$

Subtracting (13.63) from (13.68) we find

$$\delta\mu(T = 0) = \mu(N + \delta N, T = 0) - \mu(N, T = 0) = \frac{\pi^2 \hbar^3}{V m^* p_F} \delta N + \delta N \int \frac{d\Omega}{4\pi} f(p_F, \theta), \tag{13.69}$$

so that

$$\left(\frac{\partial \mu}{\partial N}\right)_{V,T} = \frac{\pi^2 \hbar^3}{V m^* p_F} + \int \frac{d\Omega}{4\pi} f(p_F, \theta). \tag{13.70}$$

Thus, (13.62) becomes

$$\begin{aligned} u^2 = \frac{N}{m}\left(\frac{\partial \mu}{\partial N}\right)_{V,T} &= \frac{V p_F^3}{3\pi^2 \hbar^3} \frac{1}{m} \left[\frac{\pi^2 \hbar^3}{V m^* p_F} + \int \frac{d\Omega}{4\pi} f(p_F, \theta)\right] \\ &= \frac{p_F^2}{3 m m^*} + \frac{2 V p_F^3}{3 (2\pi\hbar)^3} \frac{1}{m} \int d\Omega f(p_F, \theta). \end{aligned} \tag{13.71}$$

To eliminate m^* we use equation (13.54) in the form

$$\frac{1}{m^*} = \frac{1}{m} - \frac{D(\varepsilon_F)}{m^*} \int \frac{d\Omega}{4\pi} f(p_F, \theta) \cos\theta = \frac{1}{m} - \frac{2 V p_F}{(2\pi\hbar)^3} \int d\Omega\, f(p_F, \theta) \cos\theta, \tag{13.72}$$

where we have used the result

$$D(\varepsilon) = \frac{8\pi V m^*}{(2\pi\hbar)^3} p, \tag{13.73}$$

which, since $d\varepsilon = d(p^2/2m^*) = p dp/m^*$, follows from comparison of the second and final expressions in the following chain:

$$\begin{aligned} \sum_{\text{states}} f(p_F, \theta) &= \int d\varepsilon D(\varepsilon) \int \frac{d\Omega}{4\pi} f(p_F, \theta) = 2 \sum_{\boldsymbol{p}} f(p_F, \theta) \\ &\approx 2 \left(\frac{L}{2\pi\hbar}\right)^3 \int d^3\boldsymbol{p} f(p_F, \theta) = \frac{2V}{(2\pi\hbar)^3} \int p^2 dp \int d\Omega\, f(p_F, \theta). \end{aligned}$$

Putting (13.72) into (13.71) we find for the speed of sound:

$$u^2 = \frac{p_F^2}{3m^2} + \frac{2V}{3m}\frac{p_F^3}{(2\pi\hbar)^3}\int d\Omega f(p_F,\theta)(1-\cos\theta). \tag{13.74}$$

In the case of an **ideal** Fermi gas ($f = 0$), (13.74) gives for the speed of sound

$$u = \frac{1}{\sqrt{3}}\frac{p_F}{m}. \tag{13.75}$$

The result (13.74) is of theoretical interest only, since at $T = 0$ ordinary sound cannot propagate in a Fermi liquid. This is because the attenuation coefficient (damping coefficient) γ becomes infinite at zero temperature by virtue of the fact that $\gamma \propto \omega^2\eta$, where ω is the frequency of the sound and η is the viscosity, which was shown by Pomeranchuk to be proportional to T^{-2} in a Fermi liquid. However, another type of sound can propagate in a Fermi liquid at $T = 0$. This is known as **zero sound**, and is discussed in the next subsection.

13.6 Collective excitations of a Fermi liquid

13.6.1 Zero sound in a neutral Fermi gas with repulsive interactions

This is a collective Bose-type mode in which we have a density-fluctuation wave oscillating in time, the restoring force on any particle being the averaged force field of all the other particles. In the case of relatively weak repulsion, the mode lies just above the quasiparticle–quasihole continuum (see the dashed curve in figure 13.7). Since collisions would disrupt the restoring force, the zero-sound mode can exist only when the collision time τ is much longer than the period of the wave, that is, when $\omega\tau \gg 1$. (For this reason, zero sound is sometimes called high-frequency sound.) For the propagation of ordinary, first sound, on the other hand, frequent collisions are necessary ($\omega\tau \ll 1$), in order to return the particle-velocity distribution to its thermal-equilibrium form. Liquid helium-3 provides an example of a system in which it is possible to observe the transition from ordinary (first) sound to zero sound as the temperature is lowered and τ becomes longer.

13.6.2 Plasma oscillations in a charged Fermi liquid

As an example of a charged Fermi liquid we may consider an electron gas. Here, as a result of the Coulomb repulsion, the electron density in the immediate surroundings of a given electron tends to be lower than would be the case if the electrons were distributed uniformly throughout the volume. In other words, each

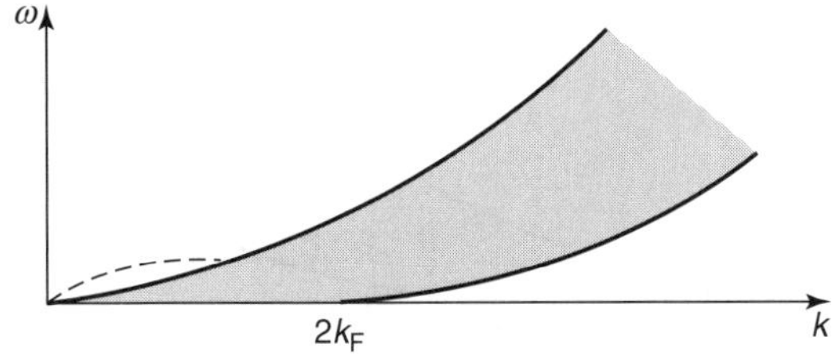

Figure 13.7 Dispersion curve (dashed curve) of the zero-sound mode in a Fermi liquid.

electron tends to be surrounded by a screening cloud of effective positive charge (relative to the mean charge that would obtain in the case of a uniform electron density). Such deviations of the charge density from uniformity are called charge-density fluctuations and, using Fourier expansion, may be regarded as composed of charge-density-deviation waves, or polarization waves, with characteristic wavelengths $2\pi/k$ (k is the wave number). These polarization waves propagate with a frequency ω_{pl} (the plasma frequency) that turns out to be practically independent of k:

$$\omega_{\text{pl}} = \sqrt{\frac{ne^2}{\varepsilon_0 m}}, \tag{13.76}$$

where n is the electron-number density, e is the magnitude of the electron charge, m is the electron mass, and ε_0 is the dielectric permittivity in vacuo. The polarization waves, in analogy with other harmonic oscillations, can receive and give up energy only in the form of quanta of magnitude $\hbar\omega_{\text{pl}}$, known as **plasmons**. A polarization wave of given wave vector $\boldsymbol{k}$ can have any positive integer number (including zero) of these quanta of energy (plasmons), that is, plasmons are bosons. The values of n in metals are such that (13.76) yields plasmon energies $\hbar\omega_{\text{pl}} \approx 5\text{ eV} - 30\text{ eV}$ and since, at room temperature, these energies are very much greater than $k_{\text{B}}T$, plasmons do not have much effect on the thermodynamics of metals at ordinary temperatures. The plasmon dispersion curve is indicated by the dashed curve in figure 13.8.

A charged Fermi liquid with even just one plasmon excited is not in an exact energy eigenstate (stationary state) of the hamiltonian of the system, that is, the plasmon is not an "exact excitation", of infinite lifetime, but rather has a finite lifetime. The corresponding damping constant for a plasmon corresponding to wave number k depends on k as:

$$\gamma_{\text{pl}} \propto \left(\frac{k}{k_{\text{F}}}\right)^2 \omega_{\text{pl}}, \tag{13.77}$$

and so the plasmon is a better-defined, longer-lived excitation for lower k ($k < k_{\text{F}}$), that is, for longer wavelengths. For larger values of k, as can be seen from the figure 13.8, the plasmon mode merges into the quasiparticle–quasihole continuum, and, in fact, the plasmon becomes unstable with respect to decay into quasiparticle–quasihole excitations.

Like zero sound, of which it is an exact analogue, polarization waves (plasma waves) can exist only if $\omega\tau \gg 1$. From the dispersion curve it can be seen that at $k = 0$ the plasma wave has infinite phase velocity ω/k and zero group velocity $\mathrm{d}\omega/\mathrm{d}k$.

Finally, it should be noted that, unlike the quasiparticle excitations considered previously, the plasmon and zero-sound excitations have no direct counterpart in the noninteracting Fermi gas.

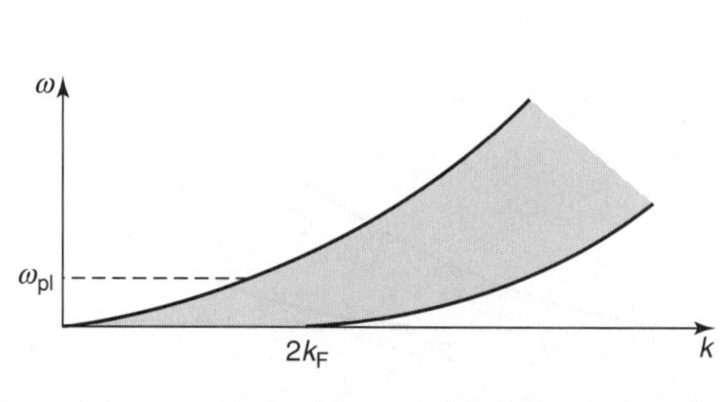

Figure 13.8 Dispersion curve (dashed curve) of the plasmon mode in a Fermi liquid.

13.7 Phonons and other excitations

We have considered, above, the quasiparticle and collective excitations of a gas of interacting fermions. In this section we consider other systems and their elementary excitations.

13.7.1 Phonons in crystals

The **crystal structure** of a physical crystal is defined as a three-dimensional Bravais lattice plus a basis. A three-dimensional **Bravais lattice** is defined as any (infinite) array of points such that all their position vectors $\boldsymbol{R}$ are of the form

$$\boldsymbol{R} = n_1\boldsymbol{a}_1 + n_2\boldsymbol{a}_2 + n_3\boldsymbol{a}_3, \tag{13.78}$$

where $\boldsymbol{a}_1$, $\boldsymbol{a}_2$ and $\boldsymbol{a}_3$ are three given vectors not in the same plane, and n_1, n_2 and n_3 take **all** integer values (negative and zero, as well as positive). The vectors $\boldsymbol{a}_i$ are called **primitive vectors** of the lattice and are said to generate or span the lattice. A **basis** is a physical unit (such as an atom or a group of atoms with some definite arrangement and orientation), such that, in any particular crystal structure, an identical copy of it is located at each lattice point.

In this section we shall consider the vibrational properties of a crystal whose structure is a Bravais lattice plus a monatomic basis (that is, a Bravais lattice that has one atom, of mass M, at each lattice point), and we shall assume that the atoms interact through simple-harmonic forces. Unlike a Bravais lattice, which has an infinite number of points (and hence infinite volume), a real crystal has finite volume, and so may be thought of a occupying a subvolume of a Bravais lattice. We shall assume that this subvolume contains the $N = N_1N_2N_3$ lattice points

$$\boldsymbol{R} = n_1\boldsymbol{a}_1 + n_2\boldsymbol{a}_2 + n_3\boldsymbol{a}_3, \text{ with } n_i = 1, 2, \ldots, N_i. \tag{13.79}$$

In order that the bulk vibrational properties are not unduly influenced by surface effects, we shall also assume that the number N of lattice points (N is also the number of atoms, in this case) is large and that the crystal is not of irregular shape, that is, that all three N_i are of order $N^{1/3}$.

To find the vibrational properties we first find the equations of motion of the atoms. For this we first write the lagrangian $L(\{q_i\}, \{\dot{q}_i\})$ of the system, taking as the set $\{q_i\}$ of generalized coordinates q_i the $3N$ quantities $u^\mu(\boldsymbol{R})$ [the μ-component ($\mu = 1, 2, 3$) of the vector displacement $\boldsymbol{u}(\boldsymbol{R})$ of the atom from its equilibrium position at the lattice point at vector position $\boldsymbol{R}$]. The lagrangian is the difference of the kinetic energy T and potential energy V^{harmonic}:

$$L = T - V^{\text{harmonic}} = \frac{1}{2}\sum_{\boldsymbol{R},\mu} M(\dot{u}^\mu(\boldsymbol{R}))^2 - \frac{1}{2}\sum_{\boldsymbol{R},\boldsymbol{R}',\mu,\nu} D_{\mu\nu}(\boldsymbol{R} - \boldsymbol{R}')u^\mu(\boldsymbol{R})u^\nu(\boldsymbol{R}'). \tag{13.80}$$

The Euler–Lagrange equations

$$\frac{\mathrm{d}}{\mathrm{d}t}\left(\frac{\partial L}{\partial \dot{q}_i}\right) = \left(\frac{\partial L}{\partial q_i}\right) \tag{13.81}$$

then take the form of the following $3N$ equations of motion:

$$M\ddot{u}^{\mu}(\boldsymbol{R}) = -\frac{\partial V^{\text{harmonic}}}{\partial u^{\mu}(\boldsymbol{R})} = -\sum_{\boldsymbol{R}',\nu} D_{\mu\nu}(\boldsymbol{R}-\boldsymbol{R}')u^{\nu}(\boldsymbol{R}'), \tag{13.82}$$

which can be written in the form of N matrix equations (one for each value of $\boldsymbol{R}$, with each equation involving the displacements at all sites):

$$M\ddot{\mathsf{u}}(\boldsymbol{R}) = -\sum_{\boldsymbol{R}'} \mathsf{D}(\boldsymbol{R}-\boldsymbol{R}')\mathsf{u}(\boldsymbol{R}'), \tag{13.83}$$

where $\mathsf{u}(\boldsymbol{R})$ is the 3×1 matrix (a column matrix) whose elements are the components $u^{\mu}(\boldsymbol{R})$ of the vector $\boldsymbol{u}(\boldsymbol{R})$, and $\mathsf{D}(\boldsymbol{R}-\boldsymbol{R}')$ is the 3×3 matrix with elements $D_{\mu\nu}(\boldsymbol{R}-\boldsymbol{R}')$.

The matrix $\mathsf{D}(\boldsymbol{R}-\boldsymbol{R}')$ has certain symmetries that will be useful in what follows. The first is

$$D_{\mu\nu}(\boldsymbol{R}-\boldsymbol{R}') = D_{\nu\mu}(\boldsymbol{R}'-\boldsymbol{R}), \tag{13.84}$$

which follows from the fact that the potential energy V^{harmonic} [see (13.80)] is unchanged by this replacement, or, alternatively, from the fact that $D_{\mu\nu}(\boldsymbol{R}-\boldsymbol{R}')$ is a second derivative of V^{harmonic} with respect to $u^{\mu}(\boldsymbol{R})$ and $u^{\nu}(\boldsymbol{R}')$ and this derivative is independent of the order of the differentiations.

The second symmetry of $\mathsf{D}(\boldsymbol{R}-\boldsymbol{R}')$ is

$$D_{\mu\nu}(\boldsymbol{R}-\boldsymbol{R}') = D_{\mu\nu}(\boldsymbol{R}'-\boldsymbol{R}), \quad \text{i.e.,} \quad \mathsf{D}(\boldsymbol{R}) = \mathsf{D}(-\boldsymbol{R}), \tag{13.85}$$

which, taken with (13.84), gives

$$D_{\mu\nu}(\boldsymbol{R}-\boldsymbol{R}') = D_{\nu\mu}(\boldsymbol{R}-\boldsymbol{R}'), \tag{13.86}$$

that is, D is a symmetric matrix. [The symmetry (13.85) follows from the fact that, by virtue of the inversion symmetry always present in a Bravais lattice (as can be seen from (13.88), for every lattice site $\boldsymbol{R}$ there is a lattice site $-\boldsymbol{R}$), replacing the displacement $\boldsymbol{u}(\boldsymbol{R})$ at every lattice point $\boldsymbol{R}$ (where the origin of $\boldsymbol{R}$ is taken to be the centre of inversion of the Bravais lattice subvolume occupied by the crystal) by minus the displacement at the lattice point $-\boldsymbol{R}$ [$\boldsymbol{u}(\boldsymbol{R}) \to -\boldsymbol{u}(-\boldsymbol{R})$] leaves the potential energy V^{harmonic} of the crystal unchanged.]

The third symmetry of $\mathsf{D}(\boldsymbol{R}-\boldsymbol{R}')$ is

$$\sum_{\boldsymbol{R}} D_{\mu\nu}(\boldsymbol{R}) = 0, \quad \text{i.e.,} \quad \sum_{\boldsymbol{R}} \mathsf{D}(\boldsymbol{R}) = 0, \tag{13.87}$$

where 0 is the 3×3 null matrix. This symmetry follows from the fact that the harmonic term in the potential energy V^{harmonic} in (13.80) must give zero in the case when the vector displacements are chosen to be the same at all lattice points, corresponding to displacement of the crystal as a whole, with no internal distortion.

Equation (13.83) for the force (in column-matrix form) on the particle at lattice point $\boldsymbol{R}$ contains in its right-hand side the vector displacements (in column-matrix form) of the atoms at **all** other lattice points $\boldsymbol{R}'$. Therefore, every possible physical solution of equation (13.83) must take the form of the set of the vector displacements (in column-matrix form) of the atoms at all lattice points $\boldsymbol{R}$, that is, must take the form of a **real** column-matrix function of $\boldsymbol{R}$ (and t) – a "displacement field" $\mathsf{u}(\boldsymbol{R}, t)$. For mathematical convenience, however, we shall first seek elementary **complex** solutions of equation (13.83) in the form of normal modes, where by a **normal mode** we mean a mode of motion in which the atoms at all lattice points $\boldsymbol{R}$ oscillate about their equilibrium positions with the same frequency. We try solutions $\mathsf{u}_{\boldsymbol{k}}(\boldsymbol{R}, t)$ that each take the form of a plane wave with wave vector $\boldsymbol{k}$ and corresponding angular frequency $\omega_{\boldsymbol{k}}$:

$$\mathsf{u}_{\boldsymbol{k}}(\boldsymbol{R}, t) = \mathsf{u}_{\boldsymbol{k}} \mathrm{e}^{i(\boldsymbol{k}\cdot\boldsymbol{R} - \omega_{\boldsymbol{k}} t)}, \tag{13.88}$$

where $\mathsf{u}_{\boldsymbol{k}}$ is a 3×1 (column) matrix to be determined. In the end we shall use the complex normal-mode solutions of the type (13.88) to construct real physical displacement fields $\mathsf{u}(\boldsymbol{R}, t)$.

In (13.88), the wave vector $\boldsymbol{k}$ takes values permitted by the boundary conditions that we choose to impose on the displacements. If the crystal were a one-dimensional chain of N atoms at lattice points R with spacing a (so that the chain has length $L = Na$), the unimportance of the "surface" of the crystal (that is, the two ends of the chain of atoms) could be emphasized by disposing of this "surface" entirely by joining the two ends together, replacing the chain of length L by a circle of circumference L. We would then have the corresponding macroscopic periodic boundary condition $\boldsymbol{u}(R + L) = \boldsymbol{u}(R)$, for every lattice point R. Such a condition is known as a Born–von Kàrmàn boundary condition. In the analogous condition for a three-dimensional crystal with lattice points (13.79) we join each of the three pairs of opposite faces of this crystal together. Although this is topologically impossible in a three-dimensional space, we can nonetheless write the analytical form of this condition:

$$\boldsymbol{u}(\boldsymbol{R} + N_1\boldsymbol{a}_1) = \boldsymbol{u}(\boldsymbol{R}), \quad \boldsymbol{u}(\boldsymbol{R} + N_2\boldsymbol{a}_2) = \boldsymbol{u}(\boldsymbol{R}), \quad \boldsymbol{u}(\boldsymbol{R} + N_3\boldsymbol{a}_3) = \boldsymbol{u}(\boldsymbol{R}), \tag{13.89}$$

where $N_1N_2N_3 = N$, the number of atoms in the crystal. Solutions of the form (13.88) will satisfy condition (13.89) only if the wave vector $\boldsymbol{k}$ is restricted to be of the form

$$\boldsymbol{k} = \frac{n_1}{N_1}\boldsymbol{b}_1 + \frac{n_2}{N_2}\boldsymbol{b}_2 + \frac{n_3}{N_3}\boldsymbol{b}_3, \tag{13.90}$$

where the vectors $\boldsymbol{b}_i$ ($i = 1, 2, 3$) are the primitive **reciprocal-lattice** vectors satisfying $\boldsymbol{b}_i \cdot \boldsymbol{a}_j = 2\pi\delta_{ij}$:

$$\boldsymbol{b}_1 = 2\pi\frac{\boldsymbol{a}_2 \times \boldsymbol{a}_3}{\boldsymbol{a}_1 \cdot (\boldsymbol{a}_2 \times \boldsymbol{a}_3)}, \quad \boldsymbol{b}_2 = 2\pi\frac{\boldsymbol{a}_3 \times \boldsymbol{a}_1}{\boldsymbol{a}_2 \cdot (\boldsymbol{a}_3 \times \boldsymbol{a}_1)}, \quad \boldsymbol{b}_3 = 2\pi\frac{\boldsymbol{a}_1 \times \boldsymbol{a}_2}{\boldsymbol{a}_3 \cdot (\boldsymbol{a}_1 \times \boldsymbol{a}_2)}, \tag{13.91}$$

and the n_i are integers. If we restrict ourselves to the following values of the integers n_i: $n_1 = 0, 1, \ldots, N_1 - 1$, $n_2 = 0, 1, \ldots, N_2 - 1$, and $n_3 = 0, 1, \ldots, N_3 - 1$, corresponding to $N_1N_2N_3 = N$ distinct values of $\boldsymbol{k}$ (13.90), these values of $\boldsymbol{k}$ clearly all lie inside the parallelepiped spanned by the primitive vectors $\boldsymbol{b}_1, \boldsymbol{b}_2, \boldsymbol{b}_3$ (a primitive cell of the reciprocal lattice). Adding any reciprocal-lattice vector ((that is, a vector of the form $\boldsymbol{K} = n'_1\boldsymbol{b}_1 + n'_2\boldsymbol{b}_2 + n'_3\boldsymbol{b}_3$, where the n'_i are integers) to a given $\boldsymbol{k}$ (13.90) within this primitive cell does not give a distinct solution of the form (13.88), because of the basic property $\mathrm{e}^{i\boldsymbol{K}\cdot\boldsymbol{R}} = 1$ of reciprocal-lattice vectors. Thus, there are just N nonequivalent values of $\boldsymbol{k}$ of the form (13.90), which can all be chosen to lie within **any** primitive cell of the reciprocal lattice. It is generally convenient to take that cell to be the first **Brillouin zone**. This zone in $\boldsymbol{k}$-space (reciprocal space) is defined as the **Wigner–Seitz primitive cell** of the reciprocal lattice, that is, the set of values of $\boldsymbol{k}$ that lie closer to the reciprocal-lattice vector $\boldsymbol{K} = 0$ than to any other reciprocal-lattice vector $\boldsymbol{K}$.

Substituting (13.88) into (13.83), we find the matrix eigenvalue equation

$$\mathsf{D}(\boldsymbol{k})\mathsf{u}_{\boldsymbol{k}} = M\omega_{\boldsymbol{k}}^2\mathsf{u}_{\boldsymbol{k}}, \tag{13.92}$$

where the 3×3 matrix $\mathsf{D}(\boldsymbol{k})$ (called the **dynamical matrix**) is given by

$$\mathsf{D}(\boldsymbol{k}) = \sum_{\boldsymbol{R}'} \mathsf{D}(\boldsymbol{R} - \boldsymbol{R}')\mathrm{e}^{-i\boldsymbol{k}\cdot(\boldsymbol{R}-\boldsymbol{R}')} = \sum_{\boldsymbol{R}} \mathsf{D}(\boldsymbol{R})\mathrm{e}^{-i\boldsymbol{k}\cdot\boldsymbol{R}}. \tag{13.93}$$

Here, the second form of the sum follows from the fact that every vector $\boldsymbol{R} - \boldsymbol{R}'$ in the first sum over all lattice vectors $\boldsymbol{R}'$ is itself a lattice vector, so that the first sum can be written alternatively as a sum over all values of $\boldsymbol{R} - \boldsymbol{R}'$ (relabelled as $\boldsymbol{R}$). Using the symmetries (13.85) and (13.87), we now rewrite the matrix $\mathsf{D}(\boldsymbol{k})$ (13.93) in the form

$$\mathsf{D}(\boldsymbol{k}) = \frac{1}{2}\sum_{\boldsymbol{R}} \mathsf{D}(\boldsymbol{R})\left[\mathrm{e}^{-i\boldsymbol{k}\cdot\boldsymbol{R}} + \mathrm{e}^{i\boldsymbol{k}\cdot\boldsymbol{R}} - 2\right] = \sum_{\boldsymbol{R}} \mathsf{D}(\boldsymbol{R})\left[\cos(\boldsymbol{k}\cdot\boldsymbol{R}) - 1\right] = -2\sum_{\boldsymbol{R}} \mathsf{D}(\boldsymbol{R})\sin^2\left(\frac{\boldsymbol{k}\cdot\boldsymbol{R}}{2}\right). \tag{13.94}$$

From this it is clear that the dynamical matrix $\mathsf{D}(\boldsymbol{k})$ is an even function of $\boldsymbol{k}$, and is also real. Moreover, because the matrix $\mathsf{D}(\boldsymbol{R})$ is symmetric [see (13.86)], so too is the matrix $\mathsf{D}(\boldsymbol{k})$. But, by a well-known theorem of matrix algebra, real symmetric 3×3 matrices have three linearly independent eigenvectors that can be chosen to be real (that is, the elements in the column matrix describing any given eigenvector have the same complex phase factor $\mathrm{e}^{i\theta}$, which can be chosen to be unity). Therefore, for every distinct value of $\boldsymbol{k}$ equation (13.92) has three linearly independent real eigenvector solutions $\mathsf{u}_{\boldsymbol{k}s}$ $(s = 1, 2, 3)$, with corresponding eigenvalues $M\omega_{\boldsymbol{k}s}^2$. The index s is said to characterize a **branch** of the vibration spectrum and, for a given value of the branch index s, the eigenvalues of (13.92) for all the allowed values of $\boldsymbol{k}$ allow one to determine the (real and positive) frequency $\omega_{\boldsymbol{k}s}$ of that branch as a function of the wave vector $\boldsymbol{k}$, that is, they yield the **dispersion relation** of the branch.

Solutions of the matrix eigenvalue equation (13.92) can clearly each be multiplied by a constant of arbitrary magnitude and physical dimensions and still remain a solution. In what follows it will be useful to work with eigenvector solutions $\varepsilon_{\boldsymbol{k}s}$ of (13.92) that are dimensionless [unlike the $\mathsf{u}_{\boldsymbol{k}}$ in (13.88), which clearly has the dimensions of length] and normalized so that the corresponding vectors $\boldsymbol{\varepsilon}_{\boldsymbol{k}s}$ satisfy

$$\boldsymbol{\varepsilon}_{\boldsymbol{k}s} \cdot \boldsymbol{\varepsilon}_{\boldsymbol{k}s'} = \delta_{ss'} \quad (s, s' = 1, 2, 3). \tag{13.95}$$

Because $\mathsf{D}(\boldsymbol{k})$ is an even function of $\boldsymbol{k}$, replacement of $\boldsymbol{k}$ by $-\boldsymbol{k}$ in equation (13.92) gives an unchanged matrix eigenvalue problem, whose solutions must be the same as those of (13.92). Thus, we have

$$\boldsymbol{\varepsilon}_{-\boldsymbol{k},s} = \boldsymbol{\varepsilon}_{\boldsymbol{k}s} \quad \text{and} \quad \omega_{-\boldsymbol{k},s} = \omega_{\boldsymbol{k}s}. \tag{13.96}$$

It is clear from (13.92) that the three vectors $\boldsymbol{\varepsilon}_{\boldsymbol{k}s}$ (for any allowed $\boldsymbol{k}$) that satisfy (13.95) and whose corresponding column-matrix forms $\varepsilon_{\boldsymbol{k}s}$ satisfy (13.92) lie in the directions of motion of the atoms in the three modes $\boldsymbol{k}$, s $(s = 1, 2, 3)$ [the same directions, according to (13.96), as those of the motion of the atoms in the three modes $-\boldsymbol{k}$, s $(s = 1, 2, 3)$, respectively]. For this reason, the vectors $\boldsymbol{\varepsilon}_{\boldsymbol{k}s}$ $(s = 1, 2, 3)$ are called the **polarization vectors** of the mode $\boldsymbol{k}$, s.

It should be noted that only in certain special cases do the three polarization vectors $\boldsymbol{\varepsilon}_{\boldsymbol{k}s}$ for a given allowed $\boldsymbol{k}$ bear a simple relation to the direction of $\boldsymbol{k}$. An example of such a case is when the lattice is simple-cubic and when $\boldsymbol{k}$ is in the direction of one of the main axes. Then one of the three polarization vectors $\boldsymbol{\varepsilon}_{\boldsymbol{k}s}$ (say, $\boldsymbol{\varepsilon}_{\boldsymbol{k}1}$) is parallel to this axis and hence to $\boldsymbol{k}$ (we call this a **longitudinal mode**, with associated frequency $\omega_{\boldsymbol{k}1} \equiv \omega_{\boldsymbol{k}\ell}$), while the other two polarization vectors $\boldsymbol{\varepsilon}_{\boldsymbol{k}2}$ and $\boldsymbol{\varepsilon}_{\boldsymbol{k}3}$ are parallel to the other two main axes, and hence perpendicular to $\boldsymbol{k}$ (we call these **transverse modes**, and their associated frequencies are equal and smaller than $\omega_{\boldsymbol{k}\ell}$: $\omega_{\boldsymbol{k}2} = \omega_{\boldsymbol{k}3} \equiv \omega_{\boldsymbol{k}t} < \omega_{\boldsymbol{k}1} \equiv \omega_{\boldsymbol{k}\ell}$).

Another special case is the case $\boldsymbol{k} = 0$ (modes of infinite wavelength). As can be seen from the normal-mode solution (13.88), the displacement vector in this case is independent of the site position vector $\boldsymbol{R}$, that is, in each of the three modes ($s = 1, 2, 3$) there is a uniform displacement of all atoms in the same direction. Setting $\boldsymbol{k} = 0$ in the dynamical matrix (13.93), and using the second relation (13.87), we find

$$\mathrm{D}(\boldsymbol{k} = 0) = \sum_{\boldsymbol{R}} \mathrm{D}(\boldsymbol{R}) = 0, \tag{13.97}$$

and so the $\boldsymbol{k} = 0$ version of the eigenvalue equation (13.92) tells us that the three modes $\boldsymbol{\varepsilon}_{\boldsymbol{k}s}$ have frequencies $\omega_{\boldsymbol{k},s}$ that all tend to zero as $\boldsymbol{k} \to 0$. These modes are called **acoustic modes**, since at low k their frequencies obey the dispersion relation for sound waves; if, for any given $\boldsymbol{k}$ and s, we expand $\omega_{\boldsymbol{k}s}^2$ in powers of $\boldsymbol{k}$ about $\boldsymbol{k} = 0$, then, because the constant term vanishes (as just proved) and the linear term vanishes by virtue of the fact that $-\boldsymbol{k}$ and $\boldsymbol{k}$ belong to the same frequency [see (13.96)], the leading term in $\omega_{\boldsymbol{k}s}^2$ is proportional to k^2. Hence, we have $\omega_{\boldsymbol{k}s} = v_{\hat{\boldsymbol{k}}s} k$ (the sound-wave dispersion law), where $v_{\hat{\boldsymbol{k}}s}$, the velocity of sound for the mode, is independent of k (and hence of the wavelength $2\pi/k$) but depends on the direction of $\boldsymbol{k}$, that is, on the unit vector $\hat{\boldsymbol{k}}$, and on the mode index s.

In this section we have been describing only the case $r = 1$ (a monatomic basis), for which there are just three modes, of the acoustic type described above. It is worth noting, however, that for the case of general r (a polyatomic basis, with $r > 1$ atoms per primitive cell) there are $3r$ modes, three of which are again acoustic. (In these three modes, as in the monatomic-basis case, the $\boldsymbol{k} = 0$ limits correspond to uniform displacements of all the atoms of the crystal, so that the restoring forces on the atoms are zero and the corresponding mode frequencies also tend to zero.) Besides these three acoustic modes there are then another $3r - 3$ modes, whose $\boldsymbol{k} = 0$ limits describe a situation in which the displacements of corresponding atoms in different primitive cells are the same, but the atoms in the same cell move relatively to each other. Because of this relative motion within each cell the restoring forces are nonzero, leading to nonzero frequencies for these modes even at $\boldsymbol{k} = 0$. Thus, in the case $r = 2$, orthogonality to the acoustic modes, which in the case of general r is expressed by a relation analogous to (13.95), requires that the two atoms in the unit cell move in opposite directions. In the ionic solid sodium chloride (NaCl), for example, this mode involves an oscillating electric dipole moment, which interacts strongly with the electric field of a light wave of the corresponding frequency. Vibrational modes of this kind are called **optical modes**.

We return to the case $r = 1$, and consider now the hamiltonian $H = H(\{q_i\}, \{p_i\})$ of the crystal, where the set $\{q_i\}$ are the same generalized coordinates $u^\mu(\boldsymbol{R})$ [components of the displacement vector $\boldsymbol{u}(\boldsymbol{R})$] as were used in the lagrangian (13.80), and the set $\{p_i\}$ are the corresponding conjugate momenta $p_i = \partial L / \partial \dot{q}_i$, which we shall denote by $P^\mu(\boldsymbol{R})$ [the corresponding components of the momentum $\boldsymbol{P}(\boldsymbol{R})$ of the atom whose equilibrium position is the lattice point $\boldsymbol{R}$]. Thus, the hamiltonian can be written as

$$H = T + V^{\text{harmonic}} = \sum_{\boldsymbol{R},\mu} \frac{1}{2M} (P^\mu(\boldsymbol{R}))^2 + \frac{1}{2} \sum_{\boldsymbol{R},\boldsymbol{R}',\mu,\nu} D_{\mu\nu}(\boldsymbol{R} - \boldsymbol{R}') u^\mu(\boldsymbol{R}) u^\nu(\boldsymbol{R}'), \tag{13.98}$$

To find the allowed vibrational energies (and the corresponding stationary quantum states) of the crystal we should solve the time-independent Schrödinger equation with the hamiltonian operator $\hat{H}$ obtained from H (13.98) by replacing the quantities $u^\mu(\boldsymbol{R})$ and $P^\mu(\boldsymbol{R})$ by the quantum-mechanical operators $\hat{u}^\mu(\boldsymbol{R})$ [$= u^\mu(\boldsymbol{R})$] and $\hat{P}^\mu(\boldsymbol{R})$ [$= -i\hbar\partial / \partial u^\mu(\boldsymbol{R})$]. To solve this problem we shall first show how the hamiltonian (13.98) can be re-expressed as a sum over modes $\boldsymbol{k},s$. For this, we introduce new operators $\hat{b}_{\boldsymbol{k}s}$ (one for each mode $\boldsymbol{k},s$), **defined** by

$$\hat{b}_{\boldsymbol{k}s} = \frac{1}{\sqrt{N}} \sum_{\boldsymbol{R}} \mathrm{e}^{-i\boldsymbol{k}\cdot\boldsymbol{R}} \boldsymbol{\varepsilon}_{\boldsymbol{k}s} \cdot \left(\sqrt{\frac{M\omega_{\boldsymbol{k}s}}{2\hbar}} \hat{\boldsymbol{u}}(\boldsymbol{R}) + i \sqrt{\frac{1}{2\hbar M \omega_{\boldsymbol{k}s}}} \hat{\boldsymbol{P}}(\boldsymbol{R}) \right), \tag{13.99}$$

the adjoint of which is

$$\hat{b}^{\dagger}_{\boldsymbol{k}s} = \frac{1}{\sqrt{N}} \sum_{\boldsymbol{R}} \mathrm{e}^{i\boldsymbol{k}\cdot\boldsymbol{R}} \boldsymbol{\varepsilon}_{\boldsymbol{k}s} \cdot \left(\sqrt{\frac{M\omega_{\boldsymbol{k}s}}{2\hbar}} \hat{\boldsymbol{u}}(\boldsymbol{R}) - i \sqrt{\frac{1}{2\hbar M \omega_{\boldsymbol{k}s}}} \hat{\boldsymbol{P}}(\boldsymbol{R}) \right), \tag{13.100}$$

where we have used the fact that the vector operators $\hat{\boldsymbol{u}}(\boldsymbol{R})$ and $\hat{\boldsymbol{P}}(\boldsymbol{R})$ are, of course, self-adjoint (hermitian). In the expression in the brackets in (13.99) the imaginary part is the momentum-vector operator $\hat{\boldsymbol{P}}(\boldsymbol{R})$ made dimensionless by dividing by the characteristic (for the mode $\boldsymbol{k},s$) momentum $\sqrt{2M\hbar\omega_{\boldsymbol{k}s}}$, while the real part is the displacement-vector operator $\hat{\boldsymbol{u}}(\boldsymbol{R})$ made dimensionless by dividing by the characteristic (for the mode $\boldsymbol{k},s$) length $2\hbar/\sqrt{2M\hbar\omega_{\boldsymbol{k}s}} = \sqrt{2\hbar/M\omega_{\boldsymbol{k}s}}$. Since the polarization vectors $\boldsymbol{\varepsilon}_{\boldsymbol{k}s}$ are also dimensionless, the operators $\hat{b}_{\boldsymbol{k}s}$ (and their adjoints $\hat{b}^{\dagger}_{\boldsymbol{k}s}$) are dimensionless operators.

Using the canonical commutation relations

$$[\hat{u}^{\mu}(\boldsymbol{R}), \hat{P}^{\nu}(\boldsymbol{R}')] = i\hbar\delta_{\mu\nu}\delta_{\boldsymbol{R},\boldsymbol{R}'} \quad \text{and} \quad [\hat{u}^{\mu}(\boldsymbol{R}), \hat{u}^{\nu}(\boldsymbol{R}')] = [\hat{P}^{\mu}(\boldsymbol{R}), \hat{P}^{\nu}(\boldsymbol{R}')] = 0, \tag{13.101}$$

the identity

$$\sum_{\boldsymbol{R}} \mathrm{e}^{i\boldsymbol{k}\cdot\boldsymbol{R}} = \begin{cases} 0 & \text{when } \boldsymbol{k} \text{ is not a reciprocal-lattice vector,} \\ N & \text{when } \boldsymbol{k} \text{ is a reciprocal-lattice vector,} \end{cases} \tag{13.102}$$

and the orthonormalization (13.95) of the polarization vectors $\boldsymbol{\varepsilon}_{\boldsymbol{k}s}$, we easily find the commutation relations for the operators (13.99) and (13.100):

$$\begin{aligned}
[\hat{b}_{\boldsymbol{k}s}, \hat{b}^{\dagger}_{\boldsymbol{k}'s'}] &= \frac{1}{N} \sum_{\boldsymbol{R},\boldsymbol{R}'} \mathrm{e}^{-i(\boldsymbol{k}\cdot\boldsymbol{R}-\boldsymbol{k}'\cdot\boldsymbol{R}')} \sum_{\mu,\nu} \varepsilon^{\mu}_{\boldsymbol{k}s}\varepsilon^{\nu}_{\boldsymbol{k}'s'} \left(\frac{-i}{2\hbar}\right) \sqrt{\frac{\omega_{\boldsymbol{k}s}}{\omega_{\boldsymbol{k}'s'}}} \left([u^{\mu}(\boldsymbol{R}), P^{\nu}(\boldsymbol{R}')] - [P^{\mu}(\boldsymbol{R}), u^{\nu}(\boldsymbol{R}')]\right) \\
&= \frac{1}{N} \sum_{\boldsymbol{R},\boldsymbol{R}'} \mathrm{e}^{-i(\boldsymbol{k}\cdot\boldsymbol{R}-\boldsymbol{k}'\cdot\boldsymbol{R}')} \sum_{\mu,\nu} \varepsilon^{\mu}_{\boldsymbol{k}s}\varepsilon^{\nu}_{\boldsymbol{k}'s'} \left(\frac{-i}{2\hbar}\right) \sqrt{\frac{\omega_{\boldsymbol{k}s}}{\omega_{\boldsymbol{k}'s'}}} 2i\hbar\delta_{\mu\nu}\delta_{\boldsymbol{R}\boldsymbol{R}'} \\
&= \frac{1}{N} \sum_{\boldsymbol{R}} \mathrm{e}^{-i(\boldsymbol{k}-\boldsymbol{k}')\cdot\boldsymbol{R}} \sum_{\mu} \varepsilon^{\mu}_{\boldsymbol{k}s}\varepsilon^{\mu}_{\boldsymbol{k}'s'} \sqrt{\frac{\omega_{\boldsymbol{k}s}}{\omega_{\boldsymbol{k}'s'}}} \\
&= \delta_{\boldsymbol{k}\boldsymbol{k}'}(\boldsymbol{\varepsilon}_{\boldsymbol{k}s}\cdot\boldsymbol{\varepsilon}_{\boldsymbol{k}'s'})\sqrt{\frac{\omega_{\boldsymbol{k}s}}{\omega_{\boldsymbol{k}'s'}}} = \delta_{\boldsymbol{k}\boldsymbol{k}'}(\boldsymbol{\varepsilon}_{\boldsymbol{k}s}\cdot\boldsymbol{\varepsilon}_{\boldsymbol{k}s'})\sqrt{\frac{\omega_{\boldsymbol{k}s}}{\omega_{\boldsymbol{k}s'}}} = \delta_{\boldsymbol{k}\boldsymbol{k}'}\delta_{ss'}\sqrt{\frac{\omega_{\boldsymbol{k}s}}{\omega_{\boldsymbol{k}s'}}} = \delta_{\boldsymbol{k}\boldsymbol{k}'}\delta_{ss'}.
\end{aligned}$$

In the corresponding derivation of $[\hat{b}_{\boldsymbol{k}s}, \hat{b}_{\boldsymbol{k}'s'}]$ the difference of the pair of commutators in the bracketed expression in the first line of the development gets replaced by the sum of the same pair of commutators, which clearly vanishes, and so $[\hat{b}_{\boldsymbol{k}s}, \hat{b}_{\boldsymbol{k}'s'}] = 0$. Similarly (or by taking the adjoint of the relation $[\hat{b}_{\boldsymbol{k}'s'}, \hat{b}_{\boldsymbol{k}s}] = 0$), we find that $[\hat{b}^\dagger_{\boldsymbol{k}s}, \hat{b}^\dagger_{\boldsymbol{k}'s'}] = 0$. Thus, we have

$$[\hat{b}_{\boldsymbol{k}s}, \hat{b}^\dagger_{\boldsymbol{k}'s'}] = \delta_{\boldsymbol{k}\boldsymbol{k}'}\delta_{ss'},$$
$$[\hat{b}_{\boldsymbol{k}s}, \hat{b}_{\boldsymbol{k}'s'}] = [\hat{b}^\dagger_{\boldsymbol{k}s}, \hat{b}^\dagger_{\boldsymbol{k}'s'}] = 0. \tag{13.103}$$

The definition (13.99) and its adjoint (13.100) can be inverted to express the displacement-vector operators $\hat{\boldsymbol{u}}(\boldsymbol{R})$ and momentum-vector operators $\hat{\boldsymbol{P}}(\boldsymbol{R})$ for each lattice site $\boldsymbol{R}$ in terms of the operators $\hat{b}_{\boldsymbol{k}s}$ and $\hat{b}^\dagger_{\boldsymbol{k}s}$. The resulting expressions are

$$\begin{aligned}\hat{\boldsymbol{u}}(\boldsymbol{R}) &= \frac{1}{\sqrt{N}}\sum_{\boldsymbol{k},s}\sqrt{\frac{\hbar}{2M\omega_{\boldsymbol{k}s}}}(\hat{b}_{\boldsymbol{k}s} + \hat{b}^\dagger_{-\boldsymbol{k}s})\boldsymbol{\varepsilon}_{\boldsymbol{k}s}\mathrm{e}^{i\boldsymbol{k}\cdot\boldsymbol{R}},\\ \hat{\boldsymbol{P}}(\boldsymbol{R}) &= \frac{-i}{\sqrt{N}}\sum_{\boldsymbol{k},s}\sqrt{\frac{\hbar M\omega_{\boldsymbol{k}s}}{2}}(\hat{b}_{\boldsymbol{k}s} - \hat{b}^\dagger_{-\boldsymbol{k}s})\boldsymbol{\varepsilon}_{\boldsymbol{k}s}\mathrm{e}^{i\boldsymbol{k}\cdot\boldsymbol{R}}.\end{aligned} \tag{13.104}$$

We verify the first of these by direct substitution of (13.99) and (13.100) into it:

$$\begin{aligned}\frac{1}{\sqrt{N}}\sum_{\boldsymbol{k},s}\sqrt{\frac{\hbar}{2M\omega_{\boldsymbol{k}s}}}(\hat{b}_{\boldsymbol{k}s} + \hat{b}^\dagger_{-\boldsymbol{k}s})\boldsymbol{\varepsilon}_{\boldsymbol{k}s}\mathrm{e}^{i\boldsymbol{k}\cdot\boldsymbol{R}} &= \frac{1}{2N}\sum_{\boldsymbol{R}'}\sum_{\boldsymbol{k},s}\mathrm{e}^{-i\boldsymbol{k}\cdot\boldsymbol{R}'}[(\boldsymbol{\varepsilon}_{\boldsymbol{k}s} + \boldsymbol{\varepsilon}_{-\boldsymbol{k}s})\cdot\hat{\boldsymbol{u}}(\boldsymbol{R}')]\boldsymbol{\varepsilon}_{\boldsymbol{k}s}\mathrm{e}^{i\boldsymbol{k}\cdot\boldsymbol{R}}\\ = \frac{1}{N}\sum_{\boldsymbol{R}'}\sum_{\boldsymbol{k},s}\mathrm{e}^{i\boldsymbol{k}\cdot(\boldsymbol{R}-\boldsymbol{R}')}\sum_{\mu,\nu}\varepsilon^\mu_{\boldsymbol{k}s}\hat{u}^\mu(\boldsymbol{R}')e_\nu\varepsilon^\nu_{\boldsymbol{k}s} &= \frac{1}{N}\sum_{\boldsymbol{R}'}\sum_{\boldsymbol{k}}\mathrm{e}^{i\boldsymbol{k}\cdot(\boldsymbol{R}-\boldsymbol{R}')}\sum_{\mu,\nu}\delta_{\mu\nu}\hat{u}^\mu(\boldsymbol{R}')e_\nu\\ = \frac{1}{N}\sum_{\boldsymbol{R}'}\sum_{\boldsymbol{k}}\mathrm{e}^{i\boldsymbol{k}\cdot(\boldsymbol{R}-\boldsymbol{R}')}\hat{\boldsymbol{u}}(\boldsymbol{R}') &= \sum_{\boldsymbol{R}'}\delta_{\boldsymbol{R}\boldsymbol{R}'}\hat{\boldsymbol{u}}(\boldsymbol{R}') = \hat{\boldsymbol{u}}(\boldsymbol{R}).\end{aligned}$$

where $\boldsymbol{e}_\nu$ is the unit vector in the ν direction. Here, we have used the fact that (in a monatomic Bravais lattice) $\omega_{\boldsymbol{k}s} = \omega_{-\boldsymbol{k}s}$ and $\boldsymbol{\varepsilon}_{\boldsymbol{k}s} = \varepsilon_{-\boldsymbol{k}s}$, the "completeness relation"

$$\sum_{s=1}^{3}\varepsilon^\mu_{\boldsymbol{k}s}\varepsilon^\nu_{\boldsymbol{k}s} = \delta_{\mu\nu} \tag{13.105}$$

(which holds for any complete set of real orthogonal vectors), and the identity

$$\sum_{\boldsymbol{k}}\mathrm{e}^{i\boldsymbol{k}\cdot(\boldsymbol{R}-\boldsymbol{R}')} = N\delta_{\boldsymbol{R}\boldsymbol{R}'}. \tag{13.106}$$

The second expression in (13.104) is verified similarly.

We now use the relations (13.104) to re-express the hamiltonian $\hat{H}$ corresponding to (13.98) in terms of the new oscillator operators $\hat{b}_{\boldsymbol{k}s}$ and $\hat{b}^\dagger_{\boldsymbol{k}s}$. The two parts of $\hat{H}$ become

$$\hat{T} = \sum_{\boldsymbol{R},\mu}\frac{1}{2M}(\hat{P}^\mu(\boldsymbol{R}))^2 = \frac{1}{2M}\sum_{\boldsymbol{R}}\boldsymbol{P}(\boldsymbol{R})\cdot\boldsymbol{P}(\boldsymbol{R})$$

$$= -\frac{1}{4N}\sum_{\boldsymbol{R}}\sum_{\boldsymbol{k},\boldsymbol{k}'}\sum_{s,s'}\hbar\sqrt{\omega_{\boldsymbol{k}s}\omega_{\boldsymbol{k}'s'}}(\hat{b}_{\boldsymbol{k}s}-\hat{b}^{\dagger}_{-\boldsymbol{k}s})(\hat{b}_{\boldsymbol{k}'s'}-\hat{b}^{\dagger}_{-\boldsymbol{k}'s'})(\boldsymbol{\varepsilon}_{\boldsymbol{k}s}\cdot\boldsymbol{\varepsilon}_{\boldsymbol{k}'s'})\mathrm{e}^{i\boldsymbol{k}\cdot\boldsymbol{R}}\mathrm{e}^{i\boldsymbol{k}'\cdot\boldsymbol{R}}$$

$$= -\frac{1}{4}\sum_{\boldsymbol{k},\boldsymbol{k}'}\sum_{s,s'}\hbar\sqrt{\omega_{\boldsymbol{k}s}\omega_{\boldsymbol{k}'s'}}(\hat{b}_{\boldsymbol{k}s}-\hat{b}^{\dagger}_{-\boldsymbol{k}s})(\hat{b}_{\boldsymbol{k}'s'}-\hat{b}^{\dagger}_{-\boldsymbol{k}'s'})\delta_{ss'}\delta_{\boldsymbol{k}',-\boldsymbol{k}}$$

$$= \frac{1}{4}\sum_{\boldsymbol{k},s}\hbar\omega_{\boldsymbol{k}s}(\hat{b}_{\boldsymbol{k}s}-\hat{b}^{\dagger}_{-\boldsymbol{k}s})(\hat{b}^{\dagger}_{\boldsymbol{k}s}-\hat{b}_{-\boldsymbol{k}s}),$$

$$\hat{V}^{\text{harmonic}} = \frac{1}{2}\sum_{\boldsymbol{R},\boldsymbol{R}',\mu,\nu} D_{\mu\nu}(\boldsymbol{R}-\boldsymbol{R}')\hat{u}^{\mu}(\boldsymbol{R})\hat{u}^{\nu}(\boldsymbol{R}')$$

$$= \frac{\hbar}{4NM}\sum_{\boldsymbol{R},\boldsymbol{R}',\mu,\nu} D_{\mu\nu}(\boldsymbol{R}-\boldsymbol{R}')\sum_{\boldsymbol{k},s}\sum_{\boldsymbol{k}',s'}\sqrt{\frac{1}{\omega_{\boldsymbol{k}s}\omega_{\boldsymbol{k}'s'}}}(\hat{b}_{\boldsymbol{k}s}+\hat{b}^{\dagger}_{-\boldsymbol{k}s})(\hat{b}_{\boldsymbol{k}'s'}+\hat{b}^{\dagger}_{-\boldsymbol{k}'s'})\varepsilon^{\mu}_{\boldsymbol{k}s}\varepsilon^{\nu}_{\boldsymbol{k}'s'}\mathrm{e}^{i\boldsymbol{k}\cdot\boldsymbol{R}}\mathrm{e}^{i\boldsymbol{k}'\cdot\boldsymbol{R}'}$$

$$= \frac{\hbar}{4NM}\sum_{\boldsymbol{R},\mu,\nu}\sum_{\boldsymbol{k},s}\sum_{\boldsymbol{k}',s'}\mathrm{e}^{i\boldsymbol{k}'\cdot\boldsymbol{R}}D_{\mu\nu}(\boldsymbol{k}')\sqrt{\frac{1}{\omega_{\boldsymbol{k}s}\omega_{\boldsymbol{k}'s'}}}(\hat{b}_{\boldsymbol{k}s}+\hat{b}^{\dagger}_{-\boldsymbol{k}s})(\hat{b}_{\boldsymbol{k}'s'}+\hat{b}^{\dagger}_{-\boldsymbol{k}'s'})\varepsilon^{\mu}_{\boldsymbol{k}s}\varepsilon^{\nu}_{\boldsymbol{k}'s'}\mathrm{e}^{i\boldsymbol{k}\cdot\boldsymbol{R}}$$

$$= \frac{\hbar}{4M}\sum_{\mu,\nu}\sum_{\boldsymbol{k},s}\sum_{\boldsymbol{k}',s'}\delta_{\boldsymbol{k}',-\boldsymbol{k}}D_{\mu\nu}(\boldsymbol{k}')\sqrt{\frac{1}{\omega_{\boldsymbol{k}s}\omega_{\boldsymbol{k}'s'}}}(\hat{b}_{\boldsymbol{k}s}+\hat{b}^{\dagger}_{-\boldsymbol{k}s})(\hat{b}_{\boldsymbol{k}'s'}+\hat{b}^{\dagger}_{-\boldsymbol{k}'s'})\varepsilon^{\mu}_{\boldsymbol{k}s}\varepsilon^{\nu}_{\boldsymbol{k}'s'}$$

$$= \frac{\hbar}{4M}\sum_{\mu,\nu}\sum_{\boldsymbol{k},s}\sum_{s'}D_{\mu\nu}(-\boldsymbol{k})\varepsilon^{\nu}_{-\boldsymbol{k}s'}\sqrt{\frac{1}{\omega_{\boldsymbol{k}s}\omega_{-\boldsymbol{k}s'}}}(\hat{b}_{\boldsymbol{k}s}+\hat{b}^{\dagger}_{-\boldsymbol{k}s})(\hat{b}_{-\boldsymbol{k}s'}+\hat{b}^{\dagger}_{\boldsymbol{k}s'})\varepsilon^{\mu}_{\boldsymbol{k}s}$$

$$= \frac{\hbar}{4M}\sum_{\mu}\sum_{\boldsymbol{k},s}\sum_{s'}M\omega^{2}_{-\boldsymbol{k}s'}\varepsilon^{\mu}_{-\boldsymbol{k}s'}\sqrt{\frac{1}{\omega_{\boldsymbol{k}s}\omega_{-\boldsymbol{k}s'}}}(\hat{b}_{\boldsymbol{k}s}+\hat{b}^{\dagger}_{-\boldsymbol{k}s})(\hat{b}_{-\boldsymbol{k}s'}+\hat{b}^{\dagger}_{\boldsymbol{k}s'})\varepsilon^{\mu}_{\boldsymbol{k}s}$$

$$= \frac{\hbar}{4}\sum_{\boldsymbol{k},s}\sum_{s'}\omega^{2}_{-\boldsymbol{k}s'}\sqrt{\frac{1}{\omega_{\boldsymbol{k}s}\omega_{-\boldsymbol{k}s'}}}(\hat{b}_{\boldsymbol{k}s}+\hat{b}^{\dagger}_{-\boldsymbol{k}s})(\hat{b}_{-\boldsymbol{k}s'}+\hat{b}^{\dagger}_{\boldsymbol{k}s'})(\boldsymbol{\varepsilon}_{\boldsymbol{k}s}\cdot\boldsymbol{\varepsilon}_{-\boldsymbol{k}s'})$$

$$= \frac{\hbar}{4}\sum_{\boldsymbol{k},s}\sum_{s'}\omega^{2}_{\boldsymbol{k}s'}\sqrt{\frac{1}{\omega_{\boldsymbol{k}s}\omega_{\boldsymbol{k}s'}}}(\hat{b}_{\boldsymbol{k}s}+\hat{b}^{\dagger}_{-\boldsymbol{k}s})(\hat{b}_{-\boldsymbol{k}s'}+\hat{b}^{\dagger}_{\boldsymbol{k}s'})\delta_{ss'} = \frac{1}{4}\sum_{\boldsymbol{k},s}\hbar\omega_{\boldsymbol{k}s}(\hat{b}_{\boldsymbol{k}s}+\hat{b}^{\dagger}_{-\boldsymbol{k}s})(\hat{b}_{-\boldsymbol{k}s}+\hat{b}^{\dagger}_{\boldsymbol{k}s}).$$

where in the third line we have used the explicit (i.e., with matrix indices) form of the first equality in (13.93), in the fourth line we have used (13.102), in the sixth line we have used the fact that the polarization vector $\boldsymbol{\varepsilon}_{-\boldsymbol{k}s'}$ is an eigenvector of the dynamical matrix $\mathsf{D}(-\boldsymbol{k})$ [see (13.92)], and in the last line we have used (13.96) and (13.95).

Thus, the full hamiltonian takes the form

$$\hat{H} = \hat{T} + \hat{V}^{\text{harmonic}}$$

$$= \frac{1}{4}\sum_{\boldsymbol{k},s}\hbar\omega_{\boldsymbol{k}s}(\hat{b}_{\boldsymbol{k}s}-\hat{b}^{\dagger}_{-\boldsymbol{k}s})(\hat{b}^{\dagger}_{\boldsymbol{k}s}-\hat{b}_{-\boldsymbol{k}s}) + \frac{1}{4}\sum_{\boldsymbol{k},s}\hbar\omega_{\boldsymbol{k}s}(\hat{b}_{\boldsymbol{k}s}+\hat{b}^{\dagger}_{-\boldsymbol{k}s})(\hat{b}_{-\boldsymbol{k}s}+\hat{b}^{\dagger}_{\boldsymbol{k}s})$$

$$= \frac{1}{2}\sum_{\boldsymbol{k},s}\hbar\omega_{\boldsymbol{k}s}(\hat{b}_{\boldsymbol{k}s}\hat{b}^{\dagger}_{\boldsymbol{k}s}+\hat{b}^{\dagger}_{\boldsymbol{k}s}\hat{b}_{\boldsymbol{k}s}).$$

Using the first of the commutation relations (13.103) we can rewrite this as

$$\hat{H} = \sum_{\mathbf{k},s} \hbar\omega_{\mathbf{k}s} \left(\hat{b}^{\dagger}_{\mathbf{k}s} \hat{b}_{\mathbf{k}s} + \frac{1}{2} \right) \equiv \sum_{\mathbf{k},s} \hat{H}_{\mathbf{k}s}. \tag{13.107}$$

This is simply a sum of $3N$ independent oscillator hamiltonians $\hat{H}_{\mathbf{k}s}$ (one for each mode $\mathbf{k}$,s, that is, one for each wave vector and polarization), which, by (13.103), commute with each other. But the eigenstates of a sum of commuting sub-hamiltonians are all simply products of the eigenstates of each of the separate sub-hamiltonians, and the corresponding eigenvalues are the sum of the eigenvalues of the sub-hamiltonians. We therefore expect the possible eigenvalues of $\hat{H}$ to take the form

$$E = \sum_{\mathbf{k},s} \hbar\omega_{\mathbf{k}s} \left(n_{\mathbf{k}s} + \frac{1}{2} \right), \tag{13.108}$$

in which $n_{\mathbf{k}s}$ is to be interpreted as the number of quanta of energy (each of magnitude $\hbar\omega_{\mathbf{k}s}$) in a given mode $\mathbf{k}$,s. The quantum of energy $\hbar\omega_{\mathbf{k}s}$ of a given normal mode of vibration of the lattice is called a **phonon**. (The term $\frac{1}{2}\hbar\omega_{\mathbf{k}s}$ in (13.108) is the **zero-point energy** of the mode.)

Comparison of (13.107) and (13.108) suggests that we seek eigenstates $|n_{\mathbf{k}s}\rangle$ of the sub-hamiltonians $\hat{H}_{\mathbf{k}s}$ such that

$$\hat{b}^{\dagger}_{\mathbf{k}s} \hat{b}_{\mathbf{k}s} |n_{\mathbf{k}s}\rangle \equiv \hat{n}_{\mathbf{k}s} |n_{\mathbf{k}s}\rangle = n_{\mathbf{k}s} |n_{\mathbf{k}s}\rangle, \tag{13.109}$$

which defines the **phonon-number operator** $\hat{n}_{\mathbf{k}s}$. The following states satisfy (13.109):

$$|n_{\mathbf{k}s}\rangle = A(\hat{b}^{\dagger}_{\mathbf{k}s})^{n_{\mathbf{k}s}} |0\rangle \tag{13.110}$$

(A is a constant) if the ground state $|0\rangle$ is chosen to satisfy $\hat{b}_{\mathbf{k}s}|0\rangle = 0$. We now check this statement:

$$\begin{aligned} \hat{b}^{\dagger}_{\mathbf{k}s} \hat{b}_{\mathbf{k}s} A(\hat{b}^{\dagger}_{\mathbf{k}s})^{n_{\mathbf{k}s}} |0\rangle &= A\hat{b}^{\dagger}_{\mathbf{k}s} \left((\hat{b}^{\dagger}_{\mathbf{k}s})^{n_{\mathbf{k}s}} \hat{b}_{\mathbf{k}s} + [\hat{b}_{\mathbf{k}s}, (\hat{b}^{\dagger}_{\mathbf{k}s})^{n_{\mathbf{k}s}}] \right) |0\rangle \\ &= A\hat{b}^{\dagger}_{\mathbf{k}s} [\hat{b}_{\mathbf{k}s}, (\hat{b}^{\dagger}_{\mathbf{k}s})^{n_{\mathbf{k}s}}] |0\rangle = n_{\mathbf{k}s} A(\hat{b}^{\dagger}_{\mathbf{k}s})^{n_{\mathbf{k}s}} |0\rangle, \end{aligned}$$

where we have used the relation (with the common subscript $\mathbf{k}s$ suppressed)

$$\begin{aligned} [\hat{b}, (\hat{b}^{\dagger})^n] &= [\hat{b}, \hat{b}^{\dagger}](\hat{b}^{\dagger})^{n-1} + \hat{b}^{\dagger}[\hat{b}, \hat{b}^{\dagger}](\hat{b}^{\dagger})^{n-2} + \ldots + (\hat{b}^{\dagger})^{n-2}[\hat{b}, \hat{b}^{\dagger}]\hat{b}^{\dagger} + (\hat{b}^{\dagger})^{n-1}[\hat{b}, \hat{b}^{\dagger}] \\ &= n(\hat{b}^{\dagger})^{n-1}, \end{aligned} \tag{13.111}$$

in which the second equality follows from the first commutation relation (13.103).

We shall require the states (13.109) (including $|0\rangle$) to be normalized to unity, that is,

$$\begin{aligned} \langle n_{\mathbf{k}s} \mid n_{\mathbf{k}s}\rangle &= AA^* \langle 0| (\hat{b}_{\mathbf{k}s})^{n_{\mathbf{k}s}} (\hat{b}^{\dagger}_{\mathbf{k}s})^{n_{\mathbf{k}s}} |0\rangle = |A|^2 \langle 0| (\hat{b}_{\mathbf{k}s})^{n_{\mathbf{k}s}-1} n_{\mathbf{k}s} (\hat{b}^{\dagger}_{\mathbf{k}s})^{n_{\mathbf{k}s}-1} |0\rangle \\ &= |A|^2 \langle 0| (\hat{b}_{\mathbf{k}s})^{n_{\mathbf{k}s}-2} n_{\mathbf{k}s}(n_{\mathbf{k}s} - 1) (\hat{b}^{\dagger}_{\mathbf{k}s})^{n_{\mathbf{k}s}-2} |0\rangle = \ldots \\ &= |A|^2 n_{\mathbf{k}s}! \langle 0 \mid 0\rangle = |A|^2 n_{\mathbf{k}s}! = 1, \end{aligned}$$

where we have used (13.111) n_{ks} times. Choosing A to be real, we then have $A = 1/\sqrt{n_{ks}!}$ and the normalized states (13.110) are

$$|n_{ks}\rangle = \frac{1}{\sqrt{n_{ks}!}}(\hat{b}^{\dagger}_{ks})^{n_{ks}}\,|0\rangle\,. \tag{13.112}$$

Equation (13.112) implies that

$$\hat{b}^{\dagger}_{ks}\,|n_{ks}\rangle = \frac{1}{\sqrt{n_{ks}!}}(\hat{b}^{\dagger}_{ks})^{n_{ks}+1}\,|0\rangle = \frac{\sqrt{n_{ks}+1}}{\sqrt{(n_{ks}+1)!}}(\hat{b}^{\dagger}_{ks})^{n_{ks}+1}\,|0\rangle = \sqrt{n_{ks}+1}\,|n_{ks}+1\rangle \tag{13.113}$$

and

$$\hat{b}_{ks}\,|n_{ks}\rangle = \frac{1}{\sqrt{n_{ks}!}}\hat{b}_{ks}(\hat{b}^{\dagger}_{ks})^{n_{ks}}\,|0\rangle = \frac{n_{ks}}{\sqrt{n_{ks}!}}(\hat{b}^{\dagger}_{ks})^{n_{ks}-1}\,|0\rangle = \sqrt{n_{ks}}\,|n_{ks}-1\rangle\,, \tag{13.114}$$

where in (13.114) we have again used (13.111) (and also (13.109) applied to the state $|0\rangle$). Since, from (13.113), $\hat{b}^{\dagger}_{ks}$ clearly increases the number of phonons belonging to the mode $\boldsymbol{k}$,s by unity, it is called a **phonon-creation operator**. Similarly, from (13.114), $\hat{b}_{ks}$ is clearly a **phonon-annihilation operator**.

Since the number n_{ks} of phonons in the mode $\boldsymbol{k}$,s can take any positive integer value (as well as the value zero), phonons are **bosons**.

When the original hamiltonian (13.98) is supplemented by **anharmonic** terms $\sim u^3$ it is no longer possible to express the hamiltonian as a sum over independent normal-mode sub-hamiltonians $\hat{H}_{ks}$, with eigenstates $|n_{ks}\rangle$ corresponding to a well defined number of phonons. In other words, states constructed as products of the states $|n_{ks}\rangle$ over all normal modes are no long exact eigenstates of the full hamiltonian of the system. If, however, the anharmonicity is weak enough, they are approximate eigenstates, that is, we can regard the states $|n_{ks}\rangle$ as having a finite lifetime, arising from phonon–phonon scattering processes. These processes can then be built into an effective hamiltonian constructed in terms of the phonon creation and annihilation operators $\hat{b}^{\dagger}_{ks}$ and $\hat{b}_{ks}$. With increase of temperature we expect the mean square displacement of each atom from its equilibrium position to increase, with a consequent increase in the importance of the anharmonic terms in the hamiltonian. Thus, only at low temperatures do we expect the phonons to be sufficiently well-defined excitations for a phonon-based description of the system to be useful.

13.7.2 Phonons in liquid helium-4

Consider a liquid with density $\rho(\boldsymbol{r})$ at some given time t. The deviation of this density from ρ_0 (the space-averaged density at this time) is

$$\delta\rho(\boldsymbol{r}) = \rho(\boldsymbol{r}) - \rho_0. \tag{13.115}$$

This "density fluctuation" can be Fourier-decomposed into harmonic waves of wave vector $\boldsymbol{k}$ (wavelength $\lambda = 2\pi/k$) and corresponding frequency $\omega(k)$. For example, in liquid helium-4 (^{4}He) at temperatures approaching absolute zero these harmonic density waves are the dominant excitations. For small k the spectrum is acoustic, that is, $\omega(k) = vk$, where v is the speed of sound. (A spectrum of this type is central to the explanation of the superfluidity of liquid helium-4 below a certain temperature T_λ.) As with the normal modes in elastic solids, quantization of the problem leads to a description in terms of which energy can be added to or taken from a given mode (with wave number k) only in the form of quanta of magnitude $\hbar\omega(k)$. As in the case of elastic solids, each such quantum of energy is called a **phonon**. Unlike the case of elastic solids with purely harmonic interatomic interaction, however, in liquid helium-4 the phonons are not exact excitations, and so in a phonon-based description phonon–phonon scatterings occur. Another difference is that, whereas in a crystal the restoring force giving rise to the elastic waves arises from nearest-neighbour interactions, in liquid helium-4 it arises from the fact that the interparticle collisions that suppress fluctuations away from thermodynamic equilibrium also tend to suppress any deviation of the density $\rho(\boldsymbol{r})$ from its thermal equilibrium value ρ_0.

13.7.3 Magnons in solids

In solids with atoms with nonzero spins we can define a spin-density field (analogous to the density field $\rho(\boldsymbol{r})$ in the case considered in Section 13.7.2), consider its deviation from the mean spin density, and Fourier-decompose this deviation (spin fluctuation) into harmonic waves of wavelength $\lambda = 2\pi/k$ and corresponding frequency $\omega(k)$. The dispersion relation depends on the type of magnetic-exchange coupling of the spins in the solid: for ferromagnetic coupling (preferential parallel alignment of neighbouring spins) we have $\omega(k) \propto k^2$, while for antiferromagnetic coupling (preferential antiparallel alignment of neighbouring spins) we have $\omega(k) \propto k$. Energy can be added to or taken from a spin-density wave of given wave number k only in the form of quanta of magnitude $\hbar\omega(k)$, called **magnons**. Like the phonons in an anharmonic solid or a liquid, the magnons are not exact excitations, that is, if we try to transform the original exchange hamiltonian into a sum over $\boldsymbol{k}$ of $\hbar\omega(k)$ multiplied by the appropriate magnon occupation number, we find that we also get terms corresponding to magnon–magnon scattering. Although the spins giving rise to magnon excitations may be spin-1/2 particles (and hence fermions), magnons are Bose-type excitations, that is, there are no restrictions on the number of magnons in a given mode $\boldsymbol{k}$.

13.7.4 Polarons and excitons

If we were able to neglect the motion of the atoms or ions in a crystal, and also neglect electron–electron interactions, the states of the electrons in the crystal would be described by **Bloch functions**, of the form

$$\langle \boldsymbol{r} \mid \boldsymbol{k}\rangle \equiv \phi_{\boldsymbol{k}}(\boldsymbol{r}) = u_{\boldsymbol{k}}(\boldsymbol{r})\mathrm{e}^{i\boldsymbol{k}\cdot\boldsymbol{r}}, \tag{13.116}$$

where the function $u_{\boldsymbol{k}}(\boldsymbol{r})$ has the periodicity of the lattice, that is, $u_{\boldsymbol{k}}(\boldsymbol{r}+\boldsymbol{R}) = u_{\boldsymbol{k}}(\boldsymbol{r})$ for any lattice vector $\boldsymbol{R}$. In a real crystal, however, the atoms or ions vibrate and the Bloch states are no longer exact energy eigenstates. However, as in the case of a Fermi liquid, under certain circumstances the Bloch wave functions provide a reasonably good description of the electrons, in that the states described by them have a reasonably long, if not infinite, lifetime. A **Bloch state** corresponding to some wave vector $\boldsymbol{k}$ has a finite energy width γ, and so overlaps other Bloch states, corresponding to other wave vectors $\boldsymbol{k}'$. Thus, the decay of this state may be described as a scattering process in which the electron moves from a state with momentum $\hbar\boldsymbol{k}$ to a state with momentum $\hbar(\boldsymbol{k}-\boldsymbol{q})$, the momentum difference being provided (or absorbed) by the crystal lattice; that is, **electron–phonon scattering** occurs. This process may be described by the diagram in figure 13.9.

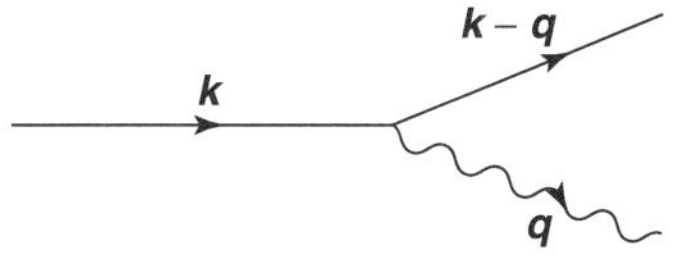

Figure 13.9 *Elementary electron–phonon interaction process.*

This description implies that the hamiltonian $\hat{H}$ of the crystal can be written as the sum of an electron part $\hat{H}_{\text{el}}$ (with a static periodic potential), a phonon part $\hat{H}_{\text{ph}}$ corresponding to a lattice with harmonic interatomic forces, and a part $\hat{H}_{\text{el-ph}}$ corresponding to the electron–phonon interaction that causes the scattering:

$$\hat{H} = \hat{H}_{\text{el}} + \hat{H}_{\text{ph}} + \hat{H}_{\text{el-ph}} \equiv \hat{H}_0 + \hat{H}_{\text{el-ph}} \tag{13.117}$$

The exact energy eigenstates of $\hat{H}_0$ (that is, when $\hat{H}_{\text{el-ph}}$ is absent) are simply products $|\boldsymbol{k};n\rangle^{(0)} \equiv |\boldsymbol{k}\rangle\,|n\rangle$ of Bloch electron states $|\boldsymbol{k}\rangle$ with phonon states $|n\rangle$, where n denotes a complete set of phonon occupation numbers for all the $3rN$ normal vibrational modes $\boldsymbol{q},s$ of the crystal.

After the term $\hat{H}_{\text{el-ph}}$ has been included in the hamiltonian, $\boldsymbol{k}$ and n are no longer good quantum numbers and the states $|\boldsymbol{k};n\rangle^{(0)}$ are no longer exact eigenstates of $\hat{H}$. If, however, $\hat{H}_{\text{el-ph}}$ is small enough, we can use perturbation theory to find approximations to the eigenstates of $\hat{H}$. For example, starting from the unperturbed state $|\boldsymbol{k};0\rangle^{(0)}$ with no phonons, we find, in first-order perturbation theory,

$$|\boldsymbol{k};0\rangle^{(1)} = |\boldsymbol{k};0\rangle^{(0)} + \sum_{\boldsymbol{q},s} \left|\boldsymbol{k}-\boldsymbol{q};1_{\boldsymbol{q},s}\right\rangle^{(0)} \frac{{}^{(0)}\langle \boldsymbol{k}-\boldsymbol{q};1_{\boldsymbol{q},s}|\hat{H}_{\text{el-ph}}|\boldsymbol{k};0\rangle^{(0)}}{\varepsilon_{\boldsymbol{k}}^{(0)} - \left(\varepsilon_{\boldsymbol{k}-\boldsymbol{q}}^{(0)} + \hbar\omega_{\boldsymbol{q}}^{(0)}\right)}, \tag{13.118}$$

where the sum represents an admixture of states $\left|\boldsymbol{k}-\boldsymbol{q};1_{\boldsymbol{q},s}\right\rangle^{(0)}$ with just one phonon in the normal mode $\boldsymbol{q},s$ and no phonons in any other normal modes, and $\varepsilon_{\boldsymbol{k}}^{(0)}$ is the energy of the electron in the unperturbed Bloch state $\boldsymbol{k}$. The expectation value $\langle n_{\text{ph}}\rangle$ of the phonon-number operator

$$\hat{n}_{\text{ph}} \equiv \sum_{\boldsymbol{q},s} \hat{n}_{\boldsymbol{q},s} = \sum_{\boldsymbol{q},s} \hat{b}_{\boldsymbol{q}s}^{\dagger}\hat{b}_{\boldsymbol{q}s} \tag{13.119}$$

in the state $|\boldsymbol{k};0\rangle^{(1)}$ gives a measure of the number of phonons in the "**phonon cloud**" associated with an electron in the Bloch state $\boldsymbol{k}$. The electron plus its phonon cloud is called a **polaron** and is a quasiparticle in the sense previously defined in our discussion of the Fermi liquid. In particular, by expressing the energy of the polaron (as derived from perturbation theory) in the free-particle form $\varepsilon_k = \hbar^2k^2/2m^*$ we can determine polaron's effective mass m^*.

Two types of polaron may be distinguished. In **covalent semiconductors**, for example, silicon and germanium, the concentration of charge carriers is low, screening effects can be neglected, and the chief contribution to $\hat{H}_{\text{el-ph}}$ comes from the long-wave acoustic phonons that give rise to dilation of the crystal on the scale of the wavelength λ. It is the change in the electron energy due to this dilation that is the most important effect here, and the corresponding electron–phonon interaction is called the **deformation potential**. In such cases the expectation value $\langle n_{\text{ph}}\rangle$ in the state $|\boldsymbol{k};0\rangle^{(1)}$ is very small ($\sim 10^{-2}$); that is, we have a small phonon cloud, validating the use of perturbation theory.

In **insulating polar crystals**, such as NaCl, AgBr, and so on, the electrons interact strongly with the longitudinal optical phonons through the long-range Coulomb field of the corresponding polarization wave. (The transverse optical phonons give rise to a smaller electric field, and the interaction of electrons with these phonons will be smaller.) The result is that in this case $\langle n_{\text{ph}}\rangle$ is large (for example, 2.8 for NaCl), so that, although this result is qualitatively useful, the use of first-order perturbation theory is not fully justified in this case.

In insulators and certain semiconductors, a further type of excitation can sometimes exist in the energy gap between the valence band and conduction band. These correspond to bound electron–hole pairs and are known as **excitons**. Excitons of small radius are known as **Frenkel excitons**, while those of large radius are called **Mott excitons**.

14
Second Quantization

14.1 The occupation-number representation

In Section 13.2 we described the states of systems of noninteracting particles by N-particle wave functions

$$\Phi_{N_1,N_2,\ldots,N_\infty}(\xi_1,\ldots,\xi_N) \qquad \left(\sum_{i=1}^{\infty} N_i = N\right).$$

In the case of bosons these are given by equations (13.5) with (13.9), and in the case of fermions they are given by (13.18) with (13.19). These wave functions in the coordinate representation can be written, in the usual way (see Chapter 2 on representation theory), in the alternative form

$$\Phi_{N_1,N_2,\ldots,N_\infty}(\xi_1,\ldots,\xi_N) \equiv \langle \xi_1,\ldots,\xi_N \mid N_1, N_2,\ldots,N_\infty\rangle, \tag{14.1}$$

and in what follows we shall derive a neater, coordinate-free description of many-particle systems in terms of the kets $|N_1, N_2,\ldots,N_\infty\rangle$; that is, we shall work in what is known as the **occupation-number representation**. Since the wave functions (14.1) are orthogonal and normalized to unity, the kets $|N_1, N_2,\ldots,N_\infty\rangle$ form an orthonormal set:

$$\langle N_1, N_2,\ldots,N_\infty|N_1', N_2',\ldots,N_\infty'\rangle = \delta_{N_1'N_1}\delta_{N_2'N_2}\ldots\delta_{N_\infty'N_\infty}. \tag{14.2}$$

Suppose that we have an operator $F^{(1)}$ in the form of a sum of single-particle operators, that is, operators each acting on the coordinates of a single particle:

$$F^{(1)} = \sum_{n=1}^{N} f^{(1)}(\xi_n) \quad \left(\text{e.g., the kinetic-energy operator} - (\hbar^2/2m)\sum_{n=1}^{N}\nabla_n^2\right). \tag{14.3}$$

What is the form of the corresponding operator acting in the Hilbert space spanned by the kets $|N_1, N_2,\ldots,N_\infty\rangle$? We shall answer this question by determining the form of the operator $\hat{F}^{(1)}$ in the

A Course in Theoretical Physics, First Edition. P. J. Shepherd.

following expression:

$$\begin{aligned}&\langle N_1', N_2', \ldots, N_\infty' | \hat{F}^{(1)} | N_1, N_2, \ldots, N_\infty \rangle \\ &\equiv \int \Phi^*_{N_1', N_2', \ldots, N_\infty'}(\xi_1, \ldots, \xi_N) F^{(1)}(\xi_1, \ldots, \xi_N) \Phi_{N_1, N_2, \ldots, N_\infty}(\xi_1, \ldots, \xi_N) \mathrm{d}\xi_1 \ldots \mathrm{d}\xi_N \\ &= \sum_{n=1}^{N} \int \Phi^*_{N_1', N_2', \ldots, N_\infty'}(\xi_1, \ldots, \xi_N) f^{(1)}(\xi_n) \Phi_{N_1, N_2, \ldots, N_\infty}(\xi_1, \ldots, \xi_N) \mathrm{d}\xi_1 \ldots \mathrm{d}\xi_N. \end{aligned} \tag{14.4}$$

We find the form of $\hat{F}^{(1)}$ first in the boson case, in which $\Phi_{N_1, N_2, \ldots, N_\infty}(\xi_1, \ldots, \xi_N)$ is given by the symmetrized sum (13.5) with (13.9). Consider first any single term in (14.4) – say, the term with $n = p$, corresponding to particle p. Because of the orthonormality (13.4) of the single-particle states, and because $f^{(1)}(\xi_p)$ is a single-particle operator, this term will always be zero unless, in (14.4), either (a) $\Phi_{N_1', N_2', \ldots, N_\infty'}(\xi_1, \ldots, \xi_N)$ and $\Phi_{N_1, N_2, \ldots, N_\infty}(\xi_1, \ldots, \xi_N)$ describe the same N-particle state ($N_i' = N_i$ for all i), or (b) $\Phi_{N_1', N_2', \ldots, N_\infty'}(\xi_1, \ldots, \xi_N)$ and $\Phi_{N_1, N_2, \ldots, N_\infty}(\xi_1, \ldots, \xi_N)$ describe two N-particle states that differ in the state of just one particle (so that, say, $N_k' = N_k + 1, \ldots, N_i' = N_i - 1$, ensuring that all the occupation numbers still sum to N).

In the case (a), the $n = p$ term in (14.4) takes the form

$$\sum_{\substack{\text{occupied}\\ i}} \left(\frac{N_1! N_2! \ldots N_\infty!}{N!} \right) S_i \int \psi^*_{\alpha_i}(\xi_p) f^{(1)}(\xi_p) \psi_{\alpha_i}(\xi_p) \mathrm{d}\xi_p, \tag{14.5}$$

where, in a given term i in the sum, the number S_i is the sum of all the unit multiple integrals (each of them a product of $N-1$ unit normalization integrals) obtained from corresponding terms in Φ (13.5) and its complex conjugate Φ^*, that is, from corresponding terms in the distinct permutations of the remaining $N-1$ particles amongst the occupied single-particle states $\psi_{\alpha_i}, \psi_{\alpha_j}, \ldots, \psi_{\alpha_l}$ when only $N_i - 1$ (rather than N_i) particles occupy the state ψ_{α_i}. Clearly,

$$S_i = \frac{(N-1)!}{N_1! N_2! \ldots (N_i - 1)! \ldots N_\infty!}, \tag{14.6}$$

so the $n = p$ term in (14.4) becomes

$$\sum_i \frac{N_i}{N} \int \psi^*_{\alpha_i}(\xi_p) f^{(1)}(\xi_p) \psi_{\alpha_i}(\xi_p) \mathrm{d}\xi_p, \tag{14.7}$$

where, because of the factor N_i, we can drop the qualification "occupied" in the sum. Clearly, all N terms in the sum (14.4) take the value (14.7), and so, for case (a), (14.4) becomes

$$\langle N_1, N_2, \ldots, N_\infty | \hat{F}^{(1)} | N_1, N_2, \ldots, N_\infty \rangle = \sum_i N_i f_{ii}^{(1)}, \tag{14.8}$$

where we have introduced the notation

$$f_{ki}^{(1)} \equiv \int \psi^*_{\alpha_k}(\xi) f^{(1)}(\xi) \psi_{\alpha_i}(\xi) \mathrm{d}\xi. \tag{14.9}$$

In the case (b), the $n = p$ term in (14.4) takes the form

$$\int \Phi^*_{N_1,N_2,\ldots,N_i-1,\ldots,N_k+1,\ldots,N'_\infty}(\xi_1,\ldots,\xi_N) f^{(1)}(\xi_p) \Phi_{N_1,N_2,\ldots,N_i,\ldots,N_k,\ldots,N_\infty}(\xi_1,\ldots,\xi_N)\mathrm{d}\xi_1\ldots\mathrm{d}\xi_N$$
$$= \sqrt{\frac{N_1!N_2!\ldots(N_i-1)!\ldots(N_k+1)!\ldots}{N!}}\sqrt{\frac{N_1!N_2!\ldots N_i!\ldots N_k!\ldots}{N!}} S_i \int \psi^*_{\alpha_k}(\xi_p) f^{(1)}(\xi_p)\psi_{\alpha_i}(\xi_p)\mathrm{d}\xi_p$$
$$= \left(\frac{N_1!N_2!\ldots(N_i-1)!\ldots N_k!\ldots}{(N-1)!}\right)\frac{\sqrt{N_i(N_k+1)}}{N} S_i f^{(1)}_{ki},$$

where the number S_i is again given by the expression (14.6), by the same arguments that led to that expression in case (a). Thus, the $n = p$ term in (14.4) becomes

$$\int \Phi^*_{N_1,N_2,\ldots,N_i-1,\ldots,N_k+1,\ldots,N'_\infty}(\xi_1,\ldots,\xi_N) f^{(1)}(\xi_p) \Phi_{N_1,N_2,\ldots,N_i,\ldots,N_k,\ldots,N_\infty}(\xi_1,\ldots,\xi_N)\mathrm{d}\xi_1\ldots\mathrm{d}\xi_N$$
$$= \frac{\sqrt{N_i(N_k+1)}}{N} f^{(1)}_{ki},$$

and, since all N terms in the sum (14.4) take this value, (14.4) becomes, in case (b),

$$\langle N_1, N_2, \ldots, N_i-1, \ldots, N_k+1, \ldots, N_\infty| \hat{F}^{(1)} |N_1, N_2, \ldots, N_i, \ldots, N_k, \ldots, N_\infty\rangle = \sqrt{N_i(N_k+1)} f^{(1)}_{ki} \tag{14.10}$$

We now assert that an operator $\hat{F}^{(1)}$ satisfying (14.8) and (14.10) can be written in the form

$$\hat{F}^{(1)} = \sum_{k,i} f^{(1)}_{ki} \hat{C}_k \hat{A}_i. \tag{14.11}$$

where the operators $\hat{A}_i$ and $\hat{C}_i$ have no effect on the occupation numbers of single-particle states other than ψ_{α_i}, but the operator $\hat{A}_i$ annihilates a particle in state ψ_{α_i} and the operator $\hat{C}_k$ creates a particle in state ψ_{α_k}:

$$\begin{aligned} \hat{A}_i |N_i\rangle &= \sqrt{N_i}\,|N_i-1\rangle, \\ \hat{C}_i |N_i\rangle &= \sqrt{N_i+1}\,|N_i+1\rangle. \end{aligned} \tag{14.12}$$

The assertion is easily proved by substituting (14.11) into (14.8) and (14.10) and making use of (14.12) and the orthonormality property (14.2). Note that, because

$$\hat{C}_i\hat{A}_i |N_i\rangle = \hat{C}_i\sqrt{N_i}\,|N_i-1\rangle = \sqrt{N_i}\sqrt{N_i-1+1}\,|N_i-1+1\rangle = N_i\,|N_i\rangle, \tag{14.13}$$

whereas

$$\hat{A}_i\hat{C}_i |N_i\rangle = \hat{A}_i\sqrt{N_i+1}\,|N_i+1\rangle = \sqrt{N_i+1}\sqrt{N_i+1}\,|N_i-1+1\rangle = (N_i+1)\,|N_i\rangle, \tag{14.14}$$

the form obtained from a modified (14.11) with the creation and annihilation operators in the opposite order does not yield the required relation (14.8) [although it does yield (14.10)].

Subtraction of (14.13) from (14.14) gives

$$(\hat{A}_i\hat{C}_i - \hat{C}_i\hat{A}_i)\,|N_1, N_2, \ldots, N_\infty\rangle = |N_1, N_2, \ldots, N_\infty\rangle \tag{14.15}$$

for all states $|N_1, N_2, \ldots, N_\infty\rangle$, that is, we have the commutation relation $[\hat{A}_i, \hat{C}_i] = 1$. Consideration of the effects of $\hat{A}_i\hat{C}_k$ and $\hat{C}_k\hat{A}_i$ on any state $|N_1, N_2, \ldots, N_\infty\rangle$ yields, similarly, $[\hat{A}_i, \hat{C}_k] = 0$ for $k \neq i$. By considering the effects of $\hat{A}_i\hat{A}_k$ and $\hat{A}_k\hat{A}_i$ on any state $|N_1, N_2, \ldots, N_\infty\rangle$ we find the result $[\hat{A}_i, \hat{A}_k] = 0$, and similar consideration of the effects of $\hat{C}_i\hat{C}_k$ and $\hat{C}_k\hat{C}_i$ yields the result $[\hat{C}_i, \hat{C}_k] = 0$. Thus, the full set of commutation relations for the creation and annihilation operators is

$$[\hat{A}_i, \hat{C}_k] = \delta_{ik}, \quad [\hat{A}_i, \hat{A}_k] = 0, \quad [\hat{C}_i, \hat{C}_k] = 0. \tag{14.16}$$

The operators $\hat{A}_i$ and $\hat{C}_i$ are mutually adjoint, as we now demonstrate. Consider the matrix element [see (14.12)]

$$\langle N_i - 1|\,\hat{A}_i\,|N_i\rangle = \sqrt{N_i}\,\langle N_i - 1 \mid N_i - 1\rangle = \sqrt{N_i}. \tag{14.17}$$

We can rewrite this as

$$\langle N_i|\,\hat{A}_i^\dagger\,|N_i - 1\rangle^* = \sqrt{N_i}, \tag{14.18}$$

or, since $\sqrt{N_i}$ is real, as

$$\langle N_i|\,\hat{A}_i^\dagger\,|N_i - 1\rangle = \sqrt{N_i}. \tag{14.19}$$

But [see (14.12)]

$$\langle N_i|\,\hat{C}_i\,|N_i - 1\rangle = \sqrt{N_i - 1 + 1}\,\langle N_i \mid N_i - 1 + 1\rangle = \sqrt{N_i}. \tag{14.20}$$

Comparing (14.19) and (14.20) we see that

$$\hat{C}_i = \hat{A}_i^\dagger \quad (\text{and so } \hat{C}_i^\dagger = (\hat{A}_i^\dagger)^\dagger = \hat{A}_i), \tag{14.21}$$

that is, $\hat{C}_i$ and $\hat{A}_i$ are mutually adjoint.

In the light of this, we introduce the standard notation $\hat{b}_i \equiv \hat{A}_i$ and $\hat{b}_i^\dagger \equiv \hat{C}_i$, that is, $\hat{b}_i$ is a boson-annihilation operator and $\hat{b}_i^\dagger$ is a boson-creation operator ($\hat{b}_i$ and $\hat{b}_i^\dagger$ annihilate and create, respectively, a boson in the single-particle state ψ_{α_i}). In this notation, the commutation relations (14.16) are

$$[\hat{b}_i, \hat{b}_k^\dagger] = \delta_{ik}, \quad [\hat{b}_i, \hat{b}_k] = 0, \quad [\hat{b}_i^\dagger, \hat{b}_k^\dagger] = 0. \tag{14.22}$$

Then (14.11) [with (14.9)] takes the form

$$\hat{F}^{(1)} = \sum_{k,i} f_{ki}^{(1)}\hat{b}_k^\dagger\hat{b}_i, \quad \text{with} \quad f_{ki}^{(1)} \equiv \int \psi_{\alpha_k}^*(\xi) f^{(1)}(\xi)\psi_{\alpha_i}(\xi)\mathrm{d}\xi \tag{14.23}$$

Still working with the boson case, our next task is to find the form of the operator $\hat{F}^{(2)}$ that is the equivalent (in the occupation-number representation) of a symmetric sum of two-particle operators $f^{(2)}(\xi_m, \xi_n)$ (in the coordinate representation), that is, the form of the operator $\hat{F}^{(2)}$ in the following expression:

$$\langle N'_1, N'_2, \ldots, N'_\infty|\hat{F}^{(2)}|N_1, N_2, \ldots, N_\infty\rangle$$
$$= \frac{1}{2}\sum_{\substack{m,n \\ (m\neq n)}} \int \Phi^*_{N'_1,N'_2,\ldots,N'_\infty}(\xi_1, \ldots, \xi_N) f^{(2)}(\xi_m, \xi_n)\Phi_{N_1,N_2,\ldots,N_\infty}(\xi_1, \ldots, \xi_N)\mathrm{d}\xi_1 \ldots \mathrm{d}\xi_N. \quad (14.24)$$

As before, this can be done by first evaluating one term in the double sum over particles in (14.24), using the wave functions $\Phi_{N_1,N_2,\ldots,N_\infty}(\xi_1, \ldots, \xi_N)$ given by the symmetrized sum (13.5) with (13.9). Clearly, the operator $f^{(2)}(\xi_m, \xi_n)$ can connect different single-particle states of no more than two particles, and the integrals over the coordinates of the remaining $N - 2$ particles in (14.24) must involve the same set of single-particle states in the factor $\Phi^*_{N'_1,N'_2\ldots}$ as in the factor $\Phi_{N_1,N_2\ldots}$. [Otherwise, the multiple integral in (14.24) would vanish by the orthogonality property (13.4).] The result is independent of the particle pair chosen (as one would expect, since the form of the fully symmetrized N-particle wave functions used has the indistinguishability of identical bosons built into it), and so multiplying it by $N(N-1)/2$ [the number of terms in the double sum in (14.24)] yields an expression from which we can deduce the form of $\hat{F}^{(2)}$ in (14.24). The result is

$$\hat{F}^{(2)} = \frac{1}{2}\sum_i\sum_k\sum_l\sum_m f^{(2)}_{ik,lm}\hat{b}^\dagger_i\hat{b}^\dagger_k\hat{b}_m\hat{b}_l. \quad (14.25)$$

where

$$f^{(2)}_{ik,lm} = \int\int \psi^*_i(\xi_1)\psi^*_k(\xi_2)f^{(2)}(\xi_1, \xi_2)\psi_l(\xi_1)\psi_m(\xi_2)\mathrm{d}\xi_1\mathrm{d}\xi_2. \quad (14.26)$$

Note that, since the boson-annihilation operators $\hat{b}_m$ and $\hat{b}_l$ commute, we could have put these operators in the "more natural" order $\hat{b}_l\hat{b}_m$ in (14.25). We have not done so, since the corresponding two-particle operator for the fermion case, as we show below, also has the form (14.25) but with annihilation operators (and creation operators) that anticommute, so that interchanging two fermion-annihilation operators pertaining to different single-particle states would introduce a minus sign into (14.25).

We now find the forms taken by the operators $\hat{F}^{(1)}$ and $\hat{F}^{(2)}$ in the fermion case. The wave function of a specific configuration of noninteracting fermions is given by equations (13.18) and (13.19):

$$\Phi_{N_1,N_2,\ldots,N_\infty}(\xi_1, \ldots, \xi_N) == \frac{1}{\sqrt{N!}}\begin{vmatrix} \psi_{\alpha_i}(\xi_1) & \psi_{\alpha_i}(\xi_2) & \cdot & \cdot & \cdot & \psi_{\alpha_i}(\xi_N) \\ \psi_{\alpha_j}(\xi_1) & \psi_{\alpha_j}(\xi_2) & \cdot & \cdot & \cdot & \psi_{\alpha_j}(\xi_N) \\ \cdot & \cdot & \cdot & \cdot & \cdot & \cdot \\ \cdot & \cdot & \cdot & \cdot & \cdot & \cdot \\ \cdot & \cdot & \cdot & \cdot & \cdot & \cdot \\ \psi_{\alpha_l}(\xi_1) & \psi_{\alpha_l}(\xi_2) & \cdot & \cdot & \cdot & \psi_{\alpha_l}(\xi_N) \end{vmatrix}. \quad (14.27)$$

Clearly, some convention is required to specify the sign of this wave function. Given that the particle coordinates appear in successive columns, from left to right, in the order $\xi_1, \xi_2, \ldots, \xi_N$, with increasing integer labels, we shall also label all the states ψ_{α_i} once and for all by identifying each label i with a different integer (quite arbitrarily). The many-fermion wave function for a given configuration is then chosen to be that in which the labels in the rows of the determinant appear in increasing order from top to bottom. No two rows can have the same state labels, and $N_i = 0$ or 1 for all values of i.

We again consider matrix elements of the following sum of one-particle operators:

$$F^{(1)} = \sum_{n=1}^{N} f^{(1)}(\xi_n). \tag{14.28}$$

Taking one term $f^{(1)}(\xi_p)$ in this sum and considering first the diagonal matrix element $\int \Phi^*_{N_1,N_2,\ldots,N_\infty}(\xi_1,\ldots,\xi_p,\ldots,\xi_N)f^{(1)}(\xi_p)\Phi_{N_1,N_2,\ldots,N_\infty}(\xi_1,\ldots,\xi_p,\ldots,\xi_N)\mathrm{d}\xi_1\ldots\mathrm{d}\xi_p\ldots\mathrm{d}\xi_N$, we obtain, for the term in which particle p is associated with a given occupied single-particle state with label i,

$$\frac{S_i}{N!}\int \psi^*_{\alpha_i}(\xi_p)f^{(1)}(\xi_p)\psi_{\alpha_i}(\xi_p)\mathrm{d}\xi_p,$$

where S_i is the number of permutations of the remaining $N-1$ particles among the remaining $N-1$ occupied single-particle states. (Each of these permutations gives a unit integral and every such integral has the same sign, since they are formed from corresponding terms in Φ and Φ^*; the cross terms vanish by orthogonality.) Thus, $S_i = (N-1)!$, and the term in which particle p is associated with single-particle state ψ_{α_i} is equal to

$$\frac{1}{N}\int \psi^*_{\alpha_i}(\xi_p)f^{(1)}(\xi_p)\psi_{\alpha_i}(\xi_p)\mathrm{d}\xi_p \equiv \frac{1}{N}f^{(1)}_{ii}. \tag{14.29}$$

There is such a term for every occupied single-particle state, and so we obtain

$$\int \Phi^*_{N_1,N_2,\ldots,N_\infty}(\xi_1,\ldots,\xi_N)f^{(1)}(\xi_p)\Phi_{N_1,N_2,\ldots,N_\infty}(\xi_1,\ldots,\xi_N)\mathrm{d}\xi_1\ldots\mathrm{d}\xi_N$$
$$= \frac{1}{N}\sum_{\substack{i\\ \text{(occupied)}}} f^{(1)}_{ii} = \frac{1}{N}\sum_i N_i f^{(1)}_{ii}. \tag{14.30}$$

Since there are N terms in the sum (14.28), each of which gives the above contribution, we obtain

$$\int \Phi^*_{N_1,N_2,\ldots,N_\infty}(\xi_1,\ldots,\xi_N)F^{(1)}\Phi_{N_1,N_2,\ldots,N_\infty}(\xi_1,\ldots,\xi_N)\mathrm{d}\xi_1\ldots\mathrm{d}\xi_N = \sum_i f^{(1)}_{ii}N_i. \tag{14.31}$$

Now consider the off-diagonal matrix elements of $F^{(1)}$, or, first, of $f^{(1)}(\xi_p)$. As in the case of bosons, in the matrix element

$$\int \Phi^*_{N'_1,N'_2,\ldots,N'_\infty}(\xi_1,\ldots,\xi_N)f^{(1)}(\xi_p)\Phi_{N_1,N_2,\ldots,N_\infty}(\xi_1,\ldots,\xi_N)\mathrm{d}\xi_1\ldots\mathrm{d}\xi_N \tag{14.32}$$

the states Φ and Φ^* can differ only in the states of one particle. For example, consider the case in which the state with quantum numbers α_j is occupied in Φ but not in Φ^*, while the state with quantum numbers α_k is occupied in Φ^* but not in Φ. Suppose too that, in the labelling of states established once and for all at the beginning, we have $k > j$. Then we can write

$$\Phi_{N_1,N_2,\ldots,N_\infty} = \frac{1}{\sqrt{N!}}\begin{vmatrix} \psi_{\alpha_i}(\xi_1) & \psi_{\alpha_i}(\xi_2) & \cdot & \cdot & \cdot & \psi_{\alpha_i}(\xi_N) \\ \cdot & \cdot & \cdot & \cdot & \cdot & \cdot \\ \psi_{\alpha_j}(\xi_1) & \psi_{\alpha_j}(\xi_2) & \cdot & \cdot & \cdot & \psi_{\alpha_j}(\xi_N) \\ \cdot & \cdot & \cdot & \cdot & \cdot & \cdot \\ \cdot & \cdot & \cdot & \cdot & \cdot & \cdot \\ \psi_{\alpha_l}(\xi_1) & \psi_{\alpha_l}(\xi_2) & \cdot & \cdot & \cdot & \psi_{\alpha_l}(\xi_N) \end{vmatrix};$$

$$\Phi^*_{N_1',N_2',\ldots,N_\infty'} = \frac{1}{\sqrt{N!}}\begin{vmatrix} \psi^*_{\alpha_i}(\xi_1) & \psi^*_{\alpha_i}(\xi_2) & \cdot & \cdot & \cdot & \psi^*_{\alpha_i}(\xi_N) \\ \cdot & \cdot & \cdot & \cdot & \cdot & \cdot \\ \cdot & \cdot & \cdot & \cdot & \cdot & \cdot \\ \psi^*_{\alpha_k}(\xi_1) & \psi^*_{\alpha_k}(\xi_2) & \cdot & \cdot & \cdot & \psi^*_{\alpha_k}(\xi_N) \\ \cdot & \cdot & \cdot & \cdot & \cdot & \cdot \\ \psi^*_{\alpha_l}(\xi_1) & \psi^*_{\alpha_l}(\xi_2) & \cdot & \cdot & \cdot & \psi^*_{\alpha_l}(\xi_N) \end{vmatrix}, \tag{14.33}$$

where in Φ there is no row with ψ_{α_k} and in Φ^* there is no row with $\psi^*_{\alpha_j}$. Apart from the complex conjugation, the other rows are the same in each determinant and, by our chosen convention, i corresponds to the smallest integer labelling the occupied states and l corresponds to the largest.

We now arrange Φ^* in such a way that rows with the same label appear in the same positions in both Φ^* and Φ. To do this, we move the row with label k up by successive interchanges of rows (thus preserving the order of all the other rows) until it is in the row position occupied by ψ_{α_j} in Φ. This process clearly multiplies Φ^* by $(-1)^y$, where y is the number of occupied states (the same in both Φ and Φ^*, of course) with labels between j and k:

$$y = \sum_{n=j+1}^{k-1} N_n. \tag{14.34}$$

Thus, the matrix element (14.32) between the states given by (14.33) is

$$\begin{aligned} &\int \Phi^*_{N_1,N_2,\ldots,0_j,\ldots,1_k,\ldots,N_\infty}(\xi_1,\ldots,\xi_N) f^{(1)}(\xi_p) \Phi_{N_1,N_2,\ldots,1_j,\ldots,0_k,\ldots,N_\infty}(\xi_1,\ldots,\xi_N)\mathrm{d}\xi_1\ldots\mathrm{d}\xi_N \\ &= \frac{C}{N!}(-1)^{\left(\sum_{n=j+1}^{k-1} N_n\right)} \int \psi^*_{\alpha_k}(\xi_p) f^{(1)}(\xi_p)\psi_{\alpha_j}(\xi_p)\mathrm{d}\xi_p, \end{aligned} \tag{14.35}$$

where C [$= (N-1)!$] is the number of permutations of the remaining $N-1$ particles among the $N-1$ single-particle states common to Φ^* and Φ. (Each permutation in Φ, together with the corresponding permutation in Φ^*, gives a product of unit integrals, and each of the latter is positive because of the corresponding ordering of the rows in Φ^* and Φ.) Thus, (14.35) becomes

$$\int \Phi^*_{N_1,N_2,\ldots,0_j,\ldots,1_k,\ldots,N_\infty}(\xi_1,\ldots,\xi_N) f^{(1)}(\xi_p) \Phi_{N_1,N_2,\ldots,1_j,\ldots,0_k,\ldots,N_\infty}(\xi_1,\ldots,\xi_N) \mathrm{d}\xi_1 \ldots \mathrm{d}\xi_N$$
$$= \frac{1}{N}(-1)^{\left(\sum\limits_{n=j+1}^{k-1} N_n\right)} f^{(1)}_{kj}. \tag{14.36}$$

Therefore, for the matrix element of $F^{(1)}$ (14.28) we have

$$\int \Phi^*_{N_1,N_2,\ldots,0_j,\ldots,1_k,\ldots,N_\infty}(\xi_1,\ldots,\xi_N) F^{(1)} \Phi_{N_1,N_2,\ldots,1_j,\ldots,0_k,\ldots,N_\infty}(\xi_1,\ldots,\xi_N) \mathrm{d}\xi_1 \ldots \mathrm{d}\xi_N$$
$$= (-1)^{\left(\sum\limits_{n=j+1}^{k-1} N_n\right)} f^{(1)}_{kj}. \tag{14.37}$$

From (14.30) and (14.37) we can now make the following identifications: for a diagonal matrix element,

$$\langle N_1, N_2, \ldots, N_\infty | \hat{F}^{(1)} | N_1, N_2, \ldots, N_\infty \rangle = \sum_i N_i f^{(1)}_{ii}, \tag{14.38}$$

while for an off-diagonal (in respect of the state of one particle) matrix element,

$$\left\langle N_1, N_2, \ldots, 0_j, \ldots, 1_k, \ldots, N_\infty \right| \hat{F}^{(1)} \left| N_1, N_2, \ldots, 1_j, \ldots, 0_k, \ldots, N_\infty \right\rangle = (-1)^{\left(\sum\limits_{n=j+1}^{k-1} N_n\right)} f^{(1)}_{kj}, \tag{14.39}$$

where all the occupation numbers are 0 or 1.

As we did in the boson case, for $\hat{F}^{(1)}$ we postulate the form [see (14.11)]

$$\hat{F}^{(1)} = \sum_{k,j} f^{(1)}_{kj} \hat{C}_k \hat{A}_j, \tag{14.40}$$

where $\hat{A}_j$ <u>a</u>nnihilates a particle in state ψ_{α_j} and $\hat{C}_k$ <u>c</u>reates a particle in state ψ_{α_k}:

$$\hat{C}_k |\ldots 0_k \ldots\rangle = \gamma_k |\ldots 1_k \ldots\rangle, \quad \hat{A}_j \left|\ldots 1_j \ldots\right\rangle = \alpha_j \left|\ldots 0_j \ldots\right\rangle, \tag{14.41}$$

in which γ_k and α_j are to be determined. The power of –1 in the right-hand side of (14.39) shows that the effect of $\hat{C}_k\hat{A}_j$ in (14.40) must depend on the occupation of other states besides k and j, and so it is reasonable to suppose that so too will the effects of $\hat{C}_k$ and $\hat{A}_j$ separately. Therefore, in particular, we must allow for the possibility that γ_k in (14.41) may depend on the occupation number N_j [we shall see from the result (14.44) below that, because $j < k$, indeed it does], and for the possibility that α_j may depend on the occupation number N_k [although the same result (14.44) shows that, because $j < k$, it does not]. Thus, since the occupation N_j of state j when $\hat{C}_k$ acts depends on whether or not $\hat{A}_j$ has acted already, the results of acting with $\hat{C}_k\hat{A}_j$ and with $\hat{A}_j\hat{C}_k$ may be different. To determine which order to use in $\hat{F}^{(1)}$ we find the form that gives a clear factorization of the power of –1 in the right-hand side of (14.39).

Consider first

$$\hat{C}_k\hat{A}_j\left|\ldots 1_j\ldots 0_k\ldots\right\rangle = \hat{C}_k\alpha_j(N_k=0)\left|\ldots 0_j\ldots 0_k\ldots\right\rangle$$
$$= \alpha_j(N_k=0)\gamma_k(N_j=0)\left|\ldots 0_j\ldots 1_k\ldots\right\rangle. \tag{14.42}$$

For the operators in the opposite order,

$$\hat{A}_j\hat{C}_k\left|\ldots 1_j\ldots 0_k\ldots\right\rangle = \hat{A}_j\gamma_k(N_j=1)\left|\ldots 1_j\ldots 1_k\ldots\right\rangle$$
$$= \gamma_k(N_j=1)\alpha_j(N_k=1)\left|\ldots 0_j\ldots 1_k\ldots\right\rangle. \tag{14.43}$$

The power of –1 in the right-hand side of (14.39) can be rewritten as

$$(-1)^{\left(\sum\limits_{n=j+1}^{k-1} N_n\right)} = (-1)^{\left(\sum\limits_{n=1}^{j-1} N_n\right)} \times (-1)^{\left(\sum\limits_{n=1}^{j-1} N_n\right)}(-1)^{\left(\sum\limits_{n=j+1}^{k-1} N_n\right)}.$$

This can this be rewritten as the product

$$(-1)^{\left(\sum\limits_{n=j+1}^{k-1} N_n\right)} = (-1)^{\left(\sum\limits_{n=1}^{j-1} N_n\right)} \times (-1)^{\left(\sum\limits_{n=1}^{k-1} N_n\right)}$$

if and only if $N_j = 0$, and so can be identified with the product $\alpha_j(N_k = 0)\gamma_k(N_j = 0)$ in (14.42), with α_j and γ_j given by the general expressions

$$\alpha_j = \gamma_j = (-1)^{\left(\sum\limits_{n=1}^{j-1} N_n\right)}. \tag{14.44}$$

Thus, the operator order used in (14.40) is the correct one, and equations (14.41) become

$$\hat{C}_k\left|\ldots 0_k\ldots\right\rangle = (-1)^{\left(\sum\limits_{n=1}^{k-1} N_n\right)}\left|\ldots 1_k\ldots\right\rangle, \tag{14.45}$$

$$\hat{A}_k\left|\ldots 1_k\ldots\right\rangle = (-1)^{\left(\sum\limits_{n=1}^{k-1} N_n\right)}\left|\ldots 0_k\ldots\right\rangle. \tag{14.46}$$

The fact that, by the Pauli exclusion principle, no state can result from creating a fermion in an already occupied single-particle state, and the fact that no state can result from the annihilation of a nonexistent particle, can be expressed by the additional relations

$$\hat{C}_k\left|\ldots 1_k\ldots\right\rangle = 0, \quad \hat{A}_k\left|\ldots 0_k\ldots\right\rangle = 0. \tag{14.47}$$

By an argument similar to the one given for the boson case, we can show that the operators $\hat{A}_k$ and $\hat{C}_k$ are mutually adjoint. Indeed, premultiply (14.45) by $\langle\ldots 1_k\ldots|$. Since the state vectors are normalized ($\langle\ldots 1_k\ldots \mid \ldots 1_k\ldots\rangle = 1$), we obtain

$$(-1)^{\left(\sum\limits_{n=1}^{k-1} N_n\right)} = \langle\ldots 1_k\ldots|\,\hat{C}_k\,|\ldots 0_k\ldots\rangle = \langle\ldots 0_k\ldots|\,\hat{C}_k^\dagger\,|\ldots 1_k\ldots\rangle^* = \langle\ldots 0_k\ldots|\,\hat{C}_k^\dagger\,|\ldots 1_k\ldots\rangle, \tag{14.48}$$

where, in the second equality, we have used the definition of the adjoint operator $\hat{C}_k^\dagger$ and, in the third, the fact that the first member of (14.48) is manifestly real (1 or -1). Comparing (14.48) with the expression obtained from (14.46) by premultiplying it by $\langle \ldots 0_k \ldots |$, we see that we can identify $\hat{C}_k^\dagger$ with $\hat{A}_k$ and hence $\hat{C}_k$ with $\hat{A}_k^\dagger$. Accordingly, we introduce the following standard notation

$$\hat{a}_k^\dagger \equiv \hat{C}_k, \quad \hat{a}_k \equiv \hat{A}_k, \tag{14.49}$$

that is, $\hat{a}_k$ is a fermion-annihilation operator and $\hat{a}_k^\dagger$ is a fermion-creation operator ($\hat{a}_k$ and $\hat{a}_k^\dagger$ annihilate and create, respectively, a fermion in the single-particle state ψ_{α_k}). In this notation, equations (14.45)–(14.47) become

$$\hat{a}_k^\dagger \, |\ldots 0_k \ldots\rangle = (-1)^{\left(\sum_{n=1}^{k-1} N_n\right)} |\ldots 1_k \ldots\rangle\,, \quad \hat{a}_k^\dagger \, |\ldots 1_k \ldots\rangle = 0, \tag{14.50}$$

$$\hat{a}_k \, |\ldots 1_k \ldots\rangle = (-1)^{\left(\sum_{n=1}^{k-1} N_n\right)} |\ldots 0_k \ldots\rangle\,, \quad \hat{a}_k \, |\ldots 0_k \ldots\rangle = 0, \tag{14.51}$$

and the operator (14.40) becomes

$$\hat{F}^{(1)} = \sum_{k,j} f_{kj}^{(1)} \hat{a}_k^\dagger \hat{a}_j. \tag{14.52}$$

In a diagonal matrix element $\langle N_1, N_2, \ldots, N_\infty | \, \hat{F}^{(1)} \, | N_1, N_2, \ldots, N_\infty\rangle$ the only terms of the operator (14.52) that give a nonzero contribution are those in which the creation and annihilation operator refer to the same single-particle state, and so we have

$$\langle N_1, N_2, \ldots, N_\infty | \, \hat{F}^{(1)} \, | N_1, N_2, \ldots, N_\infty\rangle = \langle N_1, N_2, \ldots, N_\infty | \left(\sum_i f_{ii}^{(1)} \hat{a}_i^\dagger \hat{a}_i \right) | N_1, N_2, \ldots, N_\infty\rangle\,. \tag{14.53}$$

But it follows from (14.50) and (14.51) that $\hat{a}_i^\dagger \hat{a}_i \, |\ldots 1_i \ldots\rangle = 1 \times |\ldots 1_i \ldots\rangle \, (\equiv N_i \, |\ldots 1_i \ldots\rangle)$ and $\hat{a}_i^\dagger \hat{a}_i \, |\ldots 0_i \ldots\rangle = 0 \times |\ldots 0_i \ldots\rangle \, (\equiv N_i \, |\ldots 0_i \ldots\rangle)$, and so (14.53) yields the desired result (14.38), as it should.

Similarly, using (14.50) and (14.51) we find

$$(\hat{a}_i^\dagger \hat{a}_i + \hat{a}_i \hat{a}_i^\dagger) \, |\ldots 0_i \ldots\rangle = |\ldots 0_i \ldots\rangle \quad \text{and} \quad (\hat{a}_i^\dagger \hat{a}_i + \hat{a}_i \hat{a}_i^\dagger) \, |\ldots 1_i \ldots\rangle = |\ldots 1_i \ldots\rangle\,,$$

so that

$$\hat{a}_i^\dagger \hat{a}_i + \hat{a}_i \hat{a}_i^\dagger = 1. \tag{14.54}$$

Also,

$$(\hat{a}_i^\dagger\hat{a}_k+\hat{a}_k\hat{a}_i^\dagger)\,|\ldots 0_i\ldots 1_k\ldots\rangle=\hat{a}_i^\dagger(-1)^{\left(\sum\limits_{n=1}^{k-1}N_n(N_i=0)\right)}|\ldots 0_i\ldots 0_k\ldots\rangle+\hat{a}_k(-1)^{\left(\sum\limits_{n=1}^{i-1}N_n\right)}|\ldots 1_i\ldots 1_k\ldots\rangle$$
$$=(-1)^{\left(\sum\limits_{n=1}^{k-1}N_n(N_i=0)\right)}(-1)^{\left(\sum\limits_{n=1}^{i-1}N_n\right)}|\ldots 1_i\ldots 0_k\ldots\rangle+(-1)^{\left(\sum\limits_{n=1}^{i-1}N_n\right)}(-1)^{\left(\sum\limits_{n=1}^{k-1}N_n(N_i=1)\right)}|\ldots 1_i\ldots 0_k\ldots\rangle$$
$$=(-1)^{\left(\sum\limits_{n=1}^{k-1}N_n(N_i=0)\right)}(-1)^{\left(\sum\limits_{n=1}^{i-1}N_n\right)}(1-1)\,|\ldots 1_i\ldots 0_k\ldots\rangle=0,$$

so that

$$\hat{a}_i^\dagger\hat{a}_k+\hat{a}_k\hat{a}_i^\dagger=0\quad(i\neq k). \tag{14.55}$$

Thus, from (14.54) and (14.55) we have

$$\hat{a}_i^\dagger\hat{a}_k+\hat{a}_k\hat{a}_i^\dagger=\delta_{ik}. \tag{14.56}$$

Similarly, it can be shown that, for all i and k,

$$\hat{a}_i^\dagger\hat{a}_k^\dagger+\hat{a}_k^\dagger\hat{a}_i^\dagger=0\quad\text{and}\quad\hat{a}_i\hat{a}_k+\hat{a}_k\hat{a}_i=0. \tag{14.57}$$

To summarize, a sum of single-fermion operators of the form (14.28) takes, in the occupation-number representation, the form

$$\hat{F}^{(1)}=\sum_{k,i}f_{ki}^{(1)}\hat{a}_k^\dagger\hat{a}_i,\quad\text{with}\quad f_{ki}^{(1)}\equiv\int\psi_{\alpha_k}^*(\xi)f^{(1)}(\xi)\psi_{\alpha_i}(\xi)\mathrm{d}\xi, \tag{14.58}$$

where the fermion creation and annihilation operators satisfy the relations (14.56) and (14.57), which (if we use curly brackets for an anticommutator) can be written in the form

$$\{\hat{a}_i^\dagger,\hat{a}_k\}=\delta_{ik},\quad\{\hat{a}_i^\dagger,\hat{a}_k^\dagger\}=0,\quad\{\hat{a}_i,\hat{a}_k\}=0. \tag{14.59}$$

Similarly, but with somewhat greater effort, we can find the operator $\hat{F}^{(2)}$ that is the equivalent (in the occupation-number representation) of a symmetric sum of two-particle operators $f^{(2)}(\xi_m,\xi_n)$ (in the coordinate representation), that is, we can find the form of the operator $\hat{F}^{(2)}$ in the expression

$$\langle N_1',N_2',\ldots,N_\infty'|\hat{F}^{(2)}|N_1,N_2,\ldots,N_\infty\rangle$$
$$=\frac{1}{2}\sum_{\substack{m,n\\(m\neq n)}}\int\Phi_{N_1',N_2',\ldots,N_\infty'}^*(\xi_1,\ldots,\xi_N)f^{(2)}(\xi_m,\xi_n)\Phi_{N_1,N_2,\ldots,N_\infty}(\xi_1,\ldots,\xi_N)\mathrm{d}\xi_1\ldots\mathrm{d}\xi_N. \tag{14.60}$$

The result is

$$\hat{F}^{(2)}=\frac{1}{2}\sum_i\sum_k\sum_l\sum_m f_{ik,lm}^{(2)}\hat{a}_i^\dagger\hat{a}_k^\dagger\hat{a}_m\hat{a}_l, \tag{14.61}$$

with

$$f^{(2)}_{ik,lm} = \iint \psi_i^*(\xi_1)\psi_k^*(\xi_2)f^{(2)}(\xi_1,\xi_2)\psi_l(\xi_1)\psi_m(\xi_2)\mathrm{d}\xi_1\mathrm{d}\xi_2. \tag{14.62}$$

Note the order of the last two operators in (14.61). Reversing this order would change the sign of $\hat{F}^{(2)}$ by virtue of the third anticommutation relation in (14.59).

14.2 Particle-field operators

We now introduce new operators $\hat{\psi}(\xi)$ and $\hat{\psi}^\dagger(\xi)$, in the form of the following linear combinations:

$$\hat{\psi}(\xi) = \sum_k \psi_k(\xi)\hat{c}_k, \quad \hat{\psi}^\dagger(\xi) = \sum_k \psi_k^*(\xi)\hat{c}_k^\dagger, \tag{14.63}$$

where $\psi_k(\xi)$ is the single-particle wave function (of the space–spin coordinates ξ) corresponding to the set of quantum numbers k (for example, $k \equiv \boldsymbol{k}, \lambda$) and $\hat{c}_k$ ($\hat{c}_k^\dagger$) annihilates (creates) a particle in the single-particle state k (for bosons we write $\hat{c}_k$ as $\hat{b}_k$, while for fermions we write $\hat{c}_k$ as $\hat{a}_k$).

Let us evaluate the following commutator (in the case when the particles are bosons) or anticommutator (in the case when the particles are fermions):

$$\begin{aligned}[\hat{\psi}(\xi), \hat{\psi}^\dagger(\xi')]_\mp &= \sum_k \sum_{k'} \psi_k(\xi)\psi_{k'}^*(\xi')[\hat{c}_k, \hat{c}_{k'}^\dagger]_\mp \\ &= \sum_k \sum_{k'} \psi_k(\xi)\psi_{k'}^*(\xi')\delta_{k,k'} = \sum_k \psi_k(\xi)\psi_k^*(\xi') = \delta(\xi - \xi'),\end{aligned} \tag{14.64}$$

by the closure relation (see section 2.2). Similarly, for bosons we have the commutation relations (for fermions, anticommutation relations)

$$[\hat{\psi}(\xi), \hat{\psi}(\xi')]_\mp = 0 \quad \text{and} \quad [\hat{\psi}^\dagger(\xi), \hat{\psi}^\dagger(\xi')]_\mp = 0. \tag{14.65}$$

To establish a commonly used notation, and to give a clear physical interpretation of the operators $\hat{\psi}(\xi)$ and $\hat{\psi}^\dagger(\xi)$, we shall consider these operators and their commutation (anticommutation) relations for a particular case, namely, free particles with spin. Then $\xi \equiv \boldsymbol{r}, \tau$ (where τ is the spin coordinate, which we may choose to be the quantum number of the z-component of spin, although this is not necessary to the argument) and $k \equiv \boldsymbol{k}, \lambda$ (where λ is the quantum number of the z-component of spin). In this case,

$$\hat{\psi}(\xi) = \hat{\psi}(\boldsymbol{r}, \tau) \equiv \hat{\psi}_\tau(\boldsymbol{r}) = \sum_{\boldsymbol{k}} \sum_\lambda \psi_{\boldsymbol{k}\lambda}(\boldsymbol{r}, \tau)\hat{c}_{\boldsymbol{k}\lambda}, \tag{14.66}$$

which introduces the notation $\hat{\psi}_\tau(\boldsymbol{r})$. Writing the adjoint of this operator in the form

$$\hat{\psi}_{\tau'}^\dagger(\boldsymbol{r}') = \sum_{\boldsymbol{k}'} \sum_{\lambda'} \psi_{\boldsymbol{k}'\lambda'}^*(\boldsymbol{r}', \tau')\hat{c}_{\boldsymbol{k}'\lambda'}^\dagger,$$

we find the commutation (anticommutation) relations for bosons (fermions):

$$[\hat{\psi}_\tau(\boldsymbol{r}), \hat{\psi}^\dagger_{\tau'}(\boldsymbol{r}')]_\mp = \delta(\xi - \xi') = \delta_{\tau\tau'}\delta(\boldsymbol{r} - \boldsymbol{r}'). \tag{14.67}$$

Suppose that in (14.66) the spin coordinate τ is given a particular value α (one of the possible values of the quantum number of the z-component of spin). Then (14.66) becomes

$$\hat{\psi}_\alpha(\boldsymbol{r}) = \sum_{\boldsymbol{k}}\sum_{\lambda} \psi_{\boldsymbol{k}\lambda}(\boldsymbol{r}, \alpha)\hat{c}_{\boldsymbol{k}\lambda} \equiv \sum_{\boldsymbol{k}}\sum_{\lambda} \langle \boldsymbol{r}|\boldsymbol{k}\rangle \langle \alpha|\lambda\rangle\, \hat{c}_{\boldsymbol{k}\lambda} = \sum_{\boldsymbol{k}} \psi_{\boldsymbol{k}}(\boldsymbol{r})\hat{c}_{\boldsymbol{k}\alpha}. \tag{14.68}$$

Thus, $\hat{\psi}_\alpha(\boldsymbol{r})$ is the sum over all $\boldsymbol{k}$ of $\hat{c}_{\boldsymbol{k}\alpha}$ multiplied by the probability amplitude for finding a particle (in state $\boldsymbol{k}$) at position $\boldsymbol{r}$. It thus represents an operator destroying a particle with spin α at $\boldsymbol{r}$. Similarly, $\hat{\psi}^\dagger_\alpha(\boldsymbol{r})$ creates a particle with spin α at $\boldsymbol{r}$. (Here, and henceforth, the phrase "with spin α" is always to be interpreted as "being in an eigenstate of the z-component of spin, with quantum number $m_s = \alpha$".) In line with this interpretation, the operators $\hat{\psi}_\alpha(\boldsymbol{r})$ and $\hat{\psi}^\dagger_\alpha(\boldsymbol{r})$ are often called **particle-field operators**.

Using the results (14.23), (14.25), (14.26) (for bosons), and (14.58), (14.61), (14.62) (for fermions), we can rewrite the following hamiltonian (for either bosons or fermions):

$$H = T + V = \sum_a t^{(1)}(\xi_a) + \frac{1}{2}\sum_{a}\sum_{\substack{b \\ (a \neq b)}} v^{(2)}(\xi_a, \xi_b), \tag{14.69}$$

firstly in terms of the $\hat{c}$ operators:

$$H \to \hat{H} = \sum_k\sum_i t^{(1)}_{ki}\hat{c}^\dagger_k\hat{c}_i + \frac{1}{2}\sum_i\sum_k\sum_l\sum_m v^{(2)}_{ik,lm}\hat{c}^\dagger_i\hat{c}^\dagger_k\hat{c}_m\hat{c}_l, \tag{14.70}$$

where

$$t^{(1)}_{ki} = \int \mathrm{d}\xi\, \psi^*_k(\xi)t^{(1)}(\xi)\psi_i(\xi) \tag{14.71}$$

and

$$v^{(2)}_{ik,lm} = \iint \mathrm{d}\xi_1\mathrm{d}\xi_2\psi^*_i(\xi_1)\psi^*_k(\xi_2)v^{(2)}(\xi_1, \xi_2)\psi_l(\xi_1)\psi_m(\xi_2). \tag{14.72}$$

In terms of the $\hat{\psi}$ operators (14.63), the hamiltonian can now be written as

$$\hat{H} = \int \mathrm{d}\xi\, \hat{\psi}^\dagger(\xi)t^{(1)}(\xi)\hat{\psi}(\xi) + \frac{1}{2}\iint \mathrm{d}\xi_1\mathrm{d}\xi_2\hat{\psi}^\dagger(\xi_1)\hat{\psi}^\dagger(\xi_2)v^{(2)}(\xi_1, \xi_2)\hat{\psi}(\xi_2)\hat{\psi}(\xi_1). \tag{14.73}$$

(Note that, as it should be, this form is hermitian, that is, $\hat{H}^\dagger = H$.)

In the case of particles with spin, we have $\xi \equiv \boldsymbol{r}, \lambda$ (where we have chosen the spin coordinates τ to be the quantum numbers λ) and, in the notation introduced above,

$$\hat{H} = \sum_{\lambda} \int \mathrm{d}^3 \boldsymbol{r} \hat{\psi}_{\lambda}^{\dagger}(\boldsymbol{r}) t^{(1)}(\boldsymbol{r}) \hat{\psi}_{\lambda}(\boldsymbol{r}) + \frac{1}{2} \sum_{\lambda_1} \sum_{\lambda_2} \iint \mathrm{d}^3 \boldsymbol{r}_1 \mathrm{d}^3 \boldsymbol{r}_2 \hat{\psi}_{\lambda_1}^{\dagger}(\boldsymbol{r}_1) \hat{\psi}_{\lambda_2}^{\dagger}(\boldsymbol{r}_2) v^{(2)}(\boldsymbol{r}_1, \boldsymbol{r}_2) \hat{\psi}_{\lambda_2}(\boldsymbol{r}_2) \hat{\psi}_{\lambda_1}(\boldsymbol{r}_1). \tag{14.74}$$

The two terms in this expression have a strong formal resemblance to matrix elements of the Schrödinger-representation hamiltonian between wave functions (although the $\hat{\psi}$s are not wave functions but particle-field operators). This formal similarity suggests the name **second quantization**. In the latter, the fields are the operators and the coefficients $t^{(1)}(\boldsymbol{r})$ and $v^{(2)}(\boldsymbol{r}_1, \boldsymbol{r}_2)$ are the classical expressions for the kinetic and potential energies, respectively, as in (14.69).

For single-particle operators [in the sense of (14.3)] we have

$$F^{(1)} = \sum_{n=1}^{N} f^{(1)}(\xi_n) \to \hat{F} = \sum_{\lambda} \int \mathrm{d}^3 \boldsymbol{r}_n \hat{\psi}_{\lambda}^{\dagger}(\boldsymbol{r}_n) f^{(1)}(\boldsymbol{r}_n) \hat{\psi}_{\lambda}(\boldsymbol{r}_n). \tag{14.75}$$

As an example, consider the number-density operator

$$\rho(\boldsymbol{r}) = \sum_{n=1}^{N} \delta(\boldsymbol{r} - \boldsymbol{r}_n) \equiv \sum_{n=1}^{N} f^{(1)}(\boldsymbol{r}_n) \tag{14.76}$$

(upon integration over the volume, this clearly gives the total number N of particles). This becomes

$$\begin{aligned} \hat{\rho}(\boldsymbol{r}) &= \sum_{\lambda} \int \mathrm{d}^3 \boldsymbol{r}_n \hat{\psi}_{\lambda}^{\dagger}(\boldsymbol{r}_n) \delta(\boldsymbol{r} - \boldsymbol{r}_n) \hat{\psi}_{\lambda}(\boldsymbol{r}_n) = \sum_{\lambda} \hat{\psi}_{\lambda}^{\dagger}(\boldsymbol{r}) \hat{\psi}_{\lambda}(\boldsymbol{r}) \\ &= \sum_{\lambda} \sum_{i,j} \psi_i^*(\boldsymbol{r}, \lambda) \psi_j(\boldsymbol{r}, \lambda) \hat{c}_i^{\dagger} \hat{c}_j, \end{aligned} \tag{14.77}$$

where the last line follows from (14.68).

The corresponding operator for the total number of particles is

$$\begin{aligned} \hat{N} &= \int \mathrm{d}^3 \boldsymbol{r} \hat{\rho}(\boldsymbol{r}) = \sum_{\lambda} \int \mathrm{d}^3 \boldsymbol{r} \hat{\psi}_{\lambda}^{\dagger}(\boldsymbol{r}) \hat{\psi}_{\lambda}(\boldsymbol{r}) = \sum_{\lambda} \sum_{i,j} \int \mathrm{d}^3 \boldsymbol{r} \psi_i^*(\boldsymbol{r}, \lambda) \psi_j(\boldsymbol{r}, \lambda) \hat{c}_i^{\dagger} \hat{c}_j \\ &\equiv \sum_{i,j} \int \mathrm{d}\xi \, \psi_i^*(\xi) \psi_j(\xi) \hat{c}_i^{\dagger} \hat{c}_j = \sum_{i,j} \delta_{ij} \hat{c}_i^{\dagger} \hat{c}_j = \sum_i \hat{c}_i^{\dagger} \hat{c}_i \equiv \sum_i \hat{n}_i, \end{aligned} \tag{14.78}$$

where $\hat{n}_i$ is the operator of the number of particles in single-particle state i. Writing $\hat{N}$ and $\hat{H}$ in terms of the $\hat{\psi}$s, we can easily show that $[\hat{N}, \hat{H}] = 0$ for this hamiltonian. This means that, for this hamiltonian, according to the **Ehrenfest theorem** (1.32), the number of particles is a **constant of the motion**.

Another operator that we shall need in second-quantized form is the **density-fluctuation operator**, that is, the operator corresponding to the Fourier transform ρ_k of $\rho(\boldsymbol{r})$:

$$\rho_k \equiv \rho(\boldsymbol{k}) = \int \mathrm{d}^3\boldsymbol{r}\rho(\boldsymbol{r})\mathrm{e}^{-i\boldsymbol{k}\cdot\boldsymbol{r}} = \int \mathrm{d}^3\boldsymbol{r}\sum_{n=1}^{N}\delta(\boldsymbol{r}-\boldsymbol{r}_n)\mathrm{e}^{-i\boldsymbol{k}\cdot\boldsymbol{r}} = \sum_{n=1}^{N}\mathrm{e}^{-i\boldsymbol{k}\cdot\boldsymbol{r}_n}, \tag{14.79}$$

which is clearly a sum of single-particle operators in the sense of (14.3). If we have a system of uniform density, that is, a continuum with $\rho(\boldsymbol{r}) = \text{const}$, then the only nonzero ρ_k is $\rho_{k=0}$, corresponding to zero-wave-number (infinite-wavelength) variations in density, that is, uniformity. If a particular ρ_k is nonzero for some $\boldsymbol{k} \neq 0$, we have a harmonic component of the density (a **density fluctuation**), of wave number $\boldsymbol{k}$ (wavelength $2\pi/|\boldsymbol{k}|$), and so general nonuniformity of density can be described by an integral over these Fourier components:

$$\rho(\boldsymbol{r}) = \frac{1}{(2\pi)^3}\int \mathrm{d}^3\boldsymbol{k}\rho_{\boldsymbol{k}}\mathrm{e}^{i\boldsymbol{k}\cdot\boldsymbol{r}}. \tag{14.80}$$

The second-quantized form of the sum (14.79) of single-particle operators is

$$\hat{\rho}_k = \sum_{\lambda}\int \mathrm{d}^3\boldsymbol{r}\hat{\psi}^{\dagger}_{\lambda}(\boldsymbol{r})\mathrm{e}^{-i\boldsymbol{k}\cdot\boldsymbol{r}}\hat{\psi}_{\lambda}(\boldsymbol{r}) = \sum_{\lambda}\sum_{i,j}\int \mathrm{d}^3\boldsymbol{r}\psi^*_i(\boldsymbol{r},\lambda)\mathrm{e}^{-i\boldsymbol{k}\cdot\boldsymbol{r}}\psi_j(\boldsymbol{r},\lambda)\hat{c}^{\dagger}_i\hat{c}_j. \tag{14.81}$$

Suppose now that the particles are free and possess spin, that is, let $i \equiv \boldsymbol{k}_1, \lambda_1$ and $j \equiv \boldsymbol{k}_2, \lambda_2$. Then

$$\psi_j(\boldsymbol{r},\lambda) \equiv \langle \boldsymbol{r} \mid \boldsymbol{k}_2\rangle\langle\lambda \mid \lambda_2\rangle = \langle \boldsymbol{r} \mid \boldsymbol{k}_2\rangle\delta_{\lambda\lambda_2} = \frac{1}{\sqrt{\Omega}}\mathrm{e}^{i\boldsymbol{k}_2\cdot\boldsymbol{r}}\delta_{\lambda\lambda_2} \tag{14.82}$$

and

$$\psi^*_i(\boldsymbol{r},\lambda) \equiv \langle \boldsymbol{k}_1 \mid \boldsymbol{r}\rangle\delta_{\lambda\lambda_1} = \frac{1}{\sqrt{\Omega}}\mathrm{e}^{-i\boldsymbol{k}_1\cdot\boldsymbol{r}}\delta_{\lambda\lambda_1}, \tag{14.83}$$

where we have used volume normalization (Ω is the volume of the system). Carrying out the sums over λ_1 and λ_2 (in the sums over i and j), we have

$$\begin{aligned}\hat{\rho}_k &= \sum_{\lambda}\sum_{\boldsymbol{k}_1,\boldsymbol{k}_2}\frac{1}{\Omega}\int \mathrm{d}^3\boldsymbol{r}\mathrm{e}^{-i(\boldsymbol{k}_1+\boldsymbol{k}-\boldsymbol{k}_2)\cdot\boldsymbol{r}}\hat{c}^{\dagger}_{\boldsymbol{k}_1\lambda}\hat{c}_{\boldsymbol{k}_2\lambda}\\ &= \sum_{\lambda}\sum_{\boldsymbol{k}_1,\boldsymbol{k}_2}\delta_{\boldsymbol{k}_2,\boldsymbol{k}_1+\boldsymbol{k}}\hat{c}^{\dagger}_{\boldsymbol{k}_1\lambda}\hat{c}_{\boldsymbol{k}_2\lambda} = \sum_{\boldsymbol{q},\lambda}\hat{c}^{\dagger}_{\boldsymbol{q}\lambda}\hat{c}_{\boldsymbol{q}+\boldsymbol{k},\lambda},\end{aligned} \tag{14.84}$$

where we have replaced $\boldsymbol{k}_1$ by $\boldsymbol{q}$. The adjoint operator is

$$\hat{\rho}^{\dagger}_k = \sum_{\boldsymbol{q},\lambda}\hat{c}^{\dagger}_{\boldsymbol{q}+\boldsymbol{k},\lambda}\hat{c}_{\boldsymbol{q}\lambda}. \tag{14.85}$$

Since $\hat{c}_{\boldsymbol{q}\lambda}$ destroys a particle of momentum $\boldsymbol{q}$ and spin λ, that is, creates a hole in the original distribution, while $\hat{c}^{\dagger}_{\boldsymbol{q}+\boldsymbol{k},\lambda}$ creates a particle of momentum $\boldsymbol{q}+\boldsymbol{k}$ and spin λ, we see that the density-fluctuation operator

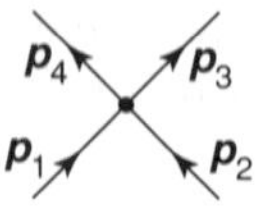

Figure 14.1 *Scattering of two particles with momenta* $\mathbf{p}_1$ *and* $\mathbf{p}_2$.

$\hat{\rho}_k^\dagger$ (we call both $\hat{\rho}_k$ and $\hat{\rho}_k^\dagger$ density-fluctuation operators) can be represented as the sum over all possible particle-hole creation operators such that the particle created has momentum equal to that of the particle destroyed plus $\boldsymbol{k}$.

To summarize this introduction to second quantization, we point out one extremely important aspect of the second-quantized form of operators, namely, that they correspond to **sums of processes**, with each process involving the creation and annihilation of particles in single-particle states of specified momentum and spin. The hamiltonian, being a sum of one-particle and two-particle operators, also reduces to sums of processes. This leads naturally to the use of diagrams (in particular, **Feynman diagrams**) in calculations in many-body problems. As a simple example, the process $f^{(2)}_{\boldsymbol{p}_4\boldsymbol{p}_3,\boldsymbol{p}_2\boldsymbol{p}_1}\hat{c}^\dagger_{\boldsymbol{p}_4}\hat{c}^\dagger_{\boldsymbol{p}_3}\hat{c}_{\boldsymbol{p}_2}\hat{c}_{\boldsymbol{p}_1}$ could be represented by the diagram shown in figure 14.1, in which the black blob represents the **amplitude** $f^{(2)}_{\boldsymbol{p}_4\boldsymbol{p}_3,\boldsymbol{p}_2\boldsymbol{p}_1}$ of the process.

15

Gas of Interacting Electrons

15.1 Hamiltonian of an electron gas

Consider a gas of interacting electrons placed in a uniformly distributed, positively charged background, chosen to make the system electrically neutral. This may provide a model for a qualitative description of the electrons in a metal, although, since the positive charges in metals are concentrated on ions and so are not uniformly distributed, a good quantitative description cannot be expected from the model.

Let the electrons be in a large cube of volume $L^3 = \Omega$ (the limit $\Omega \to \infty$ will be taken at the end). The single-particle wave functions (of the corresponding system of noninteracting particles) are

$$\psi_{k\lambda'}(\boldsymbol{r}, \lambda) = \frac{1}{\sqrt{\Omega}} \mathrm{e}^{i\boldsymbol{k}\cdot\boldsymbol{r}} \delta_{\lambda'\lambda}, \tag{15.1}$$

where we have chosen to use the "proper representation" for the spin part (see chapter 2). Application of periodic boundary conditions yields the allowed wave numbers

$$k_i = \frac{2\pi n_i}{L} \quad (i = x, y, z; \quad n_i = 0, \pm 1, \pm 2, \ldots) \tag{15.2}$$

The total hamiltonian can be represented as a sum of three terms:

$$H = H_{\mathrm{el}} + H_{\mathrm{b}} + H_{\mathrm{el-b}} \tag{15.3}$$

where

$$H_{\mathrm{el}} = -\frac{\hbar^2}{2m} \sum_n \nabla_n^2 + \frac{1}{2} \frac{e^2}{4\pi\varepsilon_0} \sum_{\substack{n,m \\ (n \neq m)}} \frac{\mathrm{e}^{-\mu|\boldsymbol{r}_n - \boldsymbol{r}_m|}}{|\boldsymbol{r}_n - \boldsymbol{r}_m|} \tag{15.4}$$

A Course in Theoretical Physics, First Edition. P. J. Shepherd.

is the hamiltonian of the electrons,

$$H_{\rm b} = \frac{1}{2}\frac{e^2}{4\pi\varepsilon_0}\int {\rm d}^3\boldsymbol{r}{\rm d}^3\boldsymbol{r}'\frac{\rho(\boldsymbol{r})\rho(\boldsymbol{r}'){\rm e}^{-\mu|\boldsymbol{r}-\boldsymbol{r}'|}}{|\boldsymbol{r}-\boldsymbol{r}'|} \tag{15.5}$$

is the hamiltonian of the positive background whose density is $\rho(\boldsymbol{r})$, and

$$H_{\rm el-b} = -\frac{e^2}{4\pi\varepsilon_0}\sum_n\int {\rm d}^3\boldsymbol{r}\frac{\rho(\boldsymbol{r}){\rm e}^{-\mu|\boldsymbol{r}-\boldsymbol{r}_n|}}{|\boldsymbol{r}-\boldsymbol{r}_n|} \tag{15.6}$$

is the hamiltonian corresponding to the interaction of the electrons with the background. In all three contributions the factor with μ is necessary to define the integrals, which would otherwise diverge. We shall see that the μ-dependent integrals all cancel each other (apart from one that vanishes in the limit $\Omega \to \infty$), and so we can set $\mu = 0$ at the end of the calculation.

We first calculate $H_{\rm b}$, putting $\rho(\boldsymbol{r}) = N/\Omega$ (uniform distribution of the positive charge):

$$H_{\rm b} = \frac{1}{2}\frac{e^2}{4\pi\varepsilon_0}\left(\frac{N}{\Omega}\right)^2\int {\rm d}^3\boldsymbol{r}{\rm d}^3\boldsymbol{r}'\frac{{\rm e}^{-\mu|\boldsymbol{r}-\boldsymbol{r}'|}}{|\boldsymbol{r}-\boldsymbol{r}'|} = \frac{1}{2}\frac{e^2}{4\pi\varepsilon_0}\left(\frac{N}{\Omega}\right)^2\int {\rm d}^3\boldsymbol{r}{\rm d}^3\boldsymbol{R}\frac{{\rm e}^{-\mu R}}{R}, \tag{15.7}$$

where we have used the translational invariance of the infinite system to shift the origin of integration. But

$$\int {\rm d}^3\boldsymbol{R}\frac{{\rm e}^{-\mu R}}{R} = 4\pi\int R^2{\rm d}R\frac{{\rm e}^{-\mu R}}{R} = \frac{4\pi}{\mu^2} \quad \text{and} \quad \int {\rm d}^3\boldsymbol{r} = \Omega,$$

and so

$$H_{\rm b} = \frac{1}{2}\frac{e^2}{4\pi\varepsilon_0}\frac{N^2}{\Omega}\frac{4\pi}{\mu^2} \quad \text{(a } c\text{-number).} \tag{15.8}$$

The interaction term $H_{\rm el-b}$ (15.6) is, in principle, a sum of one-particle operators. However, if we perform the integration over ${\rm d}^3\boldsymbol{r}$ and again use the translational invariance of the system, we see that $H_{\rm el-b}$ is also a c-number:

$$H_{\rm el-b} = -\frac{e^2}{4\pi\varepsilon_0}\sum_{n=1}^{N}\frac{N}{\Omega}\int {\rm d}^3\boldsymbol{R}\frac{{\rm e}^{-\mu R}}{R} = -\frac{e^2}{4\pi\varepsilon_0}\frac{N^2}{\Omega}\frac{4\pi}{\mu^2}. \tag{15.9}$$

The total hamiltonian therefore becomes

$$H = -\frac{1}{2}\frac{e^2}{4\pi\varepsilon_0}\frac{N^2}{\Omega}\frac{4\pi}{\mu^2} + H_{\rm el}, \tag{15.10}$$

where $H_{\rm el}$ is the only part that is not a c-number, that is, $H_{\rm el}$ is the part that contains the interesting physics. (Note that we cannot yet set $\mu = 0$.)

To write H_{el} in second quantization, we need the matrix elements of the one-particle and two-particle operators in H_{el}. The kinetic-energy term requires evaluation of the element $t^{(1)}_{\boldsymbol{k}_1\lambda_1,\boldsymbol{k}_2\lambda_2}$:

$$
\begin{aligned}
t^{(1)}_{\boldsymbol{k}_1\lambda_1,\boldsymbol{k}_2\lambda_2} &= \int \mathrm{d}\xi\, \psi^*_{\boldsymbol{k}_1\lambda_1}(\xi) t^{(1)} \psi_{\boldsymbol{k}_2\lambda_2}(\xi) \equiv \sum_{\lambda} \int \mathrm{d}^3\boldsymbol{r}\, \psi^*_{\boldsymbol{k}_1\lambda_1}(\boldsymbol{r},\lambda) t^{(1)} \psi_{\boldsymbol{k}_2\lambda_2}(\boldsymbol{r},\lambda) \\
&= \sum_{\lambda} \delta_{\lambda\lambda_1}\delta_{\lambda\lambda_2} \int \mathrm{d}^3\boldsymbol{r}\, \psi^*_{\boldsymbol{k}_1}(\boldsymbol{r}) t^{(1)} \psi_{\boldsymbol{k}_2}(\boldsymbol{r}) = -\frac{\hbar^2}{2m}\delta_{\lambda_1\lambda_2}\frac{1}{\Omega}\int \mathrm{d}^3\boldsymbol{r}\,\mathrm{e}^{-i\boldsymbol{k}_1\cdot\boldsymbol{r}}\nabla^2 \mathrm{e}^{i\boldsymbol{k}_2\cdot\boldsymbol{r}} \\
&= \frac{\hbar^2\boldsymbol{k}_2^2}{2m}\delta_{\lambda_1\lambda_2}\frac{1}{\Omega}\int \mathrm{d}^3\boldsymbol{r}\,\mathrm{e}^{-i(\boldsymbol{k}_1-\boldsymbol{k}_2)\cdot\boldsymbol{r}} = \frac{\hbar^2\boldsymbol{k}_2^2}{2m}\delta_{\lambda_1\lambda_2}\delta_{\boldsymbol{k}_1\boldsymbol{k}_2}.
\end{aligned}
\tag{15.11}
$$

Therefore, [see (14.58)]

$$
\begin{aligned}
\hat{T} &= \sum_{\boldsymbol{k}_1\lambda_1}\sum_{\boldsymbol{k}_2\lambda_2} t^{(1)}_{\boldsymbol{k}_1\lambda_1,\boldsymbol{k}_2\lambda_2}\hat{a}^\dagger_{\boldsymbol{k}_1\lambda_1}\hat{a}_{\boldsymbol{k}_2\lambda_2} = \sum_{\boldsymbol{k}_1\lambda_1}\sum_{\boldsymbol{k}_2\lambda_2}\frac{\hbar^2 k_2^2}{2m}\delta_{\lambda_1\lambda_2}\delta_{\boldsymbol{k}_1\boldsymbol{k}_2}\hat{a}^\dagger_{\boldsymbol{k}_1\lambda_1}\hat{a}_{\boldsymbol{k}_2\lambda_2} \\
&= \sum_{\boldsymbol{k}\lambda}\frac{\hbar^2k^2}{2m}\hat{a}^\dagger_{\boldsymbol{k}\lambda}\hat{a}_{\boldsymbol{k}\lambda} = \sum_{\boldsymbol{k}\lambda}\frac{\hbar^2k^2}{2m}\hat{n}_{\boldsymbol{k}\lambda},
\end{aligned}
\tag{15.12}
$$

and we see that $\hat{T}$ is the sum over modes of the kinetic energy of each mode, multiplied by the corresponding number operator.

For the two-particle part of H_{el} we need to evaluate the matrix element $v^{(2)}_{ij,lm}$ [see (14.62]:

$$
v^{(2)}_{\boldsymbol{k}_1\lambda_1\,\boldsymbol{k}_2\lambda_2,\,\boldsymbol{k}_3\lambda_3\,\boldsymbol{k}_4\lambda_4} \equiv \int \mathrm{d}\xi_1\mathrm{d}\xi_2\,\psi^*_{\boldsymbol{k}_1\lambda_1}(\xi_1)\psi^*_{\boldsymbol{k}_2\lambda_2}(\xi_2)v^{(2)}(\xi_1,\xi_2)\psi_{\boldsymbol{k}_3\lambda_3}(\xi_1)\psi_{\boldsymbol{k}_4\lambda_4}(\xi_2).
\tag{15.13}
$$

Put $\xi_1 \equiv \boldsymbol{r}_1, \lambda$ and $\xi_2 \equiv \boldsymbol{r}_2, \lambda'$. Then

$$
\begin{aligned}
&v^{(2)}_{\boldsymbol{k}_1\lambda_1\,\boldsymbol{k}_2\lambda_2,\,\boldsymbol{k}_3\lambda_3\,\boldsymbol{k}_4\lambda_4} \\
&= \sum_{\lambda}\sum_{\lambda'}\int \mathrm{d}^3\boldsymbol{r}_1\mathrm{d}^3\boldsymbol{r}_2\,\psi^*_{\boldsymbol{k}_1}(\boldsymbol{r}_1)\delta_{\lambda_1\lambda}\psi^*_{\boldsymbol{k}_2}(\boldsymbol{r}_2)\delta_{\lambda_2\lambda'}v^{(2)}(\boldsymbol{r}_1,\boldsymbol{r}_2)\psi_{\boldsymbol{k}_3}(\boldsymbol{r}_1)\delta_{\lambda_3\lambda}\psi_{\boldsymbol{k}_4}(\boldsymbol{r}_2)\delta_{\lambda_4\lambda'} \\
&= \frac{e^2}{4\pi\varepsilon_0}\frac{1}{\Omega^2}\sum_{\lambda}\sum_{\lambda'}\delta_{\lambda_1\lambda}\delta_{\lambda_2\lambda'}\delta_{\lambda_3\lambda}\delta_{\lambda_4\lambda'}\int \mathrm{d}^3\boldsymbol{r}_1\mathrm{d}^3\boldsymbol{r}_2\,\mathrm{e}^{-i\boldsymbol{k}_1\cdot\boldsymbol{r}_1}\mathrm{e}^{-i\boldsymbol{k}_2\cdot\boldsymbol{r}_2}\frac{\mathrm{e}^{-\mu|\boldsymbol{r}_1-\boldsymbol{r}_2|}}{|\boldsymbol{r}_1-\boldsymbol{r}_2|}\mathrm{e}^{i\boldsymbol{k}_3\cdot\boldsymbol{r}_1}\mathrm{e}^{i\boldsymbol{k}_4\cdot\boldsymbol{r}_2}.
\end{aligned}
\tag{15.14}
$$

Put $\boldsymbol{r}_2 = \boldsymbol{r}$ and $\boldsymbol{r}_1 - \boldsymbol{r}_2 = \boldsymbol{R}$ (that is, $\boldsymbol{r}_1 = \boldsymbol{R} + \boldsymbol{r}$). Then the matrix element $v^{(2)}_{\boldsymbol{k}_1\lambda_1\boldsymbol{k}_2\lambda_2,\boldsymbol{k}_3\lambda_3\boldsymbol{k}_4\lambda_4}$ is

$$
\begin{aligned}
&\frac{e^2}{4\pi\varepsilon_0}\frac{1}{\Omega^2}\sum_{\lambda}\sum_{\lambda'}\delta_{\lambda_1\lambda}\delta_{\lambda_2\lambda'}\delta_{\lambda_3\lambda}\delta_{\lambda_4\lambda'}\int \mathrm{d}^3\boldsymbol{r}\,\mathrm{e}^{-i(\boldsymbol{k}_1+\boldsymbol{k}_2-\boldsymbol{k}_3-\boldsymbol{k}_4)\cdot\boldsymbol{r}}\int \mathrm{d}^3\boldsymbol{R}\,\mathrm{e}^{i(\boldsymbol{k}_3-\boldsymbol{k}_1)\cdot\boldsymbol{R}}\frac{\mathrm{e}^{-\mu R}}{R} \\
&= \frac{e^2}{4\pi\varepsilon_0}\frac{1}{\Omega^2}\Omega\,\delta_{\boldsymbol{k}_1+\boldsymbol{k}_2,\,\boldsymbol{k}_3+\boldsymbol{k}_4}\delta_{\lambda_1\lambda_3}\delta_{\lambda_2\lambda_4}I,
\end{aligned}
$$

where

$$I = \int d^3 R e^{i(k_3-k_1)\cdot R} \frac{e^{-\mu R}}{R} = 2\pi \int_0^\infty dR R^2 \frac{e^{-\mu R}}{R} \int_{-1}^{1} dx e^{i|k_3-k_1|Rx}$$
$$= 2\pi \int_0^\infty dR R e^{-\mu R} \left[\frac{e^{i|k_3-k_1|Rx}}{i\,|k_3-k_1|\,R} \right]_{x=-1}^{x=1} = \frac{2\pi}{i\,|k_3-k_1|} \left[\int_0^\infty dR e^{(-\mu+i|k_3-k_1|)R} - \int_0^\infty dR e^{(-\mu-i|k_3-k_1|)R} \right]$$
$$= \frac{2\pi}{i\,|k_3-k_1|} \left[-\frac{1}{-\mu+i\,|k_3-k_1|} + \frac{1}{-\mu-i\,|k_3-k_1|} \right] = \frac{4\pi}{|k_3-k_1|^2+\mu^2}.$$

Therefore, in second quantization, the contribution of the electron–electron interaction to H_{el} is [see (14.61)]

$$\frac{1}{2}\frac{e^2}{4\pi\varepsilon_0}\frac{1}{\Omega}\sum_{k_1\lambda_1}\sum_{k_2\lambda_2}\sum_{k_3\lambda_3}\sum_{k_4\lambda_4}\delta_{k_1+k_2,k_3+k_4}\delta_{\lambda_1\lambda_3}\delta_{\lambda_2\lambda_4}\frac{4\pi}{|k_3-k_1|^2+\mu^2}\hat{a}^\dagger_{k_1\lambda_1}\hat{a}^\dagger_{k_2\lambda_2}\hat{a}_{k_4\lambda_4}\hat{a}_{k_3\lambda_3}.$$

The factor $\delta_{k_1+k_2,k_3+k_4}$ expresses momentum conservation. We make the change of variables shown in figure 15.1, that is,

$$k_1 = k + q, \quad k_2 = p - q, \quad k_3 = k, \quad k_4 = p,$$

so that $\hbar(k_1 - k_3) = \hbar q$ is the momentum transferred in the two-particle interaction.

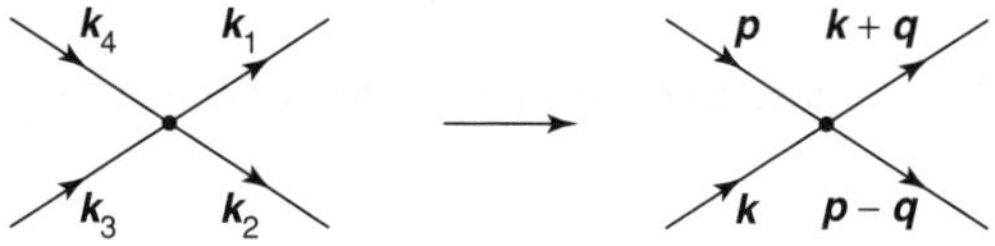

Figure 15.1 Relabelling of the momenta in electron—electron scattering.

For the contribution of the electron–electron interaction to H_{el} we then have

$$\frac{1}{2}\frac{e^2}{4\pi\varepsilon_0}\frac{1}{\Omega}\sum_{k,p,q}\sum_{\lambda_1,\lambda_2}\frac{4\pi}{q^2+\mu^2}\hat{a}^\dagger_{k+q,\lambda_1}\hat{a}^\dagger_{p-q,\lambda_2}\hat{a}_{p\lambda_2}\hat{a}_{k\lambda_1}. \tag{15.15}$$

We now show that the $q = 0$ part of this triple sum contains a part which cancels the c-number result obtained above for $H_b + H_{el\text{-}b}$. The $q = 0$ part is

$$\frac{1}{2}\frac{e^2}{4\pi\varepsilon_0}\frac{1}{\Omega}\sum_{k,p}\sum_{\lambda_1,\lambda_2}\frac{4\pi}{\mu^2}\hat{a}^\dagger_{k\lambda_1}\hat{a}^\dagger_{p\lambda_2}\hat{a}_{p\lambda_2}\hat{a}_{k\lambda_1}.$$

To simplify this we move $\hat{a}_{k\lambda_1}$ to the left through $\hat{a}^\dagger_{p\lambda_2}\hat{a}_{p\lambda_2}$. Using the anticommutation relations (14.59), we find

$$\hat{a}^\dagger_{p\lambda_2}\hat{a}_{p\lambda_2}\hat{a}_{k\lambda_1} = -\hat{a}^\dagger_{p\lambda_2}\hat{a}_{k\lambda_1}\hat{a}_{p\lambda_2} = \left(\hat{a}_{k\lambda_1}\hat{a}^\dagger_{p\lambda_2} - \{\hat{a}^\dagger_{p\lambda_2}, \hat{a}_{k\lambda_1}\}\right)\hat{a}_{p\lambda_2}$$
$$= \left(\hat{a}_{k\lambda_1}\hat{a}^\dagger_{p\lambda_2} - \delta_{p,k}\delta_{\lambda_1\lambda_2}\right)\hat{a}_{p\lambda_2}.$$

Thus, the $\boldsymbol{q} = 0$ part becomes

$$\frac{1}{2}\frac{e^2}{4\pi\varepsilon_0}\frac{1}{\Omega}\left(\sum_{\boldsymbol{k},\boldsymbol{p}}\sum_{\lambda_1,\lambda_2}\frac{4\pi}{\mu^2}\hat{a}^\dagger_{\boldsymbol{k}\lambda_1}\hat{a}_{\boldsymbol{k}\lambda_1}\hat{a}^\dagger_{\boldsymbol{p}\lambda_2}\hat{a}_{\boldsymbol{p}\lambda_2} - \sum_{\boldsymbol{k}}\sum_{\lambda_1}\frac{4\pi}{\mu^2}\hat{a}^\dagger_{\boldsymbol{k}\lambda_1}\hat{a}_{\boldsymbol{k}\lambda_1}\right)$$
$$= \frac{1}{2}\frac{e^2}{4\pi\varepsilon_0}\frac{1}{\Omega}\frac{4\pi}{\mu^2}(\hat{N}^2 - \hat{N}).$$

If we are dealing with states of fixed N, we can replace $\hat{N}$ by its eigenvalue N, thereby obtaining the c-number contribution:

$$\frac{1}{2}\frac{e^2}{4\pi\varepsilon_0}\frac{1}{\Omega}\frac{4\pi}{\mu^2}(N^2 - N). \tag{15.16}$$

Note that the first term here cancels the constant term $H_{\mathrm{b}} + H_{\mathrm{el\text{-}b}}$ in the hamiltonian (15.10).

In relation to the second term, we note that we always consider the **thermodynamic limit**, that is, the limit $N \to \infty$, $\Omega \to \infty$ with the density $\rho = N/\Omega$ held constant. In this limit we expect the total energy to be an extensive quantity, that is, proportional to N, so that the energy per particle should be independent of N (and Ω). This is clearly true of the first term in the above expression (since $N^2/\Omega = \rho N$). The energy per particle arising from the second term, however, is of order $1/\Omega$, which vanishes in the thermodynamic limit. Thus, we take the thermodynamic limit first (when the remaining constant term vanishes) and only then set $\mu = 0$ (so that the factor $4\pi/(q^2 + \mu^2) \to 4\pi/q^2$).

Therefore, the final hamiltonian for a bulk electron gas in a uniform positive background is

$$\hat{H} = \sum_{\boldsymbol{k},\lambda}\frac{\hbar^2k^2}{2m}\hat{a}^\dagger_{\boldsymbol{k}\lambda}\hat{a}_{\boldsymbol{k}\lambda} + \frac{1}{2}\frac{e^2}{4\pi\varepsilon_0}\frac{1}{\Omega}\sum_{\substack{\boldsymbol{k},\boldsymbol{p},\boldsymbol{q}\\ (\text{omit}\,\boldsymbol{q}=0)}}\sum_{\lambda_1,\lambda_2}\frac{4\pi}{q^2}\hat{a}^\dagger_{\boldsymbol{k}+\boldsymbol{q},\lambda_1}\hat{a}^\dagger_{\boldsymbol{p}-\boldsymbol{q},\lambda_2}\hat{a}_{\boldsymbol{p}\lambda_2}\hat{a}_{\boldsymbol{k}\lambda_1}, \tag{15.17}$$

where the limit $N \to \infty$, $\Omega \to \infty$ with $\rho = N/\Omega = \text{const}$ is assumed.

An alternative form of this hamiltonian is obtained by moving $\hat{a}_{\boldsymbol{k}\lambda_1}$ to the left in order to obtain an expression in terms of the second-quantized density-fluctuation operator (14.85):

$$\hat{a}^\dagger_{\boldsymbol{p}-\boldsymbol{q},\lambda_2}\hat{a}_{\boldsymbol{p}\lambda_2}\hat{a}_{\boldsymbol{k}\lambda_1} = -\hat{a}^\dagger_{\boldsymbol{p}-\boldsymbol{q},\lambda_2}\hat{a}_{\boldsymbol{k}\lambda_1}\hat{a}_{\boldsymbol{p}\lambda_2} = \left(\hat{a}_{\boldsymbol{k}\lambda_1}\hat{a}^\dagger_{\boldsymbol{p}-\boldsymbol{q},\lambda_2} - [\hat{a}^\dagger_{\boldsymbol{p}-\boldsymbol{q},\lambda_2}, \hat{a}_{\boldsymbol{k}\lambda_1}]_+\right)\hat{a}_{\boldsymbol{p}\lambda_2}$$
$$= \hat{a}_{\boldsymbol{k}\lambda_1}\hat{a}^\dagger_{\boldsymbol{p}-\boldsymbol{q},\lambda_2}\hat{a}_{\boldsymbol{p}\lambda_2} - \delta_{\boldsymbol{p}-\boldsymbol{q},\boldsymbol{k}}\delta_{\lambda_2,\lambda_1}\hat{a}_{\boldsymbol{p}\lambda_2}.$$

The interaction term in the hamiltonian thus becomes

$$\frac{1}{2}\frac{e^2}{4\pi\varepsilon_0}\frac{1}{\Omega}\sum_{\substack{\boldsymbol{q}\\ (\text{omit}\,\boldsymbol{q}=0)}}\frac{4\pi}{q^2}\left(\sum_{\boldsymbol{k},\lambda_1}\hat{a}^\dagger_{\boldsymbol{k}+\boldsymbol{q},\lambda_1}\hat{a}_{\boldsymbol{k}\lambda_1}\sum_{\boldsymbol{p},\lambda_2}\hat{a}^\dagger_{\boldsymbol{p}-\boldsymbol{q},\lambda_2}\hat{a}_{\boldsymbol{p}\lambda_2} - \sum_{\boldsymbol{k},\lambda_1}\hat{a}^\dagger_{\boldsymbol{k}+\boldsymbol{q},\lambda_1}\hat{a}_{\boldsymbol{k}+\boldsymbol{q},\lambda_1}\right)$$
$$= \frac{1}{2}\frac{e^2}{4\pi\varepsilon_0}\frac{1}{\Omega}\sum_{\substack{\boldsymbol{q}\\ (\text{omit}\,\boldsymbol{q}=0)}}\frac{4\pi}{q^2}\left(\hat{\rho}^\dagger_{\boldsymbol{q}}\hat{\rho}^\dagger_{-\boldsymbol{q}} - \hat{N}\right),$$

that is,

$$\hat{H} = \sum_{\boldsymbol{k},\lambda} \frac{\hbar^2 k^2}{2m} \hat{a}^\dagger_{\boldsymbol{k}\lambda} \hat{a}_{\boldsymbol{k}\lambda} + \frac{1}{2} \frac{e^2}{4\pi\varepsilon_0} \frac{1}{\Omega} \sum_{\substack{\boldsymbol{q} \\ (\text{omit}\, \boldsymbol{q}=0)}} \frac{4\pi}{q^2} \left(\hat{\rho}^\dagger_{\boldsymbol{q}} \hat{\rho}^\dagger_{-\boldsymbol{q}} - \hat{N} \right). \tag{15.18}$$

In most cases, however, the first form (15.17) is used.

To estimate the relative magnitudes of the kinetic-energy and potential-energy terms in $\hat{H}$, we now change to dimensionless variables. We define a length r_0 such that

$$\Omega = N \times \frac{4}{3} \pi r_0^3, \tag{15.19}$$

that is, r_0 is of the order of the interparticle spacing. The Coulomb interaction provides a second length, namely, the Bohr radius

$$a_0 = 4\pi\varepsilon_0 \frac{\hbar^2}{me^2}. \tag{15.20}$$

The dimensionless ratio

$$r_\text{s} \equiv r_0 / a_0 \tag{15.21}$$

of these two quantities characterizes the density of the system. Using r_0 as the unit of length, we define the following dimensionless quantities:

$$\bar{\Omega} \equiv \frac{\Omega}{r_0^3}, \quad \bar{\boldsymbol{k}} \equiv r_0 \boldsymbol{k}, \quad \bar{\boldsymbol{q}} \equiv r_0 \bar{\boldsymbol{q}}, \text{ etc.} \tag{15.22}$$

Then

$$\frac{\hbar^2 k^2}{2m} = \frac{\hbar^2}{2m r_0^2} \bar{k}^2 = \frac{1}{2} \frac{a_0 e^2}{4\pi\varepsilon_0} \frac{1}{a_0^2 r_\text{s}^2} \bar{k}^2 = \frac{1}{2} \frac{e^2}{4\pi\varepsilon_0} \frac{1}{a_0 r_\text{s}^2} \bar{k}^2 \tag{15.23}$$

and

$$\frac{1}{2} \frac{e^2}{4\pi\varepsilon_0} \frac{1}{\Omega} \frac{4\pi}{q^2} = \frac{1}{2} \frac{e^2}{4\pi\varepsilon_0} \frac{1}{\bar{\Omega} r_0^3} \frac{4\pi r_0^2}{\bar{q}^2} = \frac{1}{2} \frac{e^2}{4\pi\varepsilon_0} \frac{1}{a_0 r_\text{s}} \frac{4\pi}{\bar{\Omega} \bar{q}^2}, \tag{15.24}$$

so that, from (15.17),

$$\begin{aligned} \hat{H} &= \hat{T} + \hat{V} \\ &= \frac{1}{2} \frac{e^2}{4\pi\varepsilon_0} \frac{1}{a_0 r_\text{s}^2} \left(\sum_{\bar{\boldsymbol{k}},\lambda} \bar{k}^2 \hat{a}^\dagger_{\bar{\boldsymbol{k}}\lambda} \hat{a}_{\bar{\boldsymbol{k}}\lambda} + \frac{r_\text{s}}{\bar{\Omega}} \sum_{\substack{\bar{\boldsymbol{k}},\bar{\boldsymbol{p}},\bar{\boldsymbol{q}} \\ (\text{omit}\, \bar{\boldsymbol{q}}=0)}} \sum_{\lambda_1,\lambda_2} \frac{4\pi}{\bar{q}^2} \hat{a}^\dagger_{\bar{\boldsymbol{k}}+\bar{\boldsymbol{q}},\lambda_1} \hat{a}^\dagger_{\bar{\boldsymbol{p}}-\bar{\boldsymbol{q}},\lambda_2} \hat{a}_{\bar{\boldsymbol{p}}\lambda_2} \hat{a}_{\bar{\boldsymbol{k}}\lambda_1} \right). \end{aligned} \tag{15.25}$$

This important result shows that as $r_s \to 0$ (that is, as $r_0 \to 0$, the **high-density limit**) the interaction terms become a small perturbation. One might expect from this that the ground-state energy would be a power-series expansion in the small parameter r_s at high density, but detailed calculations in fact give the form

$$E = \langle 0|\hat{H}|0\rangle = \frac{1}{2}\frac{Ne^2}{4\pi\varepsilon_0}\frac{1}{a_0 r_s^2}\left(a + br_s + cr_s^2 \ln r_s + dr_s^2 + \cdots\right). \tag{15.26}$$

In the high-density limit, when r_s is small, we might expect a good result from first-order perturbation theory, that is, by finding the expectation value of the full hamiltonian in the ground state of the corresponding system of noninteracting electrons (that is, in the ground state unperturbed by the interaction). Of course, the interaction will, in fact, change the ground-state wave function, and the second and higher orders of perturbation theory correspond to finding the expectation value of $\hat{H}$ in the perturbed ground state.

So the first question is: what is the ground state of a gas of noninteracting spin-1/2 particles (a Fermi gas)? It is, of course, that determined by the Pauli principle, with two particles (with opposite values of the appropriate spin component) in each allowed momentum eigenstate, up to a maximum value (the Fermi momentum $p = \hbar k_F$), so that all allowed quantized momenta in a sphere of radius $\hbar k_F$ in momentum space correspond to doubly occupied single-particle states. We found earlier [see (13.27)] that

$$N = \frac{\Omega k_F^3}{3\pi^2}.$$

Therefore,

$$k_F = \left(\frac{3\pi^2 N}{\Omega}\right)^{1/3} = \left(\frac{3\pi^2}{\frac{4}{3}\pi r_0^3}\right)^{1/3} = \left(\frac{9\pi}{4}\right)^{1/3} r_0^{-1} \approx 1.92 r_0^{-1}, \tag{15.27}$$

so that k_F^{-1} is comparable to the interparticle spacing. We also found [see (13.29)] that for a Fermi gas (i.e., with $\hat{H} = \hat{H}_0$) we have

$$\langle 0|\hat{T}|0\rangle \equiv E_0 = \frac{3}{5}\varepsilon_F N$$

where $\varepsilon_F = \hbar^2 k_F^2/2m$ is the **Fermi energy**. In terms of r_s, we have, using (15.20), (15.27) and (15.21),

$$\begin{aligned}\langle 0|\hat{T}|0\rangle &= \frac{3}{5}\times\frac{1}{2}\times\frac{a_0 e^2}{4\pi\varepsilon_0}\times\left(\frac{9\pi}{4}\right)^{2/3}\frac{1}{r_0^2}\times N = \frac{1}{2}\frac{e^2}{4\pi\varepsilon_0 a_0}N\times\frac{3}{5}\times\left(\frac{9\pi}{4}\right)^{2/3}\frac{1}{r_s^2}\\ &= \frac{1}{2}\frac{e^2}{4\pi\varepsilon_0 a_0}N\frac{2.21}{r_s^2},\end{aligned} \tag{15.28}$$

so that, in (15.26), $a = 2.21$.

The interaction term (that is, the first-order energy shift) is [see (15.17)]:

$$\langle 0|\hat{V}|0\rangle = \frac{1}{2}\frac{e^2}{4\pi\varepsilon_0}\frac{1}{\Omega}\sum_{\substack{\boldsymbol{k},\boldsymbol{p},\boldsymbol{q}\\(\text{omit}\,\boldsymbol{q}=0)}}\sum_{\lambda_1,\lambda_2}\frac{4\pi}{q^2}\langle 0|\hat{a}^\dagger_{\boldsymbol{k}+\boldsymbol{q},\lambda_1}\hat{a}^\dagger_{\boldsymbol{p}-\boldsymbol{q},\lambda_2}\hat{a}_{\boldsymbol{p}\lambda_2}\hat{a}_{\boldsymbol{k}\lambda_1}|0\rangle.$$

The matrix element here is nonzero only if the single-particle states with quantum numbers $\boldsymbol{p}\lambda_2$ and $\boldsymbol{k}\lambda_1$ are occupied in $|0\rangle$. (Otherwise, the annihilation operators would give zero.) The two creation operators must then fill the two holes thus created. (If they created particles with other momenta, the resulting state would be orthogonal to $\langle 0|$ and the matrix element would vanish.) There are thus only two possibilities. The first, giving the so-called **direct term**, is

$$\boldsymbol{k}+\boldsymbol{q},\, \lambda_1 = \boldsymbol{k}\lambda_1 \quad \text{and} \quad \boldsymbol{p}-\boldsymbol{q},\, \lambda_2 = \boldsymbol{p}\lambda_2, \tag{15.29}$$

and implies $\boldsymbol{q}=0$. But this value of $\boldsymbol{q}$ is excluded from the sum, and so only the second possibility remains (giving the so-called **exchange term**):

$$\boldsymbol{k}+\boldsymbol{q},\, \lambda_1 = \boldsymbol{p}\lambda_2 \quad \text{and} \quad \boldsymbol{p}-\boldsymbol{q},\, \lambda_2 = \boldsymbol{k}\lambda_1. \tag{15.30}$$

The matrix element thus becomes

$$\begin{aligned}\langle 0|\hat{a}^\dagger_{\boldsymbol{k}+\boldsymbol{q},\lambda_1}\hat{a}^\dagger_{\boldsymbol{p}-\boldsymbol{q},\lambda_2}\hat{a}_{\boldsymbol{p}\lambda_2}\hat{a}_{\boldsymbol{k}\lambda_1}|0\rangle &= \delta_{\boldsymbol{k}+\boldsymbol{q},\boldsymbol{p}}\delta_{\lambda_1\lambda_2}\langle 0|\hat{a}^\dagger_{\boldsymbol{k}+\boldsymbol{q},\lambda_1}\hat{a}^\dagger_{\boldsymbol{k}\lambda_1}\hat{a}_{\boldsymbol{k}+\boldsymbol{q},\lambda_1}\hat{a}_{\boldsymbol{k}\lambda_1}|0\rangle \\ = -\delta_{\boldsymbol{k}+\boldsymbol{q},\boldsymbol{p}}\delta_{\lambda_1\lambda_2}\langle 0|\hat{n}_{\boldsymbol{k}+\boldsymbol{q},\lambda_1}\hat{n}_{\boldsymbol{k}\lambda_1}|0\rangle &= -\delta_{\boldsymbol{k}+\boldsymbol{q},\boldsymbol{p}}\delta_{\lambda_1\lambda_2}\theta(k_{\mathrm{F}}-|\boldsymbol{k}+\boldsymbol{q}|)\theta(k_{\mathrm{F}}-k),\end{aligned}$$

where in writing the second line we were able to put $\hat{a}_{\boldsymbol{k}+\boldsymbol{q},\lambda_1}$ to the left of $\hat{a}^\dagger_{\boldsymbol{k},\lambda_1}$ with only a sign change (since $\boldsymbol{q}\neq 0$), and we used the Heaviside step function, defined by

$$\theta(x) = \begin{cases} 1 & \text{for } x>0, \\ 0 & \text{for } x<0. \end{cases}$$

Then

$$\begin{aligned}\langle 0|\hat{V}|0\rangle &= -\frac{1}{2}\frac{e^2}{4\pi\varepsilon_0}\frac{1}{\Omega}\sum_{\lambda_1}\sum_{\substack{\boldsymbol{k},\boldsymbol{q}\\(\text{omit }\boldsymbol{q}=0)}}\frac{4\pi}{q^2}\theta(k_{\mathrm{F}}-|\boldsymbol{k}+\boldsymbol{q}|)\theta(k_{\mathrm{F}}-k) \\ &= -\frac{1}{2}\frac{e^2}{4\pi\varepsilon_0}\frac{1}{\Omega}\times 2\left(\frac{\Omega}{(2\pi)^3}\right)^2\int \mathrm{d}^3\boldsymbol{k}\int \mathrm{d}^3\boldsymbol{q}\frac{4\pi}{q^2}\theta(k_{\mathrm{F}}-|\boldsymbol{k}+\boldsymbol{q}|)\theta(k_{\mathrm{F}}-k),\end{aligned} \tag{15.31}$$

where the restriction $\boldsymbol{q}\neq 0$ may now be ignored, since it affects the integrand at only a single point.

We can change the step functions to a more symmetrical form by the change of variables $\boldsymbol{k}\to\boldsymbol{p}=\boldsymbol{k}+\boldsymbol{q}/2$ (so that $\boldsymbol{k}=\boldsymbol{p}-\boldsymbol{q}/2$ and $\boldsymbol{k}+\boldsymbol{q}=\boldsymbol{p}+\boldsymbol{q}/2$). Then,

$$\langle 0|\hat{V}|0\rangle = -\frac{e^2}{4\pi\varepsilon_0}\frac{\Omega}{(2\pi)^6}\int \mathrm{d}^3\boldsymbol{q}\frac{4\pi}{q^2}\int \mathrm{d}^3\boldsymbol{p}\,\theta(k_{\mathrm{F}}-|\boldsymbol{p}+\boldsymbol{q}/2|)\theta(k_{\mathrm{F}}-|\boldsymbol{p}-\boldsymbol{q}/2|). \tag{15.32}$$

The region of integration in the last integral is shown shaded in the left-hand diagram in figure 15.2.

The total shaded "volume" of wave-vector space in the left-hand diagram in figure 15.2 is just twice that in the right-hand diagram, that is, $2(I-I')$, where I is the volume of the "cone plus ice cream", and I'

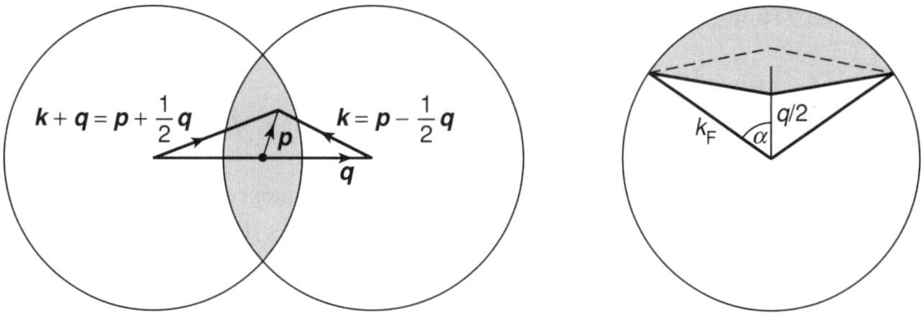

Figure 15.2 Diagrams for calculation of the integral in equation (15.32).

the volume of the flat-topped cone, in the right-hand diagram. We make the change of variable $x \equiv q/2k_\mathrm{F}$. Clearly, for $q/2 < k_\mathrm{F}$ (that is, $x < 1$), we have

$$I = \int_0^{2\pi} \mathrm{d}\varphi \int_0^{\alpha} \sin\theta \mathrm{d}\theta \int_0^{k_\mathrm{F}} k^2 \mathrm{d}k = -2\pi(\cos\alpha - 1)\frac{k_\mathrm{F}^3}{3} = \frac{2\pi}{3}\left(1 - \frac{q}{2k_\mathrm{F}}\right)k_\mathrm{F}^3 \equiv \frac{2\pi}{3}(1-x)k_\mathrm{F}^3,$$

while for $q/2 > k_\mathrm{F}$ (that is, $x > 1$), we have $I = 0$. The integral I', for $q/2 < k_\mathrm{F}$ (that is, $x < 1$), is

$$I' = \int_0^{q/2} \pi r^2 \mathrm{d}z = \frac{\pi(1-x^2)}{x^2}\int_0^{q/2} z^2 \mathrm{d}z = \frac{\pi(1-x^2)}{x^2}\frac{1}{3}\left(\frac{q}{2}\right)^3 = \frac{\pi k_\mathrm{F}^3(1-x^2)}{3x^2}x^3 = \frac{\pi k_\mathrm{F}^3(x-x^3)}{3},$$

where we used the substitution $r = z\tan\alpha = z\sqrt{1-x^2}/x$. Clearly, for $q/2 > k_\mathrm{F}$ (that is, $x > 1$), we have $I' = 0$. So, the total shaded volume of wave-vector space in the first diagram is

$$2(I - I') = 2\frac{2\pi}{3}k_\mathrm{F}^3\left(1 - x - \frac{1}{2}x + \frac{1}{2}x^3\right)\theta(1-x) = \frac{4\pi}{3}k_\mathrm{F}^3\left(1 - \frac{3}{2}x + \frac{x^3}{2}\right)\theta(1-x).$$

In the integral over $\boldsymbol{q}$ in $\langle 0|\hat{V}|0\rangle$ (15.32), substituting $q = 2k_\mathrm{F}x$ we have

$$\begin{aligned}
\langle 0|\hat{V}|0\rangle &= -\frac{e^2}{4\pi\varepsilon_0}\frac{\Omega}{(2\pi)^6}(4\pi)^2 2k_\mathrm{F}\int_0^{\infty} \mathrm{d}x \frac{4\pi}{3}k_\mathrm{F}^3\left(1 - \frac{3}{2}x + \frac{1}{2}x^3\right)\theta(1-x) \\
&= -\frac{e^2}{4\pi\varepsilon_0}\frac{\Omega}{\pi^3}\frac{2k_\mathrm{F}^4}{3}\left(1 - \frac{3}{4} + \frac{1}{8}\right) = -\frac{e^2}{4\pi\varepsilon_0}\frac{1}{4\pi^3}\Omega k_\mathrm{F}^4 \\
&= -\frac{e^2}{4\pi\varepsilon_0}\frac{1}{4\pi^3}(4/3)\pi r_\mathrm{s}^3 a_0^3 N\left(\frac{9\pi}{4}\right)^{4/3}\frac{1}{(r_\mathrm{s}a_0)^4} \\
&= -\frac{e^2}{4\pi\varepsilon_0}\frac{3}{4\pi}\left(\frac{9\pi}{4}\right)^{1/3}\frac{N}{r_\mathrm{s}a_0} = -\frac{e^2}{4\pi\varepsilon_0 a_0}N\frac{0.916}{r_\mathrm{s}},
\end{aligned} \tag{15.33}$$

where, in the penultimate line, we have used (15.19), (15.21) and (15.27).

Thus, we have $b = -0.916$ in (15.26), which becomes (as $r_s \to 0$)

$$\frac{E}{N} = \frac{1}{2}\frac{e^2}{4\pi\varepsilon_0 a_0}\left(\frac{2.21}{r_s^2} - \frac{0.916}{r_s} + \cdots\right) \tag{15.34}$$

The first term is the **kinetic energy** of a gas of interacting electrons. The second is called the **exchange energy** [recall that evaluation of the matrix element $\langle 0|\, \hat{a}^\dagger_{k+q,\lambda_1}\hat{a}^\dagger_{p-q,\lambda_2}\hat{a}_{p\lambda_2}\hat{a}_{k\lambda_1}|0\rangle$ led to two terms: the direct term (with $\boldsymbol{q} = 0$), which served to cancel $H_b + H_{el-b}$, and another term ($\boldsymbol{k}+\boldsymbol{q}, \lambda_1 = \boldsymbol{p}\lambda_2$ and $\boldsymbol{p}-\boldsymbol{q}, \lambda_2 = \boldsymbol{k}\lambda_1$), the so-called exchange term].

All terms in (15.26), beyond the first two, constitute, by definition, the **correlation energy** E_{corr}.

We thus have

$$\frac{E}{N} = \frac{2.21}{r_s^2} - \frac{0.916}{r_s} + E_{corr} \quad \text{rydberg}, \tag{15.35}$$

where

$$1\,\text{rydberg} \equiv \frac{1}{2}\frac{e^2}{4\pi\varepsilon_0 a_0}. \tag{15.36}$$

The first two terms correspond to the **Hartree–Fock approximation**, in which the only correlation taken into account in the wave functions is the "kinematic correlation", that is, the correlation arising from the Pauli exclusion principle. The "dynamic correlation", that is, that due to the forces between the electron charges, is not taken into account in the wave functions in the Hartree–Fock approximation.

The dynamic correlations obviously hinder the close approach of the electrons. (This is particularly important for electrons of antiparallel spin, since the Pauli exclusion principle will already tend to keep electrons of parallel spin apart.) This obviously reduces the energy of the system, and so we expect E_{corr} to be negative.

16

Superconductivity

16.1 Superconductors

For many metals and alloys, as the temperature is reduced the resistivity decreases and then drops discontinuously to zero at some temperature T_c (the **critical temperature**), that is, the system becomes **superconducting**. For example, in some cases a current (**supercurrent**) may persist in a ring with a decay time of 10^5 years at temperatures below the relevant T_c.

Metals and alloys in the superconducting state display the following interesting properties, which theory must attempt to explain.

(i) Below T_c, magnetic flux is totally excluded from a simply-connected superconductor (the **Meissner effect**), that is, a system in the superconducting state ($T < T_c$) is a perfect diamagnet.

(ii) Whereas in the normal state ($T > T_c$) of a metal the electronic contribution to the low-temperature specific heat is linear in the temperature T ($C_{e,\text{normal}} = \gamma T$), its temperature dependence in the superconducting state ($T < T_c$) is

$$C_{e,\text{superconductor}} = \text{const.} \exp(-\Delta / k_B T), \tag{16.1}$$

where, typically, $\Delta \approx k_B T_c$.

(iii) There is an energy gap E_g in the electron-excitation spectrum of a superconductor (manifested experimentally in the transparency of the system to photons with energy $\hbar\omega < E_g$); the gap is found experimentally to take the value $E_g = 2\Delta$. As we saw in section 13.2.2, there is no energy gap in the electron spectrum of a normal metal describable by the free-electron model (with a filled Fermi sphere for the ground state), since excitations of arbitrarily low energy are possible. In a superconductor, on the other hand, there must clearly be some attractive interaction that stabilizes the electrons near the Fermi surface in such a way that a nonzero minimum energy E_g is needed to excite them.

(iv) A clue to the nature of this interaction is provided by the so-called **isotope effect**, described by the following rough empirical relationship for the critical temperatures T_c of different isotopes of a metal:

$$T_c = \frac{\text{const}}{M^\alpha} \quad (\alpha \approx 1/2), \tag{16.2}$$

A Course in Theoretical Physics, First Edition. P. J. Shepherd.

where M is the atomic mass of the isotope. Since we expect phonon frequencies to be proportional to $M^{-1/2}$ [see (13.92)], the dependence of T_c on M suggests that the necessary attractive electron–electron interaction involves electron–phonon coupling. This is also suggested by the experimental fact that metals (such as lead) that have high resistivity in the normal state are also good superconductors, that is, have reasonably high values of T_c.

(v) As $T \to T_c$ from below, the gap $E_g = 2\Delta$ goes continuously to zero, in the absence of an external magnetic field. In this respect, Δ is analogous to the **order parameter** of a ferromagnetic material, namely, the spontaneous magnetization, which also goes to zero continuously in zero external magnetic field as the temperature approaches the critical (Curie) temperature from below.

(vi) Superconductivity can be destroyed by applying an external magnetic field. The field intensity required to bring the system into the normal state (that is, to make Δ go to zero) depends on the temperature and is denoted by $H_c(T)$. As we should expect, $H_c(T)$ is largest at $T = 0$ and is zero at temperatures $T \geq T_c$.

16.2 The theory of Bardeen, Cooper and Schrieffer

What possible interactions can cause the superconducting transition? We can rule out the Coulomb correlation energy, with a fair degree of confidence, for two reasons. One is that the observed energy gap is $E_g = 2\Delta \approx k_B T_c \approx 10^{-3}$ eV and the Coulomb correlation energy is too big to explain such a small quantitative effect. The other is that the relatively large energies associated with correlation effects occur in both the normal state and the superconducting state and may be expected to "cancel out" in the difference.

For metallic superconductors, change in the metal isotope changes the superconducting transition temperature T_c ($T_c\sqrt{M}$ = const). Spin–spin and spin–orbit interactions do not change with M (for a given element), and so can be eliminated as possible causes of the transition.

Since the phonon properties do depend on M, we may guess that superconductivity arises from electron–phonon interactions. So we try, first, a hamiltonian of the form

$$\begin{aligned}\hat{H} = &\sum_{\boldsymbol{k},\lambda} \varepsilon_{\boldsymbol{k}} \hat{a}^\dagger_{\boldsymbol{k}\lambda}\hat{a}_{\boldsymbol{k}\lambda} + \sum_{\boldsymbol{k}} \hbar\omega_{\boldsymbol{k}} \hat{b}^\dagger_{\boldsymbol{k}}\hat{b}_{\boldsymbol{k}} + \sum_{\boldsymbol{k},\boldsymbol{k}',\lambda} M_{\boldsymbol{k}\boldsymbol{k}'}\hat{b}_{\boldsymbol{k}'-\boldsymbol{k}}\hat{a}^\dagger_{\boldsymbol{k}'\lambda}\hat{a}_{\boldsymbol{k}\lambda} \\ &+ \sum_{\boldsymbol{k},\boldsymbol{k}',\lambda} M^*_{\boldsymbol{k}\boldsymbol{k}'}\hat{b}^\dagger_{\boldsymbol{k}'-\boldsymbol{k}}\hat{a}^\dagger_{\boldsymbol{k}\lambda}\hat{a}_{\boldsymbol{k}'\lambda} + \frac{1}{2}\sum_{\substack{\boldsymbol{k},\boldsymbol{k}',\boldsymbol{q}\\ (\text{omit } \boldsymbol{q}=0)}} \sum_{\lambda,\lambda'} V_{\boldsymbol{q}}\hat{a}^\dagger_{\boldsymbol{k}+\boldsymbol{q},\lambda}\hat{a}^\dagger_{\boldsymbol{k}'-\boldsymbol{q},\lambda'}\hat{a}_{\boldsymbol{k}'\lambda'}\hat{a}_{\boldsymbol{k}\lambda},\end{aligned} \tag{16.3}$$

where $\hat{a}^\dagger_{\boldsymbol{k}\lambda}$ creates an electron of momentum $\hbar\boldsymbol{k}$, $\hat{b}^\dagger_{\boldsymbol{k}}$ creates a phonon of wave vector $\boldsymbol{k}$, $\varepsilon_{\boldsymbol{k}}$ is the energy of an independent electron of momentum $\hbar\boldsymbol{k}$, $\hbar\omega_{\boldsymbol{k}}$ is the energy of an independent phonon of wave vector $\boldsymbol{k}$, and

$$V_q = \frac{e^2}{4\pi\varepsilon_0}\frac{1}{\Omega}\frac{4\pi}{q^2}. \tag{16.4}$$

For example, the fourth sum in (16.3) represents the sum of momentum-conserving processes (with amplitude $M^*_{\boldsymbol{k}\boldsymbol{k}'}$) in which an electron of momentum $\hbar\boldsymbol{k}'$ is destroyed and a phonon of wave vector $\boldsymbol{k}' - \boldsymbol{k}$ and an electron of momentum $\hbar\boldsymbol{k}$ are created. The third sum is similarly described. High $|M_{\boldsymbol{k}\boldsymbol{k}'}|$ corresponds to a high amplitude for scattering processes in which an electron is scattered from one state to another, that is, processes that lead to high resistivity at normal temperatures. That such terms are also responsible for the energy gap, and hence for superconductivity, is supported by the apparently paradoxical circumstance that metals, such as lead, with the highest resistivity at normal temperatures are often the best superconductors.

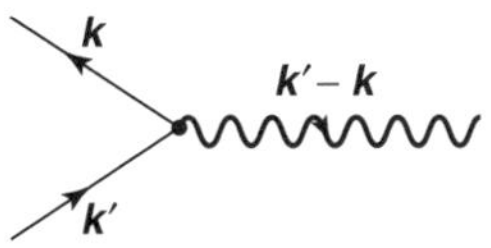

Figure 16.1 Diagrammatic representation of a term in the fourth sum in equation (16.3).

We can denote a term in the fourth sum by the diagram in figure 16.1.

There is one vertex (junction) in the diagram, implying a process with a single electron–phonon interaction. If we regard the electron–phonon interaction as a perturbation, we can ask what kind of processes can occur to second order in this perturbation, that is, to second order in $M_{kk'}$, just as the second-order correction to the energy in stationary perturbation theory contains the perturbation V bilinearly [see equation (3.14)].

The corresponding processes can obviously be represented by all possible diagrams with two vertices, such that at each vertex there are two electron lines and one phonon line. These are clearly those shown in figure 16.2.

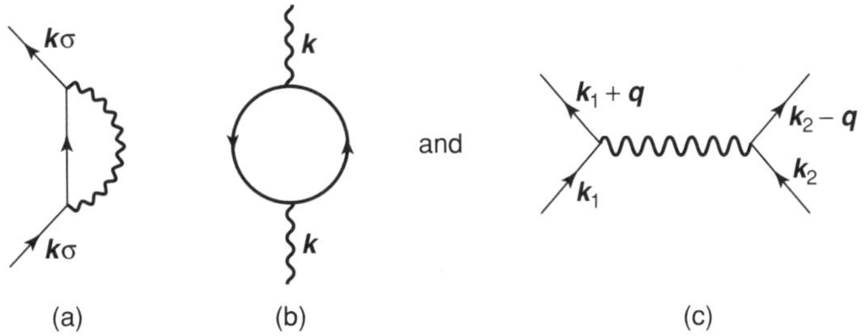

Figure 16.2 Diagrammatic representation of processes to second order in $M_{kk'}$.

Process (a), which contains an incoming and an outgoing electron line, changes the energy of the electron (by an amount larger than the gap), but does not produce the gap. Process (b) does the same for a phonon. These effects can thus be taken into account by calculating these energy changes and modifying ("renormalizing") the original single-electron and single-phonon energies (ε_k and $\hbar\omega_k$, respectively).

It will be shown below that process (c), in which one electron distorts the lattice (that is, emits a phonon) and a second electron is affected by the lattice distortion (that is, absorbs the phonon), is indeed responsible (for certain momenta of both the incoming and the outgoing electron) for superconductivity.

It is clear that the process (c), when added to the Coulomb repulsion term, gives an effective electron–electron interaction of the form

$$\frac{1}{2}\sum_{\substack{k,k',q\\ (\text{omit } q=0)}}\sum_{\lambda,\lambda'}\left(V_q + V_q^{\text{indirect}}\right)\hat{a}^\dagger_{k+q,\lambda}\hat{a}^\dagger_{k'-q,\lambda'}\hat{a}_{k'\lambda'}\hat{a}_{k\lambda}. \tag{16.5}$$

The problem is to determine V_q^{indirect} and to show that, in certain circumstances, it can be negative, so much so as to overcome the Coulomb repulsion V_q and lead to a net effective attractive interaction between electrons with the appropriate wave vectors.

Two possible processes contribute to $V_q^{\text{indirect}}\hat{a}^\dagger_{k+q,\lambda}\hat{a}^\dagger_{k'-q,\lambda'}\hat{a}_{k'\lambda'}\hat{a}_{k\lambda}$. These are illustrated in figure 16.3. We assume low (effectively zero) temperatures, so that the first part of each process is the creation rather than the absorption of a phonon, that is, there are no phonons present initially.

Note that the nature of the electron–phonon interaction is not such as to make a spin-flip ($\lambda \to \lambda' \neq \lambda$) possible (just as V_q has no power to flip spins). This means that processes of the type shown in figure 16.4 are not possible if in them $\lambda_2 \neq \lambda_1$.

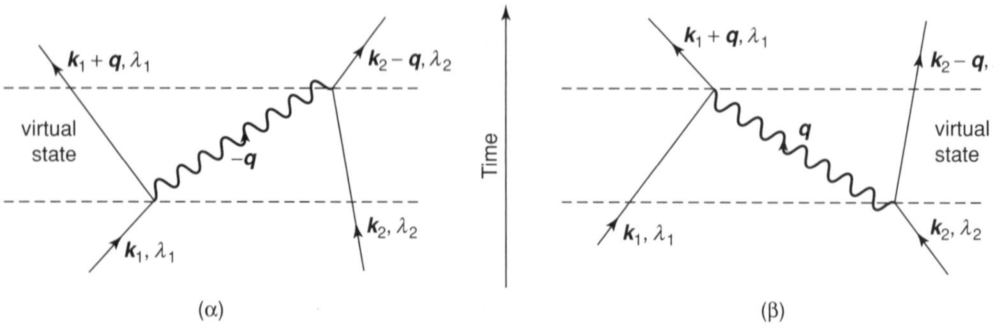

Figure 16.3 *Processes of the type* $V_{\boldsymbol{q}}^{\text{indirect}}\hat{a}^{\dagger}_{\boldsymbol{k}+\boldsymbol{q},\lambda}\hat{a}^{\dagger}_{\boldsymbol{k}'-\boldsymbol{q},\lambda'}\hat{a}_{\boldsymbol{k}'\lambda'}\hat{a}_{\boldsymbol{k}\lambda}$.

Let us now write out the electron–phonon interaction in a form suitable to display these two processes.

So far as the two-vertex process (α) is concerned, the phonon-creation part (the vertex on the left of the diagram) is described by the term

$$M^{*}_{\boldsymbol{k}_1+\boldsymbol{q},\boldsymbol{k}_1}\hat{b}^{\dagger}_{-\boldsymbol{q}}\hat{a}^{\dagger}_{\boldsymbol{k}_1+\boldsymbol{q},\lambda_1}\hat{a}_{\boldsymbol{k}_1\lambda_1} \tag{16.6}$$

which is the term with $\boldsymbol{k} = \boldsymbol{k}_1 + \boldsymbol{q}$, $\boldsymbol{k}' = \boldsymbol{k}_1$ and $\lambda = \lambda_1$ in the second sum in the electron–phonon part ($\hat{H}_{\text{ep}}$) of (16.3). The phonon-annihilation part of process (α) (the vertex on the right of the diagram) is described by the term

$$M_{\boldsymbol{k}_2,\boldsymbol{k}_2-\boldsymbol{q}}\hat{b}_{-\boldsymbol{q}}\hat{a}^{\dagger}_{\boldsymbol{k}_2-\boldsymbol{q},\lambda_2}\hat{a}_{\boldsymbol{k}_2\lambda_2}, \tag{16.7}$$

which is the term with $\boldsymbol{k} = \boldsymbol{k}_2$, $\boldsymbol{k}' = \boldsymbol{k}_2 - \boldsymbol{q}$ and $\lambda = \lambda_2$ in the first sum of $\hat{H}_{\text{ep}}$. The amplitude $V_{\boldsymbol{q}}^{\text{indirect}}$ of the effective electron–electron interaction, to second order in the electron–phonon interaction, also depends [just as in processes (a) and (b) in figure 16.2] on the energies of the intermediate states (which are virtual states – energy conservation is not required) through the energy-denominator factor $(E_{\text{initial}} - E_{\text{virtual}})^{-1}$. This factor is a standard feature in second-order perturbation theory (see section 3.1) and acts to suppress the amplitude of processes involving virtual states of high energy. For process (α),

$$E_{\text{initial}} = \varepsilon_{\boldsymbol{k}_1} + \varepsilon_{\boldsymbol{k}_2} \text{ and } E_{\text{virtual}} = \varepsilon_{\boldsymbol{k}_1+\boldsymbol{q}} + \varepsilon_{\boldsymbol{k}_2} + \hbar\omega_{\boldsymbol{q}} \tag{16.8}$$

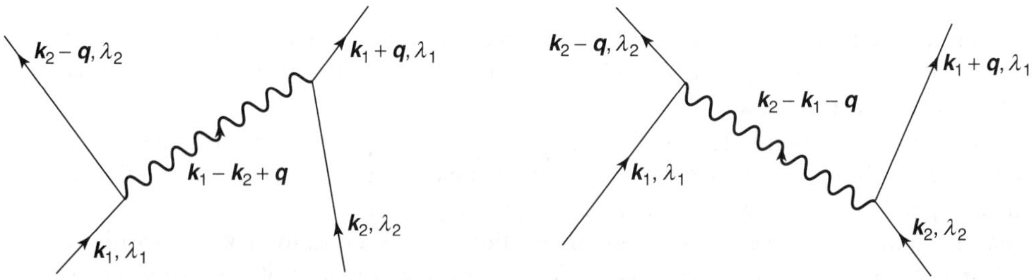

Figure 16.4 *Processes that are forbidden if* $\lambda_2 \neq \lambda_1$.

(since $\hbar\omega_{\boldsymbol{q}} = \hbar\omega_{-\boldsymbol{q}}$). Therefore [see (16.5)],

$$V_{(\alpha)}^{\text{indirect}} = M_{\boldsymbol{k}_2,\boldsymbol{k}_2-\boldsymbol{q}} \frac{1}{(\varepsilon_{\boldsymbol{k}_1} - \varepsilon_{\boldsymbol{k}_1+\boldsymbol{q}}) - \hbar\omega_{\boldsymbol{q}}} M^*_{\boldsymbol{k}_1+\boldsymbol{q},\boldsymbol{k}_1}, \tag{16.9}$$

where we have replaced $\hat{b}_{-\boldsymbol{q}}\hat{b}^\dagger_{-\boldsymbol{q}}$ by 1 (on the assumption that we are considering only zero-phonon states, that is, we consider only very low, effectively zero temperatures):

$$\hat{b}_{-\boldsymbol{q}}\hat{b}^\dagger_{-\boldsymbol{q}} = 1 - \hat{b}^\dagger_{-\boldsymbol{q}}\hat{b}_{-\boldsymbol{q}} = 1 - \hat{n}_{-\boldsymbol{q}} = 1.$$

In the case of process (β), the phonon-creation operator (the vertex on the right of the diagram) is described by the term

$$M^*_{\boldsymbol{k}_2-\boldsymbol{q},\boldsymbol{k}_2} \hat{b}^\dagger_{\boldsymbol{q}} \hat{a}^\dagger_{\boldsymbol{k}_2-\boldsymbol{q},\lambda_2} \hat{a}_{\boldsymbol{k}_2\lambda_2}, \tag{16.10}$$

which is the term with $\boldsymbol{k}' = \boldsymbol{k}_2$, $\boldsymbol{k} = \boldsymbol{k}_2 - \boldsymbol{q}$ and $\lambda = \lambda_2$ in the second sum in the electron–phonon part $\hat{H}_{\text{ep}}$ of (16.3). The phonon annihilation in process (β) arises from

$$M_{\boldsymbol{k}_1,\boldsymbol{k}_1+\boldsymbol{q}} \hat{b}_{\boldsymbol{q}} \hat{a}^\dagger_{\boldsymbol{k}_1+\boldsymbol{q},\lambda_1} \hat{a}_{\boldsymbol{k}_1\lambda_1}, \tag{16.11}$$

which is the term with $\boldsymbol{k} = \boldsymbol{k}_1$, $\boldsymbol{k}' = \boldsymbol{k}_1 + \boldsymbol{q}$ and $\lambda = \lambda_1$ in the first term in the electron–phonon part $\hat{H}_{\text{ep}}$ of (16.3). For process (β),

$$E_{\text{initial}} = \varepsilon_{\boldsymbol{k}_1} + \varepsilon_{\boldsymbol{k}_2} \quad \text{and} \quad E_{\text{virtual}} = \varepsilon_{\boldsymbol{k}_1} + \varepsilon_{\boldsymbol{k}_2-\boldsymbol{q}} + \hbar\omega_{\boldsymbol{q}}. \tag{16.12}$$

Therefore,

$$V_{(\beta)}^{\text{indirect}} = M_{\boldsymbol{k}_1,\boldsymbol{k}_1+\boldsymbol{q}} \frac{1}{\left(\varepsilon_{\boldsymbol{k}_2} - \varepsilon_{\boldsymbol{k}_2-\boldsymbol{q}}\right) - \hbar\omega_{\boldsymbol{q}}} M^*_{\boldsymbol{k}_2-\boldsymbol{q},\boldsymbol{k}_2}, \tag{16.13}$$

where, because we are considering zero-phonon states, we have again replaced $\hat{b}_{\boldsymbol{q}}\hat{b}^\dagger_{\boldsymbol{q}}$ by 1.

We consider processes in which the initial and final states of the electrons have the same total energy, that is,

$$\varepsilon_{\boldsymbol{k}_1} + \varepsilon_{\boldsymbol{k}_2} = \varepsilon_{\boldsymbol{k}_1+\boldsymbol{q}} + \varepsilon_{\boldsymbol{k}_2-\boldsymbol{q}}. \tag{16.14}$$

Therefore,

$$\varepsilon_{\boldsymbol{k}_1} - \varepsilon_{\boldsymbol{k}_1+\boldsymbol{q}} \equiv \hbar\omega_{\boldsymbol{k}_1,\boldsymbol{k}_1+\boldsymbol{q}} = -(\varepsilon_{\boldsymbol{k}_2} - \varepsilon_{\boldsymbol{k}_2-\boldsymbol{q}}). \tag{16.15}$$

We also assume that the amplitudes M depend only on the magnitude q of the momentum transfer [just as V_q was found to depend only on q – see (16.4)]. Then, from (16.9) and (16.13),

$$\begin{aligned} V^{\text{indirect}} &= V^{\text{indirect}}_{(\alpha)} + V^{\text{indirect}}_{(\beta)} = |M_q|^2 \left[\frac{1}{\hbar\omega_{k_1,k_1+q} - \hbar\omega_q} + \frac{1}{-\hbar\omega_{k_1,k_1+q} - \hbar\omega_q} \right] \\ &= -\frac{|M_q|^2}{\hbar} \frac{2\omega_q}{\omega_q^2 - \omega^2_{k_1,k_1+q}}, \end{aligned} \tag{16.16}$$

which is **negative** for $|\omega_{k_1,k_1+q}| < \omega_q$, that is, for $|\varepsilon_{k_1} - \varepsilon_{k_1+q}| < \hbar\omega_q$. Since the phonon frequencies are typically of the order of the Debye frequency ω_{D}, we may loosely define the region of attractive interactions as that in which

$$|\varepsilon_{k_1} - \varepsilon_{k_1+q}| < \hbar\omega_{\mathrm{D}}. \tag{16.17}$$

Thus, those processes in which the gain or loss of energy by an electron is less than the phonon energy lead to an effective negative (attractive) interaction between the electrons involved. Thus, it is electrons near the Fermi surface that interact via an attractive interaction.

As will be seen, it is this attractive interaction that leads to a gap in the energy spectrum, and hence to superconductivity. We therefore work, from now on, with a model hamiltonian of the form

$$\hat{H} = \sum_{k,\lambda} \varepsilon_k \hat{a}^\dagger_{k\lambda} \hat{a}_{k\lambda} + \hat{H}_{\text{int}}, \tag{16.18}$$

where

$$\hat{H}_{\text{int}} = \frac{1}{2} \sum_{\substack{k_1,k_2,q \\ (\text{omit } q=0)}} \sum_{\lambda_1\lambda_2} V(k_1+q, k_2-q; k_1, k_2) \hat{a}^\dagger_{k_1+q,\lambda_1} \hat{a}^\dagger_{k_2-q,\lambda_2} \hat{a}_{k_2\lambda_2} \hat{a}_{k_1\lambda_1}, \tag{16.19}$$

in which [see (16.4) and (16.16)]

$$V(k_1+q, k_2-q; k_1, k_2) = \begin{cases} V_q^{\text{Coulomb}} & \text{outside the region of attraction} \\ V_q^{\text{Coulomb}} - \dfrac{4|M_q|^2\omega_q}{\hbar\left(\omega_q^2 - \omega^2_{k_1,k_1+q}\right)} & \text{inside the region of attraction} \end{cases} \tag{16.20}$$

In (16.20),

$$V_q^{\text{Coulomb}} = \frac{e^2}{4\pi\varepsilon_0} \frac{1}{\Omega} \frac{4\pi}{q^2}. \tag{16.21}$$

It is clear that the terms (processes) in (16.19) that make the greatest negative contribution (corresponding to the strongest electron–electron attraction) will be those that have $\omega^2_{k_1,k_1+q} = 0$ (corresponding to a pair of electrons each with the same single-particle energy) with, at the same time, the smallest V_q^{Coulomb}, that is, the largest momentum transfer q. For electron 1 this is clearly the case for processes $k_1 \to k_1 + q = -k_1$, corresponding to scattering to the diametrically opposite point with respect to the centre of the Fermi sphere, that is, with momentum transfer $\boldsymbol{q} = -2\boldsymbol{k}_1$. At the same time, the attraction-maximizing process for electron 2 is $\boldsymbol{k}_2 \to \boldsymbol{k}_2 - \boldsymbol{q} = -\boldsymbol{k}_2$, corresponding to $\boldsymbol{q} = 2\boldsymbol{k}_2$. Comparing the expressions for the optimum $\boldsymbol{q}$ to produce electron–electron attraction we see that the most attractive terms in (16.19) have $\boldsymbol{k}_1 = -\boldsymbol{k}_2$, and so the processes corresponding to the strongest attraction are those in which two electrons of equal and opposite momenta scatter and reverse their momenta. Thus, the dominant attractive terms in the interaction hamiltonian are those containing terms of the form $\hat{a}^\dagger_{\boldsymbol{k},\lambda_1}\hat{a}^\dagger_{-\boldsymbol{k},\lambda_2}\hat{a}_{\boldsymbol{k},\lambda_2}\hat{a}_{-\boldsymbol{k},\lambda_1}$.

16.2.1 Cooper pairs

In 1957, Cooper showed that, given an attractive interaction between two electrons (and we have shown above how such an interaction can arise), two electrons interacting via this potential in the presence of a free electron gas form a state of negative energy relative to the Fermi surface, that is, form a bound state. This proves that in the presence of phonons the free-electron gas is unstable against the formation of **Cooper pairs**. The two electrons are assumed to have a fixed centre of mass, that is, fixed coordinate $\boldsymbol{R} = (\boldsymbol{r}_1 + \boldsymbol{r}_2)/2$, so that the wave function of the pair can be written as

$$\psi(\boldsymbol{r}_1, \boldsymbol{r}_2) = \psi(\boldsymbol{r}_1 - \boldsymbol{r}_2, \boldsymbol{R}) = \psi(\boldsymbol{r}_1 - \boldsymbol{r}_2) = \sum_{\boldsymbol{k}} g(\boldsymbol{k})\mathrm{e}^{i\boldsymbol{k}\cdot(\boldsymbol{r}_1-\boldsymbol{r}_2)}, \tag{16.22}$$

corresponding to a linear superposition of products $\mathrm{e}^{i\boldsymbol{k}\cdot\boldsymbol{r}_1}\mathrm{e}^{-i\boldsymbol{k}\cdot\boldsymbol{r}_2}$ of plane-wave states, or (in the occupation-number representation) a linear superposition of two-electron states $\hat{a}^\dagger_{\boldsymbol{k}}\hat{a}^\dagger_{-\boldsymbol{k}}|0\rangle$.

We now write the potential $V(\boldsymbol{k}, \boldsymbol{k}')$ as the following partial-wave expansion (θ is the angle between $\boldsymbol{k}$ and $\boldsymbol{k}'$):

$$V(\boldsymbol{k}, \boldsymbol{k}')\left(= \int \mathrm{d}^3\boldsymbol{r}\mathrm{e}^{-i\boldsymbol{k}\cdot\boldsymbol{r}}V(\boldsymbol{r})\mathrm{e}^{i\boldsymbol{k}'\cdot\boldsymbol{r}}\right) = \sum_l (2l+1)V_l(k, k')P_l(\cos\theta), \tag{16.23}$$

and consider only the $l = 0$ term (so-called s-wave scattering). Then, $\psi(\boldsymbol{r}_1, \boldsymbol{r}_2)$ is symmetric, that is,

$$\psi(\boldsymbol{r}_1, \boldsymbol{r}_2) = \psi(\boldsymbol{r}_2, \boldsymbol{r}_1), \tag{16.24}$$

so that $g(-\boldsymbol{k}) = g(\boldsymbol{k})$ [see (16.22)]. Since the pair wave function must be overall-antisymmetric, its spin part must be antisymmetric, that is, must be of the form $(\alpha(1)\beta(2) - \alpha(2)\beta(1))/\sqrt{2}$, so that the two-electron bound state corresponds to

$$\sum_{\boldsymbol{k}} g(\boldsymbol{k})\hat{a}^\dagger_{\boldsymbol{k}\uparrow}\hat{a}^\dagger_{-\boldsymbol{k}\downarrow}|0\rangle, \tag{16.25}$$

where $|0\rangle$ denotes the vacuum state.

We can use this result to motivate our choice of the ground-state wave function for a superconductor with an even number N of electrons, which we take in the form

$$|\phi_N\rangle = \sum_{\boldsymbol{k}_1}\cdots\sum_{\boldsymbol{k}_{N/2}} g(\boldsymbol{k}_1)\cdots g(\boldsymbol{k}_{N/2})\hat{a}^\dagger_{\boldsymbol{k}_1\uparrow}\hat{a}^\dagger_{-\boldsymbol{k}_1\downarrow}\cdots\hat{a}^\dagger_{\boldsymbol{k}_{N/2}\uparrow}\hat{a}^\dagger_{-\boldsymbol{k}_{N/2}\downarrow}|0\rangle, \tag{16.26}$$

Unlike (16.25), which describes a two-particle Cooper bound state, the wave function (16.26) describes an N-particle state. (If the number of electrons is odd the corrections due to the N-th particle are only of order $1/N$, which is essentially zero for a superconductor but may be important in discussing the nucleon-pairing states in nuclei.)

In fact, this wave function is rather difficult to deal with, and so Bardeen, Cooper and Schrieffer introduced the generating function

$$|\tilde{\phi}\rangle = C \prod_{\boldsymbol{k}} \left(1 + g(\boldsymbol{k}) \hat{a}^{\dagger}_{\boldsymbol{k}\uparrow} \hat{a}^{\dagger}_{-\boldsymbol{k}\downarrow}\right) |0\rangle , \tag{16.27}$$

where the product over $\boldsymbol{k}$ extends over all wave vectors allowed by the boundary conditions, and it is clear that $|\phi_N\rangle$ is that part of $|\tilde{\phi}\rangle$ which has N creation operators acting on $|0\rangle$, that is, the part of $|\tilde{\phi}\rangle$ that describes an N-particle state.

Clearly, we can write

$$|\tilde{\phi}\rangle = \sum_N \lambda_N |\phi_N\rangle, \quad \text{with} \quad \sum_N |\lambda_N|^2 = 1 \tag{16.28}$$

to ensure that $|\tilde{\phi}\rangle$ is normalized if the states $|\phi_N\rangle$ for all (even) N are normalized.

We rewrite $|\tilde{\phi}\rangle$ in a form similar to (16.27):

$$|\tilde{\phi}\rangle = \prod_{\boldsymbol{k}} \left(u_{\boldsymbol{k}} + v_{\boldsymbol{k}} \hat{a}^{\dagger}_{\boldsymbol{k}\uparrow} \hat{a}^{\dagger}_{-\boldsymbol{k}\downarrow}\right) |0\rangle, \tag{16.29}$$

with, obviously,

$$v_{-\boldsymbol{k}} = v_{\boldsymbol{k}} \text{ and } u_{-\boldsymbol{k}} = u_{\boldsymbol{k}}. \tag{16.30}$$

In fact, we shall assume full isotropy of $|\tilde{\phi}\rangle$, so that these weighting coefficients depend only on the magnitude k of the wave vector $\boldsymbol{k}$ ($u_{\boldsymbol{k}} = u_k$ and $v_{\boldsymbol{k}} = v_k$). The state $|\tilde{\phi}\rangle$ is normalized, that is,

$$\langle \tilde{\phi} \mid \tilde{\phi} \rangle = \prod_{\boldsymbol{k}} \left(u_k^2 + v_k^2\right) = 1 \tag{16.31}$$

if

$$u_k^2 + v_k^2 = 1 \quad \text{for all } \boldsymbol{k}. \tag{16.32}$$

For large N we make essentially no error by working with $|\tilde{\phi}\rangle$ instead of $|\phi_N\rangle$. This is because, when the values of u_k and v_k are chosen so as to minimize the total energy subject to the constraint that the mean number of particles in $|\tilde{\phi}\rangle$ must equal the actual number N^*, we find that these values of u_k and v_k correspond to a distribution $|\lambda_N|$ that is sharply peaked at $N = N^*$ and of width of the order of $\sqrt{N^*}$, so that the relative fluctuation of the number of particles in the energy-minimizing state $|\tilde{\phi}\rangle$ is of the order of $\sqrt{N^*}/N^* = 1/\sqrt{N^*} \approx 10^{-11}$.

Similarly, the expectation value of an arbitrary operator $\hat{F}$ in the state $|\tilde{\phi}\rangle$ is

$$\langle\tilde{\phi}|\,\hat{F}\,|\tilde{\phi}\rangle = \sum_{N,N'} \lambda_N^* \lambda_{N'} \langle\phi_N|\,\hat{F}\,|\phi_{N'}\rangle. \tag{16.33}$$

In the case when $\hat{F}$ does not change the number of particles this becomes

$$\langle\tilde{\phi}|\,\hat{F}\,|\tilde{\phi}\rangle = \sum_{N} |\lambda_N|^2 \langle\phi_N|\,\hat{F}\,|\phi_N\rangle, \tag{16.34}$$

and, unless $\langle\phi_N|\,\hat{F}\,|\phi_N\rangle$ is an extraordinarily rapidly varying function of N (we shall assume that it is not) we can replace it by its value at the peak of $|\lambda_N|$:

$$\langle\tilde{\phi}|\,\hat{F}\,|\tilde{\phi}\rangle = \langle\phi_{N^*}|\,\hat{F}\,|\phi_{N^*}\rangle \sum_{N} |\lambda_N|^2 = \langle\phi_{N^*}|\,\hat{F}\,|\phi_{N^*}\rangle\,. \tag{16.35}$$

If, instead, $\hat{F}$ is an operator whose nonzero matrix elements are $\langle\phi_{N+p}|\hat{F}|\phi_N\rangle$ for some value of $p \ll \sqrt{N^*}$, then (16.33) becomes

$$\langle\tilde{\phi}|\,\hat{F}\,|\tilde{\phi}\rangle = \sum_{N} \lambda_{N+p}^* \lambda_N \langle\phi_{N+p}|\,\hat{F}\,|\phi_N\rangle. \tag{16.36}$$

But $\lambda_{N+p} \approx \lambda_N$ for $p \ll \sqrt{N^*}$. Therefore,

$$\langle\tilde{\phi}|\,\hat{F}\,|\tilde{\phi}\rangle = \langle\phi_{N^*+p}|\,\hat{F}\,|\phi_{N^*}\rangle \sum_{N} |\lambda_N|^2 = \langle\phi_{N^*+p}|\,\hat{F}\,|\phi_{N^*}\rangle\,. \tag{16.37}$$

Equations (16.35) and (16.37) show that, in practice, $|\tilde{\phi}\rangle$ gives the same information as $|\phi_{N^*}\rangle$.

16.2.2 Calculation of the ground-state energy

If we were to use $|\phi_N\rangle$ (16.26), we would minimize $\langle\phi_N|\,\hat{H}\,|\phi_N\rangle$ directly with respect to the $g(\boldsymbol{k})$s in $|\phi_N\rangle$. To find the optimum values of the parameters u_k and v_k in $|\tilde{\phi}\rangle$, we must minimize the appropriate **free energy** in the state $|\tilde{\phi}\rangle$. Since we are concerned with the ground state, we want to find the optimum parameters at constant temperature ($T = 0$), constant volume V, and constant chemical potential μ rather than constant particle number N, that is, we must minimize the grand potential

$$\begin{aligned}\Omega(T, V, \mu) &= F - \mu N = U - TS - \mu N = U - \mu N \quad (\text{since } T = 0)\\ &= \langle E\rangle - \mu\,\langle N\rangle = \langle\tilde{\phi}|\,(\hat{H} - \mu\hat{N})\,|\tilde{\phi}\rangle\,.\end{aligned} \tag{16.38}$$

Therefore,

$$\Omega = \langle\tilde{\phi}|\,(\hat{H}_0 - \mu\hat{N})\,|\tilde{\phi}\rangle + \langle\tilde{\phi}|\,\hat{H}_{\text{int}}\,|\tilde{\phi}\rangle\,. \tag{16.39}$$

The first term in (16.39) is

$$\begin{aligned}\langle\tilde{\phi}|(\hat{H}_0-\mu\hat{N})|\tilde{\phi}\rangle &= \sum_{k,\lambda}\langle\tilde{\phi}|\,\xi_k\hat{a}^{\dagger}_{k\lambda}\hat{a}_{k\lambda}\,|\tilde{\phi}\rangle \quad (\text{with } \xi_k=\varepsilon_k-\mu)\\ &= \sum_k \xi_k\langle\tilde{\phi}|(\hat{n}_{k\uparrow}+\hat{n}_{k\downarrow})|\tilde{\phi}\rangle.\end{aligned} \tag{16.40}$$

To calculate $\langle\tilde{\phi}|\,\hat{n}_{k\uparrow}\,|\tilde{\phi}\rangle$ we note that in the operator-product expression (16.27) for $|\tilde{\phi}\rangle$, the factors with wave vectors other than this particular $\boldsymbol{k}$ simply yield unity by the normalization properties (16.31) and (16.32), and so

$$\langle\tilde{\phi}|\,\hat{n}_{k\uparrow}\,|\tilde{\phi}\rangle = v_k^2\langle\hat{a}^{\dagger}_{k\uparrow}\hat{a}^{\dagger}_{-k\downarrow}0|\hat{n}_{k\uparrow}\hat{a}^{\dagger}_{k\uparrow}\hat{a}^{\dagger}_{-k\downarrow}\,|0\rangle = v_k^2, \tag{16.41}$$

where

$$\langle\hat{a}^{\dagger}_{k\uparrow}\hat{a}^{\dagger}_{-k\downarrow}0| \equiv \left(\hat{a}^{\dagger}_{k\uparrow}\hat{a}^{\dagger}_{-k\downarrow}\,|0\rangle\right)^{\dagger}. \tag{16.42}$$

Similarly, in the calculation of $\langle\tilde{\phi}|\,\hat{n}_{k\downarrow}\,|\tilde{\phi}\rangle$, all factors in $|\tilde{\phi}\rangle$ (16.27) other than that with wave vector equal to minus this particular $\boldsymbol{k}$ again yield unity in the product, and we have

$$\langle\tilde{\phi}|\,\hat{n}_{k\downarrow}\,|\tilde{\phi}\rangle = v_k^2\langle\hat{a}^{\dagger}_{-k\uparrow}\hat{a}^{\dagger}_{k\downarrow}0|n_{k\downarrow}\hat{a}^{\dagger}_{-k\uparrow}\hat{a}^{\dagger}_{k\downarrow}\,|0\rangle = v_k^2, \tag{16.43}$$

so that (16.40) becomes

$$\langle\tilde{\phi}|(\hat{H}_0-\mu N)|\tilde{\phi}\rangle = 2\sum_k v_k^2\xi_k. \tag{16.44}$$

The second term in the grand potential Ω (16.39) can be written as

$$\langle\tilde{\phi}|\,\hat{H}_{\text{int}}\,|\tilde{\phi}\rangle = \frac{1}{2}\sum_{\substack{k_1,k_2,q\\ \lambda_1,\lambda_2}} V(\boldsymbol{k}_1+\boldsymbol{q},\boldsymbol{k}_2-\boldsymbol{q};\boldsymbol{k}_1,\boldsymbol{k}_2)\langle\tilde{\phi}|\,\hat{a}^{\dagger}_{k_1+q,\lambda_1}\hat{a}^{\dagger}_{k_2-q,\lambda_2}\hat{a}_{k_2,\lambda_2}\hat{a}_{k_1,\lambda_1}\,|\tilde{\phi}\rangle, \tag{16.45}$$

where, in the sum, the possible nonzero matrix elements are of the following four types:

(i) $\langle\tilde{\phi}|\,\hat{a}^{\dagger}_{k\uparrow}\hat{a}^{\dagger}_{k'\downarrow}\hat{a}_{k'\downarrow}\hat{a}_{k\uparrow}\,|\tilde{\phi}\rangle$

(ii) $\langle\tilde{\phi}|\,\hat{a}^{\dagger}_{k'\uparrow}\hat{a}^{\dagger}_{k\downarrow}\hat{a}_{k'\downarrow}\hat{a}_{k\uparrow}\,|\tilde{\phi}\rangle$

(iii) $\langle\tilde{\phi}|\,\hat{a}^{\dagger}_{l\uparrow}\hat{a}^{\dagger}_{-l\downarrow}\hat{a}_{-k\downarrow}\hat{a}_{k\uparrow}\,|\tilde{\phi}\rangle$

(iv) $\langle\tilde{\phi}|\,\hat{a}^{\dagger}_{-l\downarrow}\hat{a}^{\dagger}_{l\uparrow}\hat{a}_{k\uparrow}\hat{a}_{-k\downarrow}\,|\tilde{\phi}\rangle$

Those of type (i) (direct terms with $\boldsymbol{q}=0$) and those of type (ii) (exchange terms with $\boldsymbol{q}=\boldsymbol{k}_2-\boldsymbol{k}_1$) are terms in which the two electrons annihilated are replaced by a pair of electrons with the same momenta and unchanged (exchanged, respectively) spins. Those of types (iii) and (iv) (with $\boldsymbol{k}$ not equal to $\boldsymbol{l}$ or $-\boldsymbol{l}$) each correspond to a combined pair-annihilation and pair-creation process (pair-scattering process) in which the electrons in each pair have opposite momenta and antiparallel spins but the momenta of the electrons in the created pair differ from the momenta of the electrons in the annihilated pair.

It is clear that the terms of types (iii) and (iv) give a nonzero contribution to the expectation value of $\hat{H}_{\text{int}}$ in $|\tilde{\phi}\rangle$, while the expectation value of $\hat{H}_{\text{int}}$ in the free-electron ground state $|F\rangle$ has no such contribution, since $|F\rangle$ has well-defined occupancy of the single-particle momentum eigenstates.

We shall neglect the direct (i) and exchange (ii) "electron-replacing" terms, since we expect each of them to be of similar value in the normal and the superconducting state. The pair-scattering processes, however, are specific to the BCS-type state $|\tilde{\phi}\rangle$, and will be calculated. A term of type (iii) arises from the part of $|\tilde{\phi}\rangle$ in which $\boldsymbol{k}\uparrow, -\boldsymbol{k}\downarrow$ are occupied and $\boldsymbol{l}\uparrow, -\boldsymbol{l}\downarrow$ are unoccupied (this part of $|\tilde{\phi}\rangle$ contains the factor $v_k u_l$) and the part of $\langle\tilde{\phi}|$ in which $\boldsymbol{k}\uparrow, -\boldsymbol{k}\downarrow$ are unoccupied and $\boldsymbol{l}\uparrow, -\boldsymbol{l}\downarrow$ are occupied (this part of $\langle\tilde{\phi}|$ contains the factor $u_k^* v_l^*$, which is equal to $u_k v_l$, if u_k and v_l are assumed to be real).

Thus, a typical term of type (iii) is

$$u_k v_l V(\boldsymbol{l}, -\boldsymbol{l}; \boldsymbol{k}, -\boldsymbol{k}) v_k u_l.$$

Since $u_{-\boldsymbol{k}} = u_{\boldsymbol{k}} \equiv u_k$ and $v_{-\boldsymbol{k}} = v_{\boldsymbol{k}} \equiv v_k$ [see (16.30)], the terms of type (iv) are

$$u_k v_l V(-\boldsymbol{l}, \boldsymbol{l}; -\boldsymbol{k}, \boldsymbol{k}) v_k u_l.$$

Therefore,

$$\langle\tilde{\phi}|\,(\hat{H} - \mu\hat{N})\,|\tilde{\phi}\rangle = 2\sum_{\boldsymbol{k}} v_k^2 \xi_{\boldsymbol{k}} + \sum_{\boldsymbol{k},\boldsymbol{l}} V_{\boldsymbol{kl}} u_k v_k u_l v_l, \tag{16.46}$$

with

$$V_{\boldsymbol{kl}} \equiv \frac{1}{2}\left[V(\boldsymbol{l}, -\boldsymbol{l}; \boldsymbol{k}, -\boldsymbol{k}) + V(-\boldsymbol{l}, \boldsymbol{l}; -\boldsymbol{k}, \boldsymbol{k})\right] \tag{16.47}$$

Note that $V_{\boldsymbol{lk}} = V_{\boldsymbol{kl}}$, as can be checked (with appropriate relabellings) by inspecting the structure of (16.45).

To minimize this while taking the constraint (16.32) into account, it is convenient to substitute

$$u_k = \sin\theta_k, \quad v_k = \cos\theta_k. \tag{16.48}$$

Then,

$$\langle\tilde{\phi}|\,(\hat{H} - \mu N)\,|\tilde{\phi}\rangle = 2\sum_{\boldsymbol{k}} \xi_{\boldsymbol{k}} \cos^2\theta_k + \frac{1}{4}\sum_{\boldsymbol{k},\boldsymbol{l}} V_{\boldsymbol{kl}} \sin 2\theta_k \sin 2\theta_l. \tag{16.49}$$

Minimizing this with respect to each θ_i, we have

$$\begin{aligned} 0 &= \frac{\mathrm{d}}{\mathrm{d}\theta_i}\langle\tilde{\phi}|\,(\hat{H} - \mu\hat{N})\,|\tilde{\phi}\rangle = -2\xi_i \sin 2\theta_i + \frac{1}{2}\cos 2\theta_i \sum_{\boldsymbol{l}} V_{i\boldsymbol{l}} \sin 2\theta_l + \frac{1}{2}\cos 2\theta_i \sum_{\boldsymbol{k}} V_{\boldsymbol{k}i} \sin 2\theta_k \\ &= -2\xi_i \sin 2\theta_i + \cos 2\theta_i \sum_{\boldsymbol{l}} V_{i\boldsymbol{l}} \sin 2\theta_l, \end{aligned} \tag{16.50}$$

where we have put $V_{\boldsymbol{l}i} = V_{i\boldsymbol{l}}$. Therefore,

$$\xi_i \tan 2\theta_i = \frac{1}{2}\sum_{\boldsymbol{l}} V_{\boldsymbol{l}i} \sin 2\theta_l = \sum_{\boldsymbol{l}} V_{\boldsymbol{l}i} u_l v_l \equiv -\Delta_i, \tag{16.51}$$

which defines the quantity Δ_i. Therefore,

$$\tan 2\theta_i = \frac{\Delta_i}{-\xi_i}, \quad \sin 2\theta_i \ (= 2u_i v_i) = \frac{\Delta_i}{\sqrt{\Delta_i^2 + \xi_i^2}}, \quad \text{and} \quad \cos 2\theta_i \ (= v_i^2 - u_i^2) = \frac{-\xi_i}{\sqrt{\Delta_i^2 + \xi_i^2}}, \tag{16.52}$$

so that

$$\Delta_i \equiv -\sum_l V_{li} u_l v_l = -\sum_l V_{li} \frac{\Delta_l}{2\sqrt{\xi_l^2 + \Delta_l^2}}. \tag{16.53}$$

We first note that this equation always has the trivial solution $\Delta_i = 0$. From (16.52), either u_i or v_i must then equal zero (while the other equals unity). From the third expression in (16.52) it is clear that

$$v_i = \begin{cases} 1 \text{ for } \xi_i < 0 \\ 0 \text{ for } \xi_i > 0 \end{cases} \tag{16.54}$$

and the associated state $|\tilde{\phi}\rangle$ (16.29) is simply

$$|\tilde{\phi}\rangle = \prod_{\substack{\boldsymbol{k} \\ (k<k_F)}} \hat{a}^\dagger_{\boldsymbol{k}\uparrow} \hat{a}^\dagger_{-\boldsymbol{k}\downarrow} |0\rangle, \tag{16.55}$$

which is formed from all single-particle states of energy less than the Fermi energy $\varepsilon_F = \hbar^2 k_F^2 / 2m$, and is clearly thus the ground state of a gas of noninteracting electrons.

To show explicitly that there can be other solutions, we choose the simplified interaction

$$V_{il} = \begin{cases} -V & \text{if } |\xi_i|, |\xi_l| \leq \hbar\omega_D \\ 0 & \text{otherwise} \end{cases} \tag{16.56}$$

which is known as the BCS (Bardeen–Cooper–Schrieffer) interaction (V here is a positive constant).

Then, from equation (16.53),

$$\Delta_i = \begin{cases} 0 \text{ for } |\xi_i| > \hbar\omega_D \\ V \sum_l \dfrac{\Delta_l}{2\sqrt{\xi_l^2 + \Delta_l^2}} \quad \text{for } |\xi_i| \leq \hbar\omega_D \end{cases} \tag{16.57}$$

where the sum in the second line is taken (by virtue of the first line) over states wave vectors $\boldsymbol{l}$ for which $|\xi_l| \leq \hbar\omega_D$. Because this sum is clearly independent of i, we can write Δ_i for the given range of energies $|\xi_i|$ as Δ, and, since the sum is restricted to the same energy range, we can also replace Δ_i in it by Δ. Cancelling the Δ, we find

$$1 = V \sum_l \frac{1}{2\sqrt{\xi_l^2 + \Delta^2}} = VD(0) \int_{-\hbar\omega_D}^{\hbar\omega_D} \mathrm{d}\xi \frac{1}{2\sqrt{\Delta^2 + \xi^2}}, \tag{16.58}$$

where we have replaced the restricted sum over the wave vectors $\boldsymbol{l}$ by an integral over the corresponding narrow energy range, making use of the density-of-states function $D(\xi)$ (number of states per unit energy range) for a given spin direction and taking its value $D(0)$ at the midpoint $\xi = 0$ of the range, that is, at the Fermi surface.

Therefore,

$$\begin{aligned}\frac{1}{D(0)V} &= \int_0^{\hbar\omega_D} \frac{\mathrm{d}\xi}{\sqrt{\Delta^2+\xi^2}} = \int_0^{\hbar\omega_D/\Delta} \mathrm{d}x \frac{1}{\sqrt{1+x^2}} \\ &= \operatorname{arcsinh} x\big|_0^{\hbar\omega_D/\Delta} = \operatorname{arcsinh}\left(\frac{\hbar\omega_D}{\Delta}\right).\end{aligned} \tag{16.59}$$

This equation has a solution only for positive V [that is, for an attractive BCS interaction – see (16.56)]. The solution is clearly

$$\Delta = \frac{\hbar\omega_D}{\sinh(1/D(0)V)} = \frac{2\hbar\omega_D}{\exp(1/D(0)V) - \exp(-1/D(0)V)}. \tag{16.60}$$

We are interested in the weak-coupling limit $D(0)V \ll 1$, for which $\exp(-1/D(0)V) \ll \exp(1/D(0)V)$. We can, therefore, neglect the second term in the denominator, obtaining

$$\Delta = 2\hbar\omega_D \exp\left(-\frac{1}{D(0)V}\right). \tag{16.61}$$

(It will be shown that, in a simple model, $\Delta = 1.76 k_B T_c$. From measurements of T_c for various metals, we can conclude empirically that $D(0)V < 0.3$, so that the above weak-coupling assumption is justified.)

According to (16.46), the grand potential in the BCS model can be expressed as

$$\begin{aligned}\langle\tilde{\phi}|(\hat{H}-\mu\hat{N})|\tilde{\phi}\rangle &= 2\sum_{\boldsymbol{k}} v_k^2 \xi_{\boldsymbol{k}} + \sum_{\boldsymbol{k},\boldsymbol{l}} V_{\boldsymbol{k}\boldsymbol{l}} u_k v_k u_l v_l \\ &= 2\sum_{\boldsymbol{k}} v_k^2 \xi_{\boldsymbol{k}} - \sum_{\boldsymbol{k},\boldsymbol{l}} V u_k v_k u_l v_l = 2\sum_{\boldsymbol{k}} v_k^2 \xi_{\boldsymbol{k}} - \frac{\Delta^2}{V},\end{aligned} \tag{16.62}$$

where we have used the BCS interaction (16.56) and also the definition (16.51) of Δ. To find the ground-state energy we substitute the energy-minimizing value of the parameter v_k^2. From the third expression in (16.52), and the condition (16.32), we find

$$v_k^2 = \frac{1}{2}\left(1 - \frac{\xi_k}{\sqrt{\xi_k^2+\Delta_k^2}}\right), \tag{16.63}$$

from which, using the results $\Delta_k = 0$ for $|\xi_k| > \hbar\omega_\mathrm{D}$, $\Delta_k = \Delta$ for $|\xi_k| \le \hbar\omega_\mathrm{D}$ [see (16.57)], we find

$$\begin{aligned} &v_k^2 = 1 \text{ for } \xi_k < -\hbar\omega_\mathrm{D},\ v_k^2 = 0 \text{ for } \xi_k > \hbar\omega_\mathrm{D}, \\ &v_k^2 = \frac{1}{2}\left(1 - \frac{\xi_k}{\sqrt{\xi_k^2 + \Delta^2}}\right) \text{ for } |\xi_k| \le \hbar\omega_\mathrm{D}. \end{aligned} \tag{16.64}$$

The first term in (16.62) then becomes

$$\begin{aligned} 2\sum_{\boldsymbol{k}} v_k^2 \xi_{\boldsymbol{k}} &= \sum_{\substack{\boldsymbol{k} \\ (\xi_k < -\hbar\omega_\mathrm{D})}} 2\xi_k + \sum_{\substack{\boldsymbol{k} \\ (|\xi_k| < \hbar\omega_\mathrm{D})}} \left(\xi_k - \frac{\xi_k^2}{\sqrt{\xi_k^2 + \Delta^2}}\right) \\ &= \sum_{\substack{\boldsymbol{k} \\ (\xi_k < 0)}} 2\xi_k + \sum_{\substack{\boldsymbol{k} \\ (-\hbar\omega_\mathrm{D} < \xi_k < 0)}} \left(-\xi_k - \frac{\xi_k^2}{\sqrt{\xi_k^2 + \Delta^2}}\right) + \sum_{\substack{\boldsymbol{k} \\ (0 < \xi_k < \hbar\omega_\mathrm{D})}} \left(\xi_k - \frac{\xi_k^2}{\sqrt{\xi_k^2 + \Delta^2}}\right) \\ &= \sum_{\substack{\boldsymbol{k} \\ (k < k_\mathrm{F})}} 2\xi_k + \sum_{\substack{\boldsymbol{k} \\ (0 < \xi_k < \hbar\omega_\mathrm{D})}} 2\left(\xi_k - \frac{\xi_k^2}{\sqrt{\xi_k^2 + \Delta^2}}\right) \\ &= \sum_{\substack{\boldsymbol{k} \\ (k < k_\mathrm{F})}} 2\xi_k + 2D(0)\int_0^{\hbar\omega_\mathrm{D}} \mathrm{d}\xi \left(\xi - \frac{\xi^2}{\sqrt{\xi^2 + \Delta^2}}\right) \\ &= 2\sum_{\substack{\boldsymbol{k} \\ (k < k_\mathrm{F})}} \xi_k + 2D(0)\Delta^2 \int_0^{\hbar\omega_\mathrm{D}} \frac{\mathrm{d}\xi}{\sqrt{\xi^2 + \Delta^2}} + 2D(0)\int_0^{\hbar\omega_\mathrm{D}} \mathrm{d}\xi \left(\xi - \sqrt{\xi^2 + \Delta^2}\right) \\ &= 2\sum_{\substack{\boldsymbol{k} \\ (k < k_\mathrm{F})}} \xi_k + \frac{2D(0)\Delta^2}{D(0)V} + 2D(0)\frac{(\hbar\omega_\mathrm{D})^2}{2} - 2D(0)\Delta^2 I, \end{aligned} \tag{16.65}$$

where, in the last line, we have used (16.59) and the standard integral I is

$$\begin{aligned} I &= \int_0^{\hbar\omega_\mathrm{D}/\Delta} \mathrm{d}x\sqrt{x^2+1} = \frac{1}{2}\left[x\sqrt{x^2+1} + \ln\left(x + \sqrt{x^2+1}\right)\right]_0^{\hbar\omega_\mathrm{D}/\Delta} \\ &= \frac{1}{2}\left[\frac{\hbar\omega_\mathrm{D}}{\Delta}\sqrt{(\hbar\omega_\mathrm{D}/\Delta)^2 + 1} + \ln\left(\frac{\hbar\omega_\mathrm{D}}{\Delta} + \sqrt{(\hbar\omega_\mathrm{D}/\Delta)^2 + 1}\right)\right] \\ &= \frac{1}{2\Delta^2}\left[\hbar\omega_\mathrm{D}\sqrt{(\hbar\omega_\mathrm{D})^2 + \Delta^2} + \Delta^2 \ln\left\{\frac{\hbar\omega_\mathrm{D}}{\Delta}\left(1 + \sqrt{1 + (\Delta/\hbar\omega_\mathrm{D})^2}\right)\right\}\right] \\ &= \frac{1}{2\Delta^2}\left[(\hbar\omega_\mathrm{D})^2\sqrt{1 + (\Delta/\hbar\omega_\mathrm{D})^2}\right] + \Delta^2 \ln\left(\frac{\hbar\omega_\mathrm{D}}{\Delta}\right) + \Delta^2 \ln\left(1 + \sqrt{1 + (\Delta/\hbar\omega_\mathrm{D})^2}\right). \end{aligned}$$

Expanding the square roots to order $(\Delta/\hbar\omega_{\mathrm{D}})^2$ we have

$$\begin{aligned}
I &= \frac{1}{2\Delta^2}\left[(\hbar\omega_{\mathrm{D}})^2\left(1+\frac{1}{2}\frac{\Delta^2}{(\hbar\omega_{\mathrm{D}})^2}+\cdots\right)+\Delta^2\ln\left(\frac{\hbar\omega_{\mathrm{D}}}{\Delta}\right)+\Delta^2\ln\left(1+1+\frac{1}{2}\frac{\Delta^2}{(\hbar\omega_{\mathrm{D}})^2}+\cdots\right)\right]\\
&= \frac{1}{2\Delta^2}\left[(\hbar\omega_{\mathrm{D}})^2+\frac{1}{2}\Delta^2+\cdots+\Delta^2\ln\left(\frac{\hbar\omega_{\mathrm{D}}}{\Delta}\right)+\Delta^2\ln 2+\Delta^2\ln\left(1+\frac{1}{4}\frac{\Delta^2}{(\hbar\omega_{\mathrm{D}})^2}+\cdots\right)\right]\\
&= \frac{1}{2\Delta^2}\left[(\hbar\omega_{\mathrm{D}})^2+\frac{1}{2}\Delta^2+\cdots+\Delta^2\ln\left(\frac{2\hbar\omega_{\mathrm{D}}}{\Delta}\right)+\Delta^2\frac{1}{4}\frac{\Delta^2}{(\hbar\omega_{\mathrm{D}})^2}+\cdots\right].
\end{aligned}$$

But in the weak-coupling limit we have [see (16.61)]

$$\ln\left(\frac{2\hbar\omega_{\mathrm{D}}}{\Delta}\right)=\frac{1}{D(0)V}, \tag{16.66}$$

so that

$$I=\frac{1}{2\Delta^2}\left[(\hbar\omega_{\mathrm{D}})^2+\frac{1}{2}\Delta^2+\cdots+\Delta^2\frac{1}{D(0)V}+\Delta^2\frac{1}{4}\frac{\Delta^2}{(\hbar\omega_{\mathrm{D}})^2}+\cdots\right]. \tag{16.67}$$

Collecting (16.62), (16.65) and (16.67), we have

$$\begin{aligned}
\langle\tilde{\phi}|\,(\hat{H}-\mu\hat{N})\,|\tilde{\phi}\rangle &= 2\sum_{\substack{\boldsymbol{k}\\(k<k_{\mathrm{F}})}}\xi_k+\frac{2D(0)\Delta^2}{D(0)V}+2D(0)\frac{(\hbar\omega_{\mathrm{D}})^2}{2}-2D(0)\Delta^2 I-\frac{\Delta^2}{V}\\
&= 2\sum_{\substack{\boldsymbol{k}\\(k<k_{\mathrm{F}})}}\xi_k+\frac{\Delta^2}{V}+D(0)(\hbar\omega_{\mathrm{D}})^2\\
&\quad -D(0)\left[(\hbar\omega_{\mathrm{D}})^2+\frac{1}{2}\Delta^2+\cdots+\Delta^2\frac{1}{D(0)V}+\Delta^2\frac{1}{4}\frac{\Delta^2}{(\hbar\omega_{\mathrm{D}})^2}+\cdots\right]\\
&= 2\sum_{\substack{\boldsymbol{k}\\(k<k_{\mathrm{F}})}}\xi_k-\frac{1}{2}D(0)\Delta^2+O(\Delta^4).
\end{aligned}$$

Therefore, the energy difference between the minimum-free-energy state $|\tilde{\phi}\rangle$ and the ground state $|\phi_{\mathrm{n}}\rangle$ of the corresponding normal system (the system with $V=0$ and hence $\Delta=0$) is, to terms of order Δ^2,

$$\langle\tilde{\phi}|\,(\hat{H}-\mu\hat{N})\,|\tilde{\phi}\rangle-\langle\phi_{\mathrm{n}}|\,(\hat{H}-\mu\hat{N})\,|\phi_{\mathrm{n}}\rangle=-\frac{1}{2}D(0)\Delta^2. \tag{16.68}$$

Thus, the energy of the superconducting ground state is lower than that of the normal ground state. Since $D(0)$ is of the order of $N/\varepsilon_{\mathrm{F}}$ (for a free-electron gas, we found $D(0)=3N/2\varepsilon_{\mathrm{F}}$), the energy difference is of the order of $\Delta^2/\varepsilon_{\mathrm{F}}$ per particle. Thus, for a typical Δ of ~ 10 K and $\varepsilon_{\mathrm{F}}\sim 10^4$ K, we have an energy difference per particle of the order of 10^{-2} K. Although we cannot calculate the energy of the normal ground state with this accuracy, it may be expected that the unknown terms will be very nearly the same in the normal and superconducting states, so that the energy difference $-\frac{1}{2}D(0)\Delta^2$ will be very nearly correct.

What is the probability that, in the ground state $|\tilde{\phi}\rangle$, there is an electron in the single-particle state with quantum numbers $\boldsymbol{k}\lambda$? This is clearly equal to the expectation value of $\hat{n}_{\boldsymbol{k}\lambda}$ in the state $|\tilde{\phi}\rangle$ (for electrons, this expectation value must lie between 0 and 1):

$$\langle\tilde{\phi}|\,\hat{n}_{\boldsymbol{k}\lambda}\,|\tilde{\phi}\rangle = \langle\tilde{\phi}|\,\hat{a}^{\dagger}_{\boldsymbol{k}\lambda}\hat{a}_{\boldsymbol{k}\lambda}\,|\tilde{\phi}\rangle = v_k^2 = \frac{1}{2}\left(1 - \frac{\xi_k}{\sqrt{\xi_k^2 + \Delta_k^2}}\right) \tag{16.69}$$

where in the last equality we have used (16.63). As a function of k, in a superconductor it takes the form shown in figure 16.5. (We use the notation $\langle\cdots\rangle_0 \equiv \langle\tilde{\phi}|\cdots|\tilde{\phi}\rangle$.) For a normal (nonsuperconducting) system, for which $\Delta_{\boldsymbol{k}} = 0$ for all $\boldsymbol{k}$, (16.69) for the ground state gives the familiar form shown in figure 16.6.

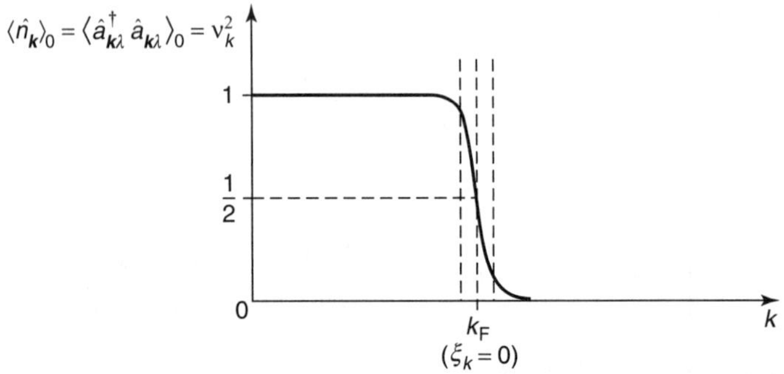

Figure 16.5 *Expectation value of $\hat{n}_{\boldsymbol{k}\lambda}$ in the ground state $|\tilde{\phi}\rangle$ of a superconductor.*

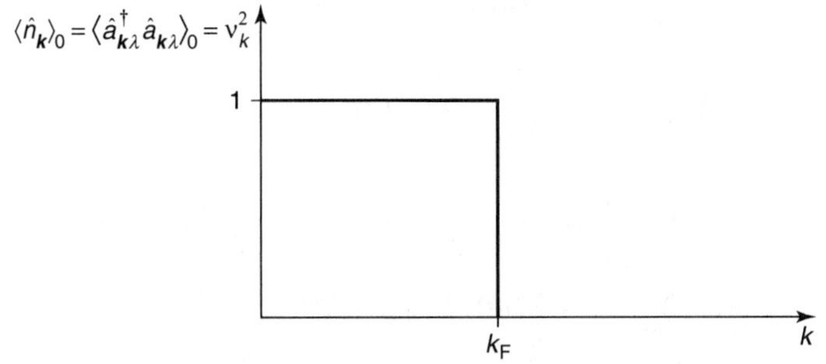

Figure 16.6 *Expectation value of $\hat{n}_{\boldsymbol{k}\lambda}$ in the ground state of a normal system.*

The width of the transition region in the superconductor can be found, in the case of the BCS interaction, from the fact that $\Delta_k = 0$ for $|\xi_k| > \hbar\omega_D$, so that (16.69) gives $v_k = 1$ for $\xi_k < -\hbar\omega_D$ and $v_k = 0$ for $\xi_k > \hbar\omega_D$. Thus, if the transition region is specified by the range $k_F - \delta k < k < k_F + \delta k$, we have

$$\hbar^2\frac{(k_F + \delta k)^2}{2m} - \hbar^2\frac{k_F^2}{2m} = \hbar\omega_D \tag{16.70}$$

so that

$$\frac{\hbar^2 k_F \delta k}{m} = \hbar\omega_D, \quad \text{i.e.,} \quad \delta k = \frac{m\omega_D}{\hbar k_F} = \frac{\omega_D}{v_F} \tag{16.71}$$

where v_F is the Fermi velocity $v_F = \hbar k_F/m$.

If we now take the N-particle state $|\phi_N\rangle$ (16.26) and try to add two electrons with quantum numbers $\boldsymbol{k}\uparrow, -\boldsymbol{k}\downarrow$, what is the probability amplitude F_k for obtaining the state $|\phi_{N+2}\rangle$? It is

$$F_k = \langle \phi_{N+2}| \hat{a}^\dagger_{\boldsymbol{k}\uparrow}\hat{a}^\dagger_{-\boldsymbol{k}\downarrow} |\phi_N\rangle = \langle \tilde{\phi}| \hat{a}^\dagger_{\boldsymbol{k}\uparrow}\hat{a}^\dagger_{-\boldsymbol{k}\downarrow} |\tilde{\phi}\rangle, \tag{16.72}$$

which clearly has a nonzero contribution only from the part of $|\tilde{\phi}\rangle$ in which the pair $\boldsymbol{k}\uparrow, -\boldsymbol{k}\downarrow$ is absent (this part has coefficient u_k) and the part of $\langle\tilde{\phi}|$ in which this pair is present (this part has coefficient v_k^*). Therefore,

$$F_k = v_k^* u_k = v_k u_k. \tag{16.73}$$

From our results for u_k and v_k, we see that F_k is nonvanishing only in the transition zone, and is maximum for $k = k_{\mathrm{F}}$. Similarly, one can show that $\langle \tilde{\phi}| \hat{a}_{-\boldsymbol{k}\downarrow}\hat{a}_{\boldsymbol{k}\uparrow} |\tilde{\phi}\rangle = u_k v_k = F_k$. Thus, plotted against k the quantity F_k has the form shown in figure 16.7.

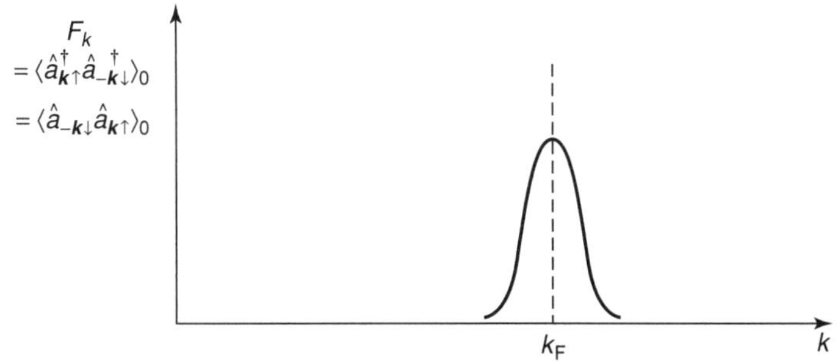

Figure 16.7 *Probability amplitude F_k for obtaining the state $|\phi_{N+2}\rangle$ from $|\phi_N\rangle$.*

16.2.3 First excited states

We have constructed a ground-state function $|\phi_N\rangle$ and found a more convenient generating function $|\tilde{\phi}\rangle$ with essentially similar matrix elements, that is, essentially the same physical consequences. We now consider the corresponding states resulting when an electron in the plane-wave state with quantum numbers $\boldsymbol{m}\lambda$ is added. From $|\phi_N\rangle$ we obtain the state

$$|\phi_{N+1,\boldsymbol{m}\lambda}\rangle \equiv \hat{a}^\dagger_{\boldsymbol{m}\lambda} |\phi_N\rangle. \tag{16.74}$$

From $|\tilde{\phi}\rangle$ we obtain the corresponding generating function

$$|\tilde{\phi}_{\boldsymbol{m}\lambda}\rangle \equiv \hat{a}^\dagger_{\boldsymbol{m}\lambda} |\tilde{\phi}\rangle = \prod_{\text{“}\boldsymbol{k}\neq\boldsymbol{m}\text{”}} \left(u_k + v_k \hat{a}^\dagger_{\boldsymbol{k}\uparrow}\hat{a}^\dagger_{-\boldsymbol{k}\downarrow}\right) \hat{a}^\dagger_{\boldsymbol{m}\lambda} |0\rangle, \tag{16.75}$$

where by “$\boldsymbol{k}\neq\boldsymbol{m}$” we mean $\boldsymbol{k}\neq\boldsymbol{m}$ if $\lambda = \uparrow$, or $\boldsymbol{k}\neq -\boldsymbol{m}$ if $\lambda = \downarrow$. We have

$$\langle \tilde{\phi}_{\boldsymbol{m}\lambda} \mid \tilde{\phi}_{\boldsymbol{m}\lambda}\rangle = \prod_{\text{“}\boldsymbol{k}\neq\boldsymbol{m}\text{”}} \left(u_k^2 + v_k^2\right) \langle 0| \hat{a}_{\boldsymbol{m}\lambda}\hat{a}^\dagger_{\boldsymbol{m}\lambda} |0\rangle = 1. \tag{16.76}$$

In the state $|\tilde{\phi}_{m\lambda}\rangle$ described by (16.75) there is unit probability of finding an electron with wave vector $\boldsymbol{m}$ (and energy ξ_m relative to the Fermi energy), whereas in the state $|\tilde{\phi}\rangle$ there was probability v_m^2 of finding an electron pair $\boldsymbol{m}\lambda$, $-\boldsymbol{m}-\lambda$ (with energy $2\xi_m$ relative to twice the Fermi energy). Therefore, for the kinetic energy of the state (16.75) with the added electron we have

$$\langle\tilde{\phi}_{m\lambda}|\,\hat{H}_0\,|\tilde{\phi}_{m\lambda}\rangle = \langle\tilde{\phi}|\,\hat{H}_0\,|\tilde{\phi}\rangle + (1-2v_m^2)\xi_m, \tag{16.77}$$

where the additional energy (the second term here) is always positive, since, from (16.63), $1-2v_m^2>0$ when $\xi_m>0$ and $1-2v_m^2<0$ when $\xi_m<0$.

To find the total energy we need to take the interaction energy into account. To do this we note that the presence of the added electron with quantum numbers $\boldsymbol{m}\lambda$ means that the processes $(\boldsymbol{k}\uparrow, -\boldsymbol{k}\downarrow)\rightarrow(\boldsymbol{m}\lambda, -\boldsymbol{m}-\lambda)$ now occur with zero probability. So, too, do processes like $(\boldsymbol{m}\lambda, -\boldsymbol{m}-\lambda)\rightarrow(\boldsymbol{l}\uparrow, -\boldsymbol{l}\downarrow)$, since there is no electron with quantum numbers $-\boldsymbol{m}$, $-\lambda$. We thus have [see (16.62) and (16.51)]

$$\langle\tilde{\phi}_{m\lambda}|\,\hat{H}_{\text{int}}\,|\tilde{\phi}_{m\lambda}\rangle = \langle\tilde{\phi}|\,\hat{H}_{\text{int}}\,|\tilde{\phi}\rangle - 2\sum_{l} V_{ml}u_m v_l u_l v_m = \langle\tilde{\phi}|\,\hat{H}_{\text{int}}\,|\tilde{\phi}\rangle + 2u_m v_m \Delta_m. \tag{16.78}$$

Thus, from (16.77) and (16.78), the total energy of the state $|\tilde{\phi}_{m\lambda}\rangle$ (with the added electron) becomes

$$\begin{aligned}\langle\tilde{\phi}_{m\lambda}|\,\hat{H}\,|\tilde{\phi}_{m\lambda}\rangle &= \langle\tilde{\phi}|\,\hat{H}\,|\tilde{\phi}\rangle + (1-2v_m^2)\xi_m + 2u_m v_m \Delta_m \\ &= E_0 + \frac{\xi_m^2}{\sqrt{\xi_m^2+\Delta_m^2}} + \frac{\Delta_m^2}{\sqrt{\xi_m^2+\Delta_m^2}} \\ &= E_0 + \sqrt{\xi_m^2+\Delta_m^2} \equiv E_0 + \varepsilon_m,\end{aligned} \tag{16.79}$$

where $E_0 = \langle\tilde{\phi}|\,\hat{H}\,|\tilde{\phi}\rangle$ is the energy of the ground state $|\tilde{\phi}\rangle$, and

$$\varepsilon_m = \sqrt{\xi_m^2+\Delta_m^2} \tag{16.80}$$

is the energy (relative to the Fermi energy) of the added electron in the plane-wave state with wave vector $\boldsymbol{m}$. If the system were normal, that is, if $\Delta_{\boldsymbol{m}}=0$ for all $\boldsymbol{m}$, the added-particle energy would be

$$\varepsilon_m = \xi_m = \frac{\hbar^2 k_m^2}{2m} - \frac{\hbar^2 k_F^2}{2m}, \tag{16.81}$$

which, as should be expected, is clearly zero when the added particle lies on the Fermi surface. In the presence of the BCS interaction, on the other hand, in a range of wave vectors about the Fermi surface Δ_m takes a nonzero value Δ, and so the corresponding energy ε_m is then nonzero. According to (16.80), the energy ε_m takes its minimum value, equal to Δ, when the added particle lies on the Fermi surface, that is, there is a gap in the excitation spectrum.

We can try to construct states with 2, 3, ..., n excitations in the same way. For example, we could try

$$|\tilde{\phi}_{m\lambda_1,n\lambda_2}\rangle \equiv \hat{a}^\dagger_{m\lambda_1}\hat{a}^\dagger_{n\lambda_2}|\tilde{\phi}\rangle = \prod_{\text{“}k\neq m,n\text{”}} \left(u_k + v_k \hat{a}^\dagger_{k\uparrow}\hat{a}^\dagger_{-k\downarrow}\right)\hat{a}^\dagger_{m\lambda_1}\hat{a}^\dagger_{n\lambda_2}|0\rangle. \tag{16.82}$$

Unfortunately, this is not necessarily orthogonal to $|\tilde{\phi}\rangle$, for example,

$$\langle \tilde{\phi} \mid \tilde{\phi}_{\boldsymbol{m}\uparrow,-\boldsymbol{m}\downarrow} \rangle = v_m, \tag{16.83}$$

which is not always zero [see (16.64)].

The way around this difficulty is to note that the required single-added-particle state $|\phi_{N+1,\boldsymbol{m}\lambda}\rangle \equiv \hat{a}^{\dagger}_{\boldsymbol{m}\lambda} |\phi_N\rangle$ (16.74) occurs not only in the generating function $\hat{a}^{\dagger}_{\boldsymbol{m}\lambda} |\tilde{\phi}\rangle$ but also in the generating function $\hat{a}_{-\boldsymbol{m},-\lambda} |\tilde{\phi}\rangle$, because we can also reach $|\phi_{N+1,\boldsymbol{m}\lambda}\rangle$ by destroying the second member of the pair $\boldsymbol{m}\lambda$, $-\boldsymbol{m} - \lambda$ in the state $|\phi_{N+2}\rangle$.

We thus try to express the generating function $|\tilde{\phi}_{\boldsymbol{m}\lambda}\rangle$ (16.75) for the first excited state in a form derived directly from the ground-state generating function $|\tilde{\phi}\rangle$ by the action of a creation operator $\hat{\gamma}^{\dagger}_{\boldsymbol{m}\lambda}$ for one elementary excitation:

$$|\tilde{\phi}_{\boldsymbol{m}\lambda}\rangle = \hat{\gamma}^{\dagger}_{\boldsymbol{m}\lambda} |\tilde{\phi}\rangle, \tag{16.84}$$

where we expect that $\hat{\gamma}^{\dagger}_{\boldsymbol{m}\lambda}$ can be expressed as a linear combination of $\hat{a}^{\dagger}_{\boldsymbol{m}\lambda}$ and $\hat{a}_{-\boldsymbol{m},-\lambda}$:

$$\hat{\gamma}^{\dagger}_{\boldsymbol{m}\uparrow} = \alpha_m \hat{a}^{\dagger}_{\boldsymbol{m}\uparrow} + \beta_m \hat{a}_{-\boldsymbol{m}\downarrow}, \quad \hat{\gamma}^{\dagger}_{\boldsymbol{m}\downarrow} = \mu_m \hat{a}^{\dagger}_{\boldsymbol{m}\downarrow} + \nu_m \hat{a}_{-\boldsymbol{m}\uparrow}, \tag{16.85}$$

with corresponding adjoint, excitation-annihilation operators

$$\hat{\gamma}_{\boldsymbol{m}\uparrow} = \alpha^*_m \hat{a}_{\boldsymbol{m}\uparrow} + \beta^*_m \hat{a}^{\dagger}_{-\boldsymbol{m}\downarrow}, \quad \hat{\gamma}_{\boldsymbol{m}\downarrow} = \mu^*_m \hat{a}_{\boldsymbol{m}\downarrow} + \nu^*_m \hat{a}^{\dagger}_{-\boldsymbol{m}\uparrow}. \tag{16.86}$$

We shall try to find operators $\hat{\gamma}^{\dagger}_{\boldsymbol{m}\lambda}$ and $\hat{\gamma}_{\boldsymbol{m}\lambda}$ such that (i) they obey the standard fermion anticommutation relations, and (ii) $\hat{\gamma}_{\boldsymbol{m}\lambda} |\tilde{\phi}\rangle = 0$, that is, $|\tilde{\phi}\rangle$ is a (generating) state with no excitations (corresponding to the ground state).

The anticommutation relations give

$$\begin{aligned} &[\hat{\gamma}_{\boldsymbol{m}\uparrow}, \hat{\gamma}^{\dagger}_{\boldsymbol{m}\uparrow}]_+ = |\alpha_m|^2 + |\beta_m|^2 = 1, \quad [\hat{\gamma}_{\boldsymbol{m}\downarrow}, \hat{\gamma}^{\dagger}_{\boldsymbol{m}\downarrow}]_+ = |\mu_m|^2 + |\nu_m|^2 = 1, \\ &[\hat{\gamma}^{\dagger}_{\boldsymbol{m}\uparrow}, \hat{\gamma}^{\dagger}_{-\boldsymbol{m}\downarrow}]_+ = \alpha_m \nu_m + \beta_m \mu_m = 0, \quad [\hat{\gamma}_{\boldsymbol{m}\uparrow}, \hat{\gamma}_{-\boldsymbol{m}\downarrow}]_+ = \alpha^*_m \nu^*_m + \beta^*_m \mu^*_m = 0. \end{aligned} \tag{16.87}$$

To apply condition (ii) we write

$$\begin{aligned} \hat{\gamma}_{\boldsymbol{m}\uparrow} |\tilde{\phi}\rangle &= \prod_{\boldsymbol{k}\neq\boldsymbol{m}} \left(u_k + v_k \hat{a}^{\dagger}_{\boldsymbol{k}\uparrow} \hat{a}^{\dagger}_{-\boldsymbol{k}\downarrow}\right) \left(\alpha^*_m \hat{a}_{\boldsymbol{m}\uparrow} + \beta^*_m \hat{a}^{\dagger}_{-\boldsymbol{m}\downarrow}\right) \left(u_m + v_m \hat{a}^{\dagger}_{\boldsymbol{m}\uparrow} \hat{a}^{\dagger}_{-\boldsymbol{m}\downarrow}\right) |0\rangle \\ &= \prod_{\boldsymbol{k}\neq\boldsymbol{m}} \left(u_k + v_k \hat{a}^{\dagger}_{\boldsymbol{k}\uparrow} \hat{a}^{\dagger}_{-\boldsymbol{k}\downarrow}\right) \left(\alpha^*_m v_m \hat{a}^{\dagger}_{-\boldsymbol{m}\downarrow} + \beta^*_m u_m \hat{a}^{\dagger}_{-\boldsymbol{m}\downarrow}\right) |0\rangle, \end{aligned}$$

which equals zero if

$$\alpha^*_m v_m + \beta^*_m u_m = 0. \tag{16.88}$$

Similarly, the condition $\hat{\gamma}_{\boldsymbol{m}\downarrow} |\tilde{\phi}\rangle = 0$ gives

$$-\mu^*_m v_m + \nu^*_m u_m = 0. \tag{16.89}$$

Equations (16.87)–(16.89) clearly have the following (real) solutions:

$$\alpha_m = u_m, \quad \beta_m = -v_m, \quad \mu_m = u_m, \quad \nu_m = v_m, \tag{16.90}$$

so the excitation-creation and excitation-annihilation operators (16.85) and (16.86) take the form

$$\hat{\gamma}^\dagger_{m\uparrow} = u_m \hat{a}^\dagger_{m\uparrow} - v_m \hat{a}_{-m\downarrow}, \quad \hat{\gamma}^\dagger_{m\downarrow} = u_m \hat{a}^\dagger_{m\downarrow} + v_m \hat{a}_{-m\uparrow} \tag{16.91}$$

and

$$\hat{\gamma}_{m\uparrow} = u_m \hat{a}_{m\uparrow} - v_m \hat{a}^\dagger_{-m\downarrow}, \quad \hat{\gamma}_{m\downarrow} = u_m \hat{a}_{m\downarrow} + v_m \hat{a}^\dagger_{-m\uparrow}. \tag{16.92}$$

All the different states obtained by applying an arbitrary number of different $\hat{\gamma}^\dagger$ operators to $|\tilde{\phi}\rangle$ are orthogonal to $|\tilde{\phi}\rangle$, orthogonal to each other, and normalized. This follows because our excitation-creation and excitation-annihilation operators have been constructed to satisfy the fermion anticommutation relations and the condition $\hat{\gamma}_{m\lambda} |\tilde{\phi}\rangle = 0$. For example, $\hat{\gamma}^\dagger_{m\alpha}\hat{\gamma}^\dagger_{n\beta} |\tilde{\phi}\rangle$ is orthogonal to $|\tilde{\phi}\rangle$:

$$\langle\tilde{\phi}| \hat{\gamma}^\dagger_{m\alpha}\hat{\gamma}^\dagger_{n\beta} |\tilde{\phi}\rangle = \langle\tilde{\phi}| \hat{\gamma}_{n\beta}\hat{\gamma}_{m\alpha} |\tilde{\phi}\rangle^* = 0. \tag{16.93}$$

The state $\hat{\gamma}^\dagger_{n\lambda} |\tilde{\phi}\rangle$ is orthogonal to $\hat{\gamma}^\dagger_{m\lambda} |\tilde{\phi}\rangle$:

$$\left(\hat{\gamma}^\dagger_{m\lambda} |\tilde{\phi}\rangle\right)^\dagger \hat{\gamma}^\dagger_{n\lambda} |\tilde{\phi}\rangle = \langle\tilde{\phi}| \hat{\gamma}_{m\lambda}\hat{\gamma}^\dagger_{n\lambda} |\tilde{\phi}\rangle = -\langle\tilde{\phi}| \hat{\gamma}^\dagger_{n\lambda}\hat{\gamma}_{m\lambda} |\tilde{\phi}\rangle = 0. \tag{16.94}$$

The state $\hat{\gamma}^\dagger_{n\lambda} |\tilde{\phi}\rangle$ is normalized:

$$\left(\hat{\gamma}^\dagger_{n\lambda} |\tilde{\phi}\rangle\right)^\dagger \hat{\gamma}^\dagger_{n\lambda} |\tilde{\phi}\rangle = \langle\tilde{\phi}| \hat{\gamma}_{n\lambda}\hat{\gamma}^\dagger_{n\lambda} |\tilde{\phi}\rangle = \langle\tilde{\phi} \mid \tilde{\phi}\rangle - \langle\tilde{\phi}| \hat{\gamma}^\dagger_{n\lambda}\hat{\gamma}_{n\lambda} |\tilde{\phi}\rangle = 1. \tag{16.95}$$

The fact that, as constructed, the operators $\hat{\gamma}^\dagger_{m\lambda}$ and $\hat{\gamma}_{m\lambda}$ satisfy the standard fermion anticommutation relations, and also ensure that $\hat{\gamma}_{m\lambda} |\tilde{\phi}\rangle = 0$, leads to the desired result that states of different excitation are automatically orthogonal to $|\tilde{\phi}\rangle$ and to each other, for example,

$$\begin{aligned} \langle\tilde{\phi}| \hat{\gamma}^\dagger_{m\lambda} |\tilde{\phi}\rangle &= \langle\hat{\gamma}_{m\lambda}\tilde{\phi} \mid \tilde{\phi}\rangle = 0, \\ \left\langle\hat{\gamma}^\dagger_{l\lambda}\tilde{\phi}\right| \hat{\gamma}^\dagger_{m\lambda} |\tilde{\phi}\rangle &= \langle\tilde{\phi}| \hat{\gamma}_{l\lambda}\hat{\gamma}^\dagger_{m\lambda} |\tilde{\phi}\rangle = (\text{if } \boldsymbol{l} \neq \boldsymbol{m}) - \langle\tilde{\phi}| \hat{\gamma}^\dagger_{m\lambda}\hat{\gamma}_{l\lambda} |\tilde{\phi}\rangle = 0. \end{aligned}$$

Thus, the operators (16.91) act on $|\tilde{\phi}\rangle$ to give states that are orthogonal to each other and to the ground state.

We say that the operator $\hat{\gamma}^\dagger_{m\lambda}$ creates a **quasiparticle** of wave vector $\boldsymbol{m}$ and spin component λ.

16.2.4 Thermodynamics of superconductors

At $T = 0$ the system is in the ground state $|\tilde{\phi}\rangle$ with certainty, and the probability of finding the system in any one of the excited states $\hat{\gamma}^\dagger_{m\lambda} |\tilde{\phi}\rangle$, $\hat{\gamma}^\dagger_{k\lambda_1}\hat{\gamma}^\dagger_{m\lambda_2} |\tilde{\phi}\rangle$, … , that is, in a state with one, two or more quasiparticles present, is zero.

At nonzero temperatures, however, the probability of finding the system in any one of these excited states is nonzero, and to determine the thermodynamic equilibrium state of a superconductor at $T \neq 0$ we shall have to take these excited states, and their probabilities, into account.

We took the ground state of a superconductor to have the form of the BCS state (16.29) (a superposition of states with differing numbers of particles), with values of u_k and v_k that minimize the grand potential of the system at $T = 0$. At any nonzero temperature, on the other hand, we determine the thermodynamic equilibrium state of the system by minimizing the grand potential of the system at $T \neq 0$. Now we shall need to find not only the u_k and v_k that minimize Ω, but also the form of the probability function f_k (the probability that a quasiparticle $\boldsymbol{k}\lambda$ is present) that minimizes Ω. The resulting f_k will be seen to have the formal structure of the Fermi–Dirac distribution, but with a more complicated temperature dependence arising from the fact that the quasiparticle-excitation energy that appears in it is itself temperature-dependent, having the form (16.80) with Δ replaced by $\Delta(T)$.

The grand potential is

$$\Omega = U - \mu N - TS, \tag{16.96}$$

where $U - \mu N = \langle \hat{H} - \mu \hat{N} \rangle$ is the thermal average energy, that is, we must first calculate

$$U - \mu N = \langle \hat{H}_0 - \mu \hat{N} \rangle + \langle \hat{V} \rangle. \tag{16.97}$$

Let $n^{(i)}_{\boldsymbol{k}\lambda}$ denote the number of quasiparticles with quantum numbers $\boldsymbol{k}\lambda$ in the many-body excited state $|i\rangle$, that is, $n^{(i)}_{\boldsymbol{k}\lambda}$ is the eigenvalue of the operator $\hat{n}_{\boldsymbol{k}\lambda} \equiv \hat{\gamma}^{\dagger}_{\boldsymbol{k}\lambda}\hat{\gamma}_{\boldsymbol{k}\lambda}$ in the many-body state $|i\rangle$. (Of course, $n^{(i)}_{\boldsymbol{k}\lambda}$ cannot exceed 1, since the $\hat{\gamma}$ are fermion operators.) Then, in a particular excited state of the many-body system:

$$|i\rangle \equiv \hat{\gamma}^{\dagger}_{\boldsymbol{m}\alpha}\hat{\gamma}^{\dagger}_{\boldsymbol{n}\beta}\cdots|\tilde{\phi}\rangle,$$

we have

$$\langle i|(\hat{H}_0 - \mu\hat{N})|i\rangle = \langle\tilde{\phi}|(\hat{H}_0 - \mu\hat{N})|\tilde{\phi}\rangle + \sum_{\boldsymbol{m},\alpha} n^{(i)}_{\boldsymbol{m}\alpha}(1 - 2v_m^2)\xi_m, \tag{16.98}$$

where we have used the result for the expectation value of the kinetic energy in a singly excited (one-quasiparticle) state. Therefore,

$$\begin{aligned}\langle \hat{H}_0 - \mu\hat{N}\rangle &= \frac{1}{\mathcal{Z}}\sum_N\sum_{i(N)} \mathrm{e}^{[\mu N - \varepsilon_i(N)]/k_B T}\langle i|(\hat{H}_0 - \mu\hat{N})|i\rangle \\ &= \langle\tilde{\phi}|(\hat{H}_0 - \mu\hat{N})|\tilde{\phi}\rangle + \sum_{\boldsymbol{m},\alpha}\langle n_{\boldsymbol{m}\alpha}\rangle(1 - 2v_m^2)\xi_m,\end{aligned} \tag{16.99}$$

where $\mathcal{Z}$ is the grand partition function and $\langle n_{\boldsymbol{m}\alpha}\rangle$ is the thermal average occupation of the quasiparticle state $\boldsymbol{m}\alpha$, which, since an occupation exceeding unity is impossible, is equal to the probability $f_{\boldsymbol{m}\alpha}$ that a quasiparticle $\boldsymbol{m}\alpha$ is present. Therefore,

$$\begin{aligned}\langle \hat{H}_0 - \mu\hat{N}\rangle &= \langle\tilde{\phi}|(\hat{H}_0 - \mu\hat{N})|\tilde{\phi}\rangle + \sum_{\boldsymbol{m},\alpha} f_{\boldsymbol{m}\alpha}(1 - 2v_m^2)\xi_m \\ &= 2\sum_{\boldsymbol{m}} v_m^2\xi_m + 2\sum_{\boldsymbol{m}} f_m(1 - 2v_m^2)\xi_m = 2\sum_{\boldsymbol{m}}\xi_m[f_m + v_m^2(1 - 2f_m)],\end{aligned} \tag{16.100}$$

where we have put $f_{m\alpha} = f_{m,-\alpha} \equiv f_m$.

We now find $\langle \hat{V} \rangle$. Once again, the only processes that come into play are the pair-scattering processes $(\boldsymbol{l}\uparrow; -\boldsymbol{l}\downarrow) \to (\boldsymbol{k}\uparrow; -\boldsymbol{k}\downarrow)$. In restricting ourselves to such processes, we are, in effect, assuming an effective hamiltonian of the form

$$\hat{H}_{\mathrm{BCS}} = \sum_{\boldsymbol{k},\lambda} \xi_k \hat{a}^\dagger_{\boldsymbol{k}\lambda}\hat{a}_{\boldsymbol{k}\lambda} + \sum_{\boldsymbol{k},\boldsymbol{l}} V_{kl}\hat{a}^\dagger_{\boldsymbol{k}\uparrow}\hat{a}^\dagger_{-\boldsymbol{k}\downarrow}\hat{a}_{-\boldsymbol{l}\downarrow}\hat{a}_{\boldsymbol{l}\uparrow}, \tag{16.101}$$

which is the Bardeen–Cooper–Schrieffer (BCS) hamiltonian.

Therefore,

$$\langle \hat{V} \rangle = \sum_{\boldsymbol{k},\boldsymbol{l}} V_{kl} \left\langle \hat{a}^\dagger_{\boldsymbol{k}\uparrow}\hat{a}^\dagger_{-\boldsymbol{k}\downarrow}\hat{a}_{-\boldsymbol{l}\downarrow}\hat{a}_{\boldsymbol{l}\uparrow} \right\rangle. \tag{16.102}$$

The simplest way to evaluate this is to express the $\hat{a}$ and $\hat{a}^\dagger$ operators in terms of the quasiparticle operators $\hat{\gamma}$ and $\hat{\gamma}^\dagger$. This we do by inverting the relations (16.91) and (16.92), which (after the replacement $\boldsymbol{m} \to -\boldsymbol{m}$ in the latter) take the form

$$\begin{pmatrix} \hat{\gamma}^\dagger_{\boldsymbol{m}\uparrow} \\ \hat{\gamma}_{-\boldsymbol{m}\downarrow} \\ \hat{\gamma}^\dagger_{\boldsymbol{m}\downarrow} \\ \hat{\gamma}_{-\boldsymbol{m}\uparrow} \end{pmatrix} = \mathrm{A} \begin{pmatrix} \hat{a}^\dagger_{\boldsymbol{m}\uparrow} \\ \hat{a}_{-\boldsymbol{m}\downarrow} \\ \hat{a}^\dagger_{\boldsymbol{m}\downarrow} \\ \hat{a}_{-\boldsymbol{m}\uparrow} \end{pmatrix}, \text{ with } \mathrm{A} = \begin{pmatrix} u_m & -v_m & 0 & 0 \\ v_m & u_m & 0 & 0 \\ 0 & 0 & u_m & v_m \\ 0 & 0 & -v_m & u_m \end{pmatrix} \text{ (so that A = 1).}$$

Inverting this relation, we have

$$\begin{pmatrix} \hat{a}^\dagger_{\boldsymbol{m}\uparrow} \\ \hat{a}_{-\boldsymbol{m}\downarrow} \\ \hat{a}^\dagger_{\boldsymbol{m}\downarrow} \\ \hat{a}_{-\boldsymbol{m}\uparrow} \end{pmatrix} = \mathrm{A}^{-1} \begin{pmatrix} \hat{\gamma}^\dagger_{\boldsymbol{m}\uparrow} \\ \hat{\gamma}_{-\boldsymbol{m}\downarrow} \\ \hat{\gamma}^\dagger_{\boldsymbol{m}\downarrow} \\ \hat{\gamma}_{-\boldsymbol{m}\uparrow} \end{pmatrix}, \text{ with } \mathrm{A}^{-1} = \begin{pmatrix} u_m & v_m & 0 & 0 \\ -v_m & u_m & 0 & 0 \\ 0 & 0 & u_m & -v_m \\ 0 & 0 & v_m & u_m \end{pmatrix}.$$

Thus, in a given many-body excited state $|i\rangle$, we have

$$\begin{aligned} &\langle i|\, \hat{a}^\dagger_{\boldsymbol{k}\uparrow}\hat{a}^\dagger_{-\boldsymbol{k}\downarrow}\hat{a}_{-\boldsymbol{l}\downarrow}\hat{a}_{\boldsymbol{l}\uparrow}\, |i\rangle \\ &\quad = \langle i|\,(u_k\hat{\gamma}^\dagger_{\boldsymbol{k}\uparrow} + v_k\hat{\gamma}_{-\boldsymbol{k}\downarrow})(u_k\hat{\gamma}^\dagger_{-\boldsymbol{k}\downarrow} - v_k\hat{\gamma}_{\boldsymbol{k}\uparrow})(-v_l\hat{\gamma}^\dagger_{\boldsymbol{l}\uparrow} + u_l\hat{\gamma}_{-\boldsymbol{l}\downarrow})(v_l\hat{\gamma}^\dagger_{-\boldsymbol{l}\downarrow} + u_l\hat{\gamma}_{\boldsymbol{l}\uparrow})\,|i\rangle\,, \\ &\quad = \langle i|\, u_k v_k[-\hat{n}_{\boldsymbol{k}\uparrow} + (1 - \hat{n}_{-\boldsymbol{k}\downarrow})]u_l v_l[-\hat{n}_{\boldsymbol{l}\uparrow} + (1 - \hat{n}_{-\boldsymbol{l}\downarrow})]\,|i\rangle\,, \end{aligned}$$

where we have used the fact that

$$\hat{n}_{\boldsymbol{k}\uparrow} \equiv \hat{\gamma}^\dagger_{\boldsymbol{k}\uparrow}\hat{\gamma}_{\boldsymbol{k}\uparrow} \quad \text{and} \quad \hat{\gamma}_{-\boldsymbol{k}\downarrow}\hat{\gamma}^\dagger_{-\boldsymbol{k}\downarrow} = -\hat{\gamma}^\dagger_{-\boldsymbol{k}\downarrow}\hat{\gamma}_{-\boldsymbol{k}\downarrow} + 1 = 1 - \hat{n}_{-\boldsymbol{k}\downarrow},$$

and the fact that $\boldsymbol{k} \neq \boldsymbol{l}$ or $-\boldsymbol{l}$. Therefore, the matrix element is

$$\langle i|\, \hat{a}^\dagger_{\boldsymbol{k}\uparrow}\hat{a}^\dagger_{-\boldsymbol{k}\downarrow}\hat{a}_{-\boldsymbol{l}\downarrow}\hat{a}_{\boldsymbol{l}\uparrow}\, |i\rangle = u_k v_k u_l v_l \left(1 - n^{(i)}_{\boldsymbol{k}\uparrow} - n^{(i)}_{-\boldsymbol{k}\downarrow}\right)\left(1 - n^{(i)}_{\boldsymbol{l}\uparrow} - n^{(i)}_{-\boldsymbol{l}\downarrow}\right)], \tag{16.103}$$

where the $n^{(i)}$ are the occupation numbers of the single-quasiparticle states $\boldsymbol{k}\uparrow$ and so on in the many-body state $|i\rangle$. Therefore

$$\langle \hat{V} \rangle = \frac{1}{\mathcal{Z}} \sum_N \sum_{i(N)} \mathrm{e}^{[\mu N - \varepsilon_i(N)]/k_B T} \langle i | \hat{V} | i \rangle = \sum_{kl} V_{kl} u_k v_k u_l v_l (1 - 2f_k)(1 - 2f_l), \tag{16.104}$$

where we have put $\langle \hat{n}_{\boldsymbol{k}\uparrow} \rangle = \langle \hat{n}_{-\boldsymbol{k}\downarrow} \rangle \equiv f_k$.

To calculate the grand potential Ω, we now need the entropy S of the "gas" of Fermi quasiparticles. This is given by equation (12.24):

$$S = -k_B \sum_{k\alpha} [f_k \ln f_k + (1 - f_k) \ln(1 - f_k)], \tag{16.105}$$

where $f_{\boldsymbol{k}\uparrow} = f_{\boldsymbol{k}\downarrow} \equiv f_k$ is the thermal average number of quasiparticles with wave vector $\boldsymbol{k}$ and "spin up" (or "spin down")

Therefore, the grand potential is

$$\begin{aligned}\Omega &= \langle H_0 - \mu N \rangle + \langle V \rangle - TS \\ &= 2 \sum_k \xi_k [f_k + v_k^2 (1 - 2f_k)] + \sum_{k,l} V_{kl} u_k v_k u_l v_l (1 - 2f_k)(1 - 2f_l) \\ &\quad + 2k_B T \sum_k [f_k \ln f_k + (1 - f_k) \ln(1 - f_k)].\end{aligned} \tag{16.106}$$

Firstly we require Ω to be stationary with respect to u_k and v_k. Again putting $u_k = \sin\theta_k$ and $v_k = \cos\theta_k$, we have

$$\Omega = 2 \sum_k \xi_k [f_k + \cos^2\theta_k (1 - 2f_k)] + \frac{1}{4} \sum_{k,l} V_{kl} \sin 2\theta_k \sin 2\theta_l (1 - 2f_k)(1 - 2f_l) - TS$$

where TS is independent of θ_k. Therefore,

$$\begin{aligned}\frac{\partial \Omega}{\partial \theta_k} = 0 &= -2\xi_k (1 - 2f_k) \sin 2\theta_k + \frac{1}{2} \sum_l V_{kl} \cos 2\theta_k \sin 2\theta_l (1 - 2f_k)(1 - 2f_l) \\ &\quad + \frac{1}{2} \sum_{k'} V_{k'k} \sin 2\theta_{k'} \cos 2\theta_k (1 - 2f_{k'})(1 - 2f_k) \\ &= -2\xi_k (1 - 2f_k) \sin 2\theta_k + \sum_l V_{kl} \cos 2\theta_k \sin 2\theta_l (1 - 2f_k)(1 - 2f_l),\end{aligned}$$

and so

$$\tan 2\theta_k = -\frac{\Delta_k(T)}{\xi_k}, \tag{16.107}$$

where

$$\Delta_k(T) = -\frac{1}{2} \sum_l V_{kl} \sin 2\theta_l (1 - 2f_l) = -\sum_l V_{kl} u_l v_l (1 - 2f_l), \tag{16.108}$$

that is, we get formally the same solution as for $T = 0$, except that Δ_k is replaced by $\Delta_k(T)$, which has the extra temperature-dependent factor $1 - 2f_l$ in the sum over l.

Returning to our expression (16.106) for Ω and minimizing it now with respect to f_k, we find

$$\frac{\partial \Omega}{\partial f_k} = 0 = 2\xi_k(1 - 2v_k^2) - 4\sum_l V_{kl} u_k v_k u_l v_l (1 - 2f_l) + 2k_\mathrm{B}T[\ln f_k + 1 - 1 - \ln(1 - f_k)],$$

that is,

$$k_\mathrm{B}T \ln\left(\frac{f_k}{1 - f_k}\right) = \xi_k(v_k^2 - u_k^2) - 2\Delta_k(T)u_k v_k. \tag{16.109}$$

But since the solutions for u_k and v_k are the same as before [see (16.52) and (16.80)] except with Δ_k replaced by $\Delta_k(T)$, we have

$$v_k^2 - u_k^2 = -\frac{\xi_k}{\varepsilon_k^\mathrm{ex}(T)} \quad \text{and} \quad 2u_k v_k = \frac{\Delta_k(T)}{\varepsilon_k^\mathrm{ex}(T)} \tag{16.110}$$

where $\varepsilon_k^\mathrm{ex}(T)$ is the temperature-dependent quasiparticle-excitation energy:

$$\varepsilon_k^\mathrm{ex}(T) = \sqrt{\xi_k^2 + \Delta_k(T)^2}. \tag{16.111}$$

Therefore,

$$k_\mathrm{B}T \ln\left(\frac{f_k}{1 - f_k}\right) = -\frac{\xi_k^2}{\varepsilon_k^\mathrm{ex}(T)} - \frac{\Delta_k(T)^2}{\varepsilon_k^\mathrm{ex}(T)} = -\varepsilon_k^\mathrm{ex}(T),$$

so that

$$\frac{f_k}{1 - f_k} = \mathrm{e}^{-\varepsilon_k^\mathrm{ex}(T)/k_\mathrm{B}T},$$

from which we obtain

$$f_k = \frac{\mathrm{e}^{-\varepsilon_k^\mathrm{ex}(T)/k_\mathrm{B}T}}{1 + \mathrm{e}^{-\varepsilon_k^\mathrm{ex}(T)/k_\mathrm{B}T}} = \frac{1}{\mathrm{e}^{\varepsilon_k^\mathrm{ex}(T)/k_\mathrm{B}T} + 1}, \tag{16.112}$$

which is formally the same as the Fermi–Dirac distribution function (of ε_k) [see (12.20)], except that here ε_k depends on T.

We now try to solve for the dependence $\Delta_k(T)$. We have, from (16.108) and (16.110),

$$\Delta_k(T) = -\frac{1}{2}\sum_l V_{kl}\frac{\Delta_l(T)(1 - 2f_l)}{\sqrt{\xi_l^2 + \Delta(T)^2}}.$$

Once again, we introduce the BCS interaction

$$V_{kl} = \begin{cases} -V \text{ if } |\xi_k|, |\xi_l| \le \hbar\omega_\mathrm{D} \\ 0 \text{ otherwise} \end{cases} \tag{16.113}$$

Therefore, $\Delta_k(T) = 0$ if $|\xi_k| > \hbar\omega_D$, and

$$\Delta_k(T) = \frac{V}{2}\sum_l \frac{\Delta_l(T)(1-2f_l)}{\sqrt{\xi_l^2 + \Delta(T)^2}} \quad \text{if} \quad |\xi_k| \leq \hbar\omega_D, \tag{16.114}$$

where the sum over $\boldsymbol{l}$ here is restricted to those $\boldsymbol{l}$ for which $|\xi_l| \leq \hbar\omega_D$. The sum in (16.114) is clearly independent of $\boldsymbol{k}$, and so we can replace $\Delta_k(T)$ by $\Delta(T)$. Therefore,

$$\begin{aligned} 1 &= \frac{D(0)V}{2}\int_{-\hbar\omega_D}^{\hbar\omega_D} \frac{d\xi[1-2f(\sqrt{\xi^2+\Delta(T)^2})]}{\sqrt{\xi^2+\Delta(T)^2}} \\ &= D(0)V\int_0^{\hbar\omega_D} \frac{d\xi[1-2f(\sqrt{\xi^2+\Delta(T)^2})]}{\sqrt{\xi^2+\Delta(T)^2}}. \end{aligned} \tag{16.115}$$

which is an implicit relation between $\Delta(T)$ and T. Note that for $T = 0$ there are no quasiparticles, so that $f = 0$ and we recover the $T = 0$ equation (16.58), which we found to have the solution (16.61). Therefore, $\Delta(T = 0) = \Delta$ (16.61). We introduce a temperature T_c (the **superconducting transition temperature**), above which we expect the gap function $\Delta(T)$ to vanish, so that the state then becomes that of a normal Fermi system. We shall postulate that $\Delta(T)$ goes continuously to zero as the temperature approaches T_c from below, so that we can find T_c by putting $T = T_c$ and $\Delta(T_c) = 0$ in equaion (16.115). We then have

$$1 - 2f(\xi) = 1 - \frac{2}{1+e^{\xi/k_B T_c}} = \tanh\frac{\xi}{2k_B T_c},$$

so that (16.115) becomes

$$1 = D(0)V\int_0^{\hbar\omega_D} \frac{d\xi}{\xi}\tanh\frac{\xi}{2k_B T_c}.$$

If $\hbar\omega_D >> k_B T_c$, the argument x in $\tanh x$ is large over the major part of the domain of integration, so that $\tanh x = (e^x - e^{-x})/(e^x + e^{-x}) \approx e^x/e^x = 1$ and we obtain

$$1 = D(0)V\left[\ln\left(\frac{\hbar\omega_D}{k_B T_c}\right) + C\right],$$

where we have changed the integration variable to $x = \xi/k_B T_c$, and the constant C arises from the lower part of the range of integration. A more detailed calculation gives $C = \ln 1.14$, and so we obtain

$$1 = D(0)V\ln\left(\frac{1.14\hbar\omega_D}{k_B T_c}\right).$$

Therefore,

$$k_B T_c = 1.14\hbar\omega_D e^{-1/D(0)V}. \tag{16.116}$$

But we found [see (16.61)]

$$\Delta = \Delta(T = 0) = 2\hbar\omega_D e^{-1/D(0)V},$$

and so

$$\Delta = (2/1.14)k_B T_c = 1.76 k_B T_c. \tag{16.117}$$

Equation (16.116) enables us to determine $D(0)V$ from experimental values of T_c and of the Debye temperature θ_D ($\equiv \hbar\omega_D/k_B$).

To summarize, starting from a generating function of pair states, we have used a variational method to find the ground state of a superconductor, and have gone on to obtain the following result for the energy (relative to the Fermi surface) of an added electron (quasiparticle) of wave vector $\boldsymbol{k}$:

$$\varepsilon_k = \sqrt{\xi_k^2 + \Delta^2}. \tag{16.118}$$

To create an excitation from the ground state of the superconductor without the addition of an odd electron from outside the system we must break a Cooper pair, that is, knock the two electrons of the pair into states in which they have the same (equal and opposite) momenta that they had previously in the pair but such that they are now no longer in a bound state. From (16.118), the energy required to create such an unbound pair is clearly a minimum when the momenta of the two electrons in the pair lie on the Fermi surface (so that $\xi_k = 0$), and is equal to $2[\varepsilon_k]_{\min} = 2\Delta$. For obvious reasons, this quantity is known as the $T = 0$ energy gap ε_{gap}:

$$\varepsilon_{\text{gap}} = 2[\varepsilon_k]_{\min} = 2\Delta, \tag{16.119}$$

where Δ is given by (16.61). At temperatures $T \neq 0$, the quantity Δ in (16.118) and (16.119) is replaced by a temperature-dependent quantity $\Delta(T)$, and so the energy gap becomes temperature-dependent:

$$\varepsilon_{\text{gap}} = 2\Delta(T). \tag{16.120}$$

Clearly, the energy gap vanishes at the temperature T_c [the superconducting transition temperature – see (16.116)] at which $\Delta(T)$ vanishes.

An important point to note is that no power series in V can reproduce equation (16.61) for Δ, and hence for the excitation spectrum. In other words, perturbation theory could not have given us this solution. The Bardeen–Cooper–Schrieffer theory is the outstanding example of the solution of a many-body problem in which perturbation theory cannot be used.

Module IV

Classical Field Theory and Relativity

17

The Classical Theory of Fields

17.1 Mathematical preliminaries

We define a **field** as a mathematical quantity that varies in a certain space (for example, in d-dimensional Euclidean space R^d, in four-dimensional Minkowski space–time M, or in some more abstract space).

Fields are classified by their **tensor character**, that is, by how they transform under transformations of the coordinates in the space on which they are defined.

Tensor fields of "rank 0" (see later) are called **scalar fields**, those of "rank 1" are called **vector fields**, and tensor fields of higher rank are also important in physics.

17.1.1 Behavior of fields under coordinate transformations

For simplicity we first consider fields defined on real two-dimensional Euclidean space R^2.

For example, suppose we approximate the Earth's mean surface as flat (a good approximation in a small enough region) and represent points in the corresponding two-dimensional Euclidean space by specifying the values of two coordinates.

We choose our origin and the orientation of the axes arbitrarily (see figure 17.1). The coordinates of the centre of Exeter in these axes are (x_1, x_2).

Consider, for example, the **temperature** at the Earth's surface at a given time. This will vary with position, (that is, it is a field), and

$$T(\text{Exeter}) = T(x_1, x_2). \tag{17.1}$$

Thus, to get the temperature at Exeter we take the function T of two arguments and set the arguments equal to Exeter's coordinates (x_1, x_2) along the axes (X_1, X_2).

We could instead have chosen another, rotated set of coordinate axes (X_1', X_2'). The same temperature field will have a different functional dependence on the new coordinates (denote the new function by T'). In the new system the coordinates of Exeter are (x_1', x_2'), and so

$$T(\text{Exeter}) = T'(x_1', x_2'). \tag{17.2}$$

A Course in Theoretical Physics, First Edition. P. J. Shepherd.

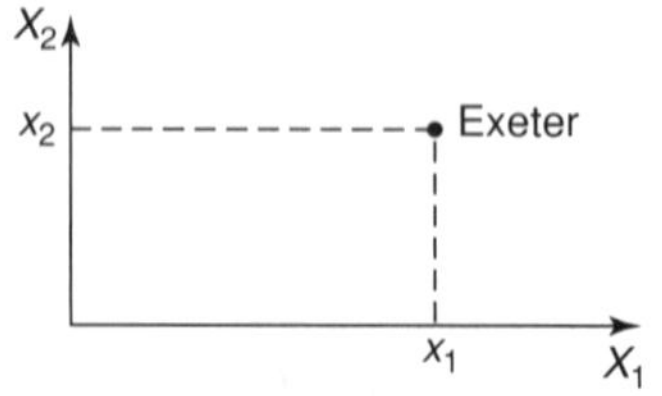

Figure 17.1 *Position of a city in arbitrary two-dimensional Cartesian coordinates (X_1, X_2).*

Comparing (17.1) and (17.2)

$$T'(x_1', x_2') = T(x_1, x_2) \tag{17.3}$$

Fields that satisfy this relation under changes of the coordinate system describing the space are called **scalar fields**. Scalar fields are described by **one** number at each point in space and that number [the result of evaluating **either** side of (17.3)] is **independent of the coordinate system chosen to describe the space**.

Consider now the **wind velocity** over the same part of the Earth's surface at a given time. This is a quantity specified by **two** numbers at each point.

For example, the wind velocity at Exeter in the coordinate system with axes (X_1, X_2) is

$$\boldsymbol{v}(\text{Exeter}) = \boldsymbol{v}(x_1, x_2) = \hat{\boldsymbol{e}}_1 v_1(x_1, x_2) + \hat{\boldsymbol{e}}_2 v_2(x_1, x_2), \tag{17.4}$$

where $\hat{\boldsymbol{e}}_1$ and $\hat{\boldsymbol{e}}_2$ are the unit vectors along the axes X_1 and X_2, and $v_1(x_1, x_2)$ and $v_2(x_1, x_2)$ are the **components** of $\boldsymbol{v}$ along these axes, evaluated at the position (x_1, x_2) of Exeter.

Expressed in the coordinate system (X_1', X_2'), the wind velocity will have a different functional dependence on the coordinates (denote the new function by $\boldsymbol{v}'$):

$$\boldsymbol{v}(\text{Exeter}) = \boldsymbol{v}'(x_1', x_2') = \hat{\boldsymbol{e}}_1' v_1'(x_1', x_2') + \hat{\boldsymbol{e}}_2' v_2'(x_1', x_2'). \tag{17.5}$$

Comparing (17.4) and (17.5), we have

$$\boldsymbol{v}(x_1, x_2) = \boldsymbol{v}'(x_1', x_2'), \tag{17.6}$$

that is,

$$\hat{\boldsymbol{e}}_1 v_1(x_1, x_2) + \hat{\boldsymbol{e}}_2 v_2(x_1, x_2) = \hat{\boldsymbol{e}}_1' v_1'(x_1', x_2') + \hat{\boldsymbol{e}}_2' v_2'(x_1', x_2'). \tag{17.7}$$

The first of these tells us that the wind velocity is unaffected by our choice of coordinate system (as it must be!). But, as the second relation shows, it is **not** true that $v_1' = v_1$ and $v_2' = v_2$, since the components along the axes must change if we change the direction of the axes. To find how they change, we express the unit vectors along (X_1', X_2') in terms of those along (X_1, X_2). Let there be an angle θ between the axes (see figure 17.2).

Then,

$$|\overrightarrow{\text{OA}}| = |\hat{\boldsymbol{e}}_1| = 1, |\overrightarrow{\text{OB}}| = |\overrightarrow{\text{OA}}| \cos\theta = \cos\theta,$$
$$|\overrightarrow{\text{BA}}| = |\overrightarrow{\text{OA}}| \sin\theta = \sin\theta, |\overrightarrow{\text{OC}}| = |\hat{\boldsymbol{e}}_2| = 1$$
$$|\overrightarrow{\text{OD}}| = |\overrightarrow{\text{OC}}| \cos\theta = \cos\theta, |\overrightarrow{\text{DC}}| = |\overrightarrow{\text{OC}}| \sin\theta = \sin\theta$$

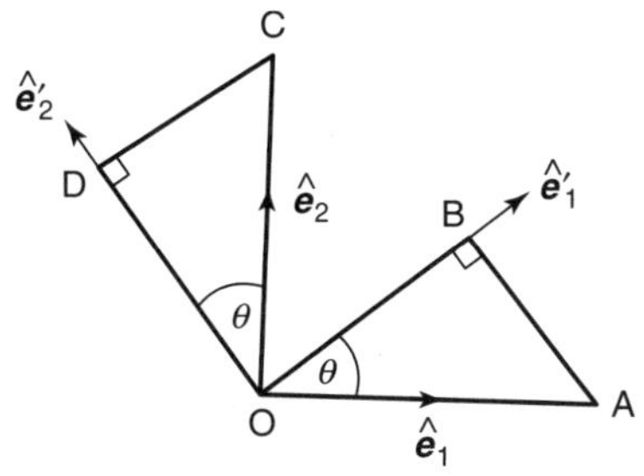

Figure 17.2 *Unit vectors in two two-dimensional coordinate frames related by rotation through angle θ.*

Therefore,

$$\hat{\boldsymbol{e}}_1 = \overrightarrow{\mathrm{OA}} = \overrightarrow{\mathrm{OB}} + \overrightarrow{\mathrm{BA}} = \hat{\boldsymbol{e}}'_1 \cos\theta - \hat{\boldsymbol{e}}'_2 \sin\theta,$$
$$\hat{\boldsymbol{e}}_2 = \overrightarrow{\mathrm{OC}} = \overrightarrow{\mathrm{DC}} + \overrightarrow{\mathrm{OD}} = \hat{\boldsymbol{e}}'_1 \sin\theta + \hat{\boldsymbol{e}}'_2 \cos\theta.$$

Put these into (17.7) and then equate the coefficients first of $\hat{\boldsymbol{e}}'_1$ and then of $\hat{\boldsymbol{e}}'_2$:

$$v'_1(x'_1, x'_2) = v_1(x_1, x_2)\cos\theta + v_2(x_1, x_2)\sin\theta,$$
$$v'_2(x'_1, x'_2) = -v_1(x_1, x_2)\sin\theta + v_2(x_1, x_2)\cos\theta.$$

Of course, if instead we consider the **position vector** of a point P on a plane in the above two coordinate systems (figure 17.3), we have

$$\boldsymbol{r}(\mathrm{P}) = \boldsymbol{r} = \hat{\boldsymbol{e}}_1 x_1 + \hat{\boldsymbol{e}}_2 x_2$$

and

$$\boldsymbol{r}(\mathrm{P}) = \boldsymbol{r}' = \hat{\boldsymbol{e}}'_1 x'_1 + \hat{\boldsymbol{e}}'_2 x'_2.$$

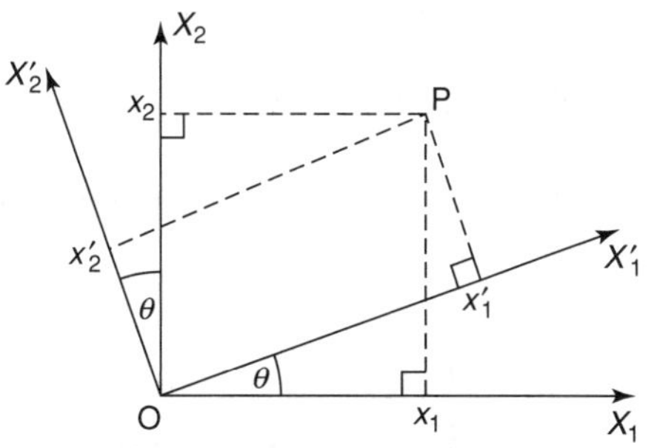

Figure 17.3 *Coordinate rotation in two dimensions.*

Again, we have

$$\boldsymbol{r} = \boldsymbol{r}'$$

with

$$\begin{aligned}\hat{\boldsymbol{e}}_1 &= \hat{\boldsymbol{e}}'_1 \cos\theta - \hat{\boldsymbol{e}}'_2 \sin\theta \\ \hat{\boldsymbol{e}}_2 &= \hat{\boldsymbol{e}}'_1 \sin\theta + \hat{\boldsymbol{e}}'_2 \cos\theta\end{aligned}$$

and

$$\begin{aligned}x'_1 &= x_1 \cos\theta + x_2 \sin\theta \\ x'_2 &= -x_1 \sin\theta + x_2 \cos\theta.\end{aligned} \tag{17.8}$$

The fact that the components v'_i are related in the **same** way to the components v_i as the components x'_i are to the components x_i is what **defines** the quantities v_i as components of a **vector** (in this case, a **vector field**).

We rewrite (17.8) as

$$\begin{aligned}x'_1 &= x_1 \cos\theta + x_2 \cos\left(\frac{\pi}{2} - \theta\right), \\ x'_2 &= x_1 \cos\left(\frac{\pi}{2} + \theta\right) + x_2 \cos\theta.\end{aligned}$$

Denoting the angle between the X'_i axis and the X_j axis by (X'_i, X_j), so that

$$\begin{aligned}x'_1 &= x_1 \cos(X'_1, X_1) + x_2 \cos(X'_1, X_2), \\ x'_2 &= x_1 \cos(X'_2, X_1) + x_2 \cos(X'_2, X_2),\end{aligned}$$

and using the abbreviation

$$\lambda_{ij} \equiv \cos(X'_i, X_j),$$

we get

$$\begin{aligned}x'_1 &= \lambda_{11}x_1 + \lambda_{12}x_2, \\ x'_2 &= \lambda_{21}x_1 + \lambda_{22}x_2.\end{aligned}$$

For a general, three-dimensional coordinate rotation we have

$$x'_1 = x_1 \cos(X'_1, X_1) + x_2 \cos(X'_1, X_2) + x_3 \cos(X'_1, X_3),$$

and similarly for x'_2 and x'_3. Thus,

$$\begin{aligned}x'_1 &= \lambda_{11}x_1 + \lambda_{12}x_2 + \lambda_{13}x_3, \\ x'_2 &= \lambda_{21}x_1 + \lambda_{22}x_2 + \lambda_{23}x_3, \\ x'_3 &= \lambda_{31}x_1 + \lambda_{32}x_2 + \lambda_{33}x_3.\end{aligned}$$

In summation notation,

$$x_i' = \sum_{j=1}^{3} \lambda_{ij} x_j \quad (i = 1, 2, 3). \tag{17.9}$$

The inverse transformation is

$$\begin{aligned} x_1 &= x_1' \cos(X_1, X_1') + x_2' \cos(X_1, X_2') + x_3' \cos(X_1, X_3') \\ &= x_1' \cos(X_1', X_1) + x_2' \cos(X_2', X_1) + x_3' \cos(X_3', X_1) \quad \text{(because } \cos\theta \text{ is even)} \\ &= \lambda_{11} x_1' + \lambda_{21} x_2' + \lambda_{31} x_3', \end{aligned}$$

and similarly for x_2 and x_3. Thus,

$$x_j = \sum_{l=1}^{3} \lambda_{lj} x_l' \quad (j = 1, 2, 3). \tag{17.10}$$

The quantities λ_{ij} form the matrix

$$\lambda = \begin{pmatrix} \lambda_{11} & \lambda_{12} & \lambda_{13} \\ \lambda_{21} & \lambda_{22} & \lambda_{23} \\ \lambda_{31} & \lambda_{32} & \lambda_{33} \end{pmatrix},$$

called the **rotation matrix**.

17.1.2 Properties of the rotation matrix

Take the partial derivative of both (17.9) and (17.10) with respect to x_k':

$$\frac{\partial x_i'}{\partial x_k'} = \sum_{j=1}^{3} \lambda_{ij} \frac{\partial x_j}{\partial x_k'}, \quad \frac{\partial x_j}{\partial x_k'} = \sum_{l=1}^{3} \lambda_{lj} \frac{\partial x_l'}{\partial x_k'}.$$

Put the second of these into the first:

$$\frac{\partial x_i'}{\partial x_k'} = \sum_{j=1}^{3} \sum_{l=1}^{3} \lambda_{ij} \lambda_{lj} \frac{\partial x_l'}{\partial x_k'}. \tag{17.11}$$

But, because the x_i' are independent variables,

$$\frac{\partial x_i'}{\partial x_k'} = \begin{cases} 1 & \text{if} \quad i = k \\ 0 & \text{if} \quad i \neq k \end{cases}$$

Use the symbol δ_{ik} for this (the **Kronecker delta**), so that (17.11) takes the form

$$\delta_{ik} = \sum_{j,l} \lambda_{ij}\lambda_{lj}\delta_{lk}.$$

Therefore,

$$\sum_{j} \lambda_{ij}\lambda_{kj} = \delta_{ik}. \tag{17.12}$$

For example, if we put $i = k = 1$ this reads

$$\lambda_{11}^2 + \lambda_{12}^2 + \lambda_{13}^2 = 1,$$

that is,

$$\cos^2(X_1', X_1) + \cos^2(X_1', X_2) + \cos^2(X_1', X_3) = 1,$$

so that the direction cosines of an arbitrary point P satisfy

$$\cos^2\alpha + \cos^2\beta + \cos^2\gamma = 1.$$

The "orthogonality relation" (17.12) could be regarded as stating that the **rows** of the rotation matrix $\boldsymbol{\lambda}$ are orthogonal to each other, and also normalized to unity (i.e., they are "orthonormal"). Writing the transpose of the matrix $\boldsymbol{\lambda}$ as $\boldsymbol{\lambda}^{\mathrm{T}}$ we can rewrite the "orthogonality relation" (17.12) as

$$\sum_{j} \lambda_{ij}(\boldsymbol{\lambda}^{\mathrm{T}})_{jk} = \delta_{ik},$$

which, by the rules of matrix multiplication, states that the *ik* element of the matrix $\boldsymbol{\lambda}\boldsymbol{\lambda}^{\mathrm{T}}$ is δ_{ik}, that is,

$$\boldsymbol{\lambda}\boldsymbol{\lambda}^{\mathrm{T}} = 1.$$

Thus,

$$\boldsymbol{\lambda}^{-1} = \boldsymbol{\lambda}^{\mathrm{T}}, \quad \text{and so} \quad \boldsymbol{\lambda}^{\mathrm{T}}\boldsymbol{\lambda} = 1.$$

(A matrix $\boldsymbol{\lambda}$ satisfying $\boldsymbol{\lambda}\boldsymbol{\lambda}^{\mathrm{T}} = \boldsymbol{\lambda}^{\mathrm{T}}\boldsymbol{\lambda} = 1$ is called an orthogonal matrix). We thus have

$$\sum_{j} (\boldsymbol{\lambda}^{\mathrm{T}})_{ij}\lambda_{jk} = \delta_{ik},$$

that is,

$$\sum_{j} \lambda_{ji}\lambda_{jk} = \delta_{ik}.$$

Thus, the **columns** of the rotation matrix $\boldsymbol{\lambda}$ are **also** orthonormal.

17.1.3 Proof that a "dot product" is a scalar

Define the dot product C of two vectors $\boldsymbol{A}$ and $\boldsymbol{B}$ by

$$C = \boldsymbol{A} \cdot \boldsymbol{B} \equiv \sum_i A_i B_i.$$

Since $\boldsymbol{A}$ and $\boldsymbol{B}$ are vectors, under a coordinate rotation described by the matrix λ their components in the rotated and unrotated frames are related by

$$A'_i = \sum_j \lambda_{ij} A_j \quad \text{and} \quad B'_i = \sum_j \lambda_{ij} B_j.$$

Therefore, the dot product $C' \equiv \mathbf{A}' \cdot \mathbf{B}'$ in the rotated frame is given by

$$\begin{aligned}
C' \equiv \boldsymbol{A}' \cdot \boldsymbol{B}' &= \sum_i A'_i B'_i \\
&= \sum_i \left(\sum_j \lambda_{ij} A_j \right) \left(\sum_k \lambda_{ik} B_k \right) = \sum_{j,k} \left(\sum_i \lambda_{ij} \lambda_{ik} \right) A_j B_k \\
&= \sum_{j,k} \delta_{jk} A_j B_k \quad \text{(since the columns of } \lambda \text{ are orthogonal)} \\
&= \sum_j A_j B_j = \boldsymbol{A} \cdot \boldsymbol{B} = C,
\end{aligned}$$

that is, $C' = C$, which is the defining relation for a scalar.

Successive orthogonal transformations (that is, transformations implemented by orthogonal matrices) produce an orthogonal transformation.

Proof: Let

$$x'_i = \sum_j \lambda_{ij} x_j \quad \text{and} \quad x''_k = \sum_i \mu_{ki} x'_i.$$

Then

$$x''_k = \sum_j \left(\sum_i \mu_{ki} \lambda_{ij} \right) x_j = \sum_j (\mu\lambda)_{kj} x_j.$$

We must show that

$$(\mu\lambda)^{\mathrm{T}} = (\mu\lambda)^{-1}.$$

For this we use the following general property of matrices:

$$(\mathrm{AB})^{\mathrm{T}} = \mathrm{B}^{\mathrm{T}} \mathrm{A}^{\mathrm{T}}$$

[proved as follows:

$$[(\mathrm{AB})^{\mathrm{T}}]_{ij} = (\mathrm{AB})_{ji} = \sum_k A_{jk} B_{ki} = \sum_k B_{ki} A_{jk}$$
$$= \sum_k (\mathrm{B}^{\mathrm{T}})_{ik} (\mathrm{A}^{\mathrm{T}})_{kj} = (\mathrm{B}^{\mathrm{T}}\mathrm{A}^{\mathrm{T}})_{ij},$$

so that $(\mathrm{AB})^{\mathrm{T}} = \mathrm{B}^{\mathrm{T}}\mathrm{A}^{\mathrm{T}}$.]

Therefore,

$$[(\mu\lambda)^{\mathrm{T}}]\mu\lambda = \lambda^{\mathrm{T}}\mu^{\mathrm{T}}\mu\lambda = \lambda^{\mathrm{T}}\lambda = 1,$$

and so

$$(\mu\lambda)^{\mathrm{T}} = (\mu\lambda)^{-1}, \text{ i.e., } \mu\lambda \text{ is an orthogonal matrix.}$$

Also, the determinants $\det\lambda$ of orthogonal transformation matrices λ are equal to ± 1.

Proof: We use

$$\det(\mathrm{AB}) = (\det \mathrm{A})(\det \mathrm{B}).$$

Therefore,

$$\det(\lambda^{\mathrm{T}}\lambda)\ [= \det 1 = 1] = (\det \lambda^{\mathrm{T}})(\det \lambda) = (\det \lambda)^2$$

and so

$$\det \lambda = \pm 1.$$

In fact, all matrices λ representing rotations starting from the original set of axes have $\det\lambda = +1$, while those for transformations involving an odd number of reflections or inversions have $\det\lambda = -1$.

For example, under reflection of the coordinate axes in the xy plane the coordinates of a fixed point P transform as follows:

$$\begin{pmatrix} x \\ y \\ z \end{pmatrix} \to \begin{pmatrix} x' \\ y' \\ z' \end{pmatrix} = \begin{pmatrix} x \\ y \\ -z \end{pmatrix} = \lambda \begin{pmatrix} x \\ y \\ z \end{pmatrix},$$

in which the transformation matrix λ has the form

$$\lambda = \begin{pmatrix} 1 & 0 & 0 \\ 0 & 1 & 0 \\ 0 & 0 & -1 \end{pmatrix},$$

which has $\det\lambda = -1$.

Under inversion of the coordinate axes,

$$\begin{pmatrix} x \\ y \\ z \end{pmatrix} \rightarrow \begin{pmatrix} x' \\ y' \\ z' \end{pmatrix} = \begin{pmatrix} -x \\ -y \\ -z \end{pmatrix} = \lambda \begin{pmatrix} x \\ y \\ z \end{pmatrix}$$

with

$$\lambda = \begin{pmatrix} -1 & 0 & 0 \\ 0 & -1 & 0 \\ 0 & 0 & -1 \end{pmatrix}$$

which also has $\det\lambda = -1$.

Orthogonal transformations whose matrices λ have $\det\lambda = +1$ are called **proper rotations**. Those with $\det\lambda = -1$ are called **improper rotations**.

17.1.4 A lemma on determinants

The determinant detA of a 3×3 matrix A can be evaluated in several ways. For example, expanding in the minors of the first column we have

$$\begin{aligned} \det A &= \begin{vmatrix} A_{11} & A_{12} & A_{13} \\ A_{21} & A_{22} & A_{23} \\ A_{31} & A_{32} & A_{33} \end{vmatrix} \\ &= A_{11}(A_{22}A_{33} - A_{32}A_{23}) + (-A_{21})(A_{12}A_{33} - A_{32}A_{13}) + A_{31}(A_{12}A_{23} - A_{22}A_{13}), \end{aligned}$$

that is,

$$\det A = \sum_{l,m,n} \varepsilon_{lmn} A_{l1} A_{m2} A_{n3},$$

where ε_{lmn} is the **Levi-Civita symbol**, defined by

$$\varepsilon_{lmn} = \begin{cases} 0 \text{ if any index equals any other index} \\ +1 \text{ if the set } \{l, m, n\} \text{ is } \{1, 2, 3\} \text{ or any even permutation of } \{1, 2, 3\} \\ -1 \text{ if the set } \{l, m, n\} \text{ is any odd permutation of } \{1, 2, 3\} \end{cases}$$

For example,

$$\begin{aligned} &\varepsilon_{112} = 0, \text{ etc.} \\ &\varepsilon_{123} = \varepsilon_{231} = \varepsilon_{312} = 1 \\ &\varepsilon_{132} = \varepsilon_{213} = \varepsilon_{321} = -1. \end{aligned}$$

If instead we expand detA in terms of the minors of the second column we obtain

$$\text{detA} = -\sum_{l,m,n} \varepsilon_{lmn} A_{l2} A_{m1} A_{n3}.$$

Using the third column, we get

$$\text{detA} = \sum_{l,m,n} \varepsilon_{lmn} A_{l3} A_{m1} A_{n2}.$$

The general expression incorporating all three of these is

$$\varepsilon_{ijk}\text{detA} = \sum_{l,m,n} \varepsilon_{lmn} A_{li} A_{mj} A_{nk}. \tag{17.13}$$

We now use (17.13) to obtain an important property of orthogonal matrices. Let A be an orthogonal-transformation matrix λ. Then (17.13) becomes

$$(\det \lambda)\varepsilon_{ijk} = \sum_{l,m,n} \varepsilon_{lmn} \lambda_{li} \lambda_{mj} \lambda_{nk}.$$

Multiply this by λ_{pi} and sum over i:

$$\begin{aligned}(\det \lambda)\sum_i \varepsilon_{ijk}\lambda_{pi} &= \sum_{l,m,n} \varepsilon_{lmn} \lambda_{mj} \lambda_{nk} \left(\sum_i \lambda_{li}\lambda_{pi}\right) \\ &= \sum_{l,m,n} \varepsilon_{lmn} \lambda_{mj} \lambda_{nk} \delta_{lp} = \sum_{m,n} \varepsilon_{pmn} \lambda_{mj} \lambda_{nk},\end{aligned}$$

that is,

$$(\det \lambda)\sum_i \varepsilon_{ijk}\lambda_{li} = \sum_{m,n} \varepsilon_{lmn} \lambda_{mj} \lambda_{nk}. \tag{17.14}$$

17.1.5 Proof that the "cross product" of two vectors is a "pseudovector"

The cross product $\boldsymbol{C} = \boldsymbol{A} \times \boldsymbol{B}$ of two vectors $\boldsymbol{A}$ and $\boldsymbol{B}$ is **defined** as a three-component quantity, with components

$$C_1 = A_2B_3 - A_3B_2, \quad C_2 = A_3B_1 - A_1B_3, \quad C_3 = A_1B_2 - A_2B_1,$$

that is,

$$C_i = \sum_{j,k} \varepsilon_{ijk} A_j B_k.$$

The corresponding quantity $\boldsymbol{C}' = \boldsymbol{A}' \times \boldsymbol{B}'$ in the rotated frame has, by definition of the cross product, components

$$C_l' = \sum_{m,n} \varepsilon_{lmn} A_m' B_n'.$$

Since $\boldsymbol{A}$ and $\boldsymbol{B}$ are vectors, this becomes

$$C_l' = \sum_{m,n} \varepsilon_{lmn} \left(\sum_j \lambda_{mj} A_j \right) \left(\sum_k \lambda_{nk} B_k \right) = \sum_{j,k} \left(\sum_{m,n} \varepsilon_{lmn} \lambda_{mj} \lambda_{nk} \right) A_j B_k.$$

Using the lemma (17.14), we have

$$C_l' = (\det \lambda) \sum_i \lambda_{li} \left(\sum_{j,k} \varepsilon_{ijk} A_j B_k \right) = (\det \lambda) \sum_i \lambda_{li} C_i,$$

that is,

$$C_l' = (\det \lambda) \sum_i \lambda_{li} C_i, \tag{17.15}$$

which is the defining relation for a **pseudovector (axial vector)**. Thus, a pseudovector transforms in the same way as a vector under proper rotations (for which $\det \lambda = +1$), but as

$$C_l' = -\sum_i \lambda_{li} C_i$$

under improper rotations (for which $\det\lambda = -1$).

17.1.6 Useful index relations

(i)

$$\sum_{i,j} \varepsilon_{ijk} \delta_{ij} = 0.$$

Proof:

$$\sum_{i,j} \varepsilon_{ijk} \delta_{ij} = \sum_i \varepsilon_{iik} = \sum_i 0 = 0.$$

(ii)

$$\sum_{j,k} \varepsilon_{ijk} \varepsilon_{ljk} = 2\delta_{il}.$$

Proof: We must have $i \neq j$ or k, and $l \neq j$ or k. Since $j \neq k$, and the indices take only three values, we must have $i = l$. For example, if $i = l = 1$ the only nonzero terms are

$$\varepsilon_{123}\varepsilon_{123} + \varepsilon_{132}\varepsilon_{132} = 1 \times 1 + (-1) \times (-1) = 2.$$

Therefore,

$$\sum_{j,k} \varepsilon_{ijk} \varepsilon_{ljk} = 2\delta_{il}.$$

(iii)
$$\sum_{i,j,k} \varepsilon_{ijk}\varepsilon_{ijk} = 6.$$

Proof: From (ii), we have

$$\sum_{i,j,k} \varepsilon_{ijk}\varepsilon_{ijk} = 2\sum_{i} \delta_{ii} = 2\sum_{i=1}^{3} 1 = 6.$$

Alternatively, there are six nonzero ε_{ijk} and each term in the sum is $1 \times 1 = 1$ or $(-1) \times (-1) = 1$.

(iv)
$$\sum_{k} \varepsilon_{ijk}\varepsilon_{lmk} = \delta_{il}\delta_{jm} - \delta_{im}\delta_{jl}.$$

Proof: For a given pair $\{ij\}(i \neq j)$ only one term in the sum over k can be nonzero – that with k equal to neither i nor j. Then the left-hand side is nonzero only when neither l nor m is equal to this value of k. But if this is so, the pair $\{lm\}$ is the same as the pair $\{ij\}$. So there are two possibilities:

(a) $l = i$ and $m = j$. Then the one nonzero term $\varepsilon_{ijk}\varepsilon_{lmk}$ in the sum is

$$\varepsilon_{ijk}\varepsilon_{ijk} = 1.$$

(b) $m = i$ and $l = j$. Then the one nonzero term $\varepsilon_{ijk}\varepsilon_{lmk}$ in the sum is

$$\varepsilon_{ijk}\varepsilon_{jik} = -\varepsilon_{ijk}\varepsilon_{ijk} = -1.$$

These results are summarized by the index relation

$$\sum_{k} \varepsilon_{ijk}\varepsilon_{lmk} = \delta_{il}\delta_{jm} - \delta_{im}\delta_{jl}.$$

17.1.7 Use of index relations to prove vector identities

As an illustration of the use of the index relations we now prove that

$$\boldsymbol{A} \times (\boldsymbol{B} \times \boldsymbol{C}) = \boldsymbol{B}(\boldsymbol{A} \cdot \boldsymbol{C}) - \boldsymbol{C}(\boldsymbol{A} \cdot \boldsymbol{B})$$

[the "bac(k) cab rule"].

The l-component is given by

$$\begin{aligned}
[\boldsymbol{A} \times (\boldsymbol{B} \times \boldsymbol{C})]_l &= \sum_{m,n} \varepsilon_{lmn} A_m (\boldsymbol{B} \times \boldsymbol{C})_n = \sum_{m,n}\sum_{r,s} \varepsilon_{lmn}\varepsilon_{nrs} A_m B_r C_s \\
&= \sum_{m,r,s}\left(\sum_{n} \varepsilon_{lmn}\varepsilon_{rsn}\right) A_m B_r C_s = \sum_{m,r,s} (\delta_{lr}\delta_{ms} - \delta_{ls}\delta_{mr}) A_m B_r C_s \\
&= B_l \sum_{m} A_m C_m - C_l \sum_{m} A_m B_m = B_l(\boldsymbol{A} \cdot \boldsymbol{C}) - C_l(\boldsymbol{A} \cdot \boldsymbol{B}).
\end{aligned}$$

Therefore,

$$\boldsymbol{A} \times (\boldsymbol{B} \times \boldsymbol{C}) = \boldsymbol{B}(\boldsymbol{A} \cdot \boldsymbol{C}) - \boldsymbol{C}(\boldsymbol{A} \cdot \boldsymbol{B}).$$

17.1.8 General definition of tensors of arbitrary rank

Under the coordinate rotation

$$x_i' = \sum_j \lambda_{ij} x_j \tag{17.16}$$

a **scalar** field specified by $\varphi(x) \equiv \varphi(x_1, x_2, x_3)$ in the frame (X_1, X_2, X_3) and by $\varphi'(x') \equiv \varphi'(x_1', x_2', x_3')$ in the rotated frame (X_1', X_2', X_3') satisfies

$$\varphi'(x') = \varphi(x).$$

A **vector** field satisfies

$$\boldsymbol{v}'(x') = \boldsymbol{v}(x)$$

with

$$v_i' = \sum_j \lambda_{ij} v_j,$$

which is the same transformation law as for the coordinates themselves [see (17.16)]. We use this to **define** a vector (in three-dimensional space) as a set of three quantities A_1, A_2, A_3 that transform as

$$A_i' = \sum_j \lambda_{ij} A_j$$

under the coordinate rotation (17.16).

Similarly, a **rank-2 tensor** in three-dimensional space is defined as a set of nine quantities $A_{ij}(x_1, x_2, x_3)$ that transform like the $3^2 = 9$ products $x_i x_j$ under the coordinate transformation (17.16). Multiplying (17.16) by itself, we have

$$x_i' x_j' = \sum_k \sum_l \lambda_{ik} \lambda_{jl} x_k x_l,$$

and so the components of a rank-2 tensor satisfy

$$A_{ij}' = \sum_k \sum_l \lambda_{ik} \lambda_{jl} A_{kl}.$$

Similarly, a **rank-3 tensor** is a set of 27 quantities B_{ijk} that transform like the $3^3 = 27$ products $x_i x_j x_k$ under the coordinate transformation (17.16):

$$B'_{ijk} = \sum_{l,m,n} \lambda_{il}\lambda_{jm}\lambda_{kn} B_{lmn}.$$

In general, a rank-n tensor in three-dimensional space is a set of $3n$ quantities transforming under (17.16) like the 3^n products $x_i x_j x_k \ldots x_l$ (with n factors).

Clearly, a vector is a tensor of rank 1 and a scalar is a tensor of rank 0.

17.2 Introduction to Einsteinian relativity

We can divide physics into quantum physics and classical physics.

Classical physics comprises **pre-Einsteinian** (for example, Newtonian) physics and **Einsteinian** physics (based on Einstein's special and general relativity).

In both pre-Einsteinian and Einsteinian classical physics a **principle of relativity** is valid, namely, that **the laws of Nature are identical in all inertial frames of reference** (i.e., frames in which a **freely** moving body proceeds with constant velocity).

In **pre-Einsteinian mechanics** it is assumed that:

(i) Interactions are instantaneous; for example, the potential energy of the interaction of two particles, at positions $\boldsymbol{r}_1$ and $\boldsymbol{r}_2$, can be expressed as $V(\boldsymbol{r}_2 - \boldsymbol{r}_1)$.

(ii) Hence, with this "action at a distance", fields need be assigned no physical reality (although they may be introduced for simplicity).

(iii) Time is absolute; for example, if two events occur simultaneously for one observer they do so for all others.

The principle of relativity plus assumption (i) constitute **Galileo's principle of relativity**.

In **Einsteinian mechanics**:

(i) There is a finite maximum velocity of propagation of interactions; that is, the effect of one body on another (its "signal" to the other) requires a finite time. Since this velocity must enter the physical laws, which must have the same form in all inertial frames, **it must have the same value in all frames**. As we shall see, it is the **velocity of light** c.

(ii) Since instantaneous "action at a distance" no longer occurs, interactions occur as follows:
   (a) the first particle creates a field;
   (b) the field interacts with the second particle.
   A change in position of one particle influences other particles only after a lapse of a certain time interval. Thus, **the field itself acquires physical reality** in Einsteinian mechanics.
   Hence the classical theory of fields is the relativistic (and nonquantum) theory of fields.

(iii) Time is no longer absolute. (If it were, velocities would be additive, and c would be different in different inertial frames – it is, of course, the same, as shown by the famous Michelson–Morley experiment.)

17.2.1 Intervals

An **event** is specified by the place and time it occurs.

The **world-line** of a particle gives its spatial coordinates at all times, that is, it is a line in a four-dimensional "space".

Consider two events:

	Emission of a photon	*Reception of the same photon*
Frame K	$x_1\ y_1\ z_1\ t_1$	$x_2\ y_2\ z_2\ t_2$
Frame K′	$x'_1\ y'_1\ z'_1\ t'_1$	$x'_2\ y'_2\ z'_2\ t'_2$

In the frame K,

$$(x_2 - x_1)^2 + (y_2 - y_1)^2 + (z_2 - z_1)^2 - c^2(t_2 - t_1)^2 = 0.$$

In the frame K′,

$$(x'_2 - x'_1)^2 + (y'_2 - y'_1)^2 + (z'_2 - z'_1)^2 - c^2(t'_2 - t'_1)^2 = 0,$$

with (of course) the same light speed c.

Now let $x_1\ y_1\ z_1\ t_1$ and $x_2\ y_2\ z_2\ t_2$ be the coordinates of **any** two events in K. Then

$$s_{12} \equiv [c^2(t_2 - t_1)^2 - (x_2 - x_1)^2 - (y_2 - y_1)^2 - (z_2 - z_1)^2]^{1/2}$$

is called the **interval** between these events.

[From the invariance of the velocity of light, if an interval is zero in one frame (i.e., the same as the interval between the emission and absorption of a photon), it is zero in all frames, i.e., if $s_{12} = 0$, then $s'_{12} = 0$.]

Now assume that the two events are infinitesimally close, that is, the interval $\mathrm{d}s$ is given by

$$\mathrm{d}s^2 = c^2\mathrm{d}t^2 - \mathrm{d}x^2 - \mathrm{d}y^2 - \mathrm{d}z^2$$

[here, $\mathrm{d}x^2 \equiv (\mathrm{d}x)^2$, etc.].

As already shown, if $\mathrm{d}s = 0$, then $\mathrm{d}s' = 0$. Also, $\mathrm{d}s^2$ and $\mathrm{d}s'^2$ are infinitesimals of the same order, that is,

$$\mathrm{d}s^2 = a\mathrm{d}s'^2,$$

where a can depend only on the absolute value $V \equiv |\boldsymbol{V}|$ of the velocity $\boldsymbol{V}$ of frame K′ relative to frame K. (It cannot depend on the coordinates or the time, since space and time are homogeneous, nor can it depend on the direction of $\boldsymbol{V}$, since space is isotropic.)

Consider three frames: K, K_1 and K_2 (K_1 has velocity $\boldsymbol{V}_1$, and K_2 has velocity $\boldsymbol{V}_2$, relative to K.) Then

$$\mathrm{d}s^2 = a(V_1)\mathrm{d}s_1^2, \quad \mathrm{d}s^2 = a(V_2)\mathrm{d}s_2^2,$$

and

$$\mathrm{d}s_1^2 = a(V_{12})\mathrm{d}s_2^2,$$

where $\boldsymbol{V}_{12} = \boldsymbol{V}_2 - \boldsymbol{V}_1$. Therefore,

$$a(V_{12}) = \frac{a(V_2)}{a(V_1)}.$$

But $V_{12} = |\boldsymbol{V}_{12}|$ depends on the angle between $\boldsymbol{V}_2$ and $\boldsymbol{V}_1$, while the right-hand side of the above expression does not. Therefore, $a(V)$ must be a constant, equal to unity by the same formula. Therefore,

$$\mathrm{d}s^2 = \mathrm{d}s'^2.$$

Building up **finite** intervals, we have

$$s = s',$$

that is, intervals are invariant under transformations from one inertial frame to another.

17.2.2 Timelike and spacelike intervals

Put $t_2 - t_1 \equiv t_{12}$ and $(x_2 - x_1)^2 + (y_2 - y_1)^2 + (z_2 - z_1)^2 \equiv l_{12}^2$. Then the interval between two events in frame K is $s_{12}^2 = c^2t_{12}^2 - l_{12}^2$, and in frame K′ is $s_{12}'^2 = c^2t_{12}'^2 - l_{12}'^2$.

The invariance of intervals then gives

$$c^2t_{12}^2 - l_{12}^2 = c^2t_{12}'^2 - l_{12}'^2.$$

If a frame K′ can be found such that $l_{12}'^2 = 0$ (so that, the two events occur at the same **spatial** point in K′), then

$$s_{12}^2 = c^2t_{12}^2 - l_{12}^2 = c^2t_{12}'^2 > 0,$$

that is, s_{12} is **real**. Real intervals are said to be **timelike.** In K′, $t_{12}' = s_{12}/c$.

If for two events there exists a frame K′ such that they occur at the same time ($t_{12}' = 0$), then

$$s_{12}^2 = c^2t_{12}^2 - l_{12}^2 = -l_{12}'^2 < 0,$$

that is, s_{12} is **imaginary**. Imaginary intervals are said to be **spacelike**. The spatial distance between the two events in this frame K′ is

$$l_{12}' = \left[l_{12}^2 - c^2t_{12}^2\right]^{1/2} = is_{12}.$$

Because intervals are invariant, if they are timelike [spacelike] in one inertial frame, they are timelike [spacelike] in all inertial frames.

17.2.3 The light cone

For convenience, consider one space dimension (x) and one time dimension (t) (see figure 17.4).

The interval between an event 1 at O and an event 2 in aOc or dOb satisfies $s_{12}^2 = c^2t_{12}^2 - x_{12}^2 > 0$, that is, is **timelike**. The region aOc with $t > 0$ is a region of events which (since the interval with respect to O is timelike) cannot occur simultaneously with O in any frame. It is the **region of absolute future** relative to O. Similarly, dOb is **the region of absolute past** relative to O. Only events in the absolute future relative to O can be influenced by an event at O, in line with the principle of causality.

Similarly, events in the region consisting of cOb and aOd have intervals $s_{12}^2 = c^2t_{12}^2 - x_{12}^2 < 0$ relative to O, that is, these intervals are **spacelike**, and this region is the **region of absolute remoteness (separation)**

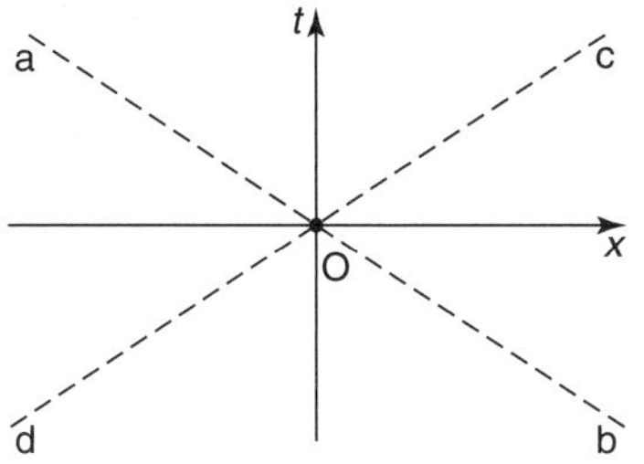

Figure 17.4 *The light cone (the lines* dc *and* ba *are photon world lines, with slopes 1/c and −1/c, respectively).*

from O. Any event in this region may occur after O in some frame, before O in others, and simultaneously with O in a certain inertial frame.

Using **three** space dimensions we get a four-dimensional "light cone" in coordinates x, y, z, t, with the t axis coinciding with the axis of the cone.

17.2.4 Variational principle for free motion

Let t_{a} and t_{b} be the times, in inertial frame K, of the beginning and end (world point a and world point b) of the motion of a particle moving relative to K with velocity $\boldsymbol{v}(t)$. Then

$$\int_{\mathrm{a}}^{\mathrm{b}} \mathrm{d}s = \int_{\mathrm{a}}^{\mathrm{b}} [c^2\mathrm{d}t^2 - \mathrm{d}x^2 - \mathrm{d}y^2 - \mathrm{d}z^2]^{1/2}$$

$$= c\int_{t_{\mathrm{a}}}^{t_{\mathrm{b}}} \mathrm{d}t\left(1 - \frac{v(t)^2}{c^2}\right)^{1/2} \quad \left[\text{since } v\,(t)^2 = \frac{\mathrm{d}x^2 + \mathrm{d}y^2 + \mathrm{d}z^2}{\mathrm{d}t^2}\right],$$

and this integral is clearly a **maximum** [equal to $c(t_{\mathrm{b}} - t_{\mathrm{a}})$] when $v(t) = 0$.

But zero velocity relative to inertial frame K corresponds to **uniform** velocity relative to any other inertial frame, which, in turn, corresponds to **free motion** (by the definition of the term "inertial frame").

We therefore have the following **variational principle**: the actual motion of a free particle is that which maximizes $\int_{\mathrm{a}}^{\mathrm{b}} \mathrm{d}s$, that is, maximizes the continuous sum of infinitesimal intervals (the integrated interval) between the start and end of the motion. (Here, of course, by the invariance of intervals, ds in the integral can be measured in **any** inertial frame.)

17.2.5 The Lorentz transformation

Let frame K′ (axes X', Y', Z') move with velocity V along the X axis relative to frame K (axes X, Y, Z), so that Y' remains parallel to Y (and Z' to Z), while X' and X coincide (apart from their origins).

In pre-Einsteinian mechanics, if x', y', z', t' are the coordinates of an event in K′, the coordinates of the same event in K are

$$x = x' + Vt, \quad y = y', \quad z = z', \quad t = t' \tag{17.17}$$

(a **Galilean transformation**).

In **Einsteinian** mechanics we find the corresponding transformation from the requirement that the interval between two events is the same in K′ as in K.

Let the first event occur at $x = x' = y = y' = z = z' = t = t' = 0$ (i.e., the origins of K and K′ coincide at $t = t' = 0$) and the second at an interval

$$s^2 = c^2t^2 - x^2 - y^2 - z^2 = s'^2 = c^2t'^2 - x'^2 - y'^2 - z'^2$$

relative to the first.

Just as a **coordinate rotation**, described by, for example,

$$\begin{aligned} x &= x' \cos\theta - y' \sin\theta \\ y &= x' \sin\theta + y' \cos\theta \\ z &= z' \\ t &= t' \end{aligned}$$

clearly satisfies $s^2 = s'^2$, and so a "rotation" in the plane of the variables ct and x, described by

$$\begin{aligned} x &= x' \cosh\psi + ct' \sinh\psi \\ ct &= x' \sinh\psi + ct' \cosh\psi \\ y &= y' \\ z &= z' \end{aligned} \tag{17.18}$$

also satisfies $s^2 = s'^2$.

The relation (17.18) is clearly the Einsteinian analogue of the Galilean transformation (17.17), that is, it corresponds to relative motion of K and K′ along X, and so ψ can depend only on the magnitude V of the relative velocity.

For example, consider the motion, in frame K, of the origin of frame K′. Then $x' = 0$ and (17.18) becomes

$$\begin{aligned} x &= ct' \sinh\psi, \\ ct &= ct' \cosh\psi. \end{aligned}$$

Therefore,

$$\frac{x}{ct} = \tanh\psi.$$

But $x/t = V$, and so

$$\tanh\psi = \frac{V}{c}.$$

Therefore,

$$\sinh\psi = \frac{V/c}{\sqrt{1 - V^2/c^2}} \quad \text{and} \quad \cosh\psi = \frac{1}{\sqrt{1 - V^2/c^2}},$$

and (17.18) becomes

$$x = \frac{x' + Vt'}{\sqrt{1 - V^2/c^2}}, \quad y = y', \quad z = z', \quad t = \frac{t' + Vx'/c^2}{\sqrt{1 - V^2/c^2}} \tag{17.19}$$

(the **Lorentz transformation**).

We can solve these for x', y', z', t' in terms of x, y, z, t. This is done most easily by interchanging the primed and unprimed coordinates and replacing V by $-V$ (the velocity of K relative to K′).

If $c \to \infty$, the Lorentz transformation (17.19) yields the Galilean transformation (17.17).

Lorentz transformations, like spatial rotations, are not commutative unless they correspond to rotations in the same "plane".

17.2.6 Length contraction and time dilation

Suppose now that a rod is at rest in frame K and lies parallel to the X axis of that frame. If the coordinates of its end points a and b are x_a and x_b, then the **proper length** of the rod (defined as the length of the rod in the frame in which it is at rest) is clearly $l_{ab} = |x_b - x_a|$. In the frame K′, relative to which our rod is moving with speed V parallel to its length and in the negative X direction, we find the length of the rod by first finding the coordinates x'_a and x'_b of its end points a and b at any chosen time $t'_a = t'_b \equiv t'_{a,b}$ (the end-point coordinates must, of course, be measured at the same time in frame K′, since the rod is in motion in this frame). Then the first of the Lorentz-transformation relations (17.19) gives:

$$x_b = \frac{x'_b + Vt'_{a,b}}{\sqrt{1 - V^2/c^2}} \quad \text{and} \quad x_a = \frac{x'_a + Vt'_{a,b}}{\sqrt{1 - V^2/c^2}}.$$

From these expressions we find that the length l'_{ab} of the rod in the frame K′ in which the rod is in motion is

$$l'_{ab} = |x'_b - x'_a| = |x_b - x_a|\sqrt{1 - V^2/c^2} = l_{ab}\sqrt{1 - V^2/c^2},$$

which is smaller than the proper length l_{ab} of the rod by the factor $\sqrt{1 - V^2/c^2}$. This is known as the **Lorentz–Fitzgerald length contraction**.

Now consider two ticks (a and b) of a clock at rest in frame K; let these occur at times t_a and t_b. Then the **proper time interval** t_{ab} between the ticks ("proper", because the clock is at rest in K, with the same space coordinates $x_a = x_b \equiv x_{a,b}$ for the events of the two ticks) is $t_{ab} = |t_b - t_a|$. As measured in frame K′ the time interval between the same two ticks is $t'_{ab} = |t'_b - t'_a|$. In K′, however, unlike in K, the clock will have moved between the two ticks. Let the space coordinates of the clock at the events of the two ticks be x'_a and x'_b. Then the fourth of the Lorentz-transformations relations (17.19) becomes, for the two tick events,

$$t_b = \frac{t'_b + Vx'_b/c^2}{\sqrt{1 - V^2/c^2}} \quad \text{and} \quad t_a = \frac{t'_a + Vx'_a/c^2}{\sqrt{1 - V^2/c^2}},$$

from which we find

$$(t_b - t_a)\sqrt{1 - V^2/c^2} = t'_b - t'_a + V(x'_b - x'_a)/c^2 = (t'_b - t'_a)\left(1 - V^2/c^2\right),$$

where we have used the fact that the change $x'_b - x'_a$ in the space coordinate of the clock in K′ between ticks a and b, divided by the time interval $t'_b - t'_a$ between these ticks in K′, can be identified with the speed $(-V)$ of the clock in K′. We therefore find

$$t'_b - t'_a = \frac{t_b - t_a}{\sqrt{1 - V^2/c^2}},$$

which shows that the time interval between two ticks of a clock is least when measured in the rest frame of the clock (this, of course, is the proper time interval between the ticks) and is longer, by the factor $(1 - V^2/c^2)^{-1/2}$, when measured in a frame moving with speed V relative to the clock. This phenomenon is called **time dilation**.

17.2.7 Transformation of velocities

Let K′ move with uniform velocity V relative to K along the X axis.

The x component of the velocity of a particle is $v_x = \mathrm{d}x/\mathrm{d}t$ in K and $v'_x = \mathrm{d}x'/\mathrm{d}t'$ in K′. We have, from (17.19),

$$\mathrm{d}x = \frac{\mathrm{d}x' + V\mathrm{d}t'}{\sqrt{1 - V^2/c^2}}, \quad \mathrm{d}y = \mathrm{d}y', \quad \mathrm{d}z = \mathrm{d}z', \quad \mathrm{d}t = \frac{\mathrm{d}t' + (V/c^2)\mathrm{d}x'}{\sqrt{1 - V^2/c^2}}.$$

Dividing the first three equations by the fourth, we have

$$\frac{\mathrm{d}x}{\mathrm{d}t} = \frac{\mathrm{d}x' + V\mathrm{d}t'}{\mathrm{d}t' + (V/c^2)\mathrm{d}x'}, \quad \frac{\mathrm{d}y}{\mathrm{d}t} = \frac{\mathrm{d}y'\sqrt{1 - V^2/c^2}}{\mathrm{d}t' + (V/c^2)\mathrm{d}x'}, \quad \frac{\mathrm{d}z}{\mathrm{d}t} = \frac{\mathrm{d}z'\sqrt{1 - V^2/c^2}}{\mathrm{d}t' + (V/c^2)\mathrm{d}x'}.$$

Now divide the numerators and denominators on the right-hand sides by $\mathrm{d}t'$:

$$v_x = \frac{v'_x + V}{1 + v'_x V/c^2}, \quad v_y = \frac{v'_y\sqrt{1 - V^2/c^2}}{1 + v'_x V/c^2}, \quad v_z = \frac{v'_z\sqrt{1 - V^2/c^2}}{1 + v'_x V/c^2}.$$

Note that as $c \to \infty$ these become the Galilean laws

$$v_x = v'_x + V, \quad v_y = v'_y, \quad v_z = v'_z.$$

17.2.8 Four-tensors

The coordinates ct, x, y, z of an event in some frame can be regarded as the components of a four-dimensional "radius vector" in a four-dimensional space. Denote the components by

$$x^0 = ct, \quad x^1 = x, \quad x^2 = y, \quad x^3 = z.$$

The "square" of this vector is defined by

$$(x^0)^2 - (x^1)^2 - (x^2)^2 - (x^3)^2$$

and does not change under any rotation (whether spatial rotation or Lorentz transformation) to a new frame.

A set of four quantities A^0, A^1, A^2, A^3 that transform like the component x^i of the radius four-vector [see (17.19)] is called a **four-vector** A^i.

For example, under a Lorentz transformation in the (x^0, x^1) [i.e., (ct, x)] plane,

$$A^0 = \frac{A'^0 + (V/c)A'^1}{\sqrt{1 - V^2/c^2}}, \quad A^1 = \frac{A'^1 + (V/c)A'^0}{\sqrt{1 - V^2/c^2}}, \quad A^2 = A'^2, \quad A^3 = A'^3. \tag{17.20}$$

The "square" of any four-vector is defined in the same way as the square of the radius four-vector:

$$(A^0)^2 - (A^1)^2 - (A^2)^2 - (A^3)^2.$$

We introduce the further notation

$$A_0 = A^0, \quad A_1 = -A^1, \quad A_2 = -A^2, \quad A_3 = -A^3.$$

The A^i $(i = 0, 1, 2, 3)$ are the called the **contravariant** components, and the A_i the **covariant** components of the four-vector.

With this notation the square of a four-vector can be written as

$$A^i A_i \equiv \sum_{i=0}^{3} A^i A_i = A^0 A_0 + A^1 A_1 + A^2 A_2 + A^3 A_3,$$

where we have used the convention that repeated indices (one upper and one lower) imply a sum. The scalar product of two four-vectors is

$$A^i B_i = A^0 B_0 + A^1 B_1 + A^2 B_2 + A^3 B_3.$$

Clearly, this is equal to $A_i B^i$.

The Lorentz-transformation laws for covariant four-vectors differ only in a sign from the Lorentz-transformation laws (17.20) for contravariant ones:

$$A_0 = \frac{A'_0 - (V/c)A'_1}{\sqrt{1 - V^2/c^2}}, \quad A_1 = \frac{A'_1 - (V/c)A'_0}{\sqrt{1 - V^2/c^2}}, \quad A_2 = A'_2, \quad A_3 = A'_3. \tag{17.21}$$

Using (17.20) and (17.21) it is easy to show that $A_i A^i$ is a Lorentz scalar (that is, an invariant under Lorentz transformations), as expected.

The component A^0 is called the **time component**, and A^1, A^2 and A^3 the **space components**, of the four-vector A^i.

In analogy with intervals, if $A_i A^i > 0$ the four-vector is said to be **timelike**, if $A_i A^i < 0$ the four-vector is said to be **spacelike**, and if $A_i A^i = 0$ the four-vector is said to be a **null** vector.

The notation A^i can indicate the vector as well as one of its components, that is,

$$A^i = (A^0, A^1, A^2, A^3),$$

where the components $A^1 \equiv A_x$, $A^2 \equiv A_y$, $A^3 \equiv A_z$ form a three-dimensional vector under spatial rotations (we can write $\hat{\mathbf{x}} A_x + \hat{\mathbf{y}} A_y + \hat{\mathbf{z}} A_z \equiv \mathbf{A}$).

The corresponding covariant vector is

$$A_i = (A^0, -A^1, -A^2, -A^3),$$

and the square of the four-vector is

$$A^i A_i = (A^0)^2 - (A^1)^2 - (A^2)^2 - (A^3)^2 \equiv (A^0)^2 - \boldsymbol{A} \cdot \boldsymbol{A}.$$

Thus, for the radius four-vector,

$$x^i = (ct, x, y, z), \;\; x_i = (ct, -x, -y, -z),$$

and

$$x^i x_i = c^2 t^2 - r^2 \quad (r^2 = \boldsymbol{r} \cdot \boldsymbol{r}).$$

For three-dimensional vectors (in Euclidean space) there is no need to distinguish between contravariant and covariant components, and we label them in both cases as A_α $(\alpha = 1, 2, 3)$; for example, we have $\boldsymbol{A} \cdot \boldsymbol{B} = A_\alpha B_\alpha$.

A four-tensor of rank 2 is a set of $4^2 = 16$ quantities A^{ik} which, under Lorentz transformations, transform like the 16 products $A^i B^k$ $(i, k = 0, 1, 2, 3)$ of the components of any two four-vectors A^i and B^k; that is, if (17.20) is put in the form

$$A^i = \sum_{j=0}^{3} \alpha^i_j A'^j \equiv \alpha^i_j A'^j,$$

then the quantities A^{ik} transform as follows:

$$A^{ik} = \alpha^i_j \alpha^k_l A'^{jl}.$$

Four-tensors of rank n transform like the 4^n products of the components of n four-vectors. For example,

A^{ik} is a contravariant rank-2 tensor;
A_{ik} is a covariant rank-2 tensor;
$A_i{}^k$ is a mixed rank-2 tensor.

Raising or lowering a space index changes the sign of the component. Raising or lowering the time index does not. For example,

$$A_{00} = A^{00} = A_0{}^0 = A^0{}_0,$$

$$A_{01} = -A^{01}, A_{11} = A^{11},$$

$$A_0{}^1 = A^{01}, A^0{}_1 = -A^{01}.$$

Under purely spatial rotations the nine quantities A^{11}, A^{12}, ... form a three-tensor, while (A^{01}, A^{02}, A^{03}) and (A^{10}, A^{20}, A^{30}) form three-vectors and A^{00} is a three-scalar (since it is unaffected by spatial rotations).

A tensor A^{ik} is said to be symmetric if

$$A^{ki} = A^{ik}$$

and antisymmetric if

$$A^{ki} = -A^{ik}.$$

In the latter case, all components with two indices equal are zero:

$$A^{00} = A^{11} = A^{22} = A^{33} = 0$$

(since $A^{00} = -A^{00}$, etc.).

For a symmetric mixed tensor,

$$A_i{}^k = A^k{}_i,$$

and so we write it as A_i^k.

In every tensor equation the two sides must contain identical and identically placed free indices (the other, repeated indices – the ones summed over – are called dummy indices).

For example, the equation

$$a_{ij}{}^k = b_{ijl}c^{lk} + d_{lmij}e^{klm}$$

could be a valid tensor equation (in fact, it represents $4^3 = 64$ equations, corresponding to the 64 choices of the free indices $\{i, j, k\}$), which would consequently take the same form for the corresponding primed components in any other inertial frame. However, the equation

$$a_{ij}{}^k = b_i{}^{jl}c^{lk} + d_{lmij}e^{klm}$$

could not, since, although it might by chance be valid in one frame, it would not remain valid on transformation to another, as a tensor equation should.

From A^{ik} one can form a scalar by lowering an index ($A^{ik} \to A^i{}_k$), setting the two indices equal, and then taking the sum

$$A^i{}_i \equiv A^0{}_0 + A^1{}_1 + A^2{}_2 + A^3{}_3.$$

This sum is called the **trace** of A^{ik} and the summation operation is **contraction**.

For example, the scalar product $A_i B^i$ is formed by contraction from the tensor $A_i B^k$. Contracting on any pair of indices reduces the rank by two.

For example, $A^i{}_{kli}$ is a rank-2 tensor, $A^i{}_k B^k$ is a four-vector (rank-1 tensor), and $A_{ik}B^{ik}$ is a scalar (rank-0 tensor).

The unit four-tensor δ_i^k satisfies

$$\delta_i^k A^i = A^k$$

for any four-vector A^i, that is, δ_i^k has components

$$\delta_i^k = \begin{cases} 1 & \text{if} \quad i = k \\ 0 & \text{if} \quad i \neq k \end{cases}$$

Its trace is $\delta_i^i = 4$.

Raising the lower index of δ_i^k gives the **contravariant metric tensor** g^{ik}. Lowering the upper index of δ_i^k gives the **covariant metric tensor** g_{ik}. These have (in flat space–time) identical components:

$$(g^{ik}) = (g_{ik}) = \begin{pmatrix} 1 & 0 & 0 & 0 \\ 0 & -1 & 0 & 0 \\ 0 & 0 & -1 & 0 \\ 0 & 0 & 0 & -1 \end{pmatrix}$$

(i labels the rows, and k the columns, in the order 0, 1, 2, 3). Clearly,

$$g_{ik}A^k = A_i$$

$$g^{ik}A_k = A^i,$$

that is, contraction with g_{ik} and g^{ik} can be used to lower and raise indices throughout a tensor equation.

For example, the scalar product

$$A^i A_i = g_{ik}A^i A^k = g^{ik}A_i A_k.$$

The tensors δ_i^k, g_{ik} and g^{ik} are special in that their components are the same in all inertial frames, as the reader may check using (17.20) and (17.21).

Another such tensor is e^{iklm} (the completely antisymmetric unit tensor of rank 4). Since it is antisymmetric in all pairs of indices, the components with two or more indices equal are zero (for example, $e^{2010} = 0$). The nonzero components are all equal to 1 or –1. We set

$$e^{0123} = +1$$

(and hence $e^{1023} = -1$ by the antisymmetry).

Clearly, the nonzero components e^{iklm} (all indices different) are $+1$ or –1, depending on whether the numbers *iklm* can be brought into the order 0123 by an even or an odd number of pair interchanges.

The number of such nonzero components is $4! = 24$. Therefore,

$$e^{iklm}e_{iklm} = -24$$

(every nonzero term in the product has one time index and three space indices, and so is equal to –1 by the index-lowering rules).

We now check the claim that a quantity e^{iklm} defined in this way is, in fact, a tensor.

For example, under a coordinate "rotation" defined by

$$x^i = \alpha^i_k x'^k,$$

where the (symmetric) matrix (α^i_k) for a Lorentz transformation from our frame K to our frame K′ is

$$(\alpha^i_k) = \begin{pmatrix} \frac{1}{\sqrt{1-V^2/c^2}} & \frac{V/c}{\sqrt{1-V^2/c^2}} & 0 & 0 \\ \frac{V/c}{\sqrt{1-V^2/c^2}} & \frac{1}{\sqrt{1-V^2/c^2}} & 0 & 0 \\ 0 & 0 & 1 & 0 \\ 0 & 0 & 0 & 1 \end{pmatrix},$$

we must expect

$$e^{iklm} = \alpha^i_p \alpha^k_q \alpha^l_r \alpha^m_s e'^{pqrs}.$$

For example,

$$e^{0123} = \alpha^0_p \alpha^1_q \alpha^2_r \alpha^3_s e'^{pqrs}.$$

From the matrix (α^i_k), only $r = 2$ and $s = 3$ survive in the sum over r and s, while p and q can each be 0 or 1, but not both 0 or both 1, since $e'^{0023} = e'^{1123} = 0$ by definition of the components of e'^{pqrs}. Therefore,

$$\begin{aligned} e^{0123} &= \alpha^0_0 \alpha^1_1 \alpha^2_2 \alpha^3_3 e'^{0123} + \alpha^0_1 \alpha^1_0 \alpha^2_2 \alpha^3_3 e'^{1023} \\ &= \frac{1}{\sqrt{1-V^2/c^2}} \times \frac{1}{\sqrt{1-V^2/c^2}} \times 1 \times 1 \times 1 + \frac{V/c}{\sqrt{1-V^2/c^2}} \times \frac{V/c}{\sqrt{1-V^2/c^2}} \times 1 \times 1 \times (-1) \\ &= 1, \end{aligned}$$

as required by the definition of e^{iklm}. The same calculation for other e^{iklm} besides e^{0123} also yields the result expected from the definition, and so e^{iklm} defined in this way is certainly a tensor (in fact, an invariant tensor) under coordinate "rotations" (including both Lorentz and spatial rotations).

However, under spatial inversion or reflection e^{iklm} does not satisfy the expected transformation law. For example, under spatial inversion, so that

$$x^i = \gamma^i_k x'^k \text{ with } (\gamma^i_k) = \begin{pmatrix} 1 & 0 & 0 & 0 \\ 0 & -1 & 0 & 0 \\ 0 & 0 & -1 & 0 \\ 0 & 0 & 0 & -1 \end{pmatrix},$$

we have

$$e^{iklm} \neq \gamma^i_p \gamma^k_q \gamma^l_r \gamma^m_s e'^{pqrs}.$$

[check: $\gamma^0_p \gamma^1_q \gamma^2_r \gamma^3_s e'^{pqrs} = \gamma^0_0 \gamma^1_1 \gamma^2_2 \gamma^3_3 e'^{0123} = -1 \neq e^{0123}$.]

Thus, e^{iklm} is a tensor under rotations (including Lorentz transformations), but not under reflection or inversion. The correct transformation law for all these types of coordinate transformation is

$$e^{iklm} = (\det\gamma) \gamma^i_p \gamma^k_q \gamma^l_r \gamma^m_s e'^{pqrs},$$

and so, in analogy with pseudovectors [see (17.15)], e^{iklm} is said to be a **pseudotensor**.

Pseudotensors of any rank (for example, pseudoscalars) behave like tensors under all coordinate transformations except those (such as reflections and inversion) that have $\det\gamma = -1$ and cannot be reduced to rotations. From any antisymmetric tensor A^{ik} we can form the pseudotensor $A^{*ik} = \frac{1}{2!}e^{iklm}A_{lm}$ **dual** to it. [Similarly, from a vector A^i we can form the antisymmetric rank-3 pseudotensor $\frac{1}{1!}e^{iklm}A_m$ dual to it.] The contraction $A^*_{ik}A^{ik} = \frac{1}{2!}e_{iklm}A^{lm}A^{ik}$ of a rank-2 pseudotensor ($\frac{1}{2!}e_{iklm}A^{lm}$) and its dual ($A^{ik}$) is a pseudoscalar.

[The concept of "duality" is not restricted to the case of four dimensions. For example, in three dimensions the familiar cross product $\boldsymbol{C} = \boldsymbol{A} \times \boldsymbol{B}$ of two vectors is a pseudovector dual to a certain antisymmetric tensor:

$$C_\alpha = \frac{1}{2!}\varepsilon_{\alpha\beta\gamma}C_{\beta\gamma}, \quad \text{with} \quad C_{\beta\gamma} = A_\beta B_\gamma - A_\gamma B_\beta \quad (= -C_{\gamma\beta}).]$$

The **four-gradient** of a four-scalar φ is the four-vector

$$\frac{\partial\varphi}{\partial x^i} = \left(\frac{1}{c}\frac{\partial\varphi}{\partial t}, \frac{\partial\varphi}{\partial x}, \frac{\partial\varphi}{\partial y}, \frac{\partial\varphi}{\partial z}\right) \equiv \partial_i\varphi \equiv \varphi_{,i},$$

that is, the derivatives here are the covariant components of a four-vector. (The defining relation for partial derivatives:

$$\mathrm{d}\varphi = \frac{\partial\varphi}{\partial x^i}\mathrm{d}x^i$$

is clearly the four-scalar product of two four-vectors.)

The four-divergence $\partial A^i/\partial x^i$ is a four-scalar.

Similarly, differentiating with respect to the "covariant coordinates" x_i, we have

$$\frac{\partial\varphi}{\partial x_i} = \left(\frac{1}{c}\frac{\partial\varphi}{\partial t}, -\frac{\partial\varphi}{\partial x}, -\frac{\partial\varphi}{\partial y}, -\frac{\partial\varphi}{\partial z}\right),$$

where the derivatives here form the contravariant components of a four-vector, that is,

$$\frac{\partial\varphi}{\partial x_i} \equiv \partial^i\varphi \equiv \varphi^{,i}.$$

17.2.9 Integration in four-space

In three-space we can integrate over curves, surfaces and volumes. In four-space, **four** types of integration are possible.

(1) **Integration over a curve in four-space**
The element of integration is the element of arc, that is, the four-vector $\mathrm{d}x^i$.

(2) **Integration over a two-dimensional surface in four-space**
Recall that in **three-space** we integrate over the vector element $\mathrm{d}\boldsymbol{S} = \mathrm{d}\boldsymbol{r} \times \mathrm{d}\boldsymbol{r}'$, which has the direction of the normal to the area element defined by $\mathrm{d}\boldsymbol{r}$ and $\mathrm{d}\boldsymbol{r}'$ and has magnitude equal to the area of this element.

If, in a given coordinate frame, $\mathrm{d}\boldsymbol{r}$ has components $\mathrm{d}x_\alpha$ and $\mathrm{d}\boldsymbol{r}'$ has components $\mathrm{d}x'_\alpha$ ($\alpha = 1, 2, 3$), then, in this frame, $\mathrm{d}\boldsymbol{S}$ has components

$$(\mathrm{d}\boldsymbol{S})_\alpha \equiv \mathrm{d}S_\alpha = \frac{1}{2!}\varepsilon_{\alpha\beta\gamma}\mathrm{d}S_{\beta\gamma},$$

where

$$\mathrm{d}S_{\beta\gamma} = \mathrm{d}x_\beta \mathrm{d}x'_\gamma - \mathrm{d}x_\gamma \mathrm{d}x'_\beta,$$

that is, $\mathrm{d}S_\alpha$ is the vector dual to the antisymmetric tensor $\mathrm{d}S_{\beta\gamma}$.

Similarly, in **four-space**, an integration over an arbitrary two-dimensional surface is represented by an integration over the tensor dual to

$$\mathrm{d}f^{ik} = \mathrm{d}x^i \mathrm{d}x'^k - \mathrm{d}x^k \mathrm{d}x'^i \quad (i, k = 0, 1, 2, 3),$$

that is, over

$$\mathrm{d}f^{*ik} = \frac{1}{2!}e^{iklm}\mathrm{d}f_{lm}$$

(an element of area equal and "normal" to $\mathrm{d}f^{ik}$).

(3) **Integration over a three-dimensional manifold in four-space**

(We say "manifold" rather than "volume", since it could have, for example, two space dimensions and one time dimension.)

Recall that in three dimensions the volume element enclosed by the parallelepiped spanned by three vectors $\mathrm{d}\boldsymbol{r}$, $\mathrm{d}\boldsymbol{r}'$ and $\mathrm{d}\boldsymbol{r}''$ is:

$$\begin{vmatrix} \mathrm{d}x & \mathrm{d}x' & \mathrm{d}x'' \\ \mathrm{d}y & \mathrm{d}y' & \mathrm{d}y'' \\ \mathrm{d}z & \mathrm{d}z' & \mathrm{d}z'' \end{vmatrix}.$$

Similarly, in **four-space** the projected three-dimensional "volumes" of the four-dimensional parallelepiped spanned by the four-vectors $\mathrm{d}x^i$, $\mathrm{d}x'^i$ and $\mathrm{d}x''^i$ are given by the determinants

$$\mathrm{d}S^{ikl} = \begin{vmatrix} \mathrm{d}x^i & \mathrm{d}x'^i & \mathrm{d}x''^i \\ \mathrm{d}x^k & \mathrm{d}x'^k & \mathrm{d}x''^k \\ \mathrm{d}x^l & \mathrm{d}x'^l & \mathrm{d}x''^l \end{vmatrix},$$

which form a rank-3 tensor, antisymmetric in all index pairs.

By analogy with the surface integration [type (2), above], it is convenient to use the **vector** $\mathrm{d}S_i$ dual to this tensor $\mathrm{d}S^{ikl}$:

$$\mathrm{d}S^i = \frac{1}{3!}e^{iklm}\mathrm{d}S_{klm}$$

(the inverse relation is $\mathrm{d}S^{ikl} = \frac{1}{1!}e^{iklm}\mathrm{d}S_m$). For example, it is easy to check that $\mathrm{d}S^0 = \mathrm{d}S_{123}$, $\mathrm{d}S^1 = \mathrm{d}S_{032}$, and so on.

Geometrically, $\mathrm{d}S^1$ is a four-vector equal in magnitude to the three-dimensional "volume" $\mathrm{d}S_{032}$ (an element of three-dimensional hypersurface in the four-dimensional hypervolume) and directed along the normal to $\mathrm{d}S_{032}$ (that is, perpendicular to all lines in this three-dimensional "volume" element).

Similarly, $\mathrm{d}S^0 = \mathrm{d}S_{123} = \mathrm{d}x\mathrm{d}y\mathrm{d}z$ (check) $= \mathrm{d}V$ is the projection of the hypervolume $c\mathrm{d}t\mathrm{d}x\mathrm{d}y\mathrm{d}z$ on to the hyperplane $x_0 =$ const and is parallel to $x = ct$.

(4) **Integration over a four-dimensional "volume" (hypervolume)**

The element is the four-volume element

$$\mathrm{d}\Omega = \mathrm{d}x^0\mathrm{d}x^1\mathrm{d}x^2\mathrm{d}x^3.$$

A hypersurface is spacelike (that is, all intervals on it are spacelike) if the normal at any of its points lies in a timelike direction, that is, inside the light cone with vertex at the same point.

If the normals lie in spacelike directions (that is, outside the corresponding light cones), the hypersurface orthogonal to these directions may have both timelike and spacelike intervals on it.

17.2.10 Integral theorems

In three dimensions we have:

(i) Gauss' theorem:

$$\oint_S \boldsymbol{A} \cdot \mathrm{d}\boldsymbol{S} = \int_V (\boldsymbol{\nabla} \cdot \boldsymbol{A})\mathrm{d}V$$

[surface integral (2 dimensions) → volume integral (3 dimensions)]
This can be symbolized as

$$\mathrm{d}S_\alpha \to \mathrm{d}V\frac{\partial}{\partial x_\alpha} \quad (\alpha = 1, 2, 3).$$

(ii) Stokes' theorem:

$$\begin{aligned}\oint_\Gamma \mathrm{d}x_\alpha A_\alpha \equiv \oint_\Gamma \mathrm{d}\boldsymbol{r} \cdot \boldsymbol{A} = \int_S \mathrm{d}\boldsymbol{S} \cdot (\boldsymbol{\nabla} \times \boldsymbol{A}) &\equiv \int_S \mathrm{d}S_\alpha(\boldsymbol{\nabla} \times \boldsymbol{A})_\alpha \\ &= \int_S \mathrm{d}S_\alpha \varepsilon_{\alpha\beta\gamma}\frac{\partial A_\gamma}{\partial x_\beta} \\ &= \int_S \mathrm{d}S_{\beta\gamma}\frac{\partial A_\gamma}{\partial x_\beta} = \int_S \mathrm{d}S_{\beta\alpha}\frac{\partial A_\alpha}{\partial x_\beta}\end{aligned}$$

[line integral (1 dimension) → surface integral (2 dimensions)]
(Here we have used the definition of $\mathrm{d}S_\alpha$ from the preceding section.) This result can be symbolized as

$$\mathrm{d}x_\alpha \to \mathrm{d}S_{\beta\alpha}\frac{\partial}{\partial x_\beta} \quad (\alpha, \beta = 1, 2, 3).$$

The analogous theorems in four-dimensional space are:

(i) $\mathrm{d}S_i \to \mathrm{d}\Omega \frac{\partial}{\partial x^i}$ $(i = 0, 1, 2, 3)$.
[hypersurface integral (3 dimensions) → hypervolume integral (4 dimensions)]

(ii) $\mathrm{d}f^*_{ik} \to \mathrm{d}S_i \frac{\partial}{\partial x^k} - \mathrm{d}S_k \frac{\partial}{\partial x^i}$ $(i, k = 0, 1, 2, 3)$.
[surface integral (2 dimensions) → hypersurface integral (3 dimensions)]

(iii) $\mathrm{d}x^i \to \mathrm{d}f^{ki} \frac{\partial}{\partial x^k}$ $(i, k = 0, 1, 2, 3)$.
[line integral (1 dimension) → surface integral (2 dimensions)]

Theorem (iii) is the four-dimensional analogue of Stokes' theorem.

17.2.11 Four-velocity and four-acceleration

Suppose that in some Lorentz frame K the spatial coordinates of a moving particle change by $\mathrm{d}x$, $\mathrm{d}y$, $\mathrm{d}z$ in time $\mathrm{d}t$. The interval $\mathrm{d}s$ between the start and end points of this infinitesimal motion is $\mathrm{d}s = c\mathrm{d}t\sqrt{1 - v^2/c^2}$, where v is the magnitude of the conventional velocity ("three-velocity") $\boldsymbol{v}$ with components $v_\alpha = \mathrm{d}x_\alpha/\mathrm{d}t$. We define the **four-velocity** of this motion of the particle in frame K as the dimensionless (!) four-vector

$$u^i = \frac{\mathrm{d}x^i}{\mathrm{d}s}$$

Putting

$$\mathrm{d}x^0 = c\mathrm{d}t, \quad \mathrm{d}x^1 = \mathrm{d}x, \quad \mathrm{d}x^2 = \mathrm{d}y, \quad \mathrm{d}x^3 = \mathrm{d}z$$

and

$$\mathrm{d}s = c\mathrm{d}t\sqrt{1 - v^2/c^2},$$

we see that u^i has components

$$u^0 = \frac{1}{\sqrt{1 - v^2/c^2}}, \quad u^\alpha = \frac{v_\alpha/c}{\sqrt{1 - v^2/c^2}} \quad (\alpha = 1, 2, 3).$$

Since $\mathrm{d}x^i\mathrm{d}x_i = \mathrm{d}s^2$, we have

$$u^i u_i = 1,$$

that is, the components are **not** independent. Thus, geometrically, u^i is a **unit** four-vector.

The **four-acceleration** is defined by

$$\frac{\mathrm{d}^2 x^i}{\mathrm{d}s^2} = \frac{\mathrm{d}u^i}{\mathrm{d}s}$$

(with dimensions of $[\text{length}]^{-1}$!).

Differentiating $u^i u_i = 1$ (expressed in the form $g_{ik}u^i u^k = 1$) with respect to s, we get

$$g_{ik}\frac{\mathrm{d}u^i}{\mathrm{d}s}u^k + g_{ik}u^i\frac{\mathrm{d}u^k}{\mathrm{d}s} = 0.$$

By the symmetry of g_{ik} these two terms are equal, and so

$$g_{ik}\frac{\mathrm{d}u^i}{\mathrm{d}s}u^k = 0, \text{ that is, } \frac{\mathrm{d}u^i}{\mathrm{d}s}u_i = 0.$$

Thus, the vectors u^i and $\mathrm{d}u^i/\mathrm{d}s$ are "mutually perpendicular".

17.3 Principle of least action

Consider the evolution of a physical system from some definite configuration (state) A at a specified initial time t_A to a new, definite configuration (state) B at some specified later time t_B. If we treat the evolution as fully deterministic, we may say that the system passes through some definite sequence of intermediate states, that is, it follows a particular "path" (the "true path") in its evolution from state A at t_A to state B at t_B. Besides the true path, other paths that also take the system from state A at t_A to state B at t_B can be considered. (Of course, such paths are not realized in classical mechanics, although they do contribute to the amplitude of processes between specified initial and final states in quantum mechanics.) For every such path (including the true one) we may construct a certain integral over the path (that is, a functional of the path) called the **action** S of the path, such that the path with the smallest value of the action is the true path (this is the **principle of least action**). Thus, infinitesimal variation of the motion (path) away from the true motion (path) gives zero first-order variation of the action; that is, for such variations we have $\delta S = 0$.

17.3.1 Free particle

The action S must be invariant under Lorentz transformations, that is, must be an integral, over the path, of a **Lorentz scalar (four-scalar)**. Also, the integrand must be a first-order differential.

The only scalar of this kind for the motion of a free particle is the interval $\mathrm{d}s$, or $\alpha\mathrm{d}s$, where α is a constant characterizing the particle.

Thus, the action integral S for the motion of a free particle between world points a and b is

$$S = -\alpha\int_\mathrm{a}^\mathrm{b} \mathrm{d}s \quad (\alpha > 0), \tag{17.22}$$

where the minus sign follows from the fact [see section 17.2.4] that $\int_\mathrm{a}^\mathrm{b}\mathrm{d}s$ is a **maximum** along a straight world line, that is, for the actual motion of a free particle.

17.3.2 Three-space formulation

The action S can also be represented as

$$S = \int_{t_\mathrm{a}}^{t_\mathrm{b}} L\mathrm{d}t,$$

where t_a and t_b are the time coordinates of the world points a and b, and $L = T - V$ is the **lagrangian** of the particle (T is the kinetic energy and V the potential energy of the particle).

Putting

$$ds = c\,dt\sqrt{1 - v^2/c^2}$$

into (17.22) and comparing our two expressions for S, we find

$$\begin{aligned} L &= -\alpha c\sqrt{1 - v^2/c^2} \\ &= -\alpha c + \frac{\alpha v^2}{2c} + \text{terms of higher order in } v^2/c^2. \end{aligned}$$

In the action $\int_{t_a}^{t_b} L\,dt$ the term $-\alpha c$ in L is constant (independent of the path), and so contributes nothing to the variation δS and can be omitted. Therefore,

$$L \approx \frac{\alpha v^2}{2c} \quad \text{if} \quad v \ll c.$$

For a free particle ($V =$ const., which we can set equal to zero) we have $L = T$, that is,

$$L \approx \frac{1}{2} m v^2 \quad \text{if} \quad v \ll c,$$

where m is the mass ("**rest mass**") of the particle, defined as the coefficient of $\frac{1}{2}v^2$ in T. Therefore,

$$\alpha = mc.$$

Thus, for the motion of a free particle between world points a and b we have

$$S = -mc \int_{a}^{b} ds,$$

and the full relativistic lagrangian for a free particle is

$$L = -mc^2\sqrt{1 - v^2/c^2}. \tag{17.23}$$

17.3.3 Momentum and energy of a free particle

The **momentum** is defined as having components $p_\alpha = \partial L/\partial v_\alpha$, and so from (17.23) we have

$$p_\alpha = \frac{m v_\alpha}{\sqrt{1 - v^2/c^2}}, \quad \text{that is,} \quad \boldsymbol{p} = \frac{m\boldsymbol{v}}{\sqrt{1 - v^2/c^2}}. \tag{17.24}$$

The **energy** $\mathcal{E}$ is defined as

$$\mathcal{E} = \boldsymbol{p} \cdot \boldsymbol{v} - L.$$

Using (17.23) and (17.24) we get

$$\mathcal{E} = \frac{mv^2}{\sqrt{1 - v^2/c^2}} + \frac{mc^2\left(1 - v^2/c^2\right)}{\sqrt{1 - v^2/c^2}} = \frac{mc^2}{\sqrt{1 - v^2/c^2}}. \tag{17.25}$$

Equations (17.24) and (17.25) are Einstein's famous expressions for the momentum and energy of a free particle. For $\boldsymbol{v} = 0$ equation (17.25) becomes

$$\mathcal{E} = mc^2,$$

while for small velocities $v \ll c$ we have

$$\mathcal{E} = mc^2 + \frac{1}{2}mv^2.$$

These derivations were not restricted to "elementary particles": they apply also to composite bodies. Thus, the total rest energy of a composite body is

$$\mathcal{E} = Mc^2,$$

where M is the coefficient of $\frac{1}{2}v^2$ in the kinetic energy of the body in the inertial frame in which the body has velocity $\boldsymbol{v} = 0$.

For example, let the body consist of N particles of masses $m^{(i)}$ $(i = 1, \ldots, N)$. By **energy** conservation we have

$$\mathcal{E} = Mc^2 = \sum_{i=1}^{N} m^{(i)}c^2 + \text{the potential energy of the interaction of the particles}$$
$$+ \text{the kinetic energy of the } \textbf{relative} \text{ motion of the particles.}$$

Therefore,

$$M \neq \sum_{i=1}^{N} m^{(i)}$$

if either interaction or relative motion of the particles is present. Thus, **the law of conservation of mass does not hold in relativistic mechanics**. For example, the mass of a nucleus is less than the sum of the masses of the constituent protons and neutrons, the difference being the **nuclear binding energy**.

For a particle moving with speed v,

$$\frac{\mathcal{E}^2}{c^2} = \frac{m^2c^2}{1 - v^2/c^2},$$

$$p^2 + m^2c^2 = \frac{m^2v^2}{1 - v^2/c^2} + m^2c^2 = \frac{m^2c^2}{1 - v^2/c^2}.$$

Therefore,

$$\frac{\mathcal{E}^2}{c^2} = p^2 + m^2c^2.$$

The energy expressed as a function of momentum (and position) is called the **hamiltonian** H, that is,

$$H = \sqrt{p^2 + m^2c^2}.$$

For low velocities (such that $p \ll mc$), we have, as expected,

$$H = mc^2 + \frac{p^2}{2m}.$$

From the expressions for $\boldsymbol{p}$ and $\mathcal{E}$, we have

$$\boldsymbol{p} = \frac{\mathcal{E}\boldsymbol{v}}{c^2}.$$

When $v \to c$ the expressions for $\boldsymbol{p}$ and $\mathcal{E}$ both $\to \infty$ if $m \neq 0$, that is, a velocity $v = c$ is impossible for particles with $m \neq 0$. If, however, $m = 0$, then motion with $v = c$ is possible, and so for particles with $v = c$ (photons, gluons, and possibly neutrinos), we clearly have

$$p = \frac{\mathcal{E}}{c},$$

which, after application of the de Broglie relation $p = h/\lambda$ and Einstein relation $\mathcal{E} = h\nu$, reduces to the familiar relation $c = \lambda\nu$ relating the wavelength λ and frequency ν of light.

17.3.4 Four-space formulation

For the variation δS of the action S under infinitesimal variation of the path of a free particle away from the true path between fixed world points a and b we have

$$\delta S = -mc\delta \int_{\mathrm{a}}^{\mathrm{b}} \mathrm{d}s = 0.$$

But

$$ds = \sqrt{dx^i dx_i} = \sqrt{g_{ik} dx^i dx^k}.$$

Therefore,

$$\delta S = -mc\delta \int_a^b \sqrt{g_{ik} dx^i dx^k} = -\frac{1}{2}mc\left[\int_a^b \frac{g_{ik} dx^i \delta dx^k}{\sqrt{g_{ik} dx^i dx^k}} + \int_a^b \frac{g_{ik} dx^k \delta dx^i}{\sqrt{g_{ik} dx^i dx^k}}\right]$$

$$= -mc \int_a^b u_i d\delta x^i \quad \left(u_i = \frac{dx_i}{ds} \text{ and } \delta dx^i = d\delta x^i\right).$$

Integrating this by parts, we find

$$\delta S = -mcu_i \delta x^i \Big|_a^b + mc \int_a^b \delta x^i \frac{du_i}{ds} ds.$$

(i) The **equations of motion** are obtained by fixing the end points of the motion [so that $(\delta x^i)_a = (\delta x^i)_b = 0$] and setting $\delta S = 0$. They are (since the variations δx^i of the world point of the particle from the true path are arbitrary)

$$\frac{du_i}{ds} = 0,$$

that is, the four-acceleration is zero (constant four-velocity u_i), as expected for a free particle.

(ii) To see **how the action varies with the coordinates of an end point**, we consider only **admissible** paths (that is, paths satisfying the equations of motion $du_i/ds = 0$) and fix the first world point: $(\delta x^i)_a = 0$. Writing $(\delta x^i)_b$ as δx^i, we then get

$$\delta S = -mcu_i \delta x^i \quad \text{(a sum over } i\text{)}.$$

But, treating the action of an admissible path as a function of the coordinates x^i of the end point of the motion, by definition of the partial derivatives $\partial S/\partial x^i$ we have

$$\delta S = \frac{\partial S}{\partial x^i}\delta x^i \quad \text{(a sum over } i\text{)}.$$

The four-vector with components $-\partial S/\partial x^i$ is called the (covariant) **four-momentum** and is denoted by p_i.

Comparing the above two forms for δS, we have

$$p_i = mcu_i.$$

Raising the indices on both sides, we get the corresponding contravariant four-momentum p^i:

$$p^i = mcu^i.$$

If we use our results for u^i, this becomes

$$p^0 = \frac{mc}{\sqrt{1 - v^2/c^2}}, \quad p^\alpha = \frac{mv_\alpha}{\sqrt{1 - v^2/c^2}} \quad (\alpha = 1, 2, 3).$$

Thus, using the results we obtained earlier for the energy $\mathcal{E}$ and three-momentum $\boldsymbol{p}$, we have

$$p^i = \left(\frac{\mathcal{E}}{c}, p_x, p_y, p_z\right), \quad p_i = \left(\frac{\mathcal{E}}{c}, -p_x, -p_y, -p_z\right).$$

Thus, **the energy and momentum are components of a single four-vector**.

Note that $p^i \equiv -\partial S/\partial x_i$ (the **four-momentum**, or **generalized momentum**) is equal to the "**kinetic four-momentum**"

$$\frac{m}{\sqrt{1 - v^2/c^2}}\left(c, v_x, v_y, v_z\right)$$

only in the free-particle case. In the presence of fields the four-momentum is **not** equal to the kinetic four-momentum (see later). Also, it is to $p_i = -\partial S/\partial x^i$ that the quantum-mechanical prescriptions $p_i \to i\hbar\partial/\partial x_i$ (that is, $\mathcal{E} \to i\hbar\partial/\partial t$, $p_x \to -i\hbar\partial/\partial x$, etc.) apply, and not to the kinetic four-momentum (except, of course, in the free-particle case).

Since p^i is a Lorentz four-vector, we can apply to it the Lorentz-transformation rule (17.20) for any contravariant four-vector A^i:

$$p^0 = \frac{p'^0 + (V/c)\,p'^1}{\sqrt{1 - V^2/c^2}}, \quad p^1 = \frac{p'^1 + (V/c)\,p'^0}{\sqrt{1 - V^2/c^2}}, \quad p^2 = p'^2, \quad p^3 = p'^3,$$

that is,

$$\mathcal{E} = \frac{\mathcal{E}' + Vp'_x}{\sqrt{1 - V^2/c^2}}, \quad p_x = \frac{p'_x + (V/c^2)\mathcal{E}'}{\sqrt{1 - V^2/c^2}}, \quad p_y = p'_y, \quad p_z = p'_z.$$

The **force four-vector** g^i is defined as the rate of change of four-momentum with interval:

$$g^i = \frac{\mathrm{d}p^i}{\mathrm{d}s} = mc\frac{\mathrm{d}u^i}{\mathrm{d}s} = mc \times \text{four-acceleration}.$$

Since $u_i \mathrm{d}u^i/\mathrm{d}s = 0$, we have $u_i g^i = 0$.

We also have

$$p^i p_i = m^2c^2u^iu_i = m^2c^2, \tag{17.26}$$

that is,

$$\frac{\mathcal{E}^2}{c^2} - p^2 = m^2c^2,$$

as before. [Equation (17.26) shows that the "length" of the four-momentum is independent of the velocity of the particle and depends only on its mass ("rest mass").]

Replacing in (17.26)

$$p_i \to -\frac{\partial S}{\partial x^i} \quad \text{and} \quad p^i = g^{ik} p_k \to -g^{ik} \frac{\partial S}{\partial x^k},$$

we get

$$g^{ik} \frac{\partial S}{\partial x^i} \frac{\partial S}{\partial x^k} = m^2 c^2,$$

that is,

$$\frac{1}{c^2}\left(\frac{\partial S}{\partial t}\right)^2 - \left(\frac{\partial S}{\partial x}\right)^2 - \left(\frac{\partial S}{\partial y}\right)^2 - \left(\frac{\partial S}{\partial z}\right)^2 = m^2 c^2,$$

which is the **relativistic Hamilton–Jacobi equation** for the action S as a function of the coordinates (ct, x, y, z) of the end point of the motion (and clearly has some similarity to the Klein–Gordon equation (19.2) in relativistic quantum theory, described later in chapter 19.)

[To get the **nonrelativistic** Hamilton–Jacobi equation, we note that

$$p_0 = \frac{\mathcal{E}}{c} = -\frac{\partial S}{\partial x^0} = -\frac{1}{c}\frac{\partial S}{\partial t},$$

that is, $\mathcal{E} = -\partial S/\partial t$, and that in relativistic mechanics this energy includes the rest energy mc^2, which does not enter into the energy in nonrelativistic mechanics. We therefore define a nonrelativistic action S' by

$$S = S' - mc^2 t,$$

so that $-\partial S'/\partial t$ does not contain the term mc^2. The relativistic Hamilton–Jacobi equation then becomes

$$\frac{1}{c^2}\left(\frac{\partial S'}{\partial t} - mc^2\right)^2 - \left(\frac{\partial S'}{\partial x}\right)^2 - \left(\frac{\partial S'}{\partial y}\right)^2 - \left(\frac{\partial S'}{\partial z}\right)^2 = m^2 c^2,$$

that is,

$$\frac{1}{2mc^2}\left(\frac{\partial S'}{\partial t}\right)^2 - \frac{\partial S'}{\partial t} - \frac{1}{2m}\left[\left(\frac{\partial S'}{\partial x}\right)^2 + \left(\frac{\partial S'}{\partial y}\right)^2 + \left(\frac{\partial S'}{\partial z}\right)^2\right] = 0,$$

which, in the limit $c \to \infty$, yields

$$\frac{\partial S'}{\partial t} + \frac{1}{2m}\left[\left(\frac{\partial S'}{\partial x}\right)^2 + \left(\frac{\partial S'}{\partial y}\right)^2 + \left(\frac{\partial S'}{\partial z}\right)^2\right] = 0$$

(the **nonrelativistic Hamilton–Jacobi equation**), which hints at the Schrödinger equation!]

This completes the consideration of **free** motion.

17.4 Motion of a particle in a given electromagnetic field

Here the action S of the motion is made up of two parts: the term $-mc\int_a^b ds$ associated with free motion and a term that takes into account the interaction of the particle with the electromagnetic field.

The interaction with the field is proportional to the "charge" q of the particle (positive, negative or zero), that is, there is just one parameter (q) characterizing the **particle** in its interaction with the field.

The quantity characterizing the **field** is a **four-vector** A_i (the **four-potential**), whose components are functions of space and time.

The simplest Lorentz-scalar action that can be constructed from A_i as an integral along the world line of the particle is

$$-q\int_a^b A_i dx^i.$$

[Actually, an even simpler Lorentz scalar would be $\int_a^b A ds$, with a **scalar** field A characterizing the electromagnetic field, but such a form does not lead to predictions in accord with experiment, that is, does not lead to the laws of electromagnetism.]

Thus, the action for the motion of a charge q in a **given** electromagnetic field is postulated to be

$$S = \int_a^b (-mc ds - qA_i dx^i).$$

The time component A^0 of the contravariant four-vector A^i is identified with φ/c, where φ is a three-scalar called the "**scalar potential**", while the space components A^α (with the corresponding unit vectors) form a three-vector $\boldsymbol{A} = \hat{\boldsymbol{x}}A_x + \hat{\boldsymbol{y}}A_y + \hat{\boldsymbol{z}}A_z$, called the "**vector potential**":

$$A^i = (A^0, A^1, A^2, A^3) = (\varphi/c, A_x, A_y, A_z),$$

$$A_i = (A_0, A_1, A_2, A_3) = \left(\varphi/c, -A_x, -A_y, -A_z\right).$$

Then S can be written as

$$S = \int_a^b (-mc ds - q\varphi dt + q\boldsymbol{A}\cdot d\boldsymbol{r}).$$

But

$$ds = c dt\sqrt{1 - v^2/c^2} \quad \text{and} \quad d\boldsymbol{r} = \boldsymbol{v} dt.$$

Therefore,

$$S = \int_{t_a}^{t_b} \left(-mc^2\sqrt{1 - v^2/c^2} - q\varphi + q\boldsymbol{A}\cdot\boldsymbol{v}\right)dt \equiv \int_{t_a}^{t_b} L dt,$$

that is, the lagrangian is

$$L = -mc^2\sqrt{1 - v^2/c^2} - q\varphi + q\boldsymbol{A} \cdot \boldsymbol{v},$$

where the first term is the free-particle lagrangian, and the second and third terms give the lagrangian describing the interaction of the charge q with the field.

The generalized momentum $\boldsymbol{P}$ has components

$$P_\alpha = \frac{\partial L}{\partial v_\alpha},$$

that is,

$$P_\alpha = \frac{mv_\alpha}{\sqrt{1 - v^2/c^2}} + qA_\alpha = p_\alpha + qA_\alpha, \tag{17.27}$$

where p_α is the ordinary (kinetic) momentum.

The hamiltonian is

$$\begin{aligned} H &= \boldsymbol{v} \cdot \boldsymbol{P} - L = \boldsymbol{v} \cdot \boldsymbol{p} + q\boldsymbol{A} \cdot \boldsymbol{v} + mc^2\sqrt{1 - v^2/c^2} + q\varphi - q\boldsymbol{A} \cdot \boldsymbol{v} \\ &= \frac{mv^2}{\sqrt{1 - v^2/c^2}} + \frac{mc^2\left(1 - v^2/c^2\right)}{\sqrt{1 - v^2/c^2}} + q\varphi \\ &= \frac{mc^2}{\sqrt{1 - v^2/c^2}} + q\varphi \end{aligned} \tag{17.28}$$

(the free-particle energy plus the energy of interaction of the charge q with the field).

From (17.27) and (17.28) above we have

$$\boldsymbol{P} - q\boldsymbol{A} = \frac{m\boldsymbol{v}}{\sqrt{1 - v^2/c^2}} \quad \text{and} \quad H - q\varphi = \frac{mc^2}{\sqrt{1 - v^2/c^2}}.$$

Therefore,

$$(H - q\varphi)^2 - (\boldsymbol{P} - q\boldsymbol{A})^2c^2 = \frac{m^2c^2(c^2 - v^2)}{1 - v^2/c^2} = m^2c^4,$$

that is, the hamiltonian is

$$H = \sqrt{(\boldsymbol{P} - q\boldsymbol{A})^2c^2 + m^2c^4} + q\varphi.$$

17.4.1 Equations of motion of a charge in an electromagnetic field

If the charge is small enough not to perturb the field, we may use the above lagrangian. The equations of motion are given by the Euler–Lagrange equations

$$\frac{\mathrm{d}}{\mathrm{d}t}\left(\frac{\partial L}{\partial v_\alpha}\right) = \frac{\partial L}{\partial x_\alpha} \quad (\alpha = 1, 2, 3). \tag{17.29}$$

But

$$\frac{\partial L}{\partial v_\alpha} = P_\alpha = p_\alpha + qA_\alpha$$

and

$$\frac{\partial L}{\partial x_\alpha} \quad [\equiv (\boldsymbol{\nabla} L)_\alpha] = q\frac{\partial}{\partial x_\alpha}(\boldsymbol{A}\cdot\boldsymbol{v}) - q\frac{\partial \varphi}{\partial x_\alpha},$$

so that (17.29) becomes

$$\begin{aligned}\frac{\mathrm{d}}{\mathrm{d}t}(\boldsymbol{p}+q\boldsymbol{A}) &= q\,\boldsymbol{\nabla}(\boldsymbol{A}\cdot\boldsymbol{v}) - q\,\boldsymbol{\nabla}\varphi \\ &= q\,[(\boldsymbol{A}\cdot\boldsymbol{\nabla})\boldsymbol{v} + (\boldsymbol{v}\cdot\boldsymbol{\nabla})\boldsymbol{A} + \boldsymbol{v}\times(\boldsymbol{\nabla}\times\boldsymbol{A}) + \boldsymbol{A}\times(\boldsymbol{\nabla}\times\boldsymbol{v})] - q\,\boldsymbol{\nabla}\varphi \\ &= q(\boldsymbol{v}\cdot\boldsymbol{\nabla})\boldsymbol{A} + q\boldsymbol{v}\times(\boldsymbol{\nabla}\times\boldsymbol{A}) - q\,\boldsymbol{\nabla}\varphi.\end{aligned}$$

But the total derivative $\mathrm{d}\boldsymbol{A}/\mathrm{d}t$ is found from

$$\begin{aligned}\mathrm{d}\boldsymbol{A} &= \mathrm{d}t\frac{\partial \boldsymbol{A}}{\partial t} + (\mathrm{d}\boldsymbol{r}\cdot\boldsymbol{\nabla})\boldsymbol{A} = \mathrm{d}t\frac{\partial \boldsymbol{A}}{\partial t} + \mathrm{d}t\left(\frac{\mathrm{d}\boldsymbol{r}}{\mathrm{d}t}\cdot\boldsymbol{\nabla}\right)\boldsymbol{A} \\ &= \mathrm{d}t\left(\frac{\partial \boldsymbol{A}}{\partial t} + (\boldsymbol{v}\cdot\boldsymbol{\nabla})\,\boldsymbol{A}\right).\end{aligned}$$

Therefore,

$$\frac{\mathrm{d}\boldsymbol{A}}{\mathrm{d}t} = \frac{\partial \boldsymbol{A}}{\partial t} + (\boldsymbol{v}\cdot\boldsymbol{\nabla})\,\boldsymbol{A},$$

and so

$$\frac{\mathrm{d}\boldsymbol{p}}{\mathrm{d}t} = -q\frac{\partial \boldsymbol{A}}{\partial t} - q\,\boldsymbol{\nabla}\varphi + q\boldsymbol{v}\times(\boldsymbol{\nabla}\times\boldsymbol{A}),$$

which is the force on the charge q in the electromagnetic field.

The first two terms are independent of $\boldsymbol{v}$ while the third depends on $\boldsymbol{v}$.

The force of the first type, per unit charge, is called the **electric intensity $\boldsymbol{E}$**:

$$\boldsymbol{E} = -\frac{\partial \boldsymbol{A}}{\partial t} - \boldsymbol{\nabla}\varphi. \tag{17.30}$$

The quantity vector-multiplying $\boldsymbol{v}$ in the force of the second type, per unit charge, is called the **magnetic induction $\boldsymbol{B}$**:

$$\boldsymbol{B} = \boldsymbol{\nabla}\times\boldsymbol{A}. \tag{17.31}$$

Thus, the equation of motion becomes

$$\frac{\mathrm{d}\boldsymbol{p}}{\mathrm{d}t} = q\boldsymbol{E} + q\boldsymbol{v} \times \boldsymbol{B} \tag{17.32}$$

(the **Lorentz force**).

We can use the relativistic energy–momentum relation to find the rate of change of the energy $\mathcal{E}$ of the particle:

$$\mathcal{E} = \sqrt{p^2c^2 + m^2c^4} = \sqrt{\boldsymbol{p} \cdot \boldsymbol{p}c^2 + m^2c^4}.$$

Therefore, the change $\mathrm{d}\mathcal{E}$ in the energy when the momentum changes by $\mathrm{d}\boldsymbol{p}$ is

$$\mathrm{d}\mathcal{E} = \frac{c^2\boldsymbol{p} \cdot \mathrm{d}\boldsymbol{p}}{\mathcal{E}} = \frac{c^2m\boldsymbol{v} \cdot \mathrm{d}\boldsymbol{p}}{\mathcal{E}\sqrt{1 - v^2/c^2}} = \boldsymbol{v} \cdot \mathrm{d}\boldsymbol{p},$$

and so

$$\frac{\mathrm{d}\mathcal{E}}{\mathrm{d}t} = \boldsymbol{v} \cdot \frac{\mathrm{d}\boldsymbol{p}}{\mathrm{d}t} = q\boldsymbol{E} \cdot \boldsymbol{v},$$

where we have used the Lorentz force law and the fact that $\boldsymbol{v} \cdot (\boldsymbol{v} \times \boldsymbol{B}) \equiv 0$. Thus, **only the electric field does work on the charge** (the magnetic field cannot, since the force it exerts is perpendicular to the motion of the charge).

The equations of motion

$$\frac{\mathrm{d}\boldsymbol{p}}{\mathrm{d}t} = q\boldsymbol{E} + q\boldsymbol{v} \times \boldsymbol{B}$$

are not altered if we make the replacements

$$t \to -t, \quad \boldsymbol{E} \to \boldsymbol{E}, \quad \boldsymbol{B} \to -\boldsymbol{B},$$

that is, the reverse motion is possible if $\boldsymbol{B}$ is reversed. (The magnetic-induction vector would, in any case, be reversed by the reversal of the currents giving rise to it.)

From the definitions of $\boldsymbol{E}$ and $\boldsymbol{B}$, the equations are invariant under the replacements

$$t \to -t, \boldsymbol{E} \to \boldsymbol{E}, \boldsymbol{B} \to -\boldsymbol{B}.$$

17.4.2 Gauge invariance

The action integral for the interaction of a charge with an electromagnetic field is (see section 17.4)

$$S_{\mathrm{int}} = -q \int_{\mathrm{a}}^{\mathrm{b}} A_i \mathrm{d}x^i .$$

Let

$$A_i \to A_i' = A_i + \frac{\partial f}{\partial x^i}, \tag{17.33}$$

where f is an arbitrary Lorentz-scalar function of the space–time coordinates x^i. Then

$$\begin{aligned} S_{\text{int}} \to S_{\text{int}}' &= -q \int_a^b A_i \mathrm{d}x^i - q \int_a^b \frac{\partial f}{\partial x^i} \mathrm{d}x^i \\ &= S_{\text{int}} - q \int_a^b \mathrm{d}f = S_{\text{int}} - q[f(\mathrm{b}) - f(\mathrm{a})] = S_{\text{int}} + \text{const.} \end{aligned}$$

But since addition of a constant to the action has no effect on the equations of motion, neither does the replacement (17.33) [a **gauge transformation**]. The transformation (17.33) can be written as

$$\frac{\varphi}{c} \to \frac{\varphi'}{c} = \frac{\varphi}{c} + \frac{\partial f}{\partial (ct)}, \qquad -\boldsymbol{A} \to -\boldsymbol{A}' = -\boldsymbol{A} + \boldsymbol{\nabla} f,$$

that is,

$$\varphi \to \varphi' = \varphi + \frac{\partial f}{\partial t}, \qquad \boldsymbol{A} \to \boldsymbol{A}' = \boldsymbol{A} - \boldsymbol{\nabla} f. \tag{17.34}$$

It is easy to check that this leaves $\boldsymbol{E}$ and $\boldsymbol{B}$ invariant:

$$\begin{aligned} \boldsymbol{E}' &= -\frac{\partial \boldsymbol{A}'}{\partial t} - \boldsymbol{\nabla}\varphi' = -\frac{\partial \boldsymbol{A}}{\partial t} + \frac{\partial}{\partial t}(\boldsymbol{\nabla} f) - \boldsymbol{\nabla}\varphi - \boldsymbol{\nabla}\left(\frac{\partial f}{\partial t}\right) = \boldsymbol{E}, \\ \boldsymbol{B}' &= \boldsymbol{\nabla} \times \boldsymbol{A}' = \boldsymbol{\nabla} \times \boldsymbol{A} = \boldsymbol{B} \quad (\text{since curl grad} \equiv 0). \end{aligned}$$

(For example, we can add an arbitrary constant vector to $\boldsymbol{A}$ and an arbitrary constant scalar to φ without changing the fields $\boldsymbol{E}$ and $\boldsymbol{B}$.)

The invariance of all equations for physical quantities under the transformation (17.34) is called **gauge invariance**.

Using (17.30) and (17.31) we can describe constant electric and magnetic fields $\boldsymbol{E}$ and $\boldsymbol{B}$ by the potentials

$$\varphi = -\boldsymbol{E} \cdot \boldsymbol{r} \quad \text{and} \quad \boldsymbol{A} = \frac{1}{2} \boldsymbol{B} \times \boldsymbol{r}.$$

17.4.3 Four-space derivation of the equations of motion

We now derive the equations of motion of a charge in a given electromagnetic field directly from the principle of least action ($\delta S = 0$), instead of from the lagrangian L and Euler–Lagrange equations, as above. We have

$$\delta S = \delta \int_a^b (-mc\mathrm{d}s - qA_i \mathrm{d}x^i) = 0.$$

Put $\mathrm{d}s = \sqrt{\mathrm{d}x^i \mathrm{d}x_i}$. Then

$$\delta S = \int_a^b \left[-mc\delta \mathrm{d}s - q\delta(A_i \mathrm{d}x^i)\right] = -\int_a^b \left[mc\frac{\mathrm{d}x_i \delta \mathrm{d}x^i}{\mathrm{d}s} + qA_i \delta \mathrm{d}x^i + q\delta A_i \mathrm{d}x^i\right]$$

$$= -\int_a^b \left[mcu_i \mathrm{d}\delta x^i + qA_i \mathrm{d}\delta x^i + q\delta A_i \mathrm{d}x^i\right]$$

$$= -\left[mcu_i \delta x^i + qA_i \delta x^i\right]\Big|_a^b + \int_a^b \left[mc\mathrm{d}u_i \delta x^i + q\mathrm{d}A_i \delta x^i - q\delta A_i \mathrm{d}x^i\right] = 0.$$

The integrated term vanishes, since $(\delta x^i)_a = (\delta x^i)_b = 0$, and we write the second term in the form $\int_a^b [\ldots]\delta x^i \mathrm{d}s$ by writing

$$\mathrm{d}u_i = \frac{\mathrm{d}u_i}{\mathrm{d}s}\mathrm{d}s, \quad \mathrm{d}A_i = \frac{\partial A_i}{\partial x^k}\mathrm{d}x^k = \frac{\partial A_i}{\partial x^k}u^k \mathrm{d}s$$

and

$$\delta A_i \mathrm{d}x^i = \frac{\partial A_i}{\partial x^k}\delta x^k \mathrm{d}x^i = \frac{\partial A_k}{\partial x^i}\delta x^i \mathrm{d}x^k = \frac{\partial A_k}{\partial x^i}u^k \mathrm{d}s \delta x^i.$$

Then

$$\delta S = \int_a^b \left[mc\frac{\mathrm{d}u_i}{\mathrm{d}s} + q\left(\frac{\partial A_i}{\partial x^k} - \frac{\partial A_k}{\partial x^i}\right)u^k\right]\delta x^i \mathrm{d}s = 0.$$

Since the variations δx^i along the path are arbitrary, we must have

$$mc\frac{\mathrm{d}u_i}{\mathrm{d}s} = q\left(\frac{\partial A_k}{\partial x^i} - \frac{\partial A_i}{\partial x^k}\right)u^k.$$

We now define the **electromagnetic-field tensor** F_{ik} as

$$F_{ik} = \frac{\partial A_k}{\partial x^i} - \frac{\partial A_i}{\partial x^k}.$$

Then the **equation of motion** derived becomes

$$mc\frac{\mathrm{d}u_i}{\mathrm{d}s} = qF_{ik}u^k.$$

Because $F_{ki} = -F_{ik}$, we have $F_{ik} = 0$ if $i = k$. The other components of F_{ik} are

$$F_{12} = \frac{\partial A_2}{\partial x^1} - \frac{\partial A_1}{\partial x^2} = \frac{\partial(-A_y)}{\partial x} - \frac{\partial(-A_x)}{\partial y} = -(\boldsymbol{\nabla} \times \boldsymbol{A})_z = -B_z,$$

that is,

$$F_{12} = -F_{21} = -B_z, \quad F_{13} = -F_{31} = B_y, \quad F_{23} = -F_{32} = -B_x.$$

Similarly,

$$F_{01} = \frac{\partial A_1}{\partial x^0} - \frac{\partial A_0}{\partial x^1} = \frac{\partial(-A_x)}{\partial(ct)} - \frac{\partial(\varphi/c)}{\partial x} = \frac{1}{c}\left(-\frac{\partial A_x}{\partial t} - \frac{\partial \varphi}{\partial x}\right) = \frac{E_x}{c},$$

that is,

$$F_{01} = -F_{10} = E_x/c, \quad F_{02} = -F_{20} = E_y/c, \quad F_{03} = -F_{30} = E_z/c.$$

Expressed in a two-dimensional array, the components of F_{ik} are

$$(F_{ik}) = \begin{pmatrix} 0 & E_x/c & E_y/c & E_z/c \\ -E_x/c & 0 & -B_z & B_y \\ -E_y/c & B_z & 0 & -B_x \\ -E_z/c & -B_y & B_x & 0 \end{pmatrix}, \tag{17.35}$$

that is, **the components of *E* and *B* are components of a single four-tensor**.

We now check that the equations

$$mc\frac{\mathrm{d}u_i}{\mathrm{d}s} = qF_{ik}u^k \quad (i, k = 0, 1, 2, 3) \tag{17.36}$$

coincide with the equations of motion obtained previously.

Put

$$\mathrm{d}s = c\mathrm{d}t\sqrt{1 - v^2/c^2}, \quad mcu_i \equiv p_i$$

and

$$u^0 = u_0 = \frac{1}{\sqrt{1 - v^2/c^2}}, \quad u^{1,2,3} = -u_{1,2,3} = \frac{v_{x,y,z}/c}{\sqrt{1 - v^2/c^2}}.$$

The $i = 1$ equation becomes

$$-\frac{1}{c\sqrt{1 - v^2/c^2}}\frac{\mathrm{d}p_x}{\mathrm{d}t} = q\left[\left(-\frac{E_x}{c}\right)\frac{1}{\sqrt{1 - v^2/c^2}} + (-B_z)\frac{v_y/c}{\sqrt{1 - v^2/c^2}} + (+B_y)\frac{v_z/c}{\sqrt{1 - v^2/c^2}}\right]$$

that is,

$$\frac{\mathrm{d}p_x}{\mathrm{d}t} = qE_x + q(\boldsymbol{v} \times \boldsymbol{B})_x, \tag{17.37}$$

as expected. Similarly, using $p_0 = \mathcal{E}/c$, we get for the $i = 0$ equation:

$$\frac{1}{c\sqrt{1-v^2/c^2}}\frac{\mathrm{d}(\mathcal{E}/c)}{\mathrm{d}t} = q\left[\frac{E_x}{c}\frac{v_x/c}{\sqrt{1-v^2/c^2}} + \frac{E_y}{c}\frac{v_y/c}{\sqrt{1-v^2/c^2}} + \frac{E_z}{c}\frac{v_z/c}{\sqrt{1-v^2/c^2}}\right],$$

that is,

$$\frac{\mathrm{d}\mathcal{E}}{\mathrm{d}t} = q\boldsymbol{E} \cdot \boldsymbol{v}, \tag{17.38}$$

as found previously. [Of course, (17.38) is a consequence of (17.37), that is, the four equations (17.36) are not independent. This can be seen by contracting both sides of (17.36) with u^i, when the left-hand side becomes zero (since $u^i \mathrm{d}u_i/\mathrm{d}s = 0$) and the right-hand side also becomes zero by virtue of the antisymmetry of F_{ik}.]

17.4.4 Lorentz transformation of the electromagnetic field

To find how $\boldsymbol{E}$ and $\boldsymbol{B}$ transform under Lorentz transformations we use the fact that F^{ik} is a four-tensor, that is, it transforms like $x^i x^k$ under Lorentz transformations. For example, for the transformation from a Lorentz frame K to a Lorentz frame K′ moving relative to K with velocity V along the X axis we have (since $x_2 = x'_2$ and $x_3 = x'_3$):

$$F_{23} = F'_{23}.$$

Also,

F_{12} transforms like x_1x_2, i.e., (in this case) like x_1;
F_{02} transforms like x_0x_2, i.e., (in this case) like x_0;
F_{13} transforms like x_1x_3, i.e., (in this case) like x_1;
F_{03} transforms like x_0x_3, i.e., (in this case) like x_0.

Thus,

$$F_{12} = \frac{F'_{12} - \frac{V}{c}F'_{02}}{\sqrt{1-V^2/c^2}}, \quad F_{02} = \frac{F'_{02} - \frac{V}{c}F'_{12}}{\sqrt{1-V^2/c^2}}, \quad F_{13} = \frac{F'_{13} - \frac{V}{c}F'_{03}}{\sqrt{1-V^2/c^2}}, \quad F_{03} = \frac{F'_{03} - \frac{V}{c}F'_{13}}{\sqrt{1-V^2/c^2}}.$$

Similarly, F_{01} transforms like x_0x_1, for which

$$x_0x_1 = \frac{x'_0 - \frac{V}{c}x'_1}{\sqrt{1-V^2/c^2}} \times \frac{x'_1 - \frac{V}{c}x'_0}{\sqrt{1-V^2/c^2}} = \frac{x'_0x'_1 + \frac{V^2}{c^2}x'_1x'_0 - \frac{V}{c}\left(x_0'^2 + x_1'^2\right)}{1-V^2/c^2},$$

that is,

$$F_{01} = \frac{F'_{01} + \frac{V^2}{c^2}F'_{10} - \frac{V}{c}\left(F'_{00} + F'_{11}\right)}{1 - V^2/c^2} = \frac{F'_{01} - \frac{V^2}{c^2}F'_{01}}{1 - V^2/c^2} = F'_{01}$$

(here we have used the fact that F_{ik} is antisymmetric). [The result $F_{01} = F'_{01}$ is a specific instance of the general theorem that an antisymmetric tensor with rank equal to the number of dimensions of the space (effectively 2 in the above instance) is invariant under rotations of the coordinates in this space – another example is our e^{ijkl}.]

Using the components of F_{ik} [see the array (17.35)], we get

$$\begin{aligned} E_x &= E'_x, \quad E_y = \frac{E'_y + VB'_z}{\sqrt{1 - V^2/c^2}}, \quad E_z = \frac{E'_z - VB'_y}{\sqrt{1 - V^2/c^2}}, \\ B_x &= B'_x, \quad B_y = \frac{B'_y - \frac{V}{c^2}E'_z}{\sqrt{1 - V^2/c^2}}, \quad B_z = \frac{B'_z + \frac{V}{c^2}E'_y}{\sqrt{1 - V^2/c^2}}. \end{aligned} \tag{17.39}$$

(As usual, all the formulas for the inverse transformation are obtained by replacing V by $-V$ and shifting the primes.)

If $V << c$ we can replace the denominators by 1 and obtain

$$\boldsymbol{E} \approx \boldsymbol{E}' + \boldsymbol{B}' \times \boldsymbol{V}, \quad \boldsymbol{B} \approx \boldsymbol{B}' - \frac{1}{c^2}\boldsymbol{E}' \times \mathbf{V}$$

(recall that $V_x = V$ and $V_y = V_z = 0$).

If there exists a velocity $\boldsymbol{V}$ such that $\boldsymbol{B}' = 0$ in the Lorentz frame K′, we always (that is, not just for $V \ll c$) have

$$\boldsymbol{B} = (\boldsymbol{V} \times \mathbf{E})/c^2 \tag{17.40}$$

in the Lorentz frame K.

[**Proof:**: By the transformation laws (17.39),

$$B_x = 0, \quad B_y = \frac{-VE'_z}{c^2\sqrt{1 - V^2/c^2}}, \quad B_z = \frac{VE'_y}{c^2\sqrt{1 - V^2/c^2}},$$

with

$$E_x = E'_x, \quad E_y = \frac{E'_y}{\sqrt{1 - V^2/c^2}}, \quad E_z = \frac{E'_z}{\sqrt{1 - V^2/c^2}}.$$

Combining these, we find

$$B_x = 0, \quad B_y = -\frac{VE_z}{c^2}, \quad B_z = \frac{VE_y}{c^2},$$

which, since $V_x = V$ and $V_y = V_z = 0$, is the component statement of (17.40).]

With regard to the direction of $\boldsymbol{B}$, equation (17.40) shows that $\boldsymbol{B}$ is perpendicular to $\boldsymbol{E}$ (and $\boldsymbol{V}$).

With regard to the magnitude of $\boldsymbol{B}$, (17.40) gives

$$\begin{aligned} B = \frac{E}{c}\frac{V}{c}\sin\theta &\leq \frac{E}{c}\frac{V}{c} \quad (\text{since } \sin\theta \leq 1) \\ &\leq \frac{E}{c} \quad (\text{since } V/c \leq 1). \end{aligned}$$

Similarly, if there exists a velocity $\boldsymbol{V}$ such that $\boldsymbol{E}' = 0$ in K′, we always have

$$\boldsymbol{E} = -\boldsymbol{V} \times \boldsymbol{B}$$

as follows from using the transformation laws (17.39).

This result shows that $\boldsymbol{E}$ is perpendicular to $\boldsymbol{B}$ in K (and to $\boldsymbol{V}$), and also that

$$\frac{E}{c} = \frac{V}{c} B \sin\theta \leq \frac{V}{c} B \leq B.$$

Summarizing, if the magnetic (respectively, electric) field at some position is zero in some inertial frame, then in all other inertial frames the magnetic and electric fields at that position are mutually perpendicular and such that $B \leq E/c$ (respectively, $E/c \leq B$).

Conversely, if $\boldsymbol{E}$ and $\boldsymbol{B}$ at some position are perpendicular in some inertial frame K, there exists an inertial frame K′ in which the field is purely electric (if $B \leq E/c$) or purely magnetic (if $E/c \leq B$).

17.4.5 Lorentz invariants constructed from the electromagnetic field

Since F_{ik} is a Lorentz four-tensor,

$$\begin{gathered} F_{ik}F^{ik} \text{ is a Lorentz invariant (scalar);} \\ e^{iklm}F_{ik}F_{lm} \text{ is a Lorentz pseudoscalar.} \end{gathered}$$

(The latter is the contraction of F_{lm} with its dual $e^{iklm}F_{ik}$.)

Using the components of the fully contravariant electromagnetic-field tensor

$$\left(F^{ik}\right) = \begin{pmatrix} 0 & -E_x/c & -E_y/c & -E_z/c \\ E_x/c & 0 & -B_z & B_y \\ E_y/c & B_z & 0 & -B_x \\ E_z/c & -B_y & B_x & 0 \end{pmatrix} \tag{17.41}$$

and the components F_{ik} [see the array (17.35)], we find

$$\begin{aligned} F_{ik}F^{ik} &= -\frac{E_x^2}{c^2} - \frac{E_y^2}{c^2} - \frac{E_z^2}{c^2} - \frac{E_x^2}{c^2} + B_z^2 + B_y^2 - \frac{E_y^2}{c^2} + B_z^2 + B_x^2 - \frac{E_z^2}{c^2} + B_y^2 + B_x^2 \\ &= 2\left(B^2 - \frac{E^2}{c^2}\right), \end{aligned}$$

that is, $B^2 - E^2/c^2$ is a **Lorentz scalar**.

Similarly,

$$\begin{aligned}
e^{iklm}F_{ik}F_{lm} &= [F_{01}F_{23} - F_{10}F_{23} - F_{01}F_{32} + F_{10}F_{32}] \quad \left(= -4\frac{E_x B_x}{c}\right) \\
&\quad + [-F_{02}F_{13} + F_{20}F_{13} + F_{02}F_{31} - F_{20}F_{31}] \quad \left(= -4\frac{E_y B_y}{c}\right) \\
&\quad + [F_{03}F_{12} - F_{30}F_{12} - F_{03}F_{21} + F_{30}F_{21}] \quad \left(= -4\frac{E_z B_z}{c}\right) \\
&= -\frac{4}{c}\boldsymbol{E}\cdot\boldsymbol{B},
\end{aligned}$$

that is, $\boldsymbol{E}\cdot\boldsymbol{B}$ is a **Lorentz pseudoscalar**. [We already knew that $\boldsymbol{E}\cdot\boldsymbol{B}$ is a pseudoscalar under purely **spatial** rotations and reflections, since $\boldsymbol{E}$ is a true (polar) three-vector and $\boldsymbol{B}$ is a pseudo- (axial) three-vector.]

Hence, if $\boldsymbol{E}$ is perpendicular to $\boldsymbol{B}$ in one inertial frame (so that $\boldsymbol{E}\cdot\boldsymbol{B} = 0$), the magnetic and electric fields remain perpendicular in all other inertial frames.

Similarly, since $B^2 - E^2/c^2$ is an invariant, if the magnitudes of $\boldsymbol{B}$ and $\boldsymbol{E}/c$ are equal in one inertial frame they are also equal in all others.

Also, if $B > E/c$ (respectively, $B < E/c$) in one Lorentz frame K, we have $B' > E'/c$ (respectively, $B' < E'/c$) in all other Lorentz frames K′.

If $\boldsymbol{E}$ and $\boldsymbol{B}$ make an acute (respectively, obtuse) angle in one frame, they make an acute (respectively, obtuse) angle in all others.

If fields $\boldsymbol{E}/c$ and $\boldsymbol{B}$ in frame K are **not** perpendicular, we can find the magnitudes E'/c and B' of the same fields as viewed in the frame K′ in which they are **parallel** by solving

$$B'^2 - \frac{E'^2}{c^2} = B^2 - \frac{E^2}{c^2}$$

and

$$\boldsymbol{B}'\cdot\boldsymbol{E}' = B'E' = \boldsymbol{B}\cdot\boldsymbol{E}$$

for B' and E'.

17.4.6 The first pair of Maxwell equations

(i) **Conventional (three-space) formulation**
From the definitions (17.30) and (17.31):

$$\boldsymbol{B} = \boldsymbol{\nabla}\times\boldsymbol{A} \quad \text{and} \quad \boldsymbol{E} = -\frac{\partial \boldsymbol{A}}{\partial t} - \boldsymbol{\nabla}\varphi,$$

we find

$$\boldsymbol{\nabla}\times\boldsymbol{E} = -\frac{\partial}{\partial t}\boldsymbol{\nabla}\times\mathbf{A} - \boldsymbol{\nabla}\times\boldsymbol{\nabla}\varphi = -\frac{\partial \boldsymbol{B}}{\partial t},$$

$$\boldsymbol{\nabla}\cdot\boldsymbol{B} = \boldsymbol{\nabla}\cdot\boldsymbol{\nabla}\times\boldsymbol{A} \equiv 0,$$

that is,

$$\nabla \times \boldsymbol{E} = -\frac{\partial \boldsymbol{B}}{\partial t}, \qquad \nabla \cdot \boldsymbol{B} = 0, \tag{17.42}$$

which are the first pair of Maxwell equations.

Therefore,

$$\int_V \nabla \cdot \boldsymbol{B} \mathrm{d}V = \oint_S \boldsymbol{B} \cdot \mathrm{d}\boldsymbol{S} = 0,$$

that is, the "flux" of the magnetic induction over any closed surface S is zero.

Similarly, by Stokes' theorem, the "circulation" of the vector E around a closed contour Γ (i.e., the electromotive "force" in the contour) is given by

$$\oint_\Gamma \boldsymbol{E} \cdot \mathrm{d}\boldsymbol{l} = \int_S (\nabla \times \boldsymbol{E}) \cdot \mathrm{d}\boldsymbol{S} = -\int_S \frac{\partial \boldsymbol{B}}{\partial t} \cdot \mathrm{d}\boldsymbol{S} = -\frac{\mathrm{d}}{\mathrm{d}t} \int_S \boldsymbol{B} \cdot \mathrm{d}\boldsymbol{S} \equiv -\frac{\mathrm{d}\Phi}{\mathrm{d}t},$$

where Φ is the magnetic flux linking any fixed surface S bounded by Γ.

(ii) **Four-space formulation**

Since

$$F_{ik} = \frac{\partial A_k}{\partial x^i} - \frac{\partial A_i}{\partial x^k},$$

it follows easily that

$$\frac{\partial F_{ik}}{\partial x^l} + \frac{\partial F_{kl}}{\partial x^i} + \frac{\partial F_{li}}{\partial x^k} = 0. \tag{17.43}$$

The expression on the left is clearly antisymmetric under interchange of any pair of indices. Thus, of the 4^3 equations (17.43), corresponding to the 4^3 possible choices of the free indices $\{i, k, l\}$, only four (corresponding to the four ways of choosing the indices i, k, l to be all different) are not of the form 0 = 0. Direct substitution of the components F_{ik} shows that these four equations are

$$\nabla \times \boldsymbol{E} = -\frac{\partial \boldsymbol{B}}{\partial t}, \qquad \nabla \cdot \boldsymbol{B} = 0$$

[the first pair of Maxwell equations (17.42)].

[Alternatively, since the contraction with e^{jikl} of each term of the fully covariant, fully antisymmetric rank-3 tensor forming the left-hand side of (17.43) gives the same result (after relabelling of the dummy indices in the second and third terms of the contraction), we find

$$e^{jikl} \frac{\partial F_{ik}}{\partial x^l} = 0,$$

which has only one free index ($j = 0, 1, 2, 3$) and is the four equations constituting the first "pair" of Maxwell equations (17.42).]

17.5 Dynamics of the electromagnetic field

In this section we examine the consequences of an action of the form

$$S = S_{\rm p} + S_{\rm pf} + S_{\rm f},$$

where the sum of

$$S_{\rm p} = -\sum_{\rm particles} mc \int {\rm d}s \quad \text{and} \quad S_{\rm pf} = -\sum_{\rm charges} q \int A_i {\rm d}x^i$$

is the many-particle form of the action (see the previous section) of the motion of a particle of mass m and charge q in a **fixed** electromagnetic field described by the four-potential A^i, and the action $S_{\rm f}$ is the new term that is required when we no longer treat the field as a fixed "background" with no intrinsic dynamics.

To take account of the dynamics of the field itself (and hence to find the remaining Maxwell equations describing the field), we must postulate a form for the action $S_{\rm f}$.

Our first requirement will be that the **principle of superposition** be valid for electric and magnetic fields, that is, that the field intensities due to several particles are the vector sums of those due to each individual particle. Quantities satisfying linear (homogeneous, of degree 1) differential equations satisfy this principle. We therefore postulate that the equation of motion of F_{ik} (F_{ik} is linear in A_i and hence in $\boldsymbol{E}$ and $\boldsymbol{B}$) is a linear differential equation.

For such an equation to be obtained by variation of the action (using the principle of least action), F_{ik} must appear **bilinearly** in the integrand of the action. (Taking the variation with respect to F_{ik} will reduce the degree in F_{ik} by unity.)

The action $S_{\rm f}$ must also be a Lorentz scalar, and so we try an integral over ${\rm d}\Omega$ [$= {\rm d}(ct){\rm d}x{\rm d}y{\rm d}z = c{\rm d}t{\rm d}V$] of some Lorentz scalar.

The only Lorentz scalar of second degree that can be formed from F_{ik} is $F_{ik}F^{ik}$ (recall that $e^{iklm}F_{ik}F_{lm}$ is a pseudoscalar).

Therefore, we try

$$S_{\rm f} = a \iint_{t_{\rm a}}^{t_{\rm b}} F_{ik}F^{ik}{\rm d}V{\rm d}t = -2a \iint_{t_{\rm a}}^{t_{\rm b}} \left(\frac{E^2}{c^2} - B^2\right){\rm d}V{\rm d}t,$$

where $\int {\rm d}V \ldots$ is over all space.

The definition of $\boldsymbol{E}$ contains the term $-\partial \boldsymbol{A}/\partial t$, and $(\partial \boldsymbol{A}/\partial t)^2$ must appear in the action with a positive sign. (Otherwise, $S_{\rm f}$ would not have a minimum, since it could be made to go to $-\infty$ by choosing a vector potential $\boldsymbol{A}$ with infinitely rapid variation in time.) We must therefore choose $a < 0$.

Choosing a value for the constant a is equivalent to choosing units of measurement of the electromagnetic field. The following choice, as we shall see later, leads to the Maxwell equations in SI units:

$$a = -\frac{\varepsilon_0 c^2}{4},$$

where ε_0 is the dielectric permittivity of free space.

Therefore,

$$S_\mathrm{f} = -\frac{\varepsilon_0 c}{4}\int F_{ik}F^{ik}\mathrm{d}\Omega,$$

or, in three-space form,

$$S_\mathrm{f} = \frac{\varepsilon_0 c^2}{2}\int\left(\frac{E^2}{c^2} - B^2\right)\mathrm{d}V\mathrm{d}t.$$

Thus, the lagrangian of the field is (recall that $S = \int L\mathrm{d}t$)

$$L_\mathrm{f} = \frac{\varepsilon_0 c^2}{2}\int\left(\frac{E^2}{c^2} - B^2\right)\mathrm{d}V.$$

The total action of the field plus particles is

$$S = -\sum_{\text{particles}} mc\int \mathrm{d}s - \sum_{\text{charges}} q\int A_i\mathrm{d}x^i - \frac{\varepsilon_0 c}{4}\int F_{ik}F^{ik}\mathrm{d}\Omega.$$

Before considering variations of this action with respect to variations of the (no longer fixed) electromagnetic field, we shall recast the second term in terms of a four-vector j^i called the **four-current**.

17.5.1 The four-current and the second pair of Maxwell equations

For a continuous distribution of charge, in some given Lorentz frame, we can introduce a charge density $\rho(\boldsymbol{r}, t)$ such that $\rho(\boldsymbol{r}, t)\mathrm{d}V$ is the charge contained in volume $\mathrm{d}V$ at position $\boldsymbol{r}$ at time t.

If the charges are pointlike, $\int\rho\mathrm{d}V$ will give the sum $\sum_n q_n$ of the charges in the integration volume. In this case, if $\boldsymbol{r}_n(t)$ is the position vector of the n-th charge at time t,

$$\rho(\boldsymbol{r}, t) = \sum_n q_n\delta^{(3)}(\boldsymbol{r} - \boldsymbol{r}_n(t)),$$

where $\delta^{(3)}(\boldsymbol{r} - \boldsymbol{r}_n(t))$ is the three-dimensional Dirac delta-function (with dimensions of $[\text{length}]^{-3}$).

The charge $\mathrm{d}q = \rho\mathrm{d}V$ in volume $\mathrm{d}V$ is a Lorentz invariant, but $\mathrm{d}V$ (which is subject to Lorentz contraction) and ρ are not. Multiply both sides of this Lorentz scalar by $\mathrm{d}x^i$:

$$\mathrm{d}q\mathrm{d}x^i = \rho\mathrm{d}V\mathrm{d}x^i = \rho\mathrm{d}V\mathrm{d}t\frac{\mathrm{d}x^i}{\mathrm{d}t}.$$

The left-hand side here is a four-vector (since $\mathrm{d}q$ is a four-scalar and $\mathrm{d}x^i$ is a four-vector), and so the right-hand side must also be a four-vector.

In the right-hand side, $\mathrm{d}V\mathrm{d}t = \mathrm{d}\Omega/c$ is a four-scalar, and so $\rho\mathrm{d}x^i/\mathrm{d}t$ must be a four-vector (although ρ is not a four-scalar and $\mathrm{d}x^i/\mathrm{d}t$ is not a four-vector). We call it the **current four-vector** or **four-current** j^i:

$$j^i = \rho\frac{\mathrm{d}x^i}{\mathrm{d}t}.$$

The $i = 0$ component of j^i is (since $x^0 = ct$)

$$j^0 = c\rho,$$

and the components j^α ($\alpha = 1, 2, 3$) form

$$\boldsymbol{j} = \rho \boldsymbol{v}$$

(the **current-density three-vector**).

The total charge in all space is

$$\int \rho \mathrm{d}V = \frac{1}{c}\int j^0 \mathrm{d}V = \frac{1}{c}\int j^0 \mathrm{d}S_0.$$

[Recall (see section 17.2.9) that $\mathrm{d}S_0 = \mathrm{d}S^0 = \mathrm{d}x\mathrm{d}y\mathrm{d}z$.]

Since the integral here is over a hyperplane $x_0 = \text{const}$, so that $\mathrm{d}S_i = (\mathrm{d}S_0, 0, 0, 0)$, we can also write this integral as

$$\int \rho \mathrm{d}V = \frac{1}{c}\int j^i \mathrm{d}S_i.$$

Just as $\frac{1}{c}\int j^0 \mathrm{d}S_0$ is the sum of the charges whose world lines pass through the hyperplane $x_0 = \text{const}$, so $\frac{1}{c}\int j^i \mathrm{d}S_i$ over an **arbitrary** spacelike hypersurface is the sum of the charges whose world lines pass through this hypersurface.

Since charge is conserved, any two such integrals over hypersurfaces covering all of three-dimensional space will be equal. The difference (0) between the two integrals $\frac{1}{c}\int j^i \mathrm{d}S_i$ can be written as $\oint j^i \mathrm{d}S_i$, where the integral $\oint \ldots$ is over the **closed** hypersurface enclosing the four-volume between the two hypersurfaces of integration in the original integrals. [We can neglect the "sides" of this four-volume at spatial infinity, since there are no charges at infinity (our hypersurfaces were chosen to be big enough to be cut by the world lines of **all** the charges). Alternatively, we can regard each hypersurface encompassing the whole of three-dimensional space as a closed (boundary-less) hypersurface, analogous to the closed two-dimensional surface of a ball.]

Therefore, using the four-dimensional Gauss' theorem, we have

$$0 = \oint j^i \mathrm{d}S_i = \int \frac{\partial j^i}{\partial x^i}\mathrm{d}\Omega.$$

We can equally well consider a group of particles carrying some specific total charge. Again, since charge is conserved (charge is neither created nor destroyed), the total charge "carried" by particle world lines entering a finite hypervolume Ω across a given finite "earlier" hypersurface (that is, occupying a given finite volume, earlier) is equal to the total charge "carried" by particle world lines leaving it through the "later" hypersurface. Thus, we have

$$\int_\Omega \frac{\partial j^i}{\partial x^i}\mathrm{d}\Omega = 0.$$

For this to be true for an arbitrary four-volume Ω of integration, we must have

$$\frac{\partial j^i}{\partial x^i} = 0,$$

which is known as the **continuity equation**.

Using

$$j^i = (c\rho, j_x, j_y, j_z) \quad \text{and} \quad x^i = (ct, x, y, z),$$

we can write the continuity equation as

$$\frac{\partial \rho}{\partial t} + \nabla \cdot \boldsymbol{j} = 0.$$

From its derivation, the continuity equation can be seen to be an expression of the conservation of charge.

Returning now to the action, we can write the interaction term $-\sum q \int A_i \mathrm{d}x^i$ in terms of the current j^i by putting $-\sum q \ldots = -\int \rho \mathrm{d}V \ldots$ and using $j^i = \rho \mathrm{d}x^i/\mathrm{d}t$. We get

$$-\sum q \int A_i \mathrm{d}x^i = -\int \rho \mathrm{d}V \int A_i \frac{\mathrm{d}x^i}{\mathrm{d}t}\mathrm{d}t = -\int A_i j^i \mathrm{d}V \mathrm{d}t = -\frac{1}{c}\int A_i j^i \mathrm{d}\Omega.$$

Therefore,

$$S = -\sum_{\text{particles}} mc \int \mathrm{d}s - \frac{1}{c}\int A_i j^i \mathrm{d}\Omega - \frac{\varepsilon_0 c}{4}\int F_{ik}F^{ik}\mathrm{d}\Omega.$$

To find the **field** equations by the principle of least action we assume the motion of the charges to be **given** and vary only the four-potential A_i (and hence the electromagnetic-field tensor F_{ik}). (To find the equations of motion of a **particle**, we varied the trajectory of the particle for a **given field**.)

Thus, the variation δS_p of the first term is zero, and in the second we must not vary j^i. Therefore,

$$\delta S = -\frac{1}{c}\int \left(j^i \delta A_i + \frac{\varepsilon_0 c^2}{2} F^{ik} \delta F_{ik} \right) \mathrm{d}\Omega = 0,$$

where we have used the fact that

$$\delta(F_{ik}F^{ik}) = 2F^{ik}\delta F_{ik} \; (= 2F_{ik}\delta F^{ik}).$$

But

$$F_{ik} = \frac{\partial A_k}{\partial x^i} - \frac{\partial A_i}{\partial x^k}.$$

Therefore,

$$\begin{aligned}
\delta S &= -\frac{1}{c}\int \left[j^i \delta A_i + \frac{\varepsilon_0 c^2}{2} F^{ik} \delta \left(\frac{\partial A_k}{\partial x^i} - \frac{\partial A_i}{\partial x^k} \right) \right] \mathrm{d}\Omega \\
&= -\frac{1}{c}\int \left[j^i \delta A_i + \frac{\varepsilon_0 c^2}{2} F^{ik} \frac{\partial}{\partial x^i}(\delta A_k) - \frac{\varepsilon_0 c^2}{2} F^{ik} \frac{\partial}{\partial x^k}(\delta A_i) \right] \mathrm{d}\Omega.
\end{aligned}$$

In the second term we interchange the dummy indices i and k, and use the antisymmetry property $F_{ki} = -F_{ik}$. Then

$$\begin{aligned}\delta S &= -\frac{1}{c}\int\left[j^i\delta A_i - \varepsilon_0 c^2 F^{ik}\frac{\partial}{\partial x^k}(\delta A_i)\right]\mathrm{d}\Omega \\ &= -\frac{1}{c}\int\left(j^i + \varepsilon_0 c^2\frac{\partial F^{ik}}{\partial x^k}\right)\delta A_i \mathrm{d}\Omega + \varepsilon_0 c\int\frac{\partial}{\partial x^k}(F^{ik}\delta A_i)\mathrm{d}\Omega.\end{aligned}$$

By the four-dimensional Gauss' theorem the last term becomes

$$\varepsilon_0 c \oint F^{ik}\delta A_i \mathrm{d}S_k$$

over a surface at the limits of the hypervolume integration (which are the "sides" of the hypervolume at spatial infinity and the hyperplanes $x^0 = ct_\mathrm{a}$ and $x^0 = ct_\mathrm{b}$, where t_a and t_b are the initial and final time values). Since $F_{ik} = 0$ at spatial infinity and $\delta A_i(t_\mathrm{a}) = \delta A_i(t_\mathrm{b}) = 0$ (the field evolves between **specified** initial and final configurations on the hyperplanes corresponding to the initial and final times), this term vanishes.

Then, since $\delta S = 0$ for arbitrary field variations δA_i (which are themselves fields defined throughout the above hypervolume), we must have

$$\frac{\partial F^{ik}}{\partial x^k} = -\frac{j^i}{\varepsilon_0 c^2} \quad \text{(four equations: } i = 0, 1, 2, 3\text{)}.$$

It is easy to show using the array (17.41) we see that these are the "second pair" of Maxwell equations:

$$\begin{aligned} i = 0 &\Rightarrow \boldsymbol{\nabla}\cdot\boldsymbol{E} = \frac{\rho}{\varepsilon_0}, \\ i = 1, 2, 3 &\Rightarrow \boldsymbol{\nabla}\times\boldsymbol{B} = \frac{1}{c^2}\frac{\partial\boldsymbol{E}}{\partial t} + \frac{1}{\varepsilon_0 c^2}\boldsymbol{j}.\end{aligned}$$

Here ε_0 is the **dielectric permittivity** of free space, and the quantity

$$\frac{1}{\varepsilon_0 c^2} \equiv \mu_0$$

is the **magnetic permeability** of free space (that is, $\varepsilon_0\mu_0 = 1/c^2$). Thus, we have

$$\boldsymbol{\nabla}\cdot\boldsymbol{E} = \frac{\rho}{\varepsilon_0}, \quad \boldsymbol{\nabla}\times\boldsymbol{B} = \frac{1}{c^2}\frac{\partial\boldsymbol{E}}{\partial t} + \mu_0\boldsymbol{j}. \tag{17.44}$$

These equations, together with the "first pair" of Maxwell equations, completely determine the electromagnetic field (if the necessary boundary conditions are given).

The flux of $\boldsymbol{E}$ through the closed surface S bounding a volume V is

$$\oint_\mathrm{S}\boldsymbol{E}\cdot\mathrm{d}\boldsymbol{S} = \int_\mathrm{V}\boldsymbol{\nabla}\cdot\boldsymbol{E}\mathrm{d}V = \frac{1}{\varepsilon_0}\oint_\mathrm{V}\rho\mathrm{d}V = \frac{Q}{\varepsilon_0},$$

where Q is the total charge in the volume V.

Similarly, by Stokes' theorem the circulation of $\boldsymbol{B}$ around a closed contour Γ is, for any surface S bounded by this contour,

$$\oint_\Gamma \boldsymbol{B} \cdot \mathrm{d}\boldsymbol{l} = \int_S (\boldsymbol{\nabla} \times \boldsymbol{B}) \cdot \mathrm{d}\boldsymbol{S} = \frac{1}{c^2} \int_S \frac{\partial \boldsymbol{E}}{\partial t} \cdot \mathrm{d}\boldsymbol{S} + \mu_0 \int_S \boldsymbol{j} \cdot \mathrm{d}\boldsymbol{S}$$
$$= \mu_0 \int_S \left(\boldsymbol{j} + \varepsilon_0 \frac{\partial \boldsymbol{E}}{\partial t} \right) \cdot \mathrm{d}\boldsymbol{S},$$

where the term $\varepsilon_0 \partial \boldsymbol{E}/\partial t$ is known as the **displacement current**.

Since div**curl** $\equiv 0$, and $\boldsymbol{\nabla} \cdot \boldsymbol{E} = \rho/\varepsilon_0$, by taking the divergence of the second equation (17.44) we get

$$0 = \frac{1}{c^2} \frac{\partial}{\partial t} \left(\frac{\rho}{\varepsilon_0} \right) + \mu_0 \boldsymbol{\nabla} \cdot \boldsymbol{j},$$

that is,

$$\frac{\partial \rho}{\partial t} + \boldsymbol{\nabla} \cdot \boldsymbol{j} = 0.$$

Alternatively, from

$$\frac{\partial F^{ik}}{\partial x^k} = -\frac{j^i}{\varepsilon_0 c^2} = -\mu_0 j^i,$$

we have

$$\frac{\partial j^i}{\partial x^i} = -\frac{1}{\mu_0} \frac{\partial^2 F^{ik}}{\partial x^i \partial x^k} = -\frac{1}{\mu_0} \frac{\partial^2 F^{ki}}{\partial x^k \partial x^i} = \frac{1}{\mu_0} \frac{\partial^2 F^{ik}}{\partial x^k \partial x^i} = \frac{1}{\mu_0} \frac{\partial^2 F^{ik}}{\partial x^i \partial x^k} = -\frac{\partial j^i}{\partial x^i} = 0,$$

where we have interchanged dummy indices in the second equality, used the antisymmetry of F^{ik} in the third, and used the commutativity of differentiations with respect to independent coordinates in the fourth. The resulting equation

$$\frac{\partial j^i}{\partial x^i} = 0$$

is the familiar four-dimensional form of the continuity equation.

17.5.2 Energy density and energy flux density of the electromagnetic field

The two "circulation" Maxwell equations are

$$\boldsymbol{\nabla} \times \boldsymbol{E} = -\frac{\partial \boldsymbol{B}}{\partial t} \quad \text{and} \quad \boldsymbol{\nabla} \times \boldsymbol{B} = \frac{1}{c^2} \frac{\partial \boldsymbol{E}}{\partial t} + \mu_0 \boldsymbol{j}.$$

Scalar-multiply the first of these by $\boldsymbol{B}$ and subtract the result from the result of scalar-multiplying the second by $\boldsymbol{E}$:

$$\frac{1}{c^2}\boldsymbol{E}\cdot\frac{\partial \boldsymbol{E}}{\partial t}+\boldsymbol{B}\cdot\frac{\partial \boldsymbol{B}}{\partial t}=-\mu_0\boldsymbol{j}\cdot\boldsymbol{E}+\boldsymbol{E}\cdot\boldsymbol{\nabla}\times\boldsymbol{B}-\boldsymbol{B}\cdot\boldsymbol{\nabla}\times\boldsymbol{E},$$

that is,

$$\frac{1}{2}\frac{\partial}{\partial t}\left(\frac{E^2}{c^2}+B^2\right)=-\mu_0\boldsymbol{j}\cdot\boldsymbol{E}-\boldsymbol{\nabla}\cdot(\boldsymbol{E}\times\boldsymbol{B}).$$

Put $1/c^2=\varepsilon_0\mu_0$, write $\boldsymbol{B}$ as $\boldsymbol{B}\equiv\mu_0\boldsymbol{H}$ (which defines the **magnetic-field strength $\boldsymbol{H}$**), and divide the resulting equation by μ_0:

$$\frac{1}{2}\frac{\partial}{\partial t}\left(\varepsilon_0E^2+\mu_0H^2\right)=-\boldsymbol{j}\cdot\boldsymbol{E}-\boldsymbol{\nabla}\cdot\boldsymbol{\mathcal{S}},$$

where the vector

$$\boldsymbol{\mathcal{S}}\equiv\frac{1}{\mu_0}\boldsymbol{E}\times\boldsymbol{B}=\boldsymbol{E}\times\boldsymbol{H}$$

is known as the **Poynting vector**.

Integrate the equation over all space and apply Gauss' theorem to the last term:

$$\frac{\mathrm{d}}{\mathrm{d}t}\int\left(\frac{\varepsilon_0E^2+\mu_0H^2}{2}\right)\mathrm{d}V=-\int\boldsymbol{j}\cdot\boldsymbol{E}\mathrm{d}V-\oint\boldsymbol{\mathcal{S}}\cdot\mathrm{d}\boldsymbol{S}.$$

The last term vanishes because the fields are zero on the surface at spatial infinity.

Also, we can put

$$\int\boldsymbol{j}\cdot\boldsymbol{E}\mathrm{d}V=\int\mathrm{d}V\rho\boldsymbol{v}\cdot\boldsymbol{E}=\sum_{\substack{\text{charged}\\ \text{particles } n}}q_n\boldsymbol{v}_n\cdot\boldsymbol{E},$$

where q_n and $\boldsymbol{v}_n$ are the charge and velocity of particle n. But we showed earlier that the time rate of change of the total energy $\mathcal{E}$ of a single charged particle of charge q moving with velocity $\boldsymbol{v}$ in an electric field of intensity $\boldsymbol{E}$ is

$$\frac{\mathrm{d}\mathcal{E}}{\mathrm{d}t}=q\boldsymbol{v}\cdot\boldsymbol{E}.$$

Therefore,

$$\int\boldsymbol{j}\cdot\boldsymbol{E}\mathrm{d}V=\sum_{\substack{\text{charged}\\ \text{particles } n}}\frac{\mathrm{d}\mathcal{E}_n}{\mathrm{d}t}=\frac{\mathrm{d}\mathcal{E}^{(\mathrm{p})}}{\mathrm{d}t},$$

where $\mathcal{E}_n$ is the total energy of charged particle n and $\mathcal{E}^{(p)}$ is the total energy of **all** the particles, charged and uncharged. (We have used the fact that there is no interaction of the uncharged particles with the field, and so that the latter particles make zero contribution to the time rate of change of the total energy.)

Therefore, for an infinite volume of integration (or for a closed system), we have

$$\frac{d}{dt}\left[\int\left(\frac{\varepsilon_0 E^2 + \mu_0 H^2}{2}\right)dV + \mathcal{E}^{(p)}\right] = 0,$$

that is, the quantity in the square brackets is conserved. Since the term $\mathcal{E}^{(p)}$ is the total energy of the particles (including their rest energy), the first term must be identified with the energy $\mathcal{E}^{(f)}$ of the field (as influenced, of course, by its interaction with the charges). Therefore, the quantity

$$W^{(f)} \equiv \frac{\varepsilon_0 E^2 + \mu_0 H^2}{2}$$

is the **energy density of the electromagnetic field** (field energy per unit volume).

If, instead, we integrate over a finite volume V, it is possible that the flux of the Poynting vector over the bounding surface S will not vanish, that is, we shall then have

$$\frac{d}{dt}\left(\int_V W^{(f)} dV + E_V^{(p)}\right) = -\oint_S \boldsymbol{\mathcal{S}} \cdot d\boldsymbol{S},$$

where $\mathcal{E}_V^{(p)}$ is the total energy of the particles inside the volume V. This equation can be written as

$$\frac{d}{dt}\left(\int_V W^{(f)} dV + \int_V W^{(p)} dV\right) = -\oint_S \mathcal{S} \cdot d\boldsymbol{S},$$

where

$$\int_V W^{(p)} dV = \mathcal{E}_V^{(p)},$$

that is, $W^{(p)}$ is the **energy density of the particles**. Defining the **total energy density** by $W = W^{(f)} + W^{(p)}$, we have

$$\frac{d}{dt}\int_V W dV = -\oint_S \boldsymbol{\mathcal{S}} \cdot d\boldsymbol{S} = -\int_V \boldsymbol{\nabla} \cdot \boldsymbol{\mathcal{S}}\, dV.$$

It is clear from this that $\oint_S \boldsymbol{\mathcal{S}} \cdot d\boldsymbol{S}$ is the flux of field energy crossing the surface S bounding the given volume V, that is, the **Poynting vector is the energy-flux density** (the energy passing through unit area of surface in unit time). [In SI units, E is in V/m, H is in A/m, and so $\mathcal{S}$ has units $VA/m^2 = W/m^2$, that is, power per unit area.]

17.6 The energy–momentum tensor

Just as the charge-conservation equation

$$\frac{\mathrm{d}}{\mathrm{d}t}\int\limits_{\mathrm{V}} \rho \mathrm{d}V = -\oint\limits_{\mathrm{S}} \boldsymbol{j}\cdot \mathrm{d}\boldsymbol{S} = -\int\limits_{\mathrm{V}} \boldsymbol{\nabla}\cdot \boldsymbol{j}\mathrm{d}V$$

in its differential form $\partial\rho/\partial t + \boldsymbol{\nabla}\cdot\boldsymbol{j} = 0$ can be written in the four-dimensional form

$$\frac{\partial j^k}{\partial x^k} = 0,$$

so the energy-conservation equation

$$\frac{\mathrm{d}}{\mathrm{d}t}\int\limits_{\mathrm{V}} W \mathrm{d}V = -\oint\limits_{\mathrm{S}} \boldsymbol{\mathcal{S}}\cdot \mathrm{d}\boldsymbol{S} = -\int\limits_{\mathrm{V}} \boldsymbol{\nabla}\cdot \boldsymbol{\mathcal{S}}\,\mathrm{d}V$$

in its differential form $\partial W/\partial t + \boldsymbol{\nabla}\cdot\boldsymbol{\mathcal{S}} = 0$ can be written in the four-dimensional form

$$\frac{\partial T^{0k}}{\partial x^k} = 0$$

with

$$T^{00} = W \quad \text{and} \quad T^{0\alpha} = \mathcal{S}_\alpha/c \quad (\alpha = 1, 2, 3).$$

The reason for the index 0 in the energy-conservation equation

$$\frac{\partial T^{0k}}{\partial x^k} = 0$$

is that this equation is just **one of four equations** (the $i = 0$ one)

$$\frac{\partial T^{ik}}{\partial x^k} = 0 \quad (i = 0, 1, 2, 3),$$

of which the other three

$$\frac{\partial T^{\alpha k}}{\partial x^k} = 0 \quad (\alpha = 1, 2, 3)$$

are the equations of **momentum conservation** (one for each component p_α).

The quantities T^{ik} form a **four-tensor** (to be proved), called, for obvious reasons, the **energy–momentum tensor**. It has the physical dimensions of energy density.

We now find a **general** expression that enables us to write down a T^{ik} with these properties.

We write the action in the form

$$S = \int L\mathrm{d}t = \int \Lambda(q, \partial q/\partial x^i)\mathrm{d}V\mathrm{d}t = \frac{1}{c}\int \Lambda \mathrm{d}\Omega,$$

where Λ is clearly the lagrangian density ($\int \Lambda \mathrm{d}V = L$).

The lagrangian density Λ depends only on the "generalized coordinates" q of the system (the components of the four-potential, in the case of the electromagnetic field), and on the first space and time derivatives of q. (For simplicity we write here only one of the qs.) The system is here assumed to be "closed", that is, the lagrangian density Λ depends on the space–time coordinates x^i only through q and $\partial q/\partial x^i$, and not explicitly.

We first find the "equations of motion" of q (that is, of A^i, in the case of the electromagnetic field) by putting $\delta S = 0$ (principle of least action):

$$\begin{aligned}\delta S = 0 &= \frac{1}{c}\int \left(\frac{\partial \Lambda}{\partial q}\delta q + \frac{\partial \Lambda}{\partial q_{,i}}\delta q_{,i}\right)\mathrm{d}\Omega \quad \left[q_{,i} \equiv \frac{\partial q}{\partial x^i}\right]\\ &= \frac{1}{c}\int \left[\frac{\partial \Lambda}{\partial q}\delta q + \frac{\partial}{\partial x^i}\left(\frac{\partial \Lambda}{\partial q_{,i}}\delta q\right) - \delta q \frac{\partial}{\partial x^i}\left(\frac{\partial \Lambda}{\partial q_{,i}}\right)\right]\mathrm{d}\Omega.\end{aligned}$$

We transform the second term by Gauss' theorem, whereupon it vanishes. Then, since δq is arbitrary, we get

$$\frac{\partial}{\partial x^i}\left(\frac{\partial \Lambda}{\partial q_{,i}}\right) - \frac{\partial \Lambda}{\partial q} = 0 \tag{17.45}$$

(the **Euler–Lagrange equation** for q).

We now find our quantity T_i^k satisfying $\partial T_i^k/\partial x^k = 0$ by first finding $\partial \Lambda/\partial x^i$:

$$\frac{\partial \Lambda}{\partial x^i} = \frac{\partial \Lambda}{\partial q}\frac{\partial q}{\partial x^i} + \frac{\partial \Lambda}{\partial q_{,k}}\frac{\partial q_{,k}}{\partial x^i} = \frac{\partial}{\partial x^k}\left(\frac{\partial \Lambda}{\partial q_{,k}}\right)\frac{\partial q}{\partial x^i} + \frac{\partial \Lambda}{\partial q_{,k}}\frac{\partial q_{,k}}{\partial x^i},$$

where in the second equality we have used equation (17.45).

Therefore, since $\partial q_{,k}/\partial x^i = \partial q_{,i}/\partial x^k$, we have

$$\frac{\partial \Lambda}{\partial x^i} = \frac{\partial}{\partial x^k}\left(\frac{\partial \Lambda}{\partial q_{,k}}q_{,i}\right).$$

If we write

$$\frac{\partial \Lambda}{\partial x^i} = \delta_i^k \frac{\partial \Lambda}{\partial x^k},$$

the above equation can be written in the form

$$\frac{\partial T_i^k}{\partial x^k} = 0$$

with

$$T_i^k = q_{,i}\frac{\partial \Lambda}{\partial q_{,k}} - \delta_i^k \Lambda,$$

which, since Λ is a Lorentz scalar and the generalized coordinates and their space–time derivatives are Lorentz tensors, is manifestly a (mixed) rank-2 Lorentz tensor.

For example, setting $i = k = 0$ and taking the integral over a volume, we have

$$\int T_0^0 \mathrm{d}V = -L + \dot{q}\frac{\partial \Lambda}{\partial \dot{q}} = H \quad \text{(the energy)}.$$

Therefore,

$$T_0^0 \ (= T^{00} = T_{00}) = W \quad \text{(the energy density)}.$$

If there are several generalized coordinates $q^{(l)}$, then

$$T_i^k = \sum_l q_{,i}^{(l)}\frac{\partial \Lambda}{\partial q_{,k}^{(l)}} - \delta_i^k \Lambda.$$

Just as the equation $\partial j^k/\partial x^k = 0$ implied that the **scalar** quantity $\int j^k \mathrm{d}S_k$ over a hypersurface containing all of three-dimensional space is conserved, so the equation $\partial T_i^k/\partial x^k = 0$ implies that the **four-vector**

$$P_i = \text{const.}\int T_i^k \mathrm{d}S_k \quad (i = 0, 1, 2, 3)$$

is conserved.

We choose the constant so that P_i is the **four-momentum** of the system. We have, for $i = 0$,

$$\begin{aligned} P_0 = \frac{\mathcal{E}}{c} &= \text{const}\int T_0^k \mathrm{d}S_k \\ &= \text{const}\int T_0^0 \mathrm{d}S_0 \quad \text{(if the hypersurface is the hyperplane } x^0 = \text{const)} \\ &= \text{const}\int T_0^0 \mathrm{d}V = \text{const} \times \mathcal{E}. \end{aligned}$$

Therefore, const $= 1/c$. Thus, the **four-momentum** of the system can be expressed in terms of T_i^k by the relation

$$P_i = \frac{1}{c}\int T_i^k \mathrm{d}S_k \quad \text{or} \quad P^i = \frac{1}{c}\int T^{ik} \mathrm{d}S_k,$$

and is conserved by virtue of the property

$$\frac{\partial T_i^k}{\partial x^k} = 0 \quad \left(\frac{\partial T^{ik}}{\partial x^k} = 0\right)$$

of the energy–momentum tensor.

To the T^{ik} defined by

$$T^{ik} = q^{,i}\frac{\partial \Lambda}{\partial q_{,k}} - g^{ik}\Lambda$$

(recall that g^{ik} is the fully contravariant form of δ_i^k) we can add any quantity $\partial \psi^{ikl}/\partial x^l$, where ψ^{ikl} is a rank-3 tensor antisymmetric in k and l, to give a new tensor

$$t^{ik} = T^{ik} + \frac{\partial \psi^{ikl}}{\partial x^l}.$$

Because

$$\frac{\partial^2 \psi^{ikl}}{\partial x^k \partial x^l} = 0$$

identically, the new tensor t^{ik} also satisfies

$$\frac{\partial t^{ik}}{\partial x^k} = 0.$$

Also, if ψ^{ikl} vanishes at spatial infinity, P^i does not change under this transformation, since

$$\int \frac{\partial \psi^{ikl}}{\partial x^l}\mathrm{d}S_k = \frac{1}{2}\int\left(\frac{\partial \psi^{ikl}}{\partial x^l}\mathrm{d}S_k - \frac{\partial \psi^{ikl}}{\partial x^k}\mathrm{d}S_l\right) = \frac{1}{2}\int \psi^{ikl}\mathrm{d}f_{kl} = 0.$$

Here we have used the fact that the latter integral is over the two-dimensional surface bounding the three-dimensional manifold (incorporating all space) over which the previous integral is taken [see section 17.2.9], and $\psi^{ikl} = 0$ on this surface at spatial infinity.

This apparent arbitrariness in T^{ik} can be removed as follows. Since

$$P^i = \frac{1}{c}\int T^{ik}\mathrm{d}S_k = \frac{1}{c}\int T^{i0}\mathrm{d}S_0 \equiv \frac{1}{c}\int T^{i0}\mathrm{d}V$$

(i.e., the integral is independent of the three-dimensional hypersurface over which it is taken, provided it includes all space), it is clear that the quantities $\frac{1}{c}T^{i0}$ are the **densities of energy–momentum**. We now introduce the **angular-momentum four-tensor**, defined for a system of point particles as

$$M^{ik} = \sum_{\text{particles}} \left(x^i p^k - x^k p^i\right).$$

This tensor (whose nonzero space components are clearly $M^{12} = L_z$, $M^{23} = L_x$, and $M^{31} = L_y$, that is, are the components of the three-dimensional angular momentum vector $\boldsymbol{L} = \boldsymbol{r} \times \boldsymbol{p}$) becomes, for a continuous macroscopic body,

$$\begin{aligned} M^{ik} &= \frac{1}{c}\int (x^i T^{k0} - x^k T^{i0})\mathrm{d}V \equiv \frac{1}{c}\int (x^i T^{k0} - x^k T^{i0})\mathrm{d}S_0 \\ &= \frac{1}{c}\int (x^i T^{kl} - x^k T^{il})\mathrm{d}S_l, \end{aligned} \tag{17.46}$$

where the latter equality follows from the fact that, because angular momentum is conserved, it does not matter which three-dimensional spacelike hypersurface we integrate over, provided it includes all space.

As usual, the conservation of M^{ik} can be expressed by equating the divergence of the integrand in (17.46) to zero:

$$\frac{\partial}{\partial x^l}\left(x^i T^{kl} - x^k T^{il}\right) = 0.$$

But

$$\frac{\partial x^i}{\partial x^l} = \delta^i_l \quad \text{and} \quad \frac{\partial T^{kl}}{\partial x^l} = 0.$$

Therefore,

$$\delta^i_l T^{kl} - \delta^k_l T^{il} = 0,$$

and so

$$T^{ki} = T^{ik}.$$

Thus, the energy–momentum tensor satisfying (17.46) must be **symmetric**, that is, to obtain a T^{ik} satisfying (17.46) we must add to

$$T^{ik} = q^{,i}\frac{\partial \Lambda}{\partial q_{,k}} - g^{ik}\Lambda$$

an expression $\partial\psi^{ikl}/\partial x^l$ that makes the result symmetric.]

From the expression

$$P^i = \frac{1}{c}\int T^{i0}\mathrm{d}V$$

we saw that $\frac{1}{c}T^{\alpha 0}$ ($\alpha = 1, 2, 3$) is the **momentum density** and T^{00} is the **energy density**.

To find the meaning of the other components of T^{ik} we write

$$\frac{\partial T^{ik}}{\partial x^k} = 0$$

in the form

$$\begin{aligned} &\frac{1}{c}\frac{\partial T^{00}}{\partial t} + \frac{\partial T^{0\alpha}}{\partial x_\alpha} = 0 \quad (\text{the } i = 0 \text{ equation}),\\ &\frac{1}{c}\frac{\partial T^{\alpha 0}}{\partial t} + \frac{\partial T^{\alpha\beta}}{\partial x_\beta} = 0 \quad (\text{the } i = \alpha = 1, 2, 3 \text{ equations}). \end{aligned} \tag{17.47}$$

We integrate these equations over a volume V in space. The first gives

$$\frac{1}{c}\frac{\mathrm{d}}{\mathrm{d}t}\int\limits_{\mathrm{V}} T^{00}\mathrm{d}V + \int\limits_{\mathrm{V}} \frac{\partial T^{0\alpha}}{\partial x_\alpha}\mathrm{d}V = 0.$$

If we use Gauss' theorem this becomes

$$\frac{\mathrm{d}}{\mathrm{d}t}\int\limits_{\mathrm{V}} T^{00}\mathrm{d}V = -c\oint\limits_{\mathrm{S}} T^{0\alpha}\mathrm{d}S_\alpha$$

(the integral on the right is over the closed surface S bounding the volume V), which is of the form

$$\frac{\mathrm{d}}{\mathrm{d}t}\int\limits_{\mathrm{V}} W\mathrm{d}V + \oint\limits_{\mathrm{S}} \mathcal{S}_\alpha \mathrm{d}S_\alpha = 0,$$

with the energy-flux density given by the components

$$\mathcal{S}_\alpha = cT^{0\alpha},$$

that is, $T^{0\alpha} = \frac{1}{c} \times$ the α-component of the energy-flux density.

Also, using the symmetry of T^{ik},

$$\mathcal{S}_\alpha = cT^{0\alpha} = c^2\left(\frac{1}{c}T^{\alpha 0}\right),$$

that is, the energy-flux density equals $c^2\times$ (the momentum density).

We still need to interpret $T^{\alpha\beta}$. The second equation (17.47) gives, similarly,

$$\frac{\mathrm{d}}{\mathrm{d}t}\int\limits_{\mathrm{V}}\left(\frac{1}{c}T^{\alpha 0}\right)\mathrm{d}V + \oint\limits_{\mathrm{S}} T^{\alpha\beta}\mathrm{d}S_\beta = 0,$$

where the first term is the rate of change of the total α-component of momentum in the volume V.

Therefore, $T^{\alpha\beta}$ is the β-component of the flux density of the α-component of momentum (that is, the amount of α-component of momentum passing in unit time through unit surface perpendicular to the x_β axis).

17.6.1 Energy–momentum tensor of the electromagnetic field

We earlier found

$$S_{\mathrm{f}} = -\frac{\varepsilon_0 c}{4}\int F_{kl}F^{kl}\mathrm{d}\Omega = -\frac{\varepsilon_0 c^2}{4}\int F_{kl}F^{kl}\mathrm{d}V\mathrm{d}t = \int \Lambda \mathrm{d}V\mathrm{d}t.$$

Therefore, the lagrangian density Λ is

$$\Lambda = -\frac{\varepsilon_0 c^2}{4}F^{kl}F_{kl},$$

where the electromagnetic-field tensor

$$F_{kl} = \frac{\partial A_l}{\partial x^k} - \frac{\partial A_k}{\partial x^l}.$$

But

$$T_i^k = \sum_l q_{,i}^{(l)} \frac{\partial \Lambda}{\partial q_{,k}^{(l)}} - \delta_i^k \Lambda,$$

where there are four $q^{(l)}$s, namely, the four-potential components A_l. Therefore,

$$T_i^k = \frac{\partial A_l}{\partial x^i} \frac{\partial \Lambda}{\partial \left(\partial A_l / \partial x^k\right)} - \delta_i^k \Lambda.$$

To calculate the derivative of Λ here, we find $\delta \Lambda$:

$$\delta \Lambda = -\frac{\varepsilon_0 c^2}{2} F^{kl} \delta F_{kl} = -\frac{\varepsilon_0 c^2}{2} F^{kl} \delta \left(\frac{\partial A_l}{\partial x^k} - \frac{\partial A_k}{\partial x^l} \right) = -\varepsilon_0 c^2 F^{kl} \delta \left(\frac{\partial A_l}{\partial x^k} \right),$$

where, in the second term, we interchanged the indices $k \leftrightarrow l$ and used the antisymmetry $F^{lk} = -F^{kl}$.
Therefore,

$$\frac{\partial \Lambda}{\partial \left(\partial A_l / \partial x^k\right)} = -\varepsilon_0 c^2 F^{kl},$$

and so

$$T_i^k = -\varepsilon_0 c^2 \frac{\partial A_l}{\partial x^i} F^{kl} + \frac{\varepsilon_0 c^2}{4} \delta_i^k F_{lm} F^{lm}.$$

The fully contravariant form is

$$T^{ik} = -\varepsilon_0 c^2 \frac{\partial A^l}{\partial x_i} F^k{}_l + \frac{\varepsilon_0 c^2}{4} g^{ik} F_{lm} F^{lm}.$$

This is not yet symmetric in i and k. We can make it so (as seen from the result below) by adding

$$\varepsilon_0 c^2 \frac{\partial A^i}{\partial x^l} F^{kl} \quad \left(= \varepsilon_0 c^2 \frac{\partial A^i}{\partial x_l} F^k{}_l \right),$$

which has the required form

$$\frac{\partial \psi^{ikl}}{\partial x^l}, \quad \text{with} \quad \psi^{ikl} = \varepsilon_0 c^2 A^i F^{kl}$$

[since $\partial F^{kl}/\partial x^l = 0$ in the absence of charges, by the second pair of Maxwell equations]. Since

$$\frac{\partial A^l}{\partial x_i} - \frac{\partial A^i}{\partial x_l} = F^{il},$$

we have, for the **energy–momentum tensor of the electromagnetic field**,

$$T^{ik} = \varepsilon_0 c^2 \left(-F^{il} F^k{}_l + \frac{1}{4} g^{ik} F_{lm} F^{lm} \right),$$

which is obviously symmetric in i and k, and also has the property

$$T_i^i = 0$$

(since $\delta_i^i = 4$).

If we now substitute for the components of the electromagnetic field tensor F_{ik} we find (using the relation $H = B/\mu_0$) the familiar results

$$T^{00} = \frac{\varepsilon_0 E^2 + \mu_0 H^2}{2} = W$$

and

$$T^{0\alpha} = \frac{1}{c}(\boldsymbol{E} \times \boldsymbol{H})_\alpha = \frac{1}{c} \mathcal{S}_\alpha,$$

and also

$$\begin{aligned} T^{11}(\equiv T_{xx}) &= \varepsilon_0 c^2 \left[-F^{1l} F^1{}_l - \frac{1}{4} \times 2 \left(B^2 - \frac{E^2}{c^2} \right) \right] \\ &= \varepsilon_0 c^2 \left(B_z^2 + B_y^2 - \frac{E_x^2}{c^2} - \frac{1}{2} B^2 + \frac{1}{2} \frac{E^2}{c^2} \right) \\ &= \frac{\mu_0}{2} \left(H_y^2 + H_z^2 - H_x^2 \right) + \frac{\varepsilon_0}{2} \left(E_y^2 + E_z^2 - E_x^2 \right), \end{aligned}$$

with similar expressions for $T^{22} = T_{yy}$ and $T^{33} = T_{zz}$. Also,

$$\begin{aligned} T^{12}\,(\equiv T_{xy}) &= \varepsilon_0 c^2 (-F^{1l} F^2{}_l) \\ &= \varepsilon_0 c^2 \left(-B_y B_x - \frac{E_x E_y}{c^2} \right) = -\mu_0 H_x H_y - \varepsilon_0 E_x E_y. \end{aligned}$$

The latter two equations can be written as

$$T_{\alpha\beta} = -\varepsilon_0 E_\alpha E_\beta - \mu_0 H_\alpha H_\beta + \frac{1}{2} \delta_{\alpha\beta} \left(\varepsilon_0 E^2 + \mu_0 H^2 \right) \quad (\alpha,\ \beta = 1, 2, 3),$$

which is a three-tensor called the **Maxwell stress tensor**.

If charged **particles** are present, the energy–momentum tensor is the sum of the above T^{ik} for the field and the T^{ik} for the particles (in the latter, the particles are assumed to be noninteracting).

17.6.2 Energy–momentum tensor of particles

Just as we introduced a **charge density**

$$\rho(\boldsymbol{r}, t) = \sum_{\text{particles } n} q_n \delta^{(3)}(\boldsymbol{r} - \boldsymbol{r}_n(t))$$

and a **current four-vector**

$$j^i = \rho \frac{\mathrm{d}x^i}{\mathrm{d}t}$$

with zeroth component $j^0 = c\rho$ (that is, $\rho = j^0/c$), so we can introduce a **mass density** (meaning, of course, rest-mass density)

$$\mu(\boldsymbol{r}, t) = \sum_{\text{particles } n} m_n \delta^{(3)}(\boldsymbol{r} - \boldsymbol{r}_n(t))$$

(m_n is the mass of particle n) and a **mass-current four-vector**

$$j^i_{\mathrm{m}} = \mu \frac{\mathrm{d}x^i}{\mathrm{d}t}$$

with zeroth component $c\mu$. The four-momentum density $\frac{1}{c}T^{i0}$ can also be written as $\mu c u^i$ (recall that $p^i = mcu^i$). Therefore,

$$T^{i0} = c\mu c u^i = \mu c u^i \frac{\mathrm{d}x^0}{\mathrm{d}t} \quad (\text{since } x^0 = ct),$$

which, since $\mu \mathrm{d}x^0/\mathrm{d}t$ is a component of a four-vector, is the $k = 0$ instance of

$$\begin{aligned} T^{ik} &= \mu c u^i \frac{\mathrm{d}x^k}{\mathrm{d}t} = c u^i \left(\mu \frac{\mathrm{d}x^k}{\mathrm{d}t}\right) \\ &\left[= \mu c \frac{\mathrm{d}x^i}{\mathrm{d}s}\frac{\mathrm{d}x^k}{\mathrm{d}t} = \mu c u^i u^k \frac{\mathrm{d}s}{\mathrm{d}t} = \mu c^2 u^i u^k \sqrt{1 - v^2/c^2}\right], \end{aligned}$$

which shows that T^{ik} is symmetric.

We now check that the sum of the energies and momenta of the field and particles is conserved, that is, that

$$\frac{\partial}{\partial x^k}\left(T^{(\mathrm{f})k}_{\ \ i} + T^{(\mathrm{p})k}_{\ \ i}\right) = 0.$$

For the field part we have

$$\frac{\partial T^{(\mathrm{f})k}_{\ \ i}}{\partial x^k} = \varepsilon_0 c^2 \left[\frac{\delta_i^k}{4} \frac{\partial}{\partial x^k} \left(F_{lm} F^{lm} \right) - \frac{\partial}{\partial x^k} \left(F_i{}^l F^k{}_l \right) \right]$$
$$= \varepsilon_0 c^2 \left(\frac{1}{2} \frac{\partial F_{lm}}{\partial x^i} F^{lm} - \frac{\partial F_i{}^l}{\partial x^k} F^k{}_l - \frac{\partial F^k{}_l}{\partial x^k} F_i{}^l \right).$$

In the first term we use the first pair of Maxwell equations

$$\frac{\partial F_{lm}}{\partial x^i} = -\frac{\partial F_{mi}}{\partial x^l} - \frac{\partial F_{il}}{\partial x^m},$$

and in the last we put $F^k{}_l = -F_l{}^k$ and use the second pair of Maxwell equations

$$\frac{\partial F_l{}^k}{\partial x^k} = -\frac{1}{\varepsilon_0 c^2} j_l.$$

Then

$$\frac{\partial T^{(\mathrm{f})k}_{\ \ i}}{\partial x^k} = \varepsilon_0 c^2 \left(-\frac{1}{2} \frac{\partial F_{mi}}{\partial x^l} F^{lm} - \frac{1}{2} \frac{\partial F_{il}}{\partial x^m} F^{lm} - \frac{\partial F_i{}^l}{\partial x^k} F^k{}_l - \frac{1}{\varepsilon_0 c^2} F_i{}^l j_l \right).$$

Relabelling dummy indices (putting $l \to k$ and $m \to l$ in the first term and $m \to k$ in the second), raising one index l in the third term while lowering the other, and using the antisymmetry property $F^{kl} = -F^{lk}$ in this term, we see that the first two terms cancel the third. Therefore,

$$\frac{\partial T^{(\mathrm{f})k}_{\ \ i}}{\partial x^k} = -F_i{}^l j_l = -F_{il} j^l.$$

For the particle part we have

$$\frac{\partial T^{(\mathrm{p})k}_{\ \ i}}{\partial x^k} = c u_i \frac{\partial}{\partial x^k} \left(\mu \frac{\mathrm{d}x^k}{\mathrm{d}t} \right) + c\mu \frac{\mathrm{d}x^k}{\mathrm{d}t} \frac{\partial u_i}{\partial x^k}.$$

The first term is zero, since it contains the four-divergence of the mass-current four-vector, which is zero by virtue of the conservation of the mass (in the present case of noninteracting particles). Therefore,

$$\frac{\partial T^{(\mathrm{p})k}_{\ \ i}}{\partial x^k} = c\mu \frac{\partial u_i}{\partial x^k} \frac{\mathrm{d}x^k}{\mathrm{d}t} = c\mu \frac{\mathrm{d}u_i}{\mathrm{d}t}.$$

But, by the equations of motion of the charges in the field, we have

$$mc \frac{\mathrm{d}u_i}{\mathrm{d}s} = q F_{ik} u^k.$$

Putting $m = \int \mu \mathrm{d}V$ and $q = \int \rho \mathrm{d}V$, we get the corresponding equations for the densities:

$$\mu c \frac{\mathrm{d}u_i}{\mathrm{d}s} = \rho F_{ik} u^k.$$

Therefore,

$$\mu c \frac{\mathrm{d}u_i}{\mathrm{d}t} = \mu c \frac{\mathrm{d}u_i}{\mathrm{d}s} \frac{\mathrm{d}s}{\mathrm{d}t} = F_{ik} \rho u^k \frac{\mathrm{d}s}{\mathrm{d}t} = F_{ik} \rho \frac{\mathrm{d}x^k}{\mathrm{d}t} = F_{ik} j^k,$$

and so

$$\frac{\partial}{\partial x^k} \left(T^{(\mathrm{f})k}_{\ \ \ i} + T^{(\mathrm{p})k}_{\ \ \ i} \right) = -F_{il} j^l + F_{ik} j^k = 0,$$

as was to be proved.

Using

$$T^{(\mathrm{p})k}_{\ \ \ i} = \mu c u_i u^k \frac{\mathrm{d}s}{\mathrm{d}t},$$

we now prove that the trace

$$T^i_i = T^{(\mathrm{f})i}_{\ \ \ i} + T^{(\mathrm{p})i}_{\ \ \ i} \geq 0.$$

We have $T^{(\mathrm{f})i}_{\ \ \ i} = 0$ (already proved). Therefore, the only contribution to the trace of the energy–momentum tensor is from the particles, and we have

$$\begin{aligned} T^i_i &= T^{(\mathrm{p})i}_{\ \ \ i} = \mu c u_i u^i \frac{\mathrm{d}s}{\mathrm{d}t} = \mu c \frac{\mathrm{d}s}{\mathrm{d}t} \quad (\text{because } u_i u^i = 1) \\ &= \mu c^2 \sqrt{1 - v^2/c^2} = \sum_{\text{particles } n} m_n c^2 \sqrt{1 - v^2/c^2} \delta^{(3)}(\boldsymbol{r} - \boldsymbol{r}_n) \\ &\geq 0 \quad (= 0, \text{ in the absence of particles}). \end{aligned}$$

17.6.3 Energy–momentum tensor of continuous media

Consider a continuous macroscopic body (solid, liquid or gas), and consider within it a volume element V.

We shall assume that the volume of the element V is sufficiently small for the energy–momentum tensor T^{ik} to be uniform (constant) throughout V, and find the components T^{ik} in the rest frame of the element, that is, in the frame in which

$$P_\alpha = \int_{\mathrm{V}} \left(\frac{1}{c} T^{\alpha 0} \right) \mathrm{d}V = 0. \tag{17.48}$$

First, we have $T^{00} = \varepsilon$, where ε is the total energy per unit **proper** volume (since we are finding T^{ik} in the **rest frame** of the element). Also, (17.48) implies that

$$T^{\alpha 0} = 0,$$

since T^{ik} is uniform throughout the element.

It remains to find $T^{\alpha\beta}$ ($\alpha, \beta = 1, 2, 3$). For this we use the fact that, if the hydrostatic pressure on the element from the surrounding matter is p, the total force on the element is

$$\boldsymbol{f} = -\oint_{\mathrm{S}} p\mathrm{d}\boldsymbol{S} = -\left(\hat{\boldsymbol{i}} \oint_{\mathrm{S}} p\mathrm{d}S_x + \hat{\boldsymbol{j}} \oint_{\mathrm{S}} p\mathrm{d}S_y + \hat{\boldsymbol{k}} \oint_{\mathrm{S}} p\mathrm{d}S_z\right),$$

which expresses **Pascal's law**: pressure is transmitted equally in all directions (we have the same p in each term) and is everywhere perpendicular to the surface S bounding the volume under consideration. (The small deviations from Pascal's law in the case of solids are negligible in a relativistic theory.) Therefore, the α-component of the force on the element is

$$f_\alpha = -\oint_{\mathrm{S}} p\mathrm{d}S_\alpha,$$

which must be equal to the rate of change of the component P_α of the three-momentum of the element:

$$\frac{\mathrm{d}P_\alpha}{\mathrm{d}t} = \frac{\mathrm{d}}{\mathrm{d}t}\int_{\mathrm{V}} \left(\frac{1}{c}T^{\alpha 0}\right)\mathrm{d}V = -\oint_{\mathrm{S}} T^{\alpha\beta}\mathrm{d}S_\beta$$

[see section 17.4.10]. Comparing, we find

$$T^{\alpha\beta}\mathrm{d}S_\beta = p\mathrm{d}S_\alpha,$$

and so

$$T^{\alpha\beta} = p\delta_{\alpha\beta}.$$

Collecting the components of T^{ik} in an array, we have, **in the frame in which the element is at rest** (that is, in which it has four-velocity $u^0 = 1, u^\alpha = 0$):

$$(T^{ik}) = \begin{pmatrix} \varepsilon & 0 & 0 & 0 \\ 0 & p & 0 & 0 \\ 0 & 0 & p & 0 \\ 0 & 0 & 0 & p \end{pmatrix}.$$

In the Lorentz frame in which the element has four-velocity u^i we must expect that the energy–momentum tensor T^{ik} will depend not only on the "intrinsic" properties ε and p of the element but also on u^i. The most general rank-2 tensor of this kind is of the form

$$T^{ik} = Xu^i u^k - Yg^{ik},$$

where X and Y depend on the intrinsic properties of the element and are easily found using the fact that in the rest frame of the element (in which $u^0 = 1$ and $u^\alpha = 0$) we have, as just shown, $T^{00} = \varepsilon$ and $T^{\alpha\alpha} = p$ ($\alpha = 1, 2, 3$). Thus, since $g^{00} = 1$ and $g^{\alpha\alpha} = -1$,

$$(T^{00})_{\text{rest}} = \varepsilon = X - Y \quad \text{and} \quad (T^{\alpha\alpha})_{\text{rest}} = p = Y,$$

that is, we have $Y = p$ and $X = \varepsilon + p$, so that the general expression for the **energy–momentum tensor T^{ik} of a macroscopic body** is

$$T^{ik} = (\varepsilon + p)u^i u^k - pg^{ik}. \tag{17.49}$$

The corresponding mixed form is

$$T_i^k = (\varepsilon + p)u_i u^k - p\delta_i^k, \tag{17.50}$$

so that **its** components in the rest frame are

$$(T_i^k) = \begin{pmatrix} \varepsilon & 0 & 0 & 0 \\ 0 & -p & 0 & 0 \\ 0 & 0 & -p & 0 \\ 0 & 0 & 0 & -p \end{pmatrix}.$$

Returning to the fully contravariant form (17.49) and substituting for the components of the four-velocity, we find

$$\begin{aligned} T^{00} &= \frac{\varepsilon + p}{1 - v^2/c^2} - p = \frac{\varepsilon + p\dfrac{v^2}{c^2}}{1 - v^2/c^2} \quad (= \varepsilon \text{ if } v = 0), \\ T^{\alpha 0} &= \frac{(\varepsilon + p)v_\alpha/c}{1 - v^2/c^2} \quad (= 0 \text{ if } v = 0), \\ T^{\alpha\beta} &= \frac{(\varepsilon + p)v_\alpha v_\beta/c^2}{1 - v^2/c^2} + p\delta_{\alpha\beta} \quad (\text{since } g^{11} = g^{22} = g^{33} = -1). \end{aligned}$$

Therefore, the momentum density is

$$\frac{1}{c}T^{\alpha 0} = \frac{(\varepsilon + p)v_\alpha}{c^2\left(1 - v^2/c^2\right)} \quad \left(\approx \frac{(\varepsilon + p)v_\alpha}{c^2} \text{ for } v \ll c\right).$$

In all these expressions, the energy density ε includes not only the rest energy of the particles but also the potential energy of their interactions and the kinetic energy of their **relative** motion (ε obviously does not include the kinetic energy of the **overall** motion, since it is the total energy density in the rest frame, that is, the frame in which the overall kinetic energy is zero.)

In the case when the velocities of the **relative** motion of the particles are $\ll c$, that is, when the system is **cold**, we can neglect all contributions to ε other than $\mu^{\text{prop}}c^2$ (μ^{prop} is the proper mass density, that is, the rest mass per unit proper volume of the element).

We can also neglect p in comparison with $\mu^{\text{prop}}c^2$ in these circumstances and (17.49) becomes

$$T^{ik} \approx \mu^{\text{prop}} c^2 u^i u^k. \tag{17.51}$$

The mass density μ in the frame relative to which the element is moving with velocity $\boldsymbol{v}$ (that is, in the frame moving with velocity $-\boldsymbol{v}$ relative to the rest frame of the element) is related to μ^{prop} by

$$m = \mu^{\text{prop}} V = \mu V \sqrt{1 - v^2/c^2}$$

because of the Lorentz contraction of the volume in the direction of motion (m here is the rest mass of the element). Thus, (17.51) becomes

$$T^{ik} \approx \mu c^2 u^i u^k \sqrt{1 - v^2/c^2}, \tag{17.52}$$

which is the "particle" T^{ik} found earlier.

[***Check*:** For the total energy of the element, (17.51) gives

$$\mathcal{E} = \int T^{00} \mathrm{d}V = \int \mathrm{d}V \frac{\mu c^2 \sqrt{1 - v^2/c^2}}{1 - v^2/c^2} = \frac{mc^2}{\sqrt{1 - v^2/c^2}},$$

as expected, while for the momentum it gives

$$P^\alpha = \int \left(\frac{1}{c} T^{\alpha 0} \right) \mathrm{d}V = \frac{1}{c} \int \mathrm{d}V \frac{\mu c^2 (v_\alpha / c) \sqrt{1 - v^2/c^2}}{\sqrt{1 - v^2/c^2}\sqrt{1 - v^2/c^2}} = \frac{m v_\alpha}{\sqrt{1 - v^2/c^2}},$$

again as expected.]

From (17.50) we find the trace

$$T_i^i = (\varepsilon + p) - 4p = \varepsilon - 3p.$$

But we showed earlier that the trace $T_i^i > 0$ if matter is present. Therefore,

$$p < \frac{\varepsilon}{3},$$

that is, even for a very hot body the pressure p can never exceed $\varepsilon/3$. (In the very early, radiation-dominated universe, p was close to $\varepsilon/3$.)

The amount by which ε exceeds $3p$ is given by

$$\varepsilon - 3p = T_i^i = \sum_{\text{particles } n} m_n c^2 \sqrt{1 - v_n^2/c^2}\,\delta^{(3)}(\boldsymbol{r} - \boldsymbol{r}_n(t)),$$

where the sum is over the particles in the volume element V [see the end of the preceding section].

Finally, the rest frame components of the T^{ik} of a macroscopic body, which we found to be

$$(T^{ik}) = \begin{pmatrix} \varepsilon & 0 & 0 & 0 \\ 0 & p & 0 & 0 \\ 0 & 0 & p & 0 \\ 0 & 0 & 0 & p \end{pmatrix},$$

are given (as proved below) by the following expressions:

$$T^{00}_{\text{macro}} = \varepsilon = \mu c^2 \overline{\left(1 - v^2/c^2\right)^{-1/2}} \tag{17.53}$$

and

$$T^{\alpha\alpha}_{\text{macro}} = p = \frac{1}{3}\mu \overline{v^2\left(1 - v^2/c^2\right)^{-1/2}}. \tag{17.54}$$

Here the bars denote averaging over the velocities (in the rest frame of the macroscopic element) of the particles constituting the element.

We now give a proof of equations (17.53) and (17.54). We have

$$T^{ik}(\text{particle}\, n) = m_n c^2 \delta^{(3)}(\boldsymbol{r} - \boldsymbol{r}_n) u^i_n u^k_n \sqrt{1 - v_n^2/c^2},$$

where $\boldsymbol{r}_n$ is the position of particle n and u^i_n is its four-velocity relative to the rest frame of the element. Therefore,

$$T^{ik}(N\ \text{particles}) = \sum_{n=1}^{N} m_n c^2 \delta^{(3)}(\boldsymbol{r} - \boldsymbol{r}_n) u^i_n u^k_n \sqrt{1 - v_n^2/c^2}.$$

Averaging this over the velocities, we get

$$\begin{aligned} \overline{T^{ik}} \equiv T^{ik}_{\text{macro}} &= \sum_{n=1}^{N} m_n c^2 \delta^{(3)}(\boldsymbol{r} - \boldsymbol{r}_n) \overline{u^i u^k \sqrt{1 - v^2/c^2}} \\ &= \mu c^2 \overline{u^i u^k \sqrt{1 - v^2/c^2}}. \end{aligned}$$

Therefore,

$$\overline{T^{00}} \equiv T^{00}_{\text{macro}} = \varepsilon = \mu c^2 \overline{(1 - v^2/c^2)^{-1/2}} \tag{17.55}$$

and

$$\begin{aligned} \overline{T^{\alpha\alpha}} \equiv T^{\alpha\alpha}_{\text{macro}} = p &= \mu c^2 \overline{\left(\frac{v_\alpha^2 \sqrt{1 - v^2/c^2}}{c^2(1 - v^2/c^2)}\right)} \\ &= \frac{1}{3}\mu \overline{v^2\left(1 - v^2/c^2\right)^{-1/2}}. \end{aligned} \tag{17.56}$$

In the cold-body (nonrelativistic) limit, when $v^2 \ll c^2$, (17.55) and (17.56) become

$$\varepsilon \approx \mu c^2 + \frac{1}{2}\mu\overline{v^2} \approx \mu c^2 \quad \text{and} \quad p \approx \frac{1}{3}\mu\overline{v^2}$$

as expected from standard (nonrelativistic) kinetic theory. From these expressions, we find

$$\frac{3p}{\varepsilon} \approx \frac{\overline{v^2}}{c^2},$$

that is, $3p \ll \varepsilon$ in the cold-body limit.

18

General Relativity

18.1 Introduction

We begin by stating the **basic property of gravitational fields**: in a gravitational field all bodies move in the same manner, whatever their mass or charge, if their initial position and initial velocity are the same.

The acceleration (relative to an inertial frame) of a body "acted upon" by a **uniform** gravitational field can be interpreted instead as **free motion** (uniform velocity) but viewed from a **noninertial frame** moving (relative to the inertial frame) with an equal and opposite acceleration. In other words, this noninertial reference system is equivalent to the gravitational field (**principle of equivalence**).

Therefore, if the gravitational field is **nonuniform**, at each point in space a **different** noninertial frame is equivalent to it.

In **nonrelativistic mechanics**, the **equation of motion** of a particle of mass m in a gravitational field is determined by a lagrangian L having (in any inertial frame) the form

$$L = \frac{1}{2}mv^2 - m\phi$$

where ϕ is the gravitational potential. The Euler–Lagrange equations then yield

$$\dot{\boldsymbol{v}} = -\boldsymbol{\nabla}\phi$$

which is independent of the mass and charge of the particle, as required by the above "basic property". (Here, ϕ at a distance r from a point mass is given **empirically** by $\phi \propto 1/r$, so that the force ($\propto \dot{\boldsymbol{v}}$) on a particle in this potential is $\propto 1/r^2$.)

In the **relativistic mechanics** of the motion of a particle in a gravitational field, the analogue of $\dot{v} = -\boldsymbol{\nabla}\phi$ is the so-called **geodesic equation**, and the $1/r^2$ force law emerges as a consequence of the **Einstein field equations**, rather than empirically.

A Course in Theoretical Physics, First Edition. P. J. Shepherd.

18.2 Space–time metrics

In an inertial reference frame the interval $\mathrm{d}s$ between two infinitesimally separated events is given by

$$\mathrm{d}s^2 = c^2\mathrm{d}t^2 - \mathrm{d}x^2 - \mathrm{d}y^2 - \mathrm{d}z^2 \tag{18.1}$$

(a sum of squares) and remains in the form of a sum of squares

$$\mathrm{d}s'^2 = c^2\mathrm{d}t'^2 - \mathrm{d}x'^2 - \mathrm{d}y'^2 - \mathrm{d}z'^2 \ (=\mathrm{d}s^2)$$

when we transform to any other **inertial** frame.

If, instead, we transform to a noninertial frame, $\mathrm{d}s^2$ is, in general, no longer a simple sum of squares. For example, if we transform to a frame rotating with angular frequency ω about the Z axis, that is,

$$\begin{aligned} x &= x'\cos\omega t - y'\sin\omega t,\\ y &= x'\sin\omega t + y'\cos\omega t,\\ z &= z', \end{aligned}$$

we have

$$\begin{aligned} \mathrm{d}x &= \mathrm{d}x'\cos\omega t - x'\omega\mathrm{d}t\sin\omega t - \mathrm{d}y'\sin\omega t - y'\omega\mathrm{d}t\cos\omega t,\\ \mathrm{d}y &= \mathrm{d}x'\sin\omega t + x'\omega\mathrm{d}t\cos\omega t + \mathrm{d}y'\cos\omega t - y'\omega\mathrm{d}t\sin\omega t,\\ \mathrm{d}z &= \mathrm{d}z', \end{aligned}$$

and so

$$\mathrm{d}s^2 = [c^2 - \omega^2(x'^2 + y'^2)]\mathrm{d}t^2 - \mathrm{d}x'^2 - \mathrm{d}y'^2 - \mathrm{d}z'^2 + 2\omega y'\mathrm{d}x'\mathrm{d}t - 2\omega x'\mathrm{d}y'\mathrm{d}t. \tag{18.2}$$

Whatever the corresponding expression for $\mathrm{d}t$ in terms of $\mathrm{d}t'$, $\mathrm{d}x'$ and $\mathrm{d}y'$, this cannot be represented as a simple sum of squares, but only as the general form

$$\mathrm{d}s^2 = g_{ik}\mathrm{d}x^i\mathrm{d}x^k, \tag{18.3}$$

where the g_{ik} are **functions of the space and time coordinates**, that is, the coordinate system is **curvilinear**. [In equation (18.3) we have dropped the primes, and summation over repeated indices is understood.]

Because $\mathrm{d}s^2$ is a Lorentz scalar, and $\mathrm{d}x^i$ is a Lorentz four-vector, the quantities g_{ik} are the components of a rank-2 Lorentz four-tensor, called the **space–time metric** or **metric tensor**.

From (18.3),

$$g_{ki} = g_{ik},$$

that is, there are $4 + 3 + 2 + 1 = 10$ independent components of g_{ik}:

$$(g_{ik}) = \begin{pmatrix} \times & \times & \times & \times \\ & \times & \times & \times \\ & & \times & \times \\ & & & \times \end{pmatrix}$$

In an inertial frame, from (18.1), using

$$\mathrm{d}x^i = (\mathrm{d}x^0, \mathrm{d}x^1, \mathrm{d}x^2, \mathrm{d}x^3) = (c\mathrm{d}t, \mathrm{d}x, \mathrm{d}y, \mathrm{d}z)$$

we have

$$g_{00} = 1, \quad g_{11} = g_{22} = g_{33} = -1, \quad g_{ik} = 0 \quad (i \neq k). \tag{18.4}$$

A system of coordinates with these g_{ik} is called **Galilean**.

Gravitational fields correspond to noninertial reference frames that may differ from point to point and time to time, that is, **these fields are determined by the space–time metric g_{ik} as a function of the coordinates** x^0, x^1, x^2, x^3.

Einstein's **general theory of relativity** yields second-order differential equations (the **Einstein field equations**) for each of the ten components g_{ik} in terms of all the components T_{ik} of the energy–momentum tensor. For given boundary conditions these can (in principle!) be solved for the metric g_{ik}, that is, for the "**gravitational field**".

In the complete absence of matter/energy–momentum, that is, when $T_{ik} = 0$ everywhere, the Einstein field equations have the constant solution (18.4), and the space–time is said to be **flat**. The metric (18.2) also corresponds to a flat space–time, since a **single** simple transformation (the inverse rotation) transforms the metric back to its Galilean form **at all space–time points**, that is, transforms the interval back to (18.1) everywhere.

Space–times whose metrics cannot be transformed to the Galilean form (18.4) everywhere by a single coordinate transformation are said to be **curved**. Even in curved space–times, however, it is always possible to find a coordinate transformation that will bring g_{ik} to diagonal form [and hence, by a coordinate scaling, to the Galilean form (18.4)] **at any particular space–time point**, but, in general, a different such transformation will be needed for each space–time point.

For any **real** space–time, such a transformation at a given point always results in a metric tensor g_{ik} at that point with one positive and three negative principal values. The set of signs, for example, (+, –, –, –), is called the **signature** of the metric. Thus, we always have

$$g \equiv \det(g_{ik}) < 0$$

for a real space–time.

A curved space–time, with a non-Galilean space–time metric, implies a non-Euclidean space metric, that is, in the presence of matter/energy–momentum the geometry of space becomes non-Euclidean.

In the case of an **arbitrary** "gravitational field", such as is produced by moving bodies, the space metric is not only non-Euclidean but also varies with time. Thus, the relative positions of "test bodies" cannot remain fixed in any coordinate system.

Thus, whereas in the **special** theory of relativity we can define a reference frame by a set of bodies at rest in unchanging relative positions, such sets of bodies cannot exist in the **general** theory, and the reference frame must be specified by an infinite number of bodies filling **all** space, each with its own clock running at a rate determined by its position in space.

18.3 Curvilinear coordinates

When we restricted ourselves to inertial frames (special relativity), we considered Lorentz transformations described by

$$x^i = \alpha^i_j x'^j \tag{18.5}$$

and defined Lorentz four-tensors by their behavior under these transformations:

$$A^i = \alpha^i_j A'^j$$
$$A^{ik} = \alpha^i_j \alpha^k_l A'^{jl}, \quad \text{etc.}$$

For an **arbitrary** (that is, possibly nonlinear) transformation

$$x^i = x^i(x'_0, x'_1, x'_2, x'_3), \tag{18.6}$$

the differential change $\mathrm{d}x^i$ of x^i that results from differential changes $\mathrm{d}x'^j$ of the x'^j is given by

$$\mathrm{d}x^i = \frac{\partial x^i}{\partial x'^j}\mathrm{d}x'^j. \tag{18.7}$$

For example, in the case of the transformation (18.5), for which

$$\frac{\partial x^i}{\partial x'^j} = \alpha^i_j,$$

equation (18.7) takes the form

$$\mathrm{d}x^i = \alpha^i_j \mathrm{d}x'^j, \tag{18.8}$$

that is, $\mathrm{d}x^i$ is a contravariant four-vector under Lorentz transformations.

So, for **general** coordinate transformations [whether linear such as (18.5) or nonlinear], we now **define** a **contravariant four-vector** as a set of four quantities A^i that transform in the same way as the differentials $\mathrm{d}x^i$, that is [see (18.7)], as

$$A^i = \frac{\partial x^i}{\partial x'^j}A'^j. \tag{18.9}$$

If φ is a scalar (i.e., is invariant) under these (in general, nonlinear) transformations, that is, if

$$\varphi'(x') = \varphi(x),$$

the general rule for implicit differentiation

$$\frac{\partial \varphi}{\partial x^i} = \frac{\partial \varphi}{\partial x'^k}\frac{\partial x'^k}{\partial x^i}$$

takes the form

$$\frac{\partial \varphi}{\partial x^i} = \frac{\partial \varphi'}{\partial x'^k}\frac{\partial x'^k}{\partial x^i}, \tag{18.10}$$

that is, $\partial\varphi/\partial x^i$ is clearly **not** a contravariant four-vector. A set of four quantities A_i that transform like $\partial\varphi/\partial x^i$ under general coordinate transformations, that is, as

$$A_i = \frac{\partial x'^k}{\partial x^i} A'_k, \tag{18.11}$$

are said to form a **covariant four-vector**.

[In the case when the coordinate transformations correspond to rotations in three-dimensional Euclidean space, we have

$$\frac{\partial x^i}{\partial x'^k} = \frac{\partial x'^k}{\partial x^i}$$

(which follows from the fact that the rotation matrices are orthogonal, i.e., $\lambda^{\mathrm{T}} = \lambda^{-1}$), and so there is no distinction between contravariant and covariant vectors in this case.]

A **contravariant** rank-2 four-tensor A^{ik} (which has $4 \times 4 = 16$ components) transforms like the $4 \times 4 = 16$ products $\mathrm{d}x^i \mathrm{d}x^k$ [see (18.7)]:

$$A^{ik} = \frac{\partial x^i}{\partial x'^l}\frac{\partial x^k}{\partial x'^m} A'^{lm}, \tag{18.12}$$

while a **covariant** rank-2 four-tensor A_{ik} transforms like the 16 products $(\partial\varphi/\partial x^i)(\partial\varphi/\partial x^k)$ [see (18.10)]:

$$A_{ik} = \frac{\partial x'^l}{\partial x^i}\frac{\partial x'^m}{\partial x^k} A'_{lm}. \tag{18.13}$$

For a **mixed** rank-2 four-tensor A^i_k,

$$A^i_k = \frac{\partial x'^m}{\partial x^k}\frac{\partial x^i}{\partial x'^l} A'^l{}_m.$$

18.4 Products of tensors

The pair products of the 4^3 components of a four-tensor $A_i{}^{kj}$ with the 4^4 components of a four-tensor $B^l{}_{mnp}$ is a 4^7-component four-tensor $C_i{}^{kjl}{}_{mnp}$ of rank 7 (contravariant of rank 3 and covariant of rank 4).

18.5 Contraction of tensors

If we put a covariant index equal to a contravariant one and sum over all values of this index, we get a tensor lower in rank by 2. This is called **contraction**. For example,

$$C_i{}^{kil}{}_{mnp} = D^{kl}{}_{mnp}, \quad A^i B_i = C, \qquad \text{etc.}$$

[We can use (18.9) and (18.11) to confirm that $A^i B_i$ is indeed a scalar. Note that $A^i B^i$ is not a scalar, nor is $C_i{}^{kjl}{}_{inp}$ a tensor.]

18.6 The unit tensor

Define a quantity

$$\delta^i_k = \begin{cases} 1 & \text{for } i = k \\ 0 & \text{for } i \neq k \end{cases} \tag{18.14}$$

Contracting this with a vector A^k we get

$$A^k \delta^i_k = A^i,$$

that is, we get the same vector. Therefore, δ^i_k is a tensor, called the **unit tensor**.

18.7 Line element

We have

$$\mathrm{d}s^2 = g_{ik}\mathrm{d}x^i \mathrm{d}x^k. \tag{18.15}$$

Since $\mathrm{d}s^2$ is a scalar and the right-hand side here is the contraction of g_{ik} with the contravariant rank-2 tensor $\mathrm{d}x^i\mathrm{d}x^k$, g_{ik} is a covariant rank-2 tensor (the **covariant metric tensor**).

18.8 Tensor inverses

If two tensors A_{ik} and B^{ik} satisfy

$$A_{ik}B^{kl} = \delta^l_i,$$

they are mutual inverses.

In particular, we **define** the **contravariant metric tensor** g_{ik} as the inverse of g_{ik}:

$$g_{ik}g^{kl} = \delta^l_i. \tag{18.16}$$

18.9 Raising and lowering of indices

The contravariant and covariant forms of a vector can be related, obviously, only by the metric tensor. Indices are raised by the use of g^{ik} and lowered by the use of g_{ik}. For example,

$$\begin{aligned} g^{ik}A_k &= A^i, \\ g_{ik}A^k &= A_i. \end{aligned} \tag{18.17}$$

[Note that, although these **include** the results for flat space–time ($A_0 = A^0$, $A_\alpha = -A^\alpha$; $\alpha = 1, 2, 3$) and for four-dimensional Euclidean space ($A_i = A^i$; $i = 1, 2, 3, 4$), they apply to any space–time; for curved space–times, the components g_{ik} **will depend on the position in space–time**.]

Similarly,

$$A^i_{kl} = g_{lm}A^{im}_{k}, \quad A^{ik} = g^{il}g^{km}A_{lm}, \quad \text{etc.}$$

Note that if A_{kl} is not symmetric, we must distinguish between $A^i{}_j$ and $A_j{}^i$, that is, between the positions from which the subscript is raised.

Scalar products can be written as

$$A_i B^i = g_{ik}A^k B^i = g^{ik}A_i B_k. \tag{18.18}$$

In a product, an index can be raised in one factor and lowered in another without changing the value. For example,

$$A_{ik}B^{lk} = A_i{}^k B^l{}_k, \quad \text{etc.}$$

18.10 Integration in curved space–time

In a **flat** space–time the hypervolume element

$$\mathrm{d}\Omega \equiv \mathrm{d}x^0\mathrm{d}x^1\mathrm{d}x^2\mathrm{d}x^3$$

is a Lorentz scalar ($\mathrm{d}\Omega' = J\mathrm{d}\Omega = \mathrm{d}\Omega$), since the Jacobian J of the Lorentz transformation

$$x^i = \alpha^i_j x'^j$$

is

$$J \equiv \det\left(\frac{\partial x^i}{\partial x'^j}\right) = \det\left(\alpha^i_j\right) = 1.$$

To integrate over a hypervolume in **curved** space–time, for every hypervolume element $\mathrm{d}\Omega' \equiv \mathrm{d}x'^0\mathrm{d}x'^1\mathrm{d}x'^2\mathrm{d}x'^3$ we make that coordinate transformation which brings the local metric tensor g'_{ik} to Galilean form.

If, at a particular point, the transformation required is

$$x^i = x^i(x'^0, x'^1, x'^2, x'^3),$$

we have, by the rule (18.13) for transformation of a covariant rank-2 tensor,

$$g_{ik} = g'_{lm} \frac{\partial x'^l}{\partial x^i} \frac{\partial x'^m}{\partial x^k}, \tag{18.19}$$

where g_{ik} is Galilean, that is, $g_{00} = 1$, $g_{11} = g_{22} = g_{33} = -1$, and $g_{ik} = 0$ for $i \neq k$.

Although tensors should never be confused with matrices, equations (18.19) have, by the rules of matrix multiplication, the same form as

$$\mathrm{g} = \mathrm{J}\mathrm{g}'\mathrm{J}^{\mathrm{T}} \tag{18.20}$$

if g is a matrix with elements equal to the tensor components g_{ik}, g′ is a matrix with elements equal to the tensor components g'_{ik}, and J is a matrix with elements J_{il} equal to the tensor components $\partial x'^l/\partial x^i$.

[**Check**: The component form of equation (18.20) is

$$g_{ik} = J_{il} g'_{lm} (\mathrm{J}^{\mathrm{T}})_{mk} = J_{il} g'_{lm} J_{km}$$

which is the same as the equations (18.19).]

Taking determinants of both sides of equation (18.20), we have

$$-1 = g' J^2,$$

where $g' \equiv \det \mathrm{g}'$ and

$$J \equiv \det \mathrm{J}\ (= \det \mathrm{J}^{\mathrm{T}}) = \det \left(\frac{\partial x'^l}{\partial x^i} \right) \equiv \frac{\partial(x'^0, x'^1, x'^2, x'^3)}{\partial(x^0, x^1, x^2, x^3)}$$

is the **Jacobian** of the transformation.

Therefore,

$$J = \frac{1}{\sqrt{-g'}}. \tag{18.21}$$

But the Jacobian relates the hypervolume elements by the usual rule

$$\mathrm{d}\Omega' = J \mathrm{d}\Omega. \tag{18.22}$$

Therefore, defining a hypervolume integral by

$$\int \cdots \mathrm{d}\Omega,$$

where, in general, every hypervolume element $d\Omega$ in the region of integration pertains to a different frame (that in which the space–time metric is locally Galilean), we have

$$\int \cdots d\Omega = \int \cdots \frac{d\Omega'}{J} = \int \cdots \sqrt{-g'} d\Omega', \tag{18.23}$$

where g' and $d\Omega'$ refer to whatever curvilinear coordinate system is locally in use. Thus, in curvilinear coordinates (and dropping the primes), $\sqrt{-g}d\Omega$ **is the scalar element of integration.**

18.11 Covariant differentiation

Consider the partial derivative

$$A_{i,k} \equiv \frac{\partial A_i}{\partial x^k}$$

of a covariant vector A_i. From the transformation properties of A_i we have

$$A'_{i,k} \equiv \frac{\partial A'_i}{\partial x'^k} = \frac{\partial}{\partial x'^k}\left(\frac{\partial x^j}{\partial x'^i}A_j\right) = \frac{\partial^2 x^j}{\partial x'^i \partial x'^k}A_j + \frac{\partial x^j}{\partial x'^i}\frac{\partial x^l}{\partial x'^k}A_{j,l}.$$

If the first term on the right-hand side were to vanish, as it would in the case of a **linear** transformation $x \to x'$, this would be the transformation law for a rank-2 covariant tensor. In the **general** case it does not vanish, and so $A_{i,k}$ is **not** a tensor.

We can rewrite the transformation as

$$A'_{i,k} - \frac{\partial^2 x^j}{\partial x'^i \partial x'^k}\frac{\partial x'^m}{\partial x^j}A'_m = \frac{\partial x^j}{\partial x'^i}\frac{\partial x^l}{\partial x'^k}A_{j,l}. \tag{18.24}$$

The structure of the left-hand side here suggests that we might be able to find quantities

$$A_{i;k} = A_{i,k} - \Gamma^m_{ik} A_m$$

that **do** transform as a rank-2 covariant tensor, that is, that satisfy

$$A'_{i,k} - \Gamma'^m_{ik} A'_m = \frac{\partial x^j}{\partial x'^i}\frac{\partial x^l}{\partial x'^k}\left(A_{j,l} - \Gamma^n_{jl} A_n\right). \tag{18.25}$$

This is possible only if the quantities Γ^m_{ik} are coordinate-dependent and satisfy a transformation law that makes equations (18.25) entirely equivalent to equations (18.24). Comparing (18.24) and (18.25), having first put

$$A_n = A'_m \frac{\partial x'^m}{\partial x^n}$$

in (18.25), we find

$$A'_m \left(\Gamma'^m_{ik} - \frac{\partial x^j}{\partial x'^i} \frac{\partial x^l}{\partial x'^k} \frac{\partial x'^m}{\partial x^n} \Gamma_{jl}{}^n - \frac{\partial x'^m}{\partial x^j} \frac{\partial^2 x^j}{\partial x'^i \partial x'^k} \right) = 0,$$

which, since it must be valid for arbitrary A'_m, yields the following **nontensor** transformation law for the quantities $\Gamma_{ik}{}^m$:

$$\Gamma'^m_{ik} = \frac{\partial x'^m}{\partial x^n} \frac{\partial x^j}{\partial x'^i} \frac{\partial x^l}{\partial x'^k} \Gamma_{jl}{}^n + \frac{\partial x'^m}{\partial x^j} \frac{\partial^2 x^j}{\partial x'^i \partial x'^k}. \tag{18.26}$$

The set of quantities Γ^m_{ik} satisfying this (inhomogeneous) transformation law is called the **connection**, and the rank-2 covariant tensor

$$A_{i;k} = A_{i,k} - \Gamma_{ik}{}^m A_m \tag{18.27}$$

is called the **covariant derivative** of the vector A_i.

18.12 Parallel transport of vectors

Consider a covariant-vector field $A_i(x)$ at space–time points separated by $\mathrm{d}x^k$. The difference

$$\mathrm{d}A_i = A_i(x^k + \mathrm{d}x^k) - A_i(x^k) = A_i(x^k) + \frac{\partial A_i}{\partial x^k}\mathrm{d}x^k - A_i(x^k) \equiv A_{i,k}\mathrm{d}x^k \tag{18.28}$$

of these two vectors is **not** a vector, since, as just shown, $A_{i,k} \equiv \partial A_i / \partial x^k$ is not a tensor. The reason is that vectors at different points in space–time have different transformation properties (because of the varying metric), and so their sums and differences do not give vectors. The "difference"

$$\mathrm{D}A_i \equiv A_{i;k}\mathrm{d}x^k \tag{18.29}$$

is, however, a vector, since $A_{i;k}$ is a tensor.

We therefore **transport** the first vector from the space–time point x^k to $x^k + \mathrm{d}x^k$ in a manner such that it can be subtracted from $A_i + \mathrm{d}A_i$ to give a **vector** difference $\mathrm{D}A_i$.

In a flat space–time the required transport leaves the components of A_i unchanged (parallel transport). The corresponding operation in a curved space–time is also called **parallel transport**, but will involve a change δA_i in A_i, that is,

$$A_i \text{ at } x^k \rightarrow A_i + \delta A_i \text{ at } x^k + \mathrm{d}x^k.$$

The required difference is then

$$\mathrm{D}A_i \equiv (A_i + \mathrm{d}A_i) - (A_i + \delta A_i) = \mathrm{d}A_i - \delta A_i = A_{i;k}\mathrm{d}x^k.$$

Substituting [see (18.27)]

$$\mathrm{d}A_i = A_{i,k}\mathrm{d}x^k$$

and

$$A_{i;k} = A_{i,k} - \Gamma_{ik}^{m} A_m,$$

we find the rule for parallel transport of a covariant vector:

$$\delta A_i = \Gamma_{ik}^{m} A_m \mathrm{d}x^k. \qquad (18.30)$$

Note that δA_i is **not** a vector, since Γ_{ik}^{m} is not a tensor.

We can also define quantities $\Gamma_{i,kl}$ by

$$\Gamma_{i,kl} \equiv g_{im}\Gamma_{kl}^{m} \qquad (18.31)$$

[so that $g^{ji}\Gamma_{i,kl} = g^{ji}g_{im}\Gamma_{kl}^{m} = \delta_m^j \Gamma_{kl}^{m} = \Gamma_{kl}^{j}$].

By definition of a scalar, $\delta(\text{scalar}) = 0$, and so

$$\delta(A_i B^i) = 0.$$

Therefore,

$$A_i \delta B^i = -B^i \delta A_i = -B^i \Gamma_{ik}^{m} A_m \mathrm{d}x^k = -B^j \Gamma_{jk}^{i} A_i \mathrm{d}x^k,$$

from which we find

$$\delta B^i = -\Gamma_{jk}^{i} B^j \mathrm{d}x^k, \qquad (18.32)$$

which defines the **parallel transport of a contravariant vector**.

From (18.32) we have

$$\mathrm{D}B^i \equiv \mathrm{d}B^i - \delta B^i = \frac{\partial B^i}{\partial x^k}\mathrm{d}x^k + \Gamma_{jk}^{i} B^j \mathrm{d}x^k,$$

so that, since $\mathrm{D}B^i = B^i_{;k}\mathrm{d}x^k$, we have

$$B^i_{;k} = \frac{\partial B^i}{\partial x^k} + \Gamma_{jk}^{i} B^j. \qquad (18.33)$$

Another useful concept is that of a **geodesic**, which is defined as a curve $x^j(s)$ such that the tangent vector $\mathrm{d}x^j/\mathrm{d}s$ remains tangent to the curve when parallel-transported along it in the sense defined here in terms of the connection.

For a Riemannian space–time, that is, a space–time with interval

$$\mathrm{d}s^2 = g_{ik}\mathrm{d}x^i \mathrm{d}x^k,$$

the components of the connection Γ_{ik}^{l} are called **Christoffel symbols** and are symmetric:

$$\Gamma_{ik}^{l} = \Gamma_{ki}^{l}$$

[***Proof***: For $A_i = \partial\varphi/\partial x^i$, where φ is a scalar, the tensor

$$A_{k;i} - A_{i;k} = \left(\Gamma^l_{ik} - \Gamma^l_{ki}\right)\frac{\partial\varphi}{\partial x^l}$$

is zero in a Cartesian coordinate system and, hence (since it is a tensor), in all systems. Therefore, the result $\Gamma^l_{ik} = \Gamma^l_{ki}$ follows.]

To find the rule for covariant differentiation of tensors of arbitrary rank, consider

$$\delta(A_i B_k) = A_i \delta B_k + B_k \delta A_i = A_i \Gamma^l_{km} B_l \mathrm{d}x^m + B_k \Gamma^l_{im} A_l \mathrm{d}x^m.$$

Then, since a tensor A_{ik} transforms like $A_i B_k$, we have

$$\mathrm{D}A_{ik} = \mathrm{d}A_{ik} - \delta A_{ik} = A_{ik;m}\mathrm{d}x^m$$

with

$$A_{ik;m} = \frac{\partial A_{ik}}{\partial x^m} - \Gamma^l_{km} A_{il} - \Gamma^l_{im} A_{lk}.$$

We now apply this to

$$g_{ik;m} = 0.$$

[This states that g_{ik} is "covariantly constant" ($\mathrm{D}g_{ik} = 0$), which follows from the fact that we can express $\mathrm{D}A_i$ either as $g_{ik}\mathrm{D}A^k$ (since $\mathrm{D}A_i$ is a vector) or, since A_i is also a vector, as $\mathrm{D}(g_{ik}A^k) = g_{ik}\mathrm{D}A^k + A^k \mathrm{D}g_{ik}$.] We have

$$g_{ik;m} = 0 = \frac{\partial g_{ik}}{\partial x^m} - g_{il}\Gamma^l_{km} - g_{lk}\Gamma^l_{im} \equiv \frac{\partial g_{ik}}{\partial x^m} - \Gamma_{i,km} - \Gamma_{k,im}.$$

Therefore,

$$\frac{\partial g_{ik}}{\partial x^m} = \Gamma_{i,km} + \Gamma_{k,im},$$

$$\frac{\partial g_{im}}{\partial x^k} = \Gamma_{i,km} + \Gamma_{m,ik},$$

$$-\frac{\partial g_{km}}{\partial x^i} = -\Gamma_{k,im} - \Gamma_{m,ik},$$

where the second equation is obtained from the first by the interchange $k \leftrightarrow m$ and the third from the second by the interchange $k \leftrightarrow i$ (with the use of the symmetry property $\Gamma^l_{ik} = \Gamma^l_{ki}$ in both cases). Adding these equations and dividing by two, we get

$$\Gamma_{i,km} = \frac{1}{2}\left(\frac{\partial g_{ik}}{\partial x^m} + \frac{\partial g_{im}}{\partial x^k} - \frac{\partial g_{km}}{\partial x^i}\right). \qquad (18.34)$$

Therefore,

$$\Gamma^l_{km} = \frac{1}{2} g^{li} \left(\frac{\partial g_{ik}}{\partial x^m} + \frac{\partial g_{im}}{\partial x^k} - \frac{\partial g_{km}}{\partial x^i} \right). \tag{18.35}$$

The contraction of this is

$$\Gamma^l_{kl} = \frac{1}{2} g^{li} \left(\frac{\partial g_{ik}}{\partial x^l} + \frac{\partial g_{il}}{\partial x^k} - \frac{\partial g_{kl}}{\partial x^i} \right) = \frac{1}{2} g^{li} \frac{\partial g_{il}}{\partial x^k}. \tag{18.36}$$

We now relate this to the change $\mathrm{d}g$ in $g \equiv \det(g_{ik})$ resulting from the change

$$g_{ik}(x_i) \to g_{ik}(x_i + \mathrm{d}x_i) = g_{ik}(x_i) + \mathrm{d}g_{ik},$$

where

$$\mathrm{d}g_{ik} = \frac{\partial g_{ik}}{\partial x^l} \mathrm{d}x^l \equiv g_{ik,l} \mathrm{d}x^l$$

is **not** a tensor (just as $\mathrm{d}A_i = A_{i,l}\mathrm{d}x^l$ is not a vector). We have

$$\mathrm{d}g = \begin{vmatrix} g_{00} + \mathrm{d}g_{00} & g_{01} + \mathrm{d}g_{01} & \cdot & \cdot \\ g_{10} + \mathrm{d}g_{10} & g_{11} + \mathrm{d}g_{11} & \cdot & \cdot \\ \cdot & \cdot & \cdot & \cdot \\ \cdot & \cdot & \cdot & \cdot \end{vmatrix} - \begin{vmatrix} g_{00} & g_{01} & \cdot & \cdot \\ g_{10} & g_{11} & \cdot & \cdot \\ \cdot & \cdot & \cdot & \cdot \\ \cdot & \cdot & \cdot & \cdot \end{vmatrix}.$$

Therefore, to first order in $\mathrm{d}g_{il}$,

$$\mathrm{d}g = G^{il} \mathrm{d}g_{il}$$

(one term for each $\mathrm{d}g_{il}$), where G^{il} is the cofactor of g_{il} in g.

But the inverse of the matrix (g_{il}) can be found as $(G^{il})/g$ and, since the inverse of (g_{il}) is (g^{il}), we have, for the components,

$$\frac{G^{il}}{g} = g^{li} = g^{il}. \tag{18.37}$$

Therefore,

$$\mathrm{d}g = g g^{il} \mathrm{d}g_{il}, \tag{18.38}$$

that is,

$$g^{il} \mathrm{d}g_{il} = \frac{\mathrm{d}g}{g} \tag{18.39}$$

(which is **not** a scalar, since dg_{il} is not a tensor). (For later use, since

$$d\left(g^{il}g_{il}\right) = d\left(\delta^i_i\right) = d(4) = 0 = g^{il}dg_{il} + dg^{il}g_{il},$$

we have

$$g_{il}dg^{il} = -g^{il}dg_{il} = -\frac{dg}{g}.\Big) \tag{18.40}$$

Therefore, (18.36) becomes

$$\Gamma^l_{kl} = \frac{1}{2}g^{li}\frac{\partial g_{il}}{\partial x^k} = \frac{1}{2g}\frac{\partial g}{\partial x^k} = \frac{1}{2(-g)}\frac{\partial(-g)}{\partial x^k} = \frac{\partial \ln\sqrt{-g}}{\partial x^k}. \tag{18.41}$$

We use this in (18.33) to find an expression for the **covariant divergence** of the vector A^i:

$$A^i_{;i} = \frac{\partial A^i}{\partial x^i} + \Gamma^i_{ki}A^k = \frac{\partial A^i}{\partial x^i} + A^k\frac{\partial \ln\sqrt{-g}}{\partial x^k}$$

that is,

$$A^i_{;i} = \frac{1}{\sqrt{-g}}\frac{\partial(\sqrt{-g}A^i)}{\partial x^i}. \tag{18.42}$$

18.13 Curvature

We get a well-defined measure of the curvature of space–time at a point by finding the change ΔA_k of any vector A_k as a result of parallel transport of the vector around a closed infinitesimal contour enclosing the point. (ΔA_k is a **vector**, since it is the difference of two vectors **at the same point**.)

Using (18.30), we can write this change as

$$\Delta A_k = \oint \delta A_k = \oint \Gamma^i_{kl}A_i dx^l. \tag{18.43}$$

But we showed in chapter 17 on the classical theory of fields (see section 17.2.9) that the analogue of Stokes' theorem in a four-dimensional space–time is

$$dx^l \to df^{*ml}\frac{\partial}{\partial x^m},$$

that is,

$$\oint A_l dx^l = \int df^{*ml}\left(\frac{\partial A_l}{\partial x^m}\right)_{\text{PT}} = \frac{1}{2}\int df^{*lm}\left(\frac{\partial A_m}{\partial x^l} - \frac{\partial A_l}{\partial x^m}\right)_{\text{PT}}, \tag{18.44}$$

where we have used the antisymmetry of the tensor df^{*ml} and the subscript PT indicates that we are considering changes due to parallel transport only.

Therefore,

$$\Delta A_k = \frac{1}{2}\int\left[\left(\frac{\partial\left(\Gamma^i_{mk}A_i\right)}{\partial x^l}\right)_{\mathrm{PT}} - \left(\frac{\partial\left(\Gamma^i_{kl}A_i\right)}{\partial x^m}\right)_{\mathrm{PT}}\right]\mathrm{d}f^{*lm}$$
$$\approx \frac{1}{2}\left[\left(\frac{\partial\Gamma^i_{mk}}{\partial x^l}\right)_{\mathrm{PT}} A_i - \left(\frac{\partial\Gamma^i_{kl}}{\partial x^m}\right)_{\mathrm{PT}} A_i + \Gamma^i_{km}\left(\frac{\partial A_i}{\partial x^l}\right)_{\mathrm{PT}} - \Gamma^i_{kl}\left(\frac{\partial A_i}{\partial x^m}\right)_{\mathrm{PT}}\right]\Delta f^{*lm}, \tag{18.45}$$

since the area Δf^{*lm} is infinitesimal.

But [see (18.30)], the change of a covariant vector A_i under parallel transport is

$$\delta A_i = \left(\frac{\partial A_i}{\partial x^l}\right)_{\mathrm{PT}}\mathrm{d}x^l = \Gamma^n_{il}A_n\mathrm{d}x^l,$$

and so

$$\left(\frac{\partial A_i}{\partial x^l}\right)_{\mathrm{PT}} = \Gamma^n_{il}A_n. \tag{18.46}$$

Also, we have

$$\left(\frac{\partial\Gamma^i_{mk}}{\partial x^l}\right)_{\mathrm{PT}} = \frac{\partial\Gamma^i_{mk}}{\partial x^l}, \tag{18.47}$$

which is proved as follows. We have

$$\delta\Gamma^i_{mk} = \left(\frac{\partial\Gamma^i_{mk}}{\partial x^l}\right)_{\mathrm{PT}}\mathrm{d}x^l.$$

Alternatively, using expression (18.35) for the connection, we have

$$\delta\Gamma^i_{mk} = \delta\left\{\frac{1}{2}g^{il}\left(\frac{\partial g_{lk}}{\partial x^m} + \cdots\right)\right\}$$
$$= \frac{1}{2}\delta g^{il}\frac{\partial g_{lk}}{\partial x^m} + \frac{1}{2}g^{il}\delta\frac{\partial g_{lk}}{\partial x^m} + \cdots = \frac{1}{2}\delta g^{il}\frac{\partial g_{lk}}{\partial x^m} + \frac{1}{2}g^{il}\frac{\partial\delta g_{lk}}{\partial x^m} + \cdots.$$

But, since $\mathrm{D}g_{il} = 0$, we have $\delta g_{il} = \mathrm{d}g_{il}$, and so this becomes

$$\delta\Gamma^i_{mk} = \frac{1}{2}\mathrm{d}g^{il}\frac{\partial g_{lk}}{\partial x^m} + \frac{1}{2}g^{il}\mathrm{d}\frac{\partial g_{lk}}{\partial x^m} + \cdots = \mathrm{d}\Gamma^i_{mk} = \frac{\partial\Gamma^i_{mk}}{\partial x^l}\mathrm{d}x^l,$$

thus proving (18.47).

Putting (18.47) and (18.46) into (18.45), we have

$$\Delta A_k = \frac{1}{2}R^i_{klm}A_i\Delta f^{*lm} \tag{18.48}$$

with

$$R^i_{klm} = \frac{\partial \Gamma^i_{km}}{\partial x^l} - \frac{\partial \Gamma^i_{kl}}{\partial x^m} + \Gamma^i_{nl}\Gamma^n_{km} - \Gamma^i_{nm}\Gamma^n_{kl}, \quad (18.49)$$

which, as can be seen from the structure of (18.48), is a tensor (the **curvature tensor**), even though none of the four contributions to it is separately a tensor.

The contraction of the curvature tensor is

$$R_{ik} \equiv R^l_{ilk} = \frac{\partial \Gamma^l_{ik}}{\partial x^l} - \frac{\partial \Gamma^l_{il}}{\partial x^k} + \Gamma^l_{nl}\Gamma^n_{ik} - \Gamma^l_{nk}\Gamma^n_{il} \quad (18.50)$$

(the **Ricci tensor**).

The **scalar curvature** R is defined as

$$R \equiv g^{ik} R_{ik}. \quad (18.51)$$

18.14 The Einstein field equations

To obtain equations for the ten fields $g_{ik}(x)$ we use the **principle of least action**, varying the action with respect to the fields g_{ik} and their space–time derivatives (just as in chapter 17 on the classical theory of fields we found the second pair of Maxwell equations by varying only the electromagnetic fields A_i).

We use

$$\delta S = \delta(S_{\mathrm{m}} + S_{\mathrm{g}}) = 0, \quad (18.52)$$

where S_{m} is the action of the matter (including the electromagnetic fields) and S_{g} is the action functional of the gravitational field.

We have for the variation of the matter action

$$\delta S_{\mathrm{m}} = \frac{1}{c}\delta \int \Lambda\sqrt{-g}\mathrm{d}\Omega = \frac{1}{c}\int \left[\frac{\partial\left(\sqrt{-g}\Lambda\right)}{\partial g^{ik}}\delta g^{ik} + \frac{\partial\left(\sqrt{-g}\Lambda\right)}{\partial\left(\partial g^{ik}/\partial x^l\right)}\delta\left(\frac{\partial g^{ik}}{\partial x^l}\right)\mathrm{d}\Omega\right]. \quad (18.53)$$

Here Λ is the lagrangian density and δg^{ik} is a field describing an arbitrary infinitesimal variation (imagined, or virtual variation) of the contravariant metric tensor at all points in the space–time domain of integration. Because, at each point, it is the difference of two contravariant rank-2 tensors at that point, it is itself a tensor (unlike the δg^{ik} considered earlier, which was the change of g^{ik} produced by parallel transport to a different, infinitesimally displaced point).

In the second term of (18.53) we use the fact that

$$\delta\left(\frac{\partial g^{ik}}{\partial x^l}\right) = \frac{\partial \delta g^{ik}}{\partial x^l}$$

and integrate this term by parts, using Gauss' theorem to eliminate the exact-divergence term. The result is

$$\begin{aligned}\delta S_{\mathrm{m}} &= \frac{1}{c}\int\left[\frac{\partial\left(\sqrt{-g}\Lambda\right)}{\partial g^{ik}} - \frac{\partial}{\partial x^{l}}\left(\frac{\partial\left(\sqrt{-g}\Lambda\right)}{\partial\left(\partial g^{ik}/\partial x^{l}\right)}\right)\right]\delta g^{ik}\mathrm{d}\Omega \\ &\equiv \frac{1}{2c}\int T_{ik}\delta g^{ik}\sqrt{-g}\mathrm{d}\Omega,\end{aligned} \tag{18.54}$$

with

$$\frac{1}{2}\sqrt{-g}T_{ik} = \frac{\partial\left(\sqrt{-g}\Lambda\right)}{\partial g^{ik}} - \frac{\partial}{\partial x^{l}}\left(\frac{\partial\left(\sqrt{-g}\Lambda\right)}{\partial\left(\partial g^{ik}/\partial x^{l}\right)}\right) \tag{18.55}$$

The quantity T_{ik} defined by this expression is obviously symmetric in i and k, and, in fact, coincides with the **energy–momentum tensor**.

Check: For the example of the electromagnetic field (see section 17.5) we have

$$\Lambda = -\frac{\varepsilon_0 c^2}{4}F_{ik}F^{ik} = -\frac{\varepsilon_0 c^2}{4}F_{ik}F_{lm}g^{il}g^{km}, \tag{18.56}$$

that is,

$$\Lambda = \Lambda(g_{ik})$$

(there is no dependence of the lagrangian density on the space and time derivatives of g^{ik}). Therefore,

$$\frac{1}{2}\sqrt{-g}T_{ik} = \frac{\partial\left(\sqrt{-g}\Lambda\right)}{\partial g^{ik}} = \Lambda\frac{\partial\sqrt{-g}}{\partial g^{ik}} + \sqrt{-g}\frac{\partial\Lambda}{\partial g^{ik}}. \tag{18.57}$$

In the first term we have

$$\frac{\partial\sqrt{-g}}{\partial g^{ik}} = \frac{1}{2\sqrt{-g}}\frac{\partial(-g)}{\partial g^{ik}} = \frac{\sqrt{-g}}{2}.\frac{1}{g}.\frac{\partial g}{\partial g^{ik}} \tag{18.58}$$

The derivation that led to (18.40) for variations related to infinitesimal displacements in space–time can also be applied to the virtual variations at a point that are considered in this application of the principle of least action. So we have

$$g_{ik}\mathrm{d}g^{ik} = -\frac{\mathrm{d}g}{g}, \tag{18.59}$$

that is,

$$\frac{\partial g}{\partial g^{ik}} = -g g_{ik}, \tag{18.60}$$

and so

$$\frac{\partial\sqrt{-g}}{\partial g^{ik}} = -\frac{\sqrt{-g}}{2} g_{ik}. \tag{18.61}$$

In the second term in (18.57),

$$\begin{aligned}\frac{\partial\Lambda}{\partial g^{ik}} &= \frac{\partial}{\partial g^{ik}}\left[-\frac{\varepsilon_0 c^2}{4} F_{il}F_{km}g^{ik}g^{lm}\right] \\ &= -\frac{\varepsilon_0 c^2}{4}\left[F_{il}F_{km}g^{lm} + F_{li}F_{mk}g^{lm}\right] = -\frac{\varepsilon_0 c^2}{2} F_{il}F_k{}^l.\end{aligned} \tag{18.62}$$

Therefore,

$$\frac{1}{2}\sqrt{-g}T_{ik} = -\frac{\Lambda\sqrt{-g}}{2} g_{ik} - \frac{\varepsilon_0 c^2}{2}\sqrt{-g}F_{il}F_k{}^l, \tag{18.63}$$

and so

$$T_{ik} = \varepsilon_0 c^2\left(-F_{il}F_k{}^l + \frac{1}{4}g_{ik}F_{lm}F^{lm}\right), \tag{18.64}$$

in agreement with what we found in chapter 17 (see subsection 17.6.1). Recall also that for a macroscopic continuum we found

$$T_{ik} = (p+\varepsilon)u_i u_k - p g_{ik}, \quad \text{i.e.,} \quad T_i^k = (p+\varepsilon)u_i u^k - p\delta_i^k. \tag{18.65}$$

We now find δS_g.

The gravitational action S_g must be an integral over $\sqrt{-g}\mathrm{d}\Omega$ of a scalar field characterizing the curvature of space–time. We choose this to be minus the curvature scalar R:

$$\delta S_g = \delta\left[\frac{c^3}{16\pi G}\int(-R)\sqrt{-g}\mathrm{d}\Omega\right], \tag{18.66}$$

where G is the gravitational constant and the prefactor will fix the units of mass.

We have

$$\begin{aligned}\delta S_g &= -\frac{c^3}{16\pi G}\delta\left[\int g^{ik}R_{ik}\sqrt{-g}\mathrm{d}\Omega\right] \\ &= -\frac{c^3}{16\pi G}\left[\int \delta g^{ik}R_{ik}\sqrt{-g}\mathrm{d}\Omega + \int g^{ik}\delta R_{ik}\sqrt{-g}\mathrm{d}\Omega + \int R\delta\sqrt{-g}\mathrm{d}\Omega\right].\end{aligned} \tag{18.67}$$

But

$$\delta\sqrt{-g} = \frac{1}{2}\frac{\delta(-g)}{\sqrt{-g}} = \frac{\sqrt{-g}}{2}\frac{\delta g}{g} = -\frac{\sqrt{-g}}{2}g_{ik}\delta g^{ik}, \tag{18.68}$$

which, once again, is derived in exactly the same way as (18.40) but now for virtual variations of g_{ik} at a point. In the second term we use our expression (18.50) for the Ricci tensor:

$$\delta R_{ik} = \frac{\partial}{\partial x^l}\delta\Gamma^l_{ik} - \frac{\partial}{\partial x^k}\delta\Gamma^l_{il} + \Gamma^l_{nl}\delta\Gamma^n_{ik} + \left(\delta\Gamma^l_{nl}\right)\Gamma^n_{ik} - \Gamma^l_{nk}\delta\Gamma^n_{il} - \left(\delta\Gamma^l_{nk}\right)\Gamma^n_{il}. \tag{18.69}$$

At any given point in space–time we can find an inertial frame, with, correspondingly, a locally Galilean metric in which $\Gamma^n_{ik} = 0$. In this frame, (18.69) gives

$$g^{ik}\delta R_{ik} = g^{ik}\left[\frac{\partial}{\partial x^l}\delta\Gamma^l_{ik} - \frac{\partial}{\partial x^k}\delta\Gamma^l_{il}\right]. \tag{18.70}$$

Bringing g^{ik} inside the derivatives (since g^{ik} is constant in a locally inertial frame), and interchanging l and k in the second term, we can write this as

$$g^{ik}\delta R_{ik} = \frac{\partial}{\partial x^l}\left(g^{ik}\delta\Gamma^l_{ik}\right) - \frac{\partial}{\partial x^l}\left(g^{il}\delta\Gamma^k_{ik}\right) \equiv \frac{\partial w^l}{\partial x^l} \equiv w^l_{,l}. \tag{18.71}$$

In a noninertial frame at the given point this divergence must generalize to a covariant divergence $w^l_{;l}$, and so, using (18.42), we have

$$g^{ik}\delta R_{ik} = w^l_{;l} = \frac{1}{\sqrt{-g}}\frac{\partial}{\partial x^l}\left(\sqrt{-g}w^l\right). \tag{18.72}$$

In the integral containing this, the factors of $\sqrt{-g}$ cancel, we apply Gauss' theorem, and the term vanishes by virtue of the vanishing of w^l on the hypersurface at infinity. Therefore, from (18.67),

$$\delta S_{\mathrm{g}} = -\frac{c^3}{16\pi G}\int\left(R_{ik} - \frac{1}{2}g_{ik}R\right)\delta g^{ik}\sqrt{-g}\mathrm{d}\Omega. \tag{18.73}$$

From this and (18.54), the principle of least action $\delta(S_{\mathrm{m}} + S_{\mathrm{g}}) = 0$ gives

$$-\frac{c^3}{16\pi G}\int\left(R_{ik} - \frac{1}{2}g_{ik}R - \frac{8\pi G}{c^4}T_{ik}\right)\delta g^{ik}\sqrt{-g}\mathrm{d}\Omega = 0. \tag{18.74}$$

Since the virtual (imagined) variations δg^{ik} at every point are arbitrary, it follows that

$$R_{ik} - \frac{1}{2}g_{ik}R = \frac{8\pi G}{c^4}T_{ik}, \tag{18.75}$$

or, in mixed components,

$$R^k_i - \frac{1}{2}\delta^k_i R = \frac{8\pi G}{c^4}T^k_i. \tag{18.76}$$

Contracting this and using $\delta_i^i = 4$, we get

$$R = -\frac{8\pi G}{c^4}T \quad \left(T \equiv T_i^i\right). \tag{18.77}$$

Putting this into (18.75) we get the alternative form

$$R_{ik} = \frac{8\pi G}{c^4}\left(T_{ik} - \frac{1}{2}g_{ik}T\right). \tag{18.78}$$

Equations (18.78) for the components of the Ricci curvature tensor in terms of the energy–momentum-density tensor are the **Einstein field equations**.

From the definition (18.50) of R_{ik}, these equations are **second-order** differential equations for the metric tensor $g_{ik}(x)$. Since they are nonlinear, **the principle of superposition is not valid** (unlike in electrodynamics). (For **weak** gravitational fields, however, the equations are approximately linear, and so the principle of superposition is approximately valid.)

In an empty region of space–time, $T_{ik}(x) = 0$, and so $R_{ik}(x) = 0$. [Note that this does **not** mean that space–time is flat in a field-free vacuum. We need the stronger condition $R^i_{klm}(x) = 0$ for that.]

In regions with an electromagnetic field alone, we have $T \equiv T_i^i = 0$, and so $R = -(8\pi G/c^4)T = 0$; that is, the scalar curvature is zero in such regions.

There is a very important difference between the Einstein and the Maxwell equations. Namely, the Maxwell equations contain the equations of motion of the fields and the equation of conservation of the total charge, but **not** the equations of motion of the charges producing the field.

The Einstein equations, on the other hand, also "contain" the equations $T^k_{i;k} = 0$ (the laws of conservation of energy and momentum), which contain the equations of motion of the physical system with energy–momentum tensor T_i^k. We prove this as follows. Taking the covariant divergence of both sides of equation (18.76) we have

$$\frac{8\pi G}{c^4}T^k_{i;k} = R^k_{i;k} - \frac{1}{2}\left(\delta_i^k R\right)_{;k} = R^k_{i;k} - \frac{1}{2}R_{;i}. \tag{18.79}$$

We now show that the right-hand side of (18.79) is equal to zero. We have from (18.49)

$$R^n_{ikl} = \frac{\partial \Gamma^n_{il}}{\partial x^k} - \frac{\partial \Gamma^n_{ik}}{\partial x^l} + \Gamma^n_{jk}\Gamma^j_{il} - \Gamma^n_{jl}\Gamma^j_{ik}. \tag{18.80}$$

We find the covariant derivative of this in an inertial frame (so that, in this frame, $\Gamma^i_{kl} = 0$ and, hence, $R^n_{ikl;m} = \partial R^n_{ikl}/\partial x^m$). We then have

$$R^n_{ikl;m} = \frac{\partial R^n_{ikl}}{\partial x^m} = \frac{\partial^2 \Gamma^n_{il}}{\partial x^m \partial x^k} - \frac{\partial^2 \Gamma^n_{ik}}{\partial x^m \partial x^l}. \tag{18.81}$$

Therefore,

$$R^n_{imk;l} = \frac{\partial^2 \Gamma^n_{ik}}{\partial x^l \partial x^m} - \frac{\partial^2 \Gamma^n_{im}}{\partial x^l \partial x^k} \tag{18.82}$$

and

$$R^n_{ilm;k} = \frac{\partial^2 \Gamma^n_{im}}{\partial x^k \partial x^l} - \frac{\partial^2 \Gamma^n_{il}}{\partial x^k \partial x^m}. \tag{18.83}$$

Adding (18.81), (18.82) and (18.83), we get

$$R^n_{ikl;m} + R^n_{imk;l} + R^n_{ilm;k} = 0 \tag{18.84}$$

(the **Bianchi identity**), which, in view of its tensor character, must also be true in **all** frames.

Contracting the Bianchi identity with $g^{ik}\delta^l_n$, we get

$$R^{lk}_{kl;m} + R^{lk}_{mk;l} + R^{lk}_{lm;k} = 0. \tag{18.85}$$

But, by the antisymmetry of R_{iklm} (and hence of R^{ik}_{lm}) in the first and second (and in the third and fourth) indices, and after the interchange $l \leftrightarrow k$ in the second term, (18.85) becomes

$$-R^{kl}_{kl;m} + 2R^k_{m;k} = 0, \tag{18.86}$$

that is,

$$-R_{;m} + 2R^k_{m;k} = 0. \tag{18.87}$$

Therefore,

$$R^k_{i;k} - \frac{1}{2}R_{;i} = 0, \tag{18.88}$$

as required, and so

$$T^k_{i;k} = 0. \tag{18.89}$$

Thus, the energy–momentum conservation relations are contained in the Einstein field equations.

18.15 Equation of motion of a particle in a gravitational field

The motion of a free particle is determined, as in special relativity, by the principle of least action in the form

$$\delta S = -mc\delta \int \mathrm{d}s = 0, \tag{18.90}$$

where

$$\mathrm{d}s^2 = g_{ik}\mathrm{d}x^i\mathrm{d}x^k. \tag{18.91}$$

In an inertial frame the result is

$$\frac{\mathrm{d}u^i}{\mathrm{d}s} = 0,$$

which generalizes in an arbitrary frame to

$$\frac{Du^i}{ds} = 0, \tag{18.92}$$

that is,

$$\frac{du^i}{ds} + \Gamma^i_{kl} u^k \frac{dx^l}{ds} = 0, \tag{18.93}$$

or, since $u^i = dx^i/ds$,

$$\frac{d^2x^i}{ds^2} + \Gamma^i_{kl} \frac{dx^k}{ds}\frac{dx^l}{ds} = 0, \tag{18.94}$$

that is,

$$m\frac{du^i}{ds} = -m\Gamma^i_{kl} u^k u^l. \tag{18.95}$$

Thus, $-m\Gamma^i_{kl}u^k u^l$ is the "four-force" on a particle in a gravitational field (that is, Γ^i_{kl} is a "field intensity", expressible in terms of derivatives of the "potentials" g^{ik}).

For example, consider how the tensor g_{ik} is related to the nonrelativistic potential ϕ of the gravitational field. In nonrelativistic mechanics we have [see (18.23)]

$$L = -mc^2 + \frac{mv^2}{2} - m\phi, \tag{18.96}$$

which is the limit of the relativistic lagrangian

$$L = -mc^2\sqrt{1 - v^2/c^2} - m\phi \tag{18.97}$$

as $v/c \to 0$. Therefore, the action of the motion of a particle along a given path in the field ϕ is

$$S = \int L dt = -mc \int \left(c - \frac{v^2}{2c} + \frac{\phi}{c}\right) dt. \tag{18.98}$$

Since this can be written as $-mc\int ds$ along the path, we have

$$ds = \left(c - \frac{v^2}{2c} + \frac{\phi}{c}\right) dt. \tag{18.99}$$

Therefore (since $\boldsymbol{v}dt = d\boldsymbol{r}$),

$$ds^2 = (c^2 + 2\phi)dt^2 - dr^2 + \text{terms which vanish as } c \to \infty. \tag{18.100}$$

This tells us that, in the nonrelativistic limit, except for g_{00} all components of the metric tensor are the same as the corresponding components of the Galilean metric tensor. Inspection of (18.100) shows that $g_{00}(x)$ is related to the nonrelativistic gravitational potential $\phi(x)$ by

$$g_{00} = 1 + \frac{2\phi}{c^2}. \tag{18.101}$$

Thus, by first solving Einstein's field equations for g_{00} in the case of a weak gravitational field (that is, one in which space–time has small curvature), we should be able to use (18.101) to derive the famous Newtonian gravitational potential $\phi \propto 1/r$ at a spatial distance r from a point mass.

18.16 Newton's law of gravity

We consider the limit of nonrelativistic mechanics. When the relative velocities of the particles making up the body are small (that is, the body is cold), we found in chapter 17 [see (17.51)]:

$$T_i^k \approx \mu^{\text{prop}} c^2 u_i u^k, \tag{18.102}$$

where μ^{prop} is the proper rest-mass density, that is, the rest mass per unit proper volume in the small element under consideration, and u_i is the four-velocity of the macroscopic motion.

In the limit of slow macroscopic motion we had $u_0 = 1$ and $u_\alpha = 0$, so that the only nonzero component of T_i^k is

$$T_0^0 \approx \mu^{\text{prop}} c^2. \tag{18.103}$$

Therefore,

$$T \equiv T_i^i = T_0^0 = \mu^{\text{prop}} c^2. \tag{18.104}$$

The Einstein field equations (18.78) have the form

$$R_i^k = \frac{8\pi G}{c^4}\left(T_i^k - \frac{1}{2}\delta_i^k T\right). \tag{18.105}$$

For the component with $i = k = 0$ this becomes

$$R_0^0 = \frac{4\pi G}{c^2}\mu^{\text{prop}} \tag{18.106}$$

and the equations for all the other components (18.105) vanish identically.

From the definition (18.50) of R_{ik}, we have

$$R_{00} = \frac{\partial \Gamma_{00}^l}{\partial x^l} - \frac{\partial \Gamma_{0l}^l}{\partial x^0} + \Gamma_{nl}^l \Gamma_{00}^n + \Gamma_{n0}^l \Gamma_{0l}^n. \tag{18.107}$$

In a weak gravitational field ($\Gamma^l_{ik} \approx 0$) we can neglect the terms bilinear in Γ. Also, derivatives with respect to $x_0 = ct$ are small (by a factor c) compared with derivatives with respect to x^α. Therefore,

$$R_{00} \approx \frac{\partial \Gamma^\alpha_{00}}{\partial x^\alpha} \quad (\alpha = 1, 2, 3). \tag{18.108}$$

Thus, from (18.106),

$$R_{00} = \frac{\partial \Gamma^\alpha_{00}}{\partial x^\alpha} = \frac{4\pi G}{c^2}\mu^{\text{prop}}. \tag{18.109}$$

But, from the expression (18.35) for Γ^l_{km}, we have

$$\Gamma^\alpha_{00} = \frac{1}{2}g^{\alpha i}\left(\frac{\partial g_{i0}}{\partial x^0} + \frac{\partial g_{i0}}{\partial x^0} - \frac{\partial g_{00}}{\partial x^i}\right) \approx -\frac{1}{2}g^{\alpha\beta}\frac{\partial g_{00}}{\partial x^\beta} = \frac{1}{c^2}\frac{\partial \phi}{\partial x^\alpha}, \tag{18.110}$$

where we have used again the relative smallness of derivatives with respect to $x_0 = ct$, and also the fact that the metric is Galilean except for the component g_{00}, given by equation (18.101). Therefore,

$$R_{00} = \frac{1}{c^2}\frac{\partial^2 \phi}{\partial x^\alpha \partial x^\alpha} \equiv \frac{1}{c^2}\nabla^2\phi, \tag{18.111}$$

and so

$$\nabla^2\phi(\boldsymbol{r}) = 4\pi G\mu^{\text{prop}}(\boldsymbol{r}). \tag{18.112}$$

This equation has the solution

$$\phi(\boldsymbol{r}) = -G\int \frac{\mu^{\text{prop}}(\boldsymbol{r}')\mathrm{d}^3\boldsymbol{r}'}{|\boldsymbol{r} - \boldsymbol{r}'|}, \tag{18.113}$$

as can be seen by noting that

$$\nabla^2 \frac{1}{|\boldsymbol{r} - \boldsymbol{r}'|} = -4\pi\delta^{(3)}(\boldsymbol{r} - \boldsymbol{r}'),$$

which is a formal statement of the fact that it is equal to zero for $\boldsymbol{r} \neq \boldsymbol{r}'$ and its volume integral is -4π. Thus, the gravitational potential (18.113) produced at $\boldsymbol{r}$ by a single point particle of mass m at the origin [that is, $\mu^{\text{prop}}(\boldsymbol{r}') = m\delta^{(3)}(\boldsymbol{r}')$] is

$$\phi(\boldsymbol{r}) = -\frac{Gm}{r}, \tag{18.114}$$

so the force acting on another particle (of mass m') is

$$F = -m'\frac{\partial \phi}{\partial r} = -\frac{Gmm'}{r^2}, \tag{18.115}$$

which is **Newton's law of gravitational attraction**.

Module V

Relativistic Quantum Mechanics and Gauge Theories

19

Relativistic Quantum Mechanics

19.1 The Dirac equation

The **nonrelativistic** free-particle energy–momentum relation is

$$E = \frac{p^2}{2m}.$$

Applying the usual operator prescriptions

$$E \to i\hbar\frac{\partial}{\partial t}, \qquad \boldsymbol{p} \to -i\hbar\boldsymbol{\nabla},$$

we find the free-particle time-dependent Schrödinger equation

$$i\hbar\frac{\partial\psi}{\partial t} = -\frac{\hbar^2}{2m}\nabla^2\psi.$$

The **relativistic** free-particle energy–momentum relation is

$$E = +\sqrt{p^2c^2 + m^2c^4}.$$

This follows from the manifestly Lorentz-covariant relation [see (17.26)]

$$p^i p_i = m^2c^2, \tag{19.1}$$

where p^i is the contravariant form of a Lorentz four-vector – the four-momentum, and p_i is the corresponding covariant form:

$$p^i = (E/c, p_x, p_y, p_z), \qquad p_i = g_{ij}p^j = (E/c, -p_x, -p_y, -p_z).$$

A Course in Theoretical Physics, First Edition. P. J. Shepherd.

Again,

$$p^i \to i\hbar\frac{\partial}{\partial x_i} = i\hbar\left(\frac{\partial}{\partial(ct)}, -\frac{\partial}{\partial x}, -\frac{\partial}{\partial y}, -\frac{\partial}{\partial z}\right).$$

The covariant form of this is

$$p_i \to i\hbar\frac{\partial}{\partial x^i} = i\hbar\left(\frac{\partial}{\partial(ct)}, \frac{\partial}{\partial x}, \frac{\partial}{\partial y}, \frac{\partial}{\partial z}\right).$$

[Recall that

$$x^i = (x^0, x^1, x^2, x^3) = (ct, x, y, z) \quad \text{and} \quad x_i = g_{ij}x^j = (ct, -x, -y, -z).]$$

Equation (19.1) becomes

$$\hbar^2\frac{\partial^2\psi}{\partial x_i \partial x^i} + m^2c^2\psi = 0, \tag{19.2}$$

which is the **Klein–Gordon equation**, often written in the form

$$-\hbar^2\frac{\partial^2\psi}{\partial t^2} = (-\hbar^2c^2\nabla^2 + m^2c^4)\psi.$$

Equation (19.2) is a **second-order** differential equation, and such an equation requires **two** initial conditions for its solution. In other words, equation (19.2), unlike the Schrödinger equation, does not entirely determine the evolution of $\psi(x) = \psi(\boldsymbol{r}, t)$ from its initial form $\psi(\boldsymbol{r}, 0)$.

Is it possible to find a **first-order** differential equation for ψ that implies the truth of the Klein–Gordon equation (19.2)?

To satisfy the requirements of special relativity, we need an equation of first order in both the time and the space partial derivatives. Dirac proposed the dimensionally obvious form

$$\left(i\hbar\gamma^i\frac{\partial}{\partial x^i} - mc\right)\psi = 0, \tag{19.3}$$

where the dimensionless quantities γ^i ($i = 0, 1, 2, 3$) are to be determined from the requirement that equation (19.3) implies equation (19.2).

Let us operate on (19.3) with

$$i\hbar\gamma^j\frac{\partial}{\partial x^j} + mc.$$

The result is

$$\left(i\hbar\gamma^j\frac{\partial}{\partial x^j} + mc\right)\left(i\hbar\gamma^i\frac{\partial}{\partial x^i} - mc\right)\psi = 0,$$

which can be rewritten as

$$-\hbar^2\gamma^j\gamma^i\frac{\partial^2\psi}{\partial x^j\partial x^i} - m^2c^2\psi = 0,$$

that is,

$$-\hbar^2\left(\frac{\gamma^j\gamma^i + \gamma^i\gamma^j}{2}\right)\frac{\partial^2\psi}{\partial x^j\partial x^i} - m^2c^2\psi = 0,$$

which is the same as equation (19.2) if and only if

$$\gamma^j\gamma^i + \gamma^i\gamma^j = 2g^{ij}$$

(since $g^{ij}\partial/\partial x^j = \partial/\partial x_i$). [Recall that, in flat space–time,

$$g^{00} = 1, \quad g^{11} = g^{22} = g^{33} = -1, \quad g^{ij}(i \neq j) = 0.]$$

No **numbers** γ^i can be found that satisfy this anticommutation relation. However, it **is** possible to find **matrices** γ^i such that their anticommutator

$$\gamma^j\gamma^i + \gamma^i\gamma^j = 2g^{ij}1, \tag{19.4}$$

where 1 is the unit matrix of the same dimensionality as γ^i and γ^j.

Therefore, Dirac postulated that equation (19.3) is an $N \times N$ matrix equation

$$\left(i\hbar\gamma^i\frac{\partial}{\partial x^i} - mc1\right)\Psi(x) = 0, \tag{19.5}$$

for the N-component column-matrix wave function

$$\Psi(x) \equiv \begin{pmatrix} \psi^{(1)}(x) \\ \cdots \\ \cdots \\ \cdots \\ \psi^{(N)}(x) \end{pmatrix}. \tag{19.6}$$

To find the smallest possible value of N for which the anticommutation relations (19.4) are satisfied, we first multiply (19.5) on the left by the matrix γ^0. Using the relation $(\gamma^0)^2 = 1$ [from (19.4)], and denoting $\beta \equiv \gamma^0$ and $\alpha_\mu \equiv \gamma^0\gamma^\mu$ ($\mu = 1, 2, 3$), we write (19.5) in the form

$$i\hbar\frac{\partial\Psi}{\partial t} = \mathrm{H}\Psi \equiv \left(-i\hbar c\alpha_\mu\frac{\partial}{\partial x^\mu} + \beta mc^2\right)\Psi = \left(c\alpha_\mu\hat{p}_\mu + \beta mc^2\right)\Psi.$$

This can be written as N coupled first-order equations for the N components of (19.6):

$$i\hbar\frac{\partial\psi^{(r)}}{\partial t}=\sum_{s=1}^{N}H_{rs}\psi^{(s)}$$

with

$$H_{rs}=c(\alpha_\mu\hat{p}_\mu)_{rs}+mc^2\beta_{rs}.$$

Because the hamiltonian matrix H must be hermitian (self-adjoint), so too must be the matrices α_μ (because the linear-momentum operator $\hat{p}_\mu$ is itself hermitian) and β. But from (19.4) we have $(\gamma^0)^2=1$ and $\alpha_\mu^2\equiv\gamma^0\gamma^\mu\gamma^0\gamma^\mu=-\gamma^0\gamma^0\gamma^\mu\gamma^\mu=-\gamma^\mu\gamma^\mu=1$. The necessarily real eigenvalues of the hermitian matrices $\beta\equiv\gamma^0$ and $\alpha_\mu\equiv\gamma^0\gamma^\mu$ can therefore only be ±1.

Also, the trace (the sum of the diagonal elements) of each of the four matrices β and α_μ is zero. For example, $\mathrm{Tr}\alpha_\mu=\mathrm{Tr}\beta^2\alpha_\mu=\mathrm{Tr}\beta\alpha_\mu\beta=-\mathrm{Tr}\alpha_\mu=0$, where in the second step we have used the cyclic property of the trace (Tr ABC = Tr BCA) and in the third step we have used (19.4). Since the trace is equal to the sum of the eigenvalues, the number of positive and negative eigenvalues ±1 must be equal, and so the matrices β and α_μ are even-dimensional. For the smallest even dimensionality $N=2$, all hermitian and traceless matrices take the form $\begin{pmatrix}a & b-ic\\ b+ic & -a\end{pmatrix}$, with just three real parameters a, b and c, and clearly there are only three such independent matrices [namely, the Pauli matrices (2.37)], whereas we require four. Thus, the smallest value of N for the four matrices satisfying (19.4) is $N=4$, and this is the case we shall consider.

Acting on equation (19.5) with

$$i\hbar\gamma^j\frac{\partial}{\partial x^j}+mc1$$

gives [if equation (19.4) is satisfied]

$$\left(\hbar^2\frac{\partial^2}{\partial x_i\partial x^i}+m^2c^2\right)\Psi(x)=0,$$

that is, the Klein–Gordon equation is satisfied for each component $\psi^{(\alpha)}(\alpha=1,2,3,4)$ of Ψ, as required. The following 4×4 γ-matrices (**Dirac matrices**) satisfy equation (19.4):

$$\gamma^0=\begin{pmatrix}1 & 0\\ 0 & -1\end{pmatrix},\quad \gamma^\mu=\begin{pmatrix}0 & \sigma_\mu\\ -\sigma_\mu & 0\end{pmatrix}\quad(\mu=1,2,3), \tag{19.7}$$

in which 0 and 1 here are the 2×2 null and unit matrices, and σ_μ are the **Pauli matrices** [see (2.37)]

$$\sigma_1=\begin{pmatrix}0 & 1\\ 1 & 0\end{pmatrix},\quad \sigma_2=\begin{pmatrix}0 & -i\\ i & 0\end{pmatrix},\quad \sigma_3=\begin{pmatrix}1 & 0\\ 0 & -1\end{pmatrix}. \tag{19.8}$$

For example,

$$\gamma^1\gamma^2 + \gamma^2\gamma^1 = \begin{pmatrix} 0 & \sigma_1 \\ -\sigma_1 & 0 \end{pmatrix}\begin{pmatrix} 0 & \sigma_2 \\ -\sigma_2 & 0 \end{pmatrix} + \begin{pmatrix} 0 & \sigma_2 \\ -\sigma_2 & 0 \end{pmatrix}\begin{pmatrix} 0 & \sigma_1 \\ -\sigma_1 & 0 \end{pmatrix}$$
$$= \begin{pmatrix} -\sigma_1\sigma_2 & 0 \\ 0 & -\sigma_1\sigma_2 \end{pmatrix} + \begin{pmatrix} -\sigma_2\sigma_1 & 0 \\ 0 & -\sigma_2\sigma_1 \end{pmatrix} = -\begin{pmatrix} \sigma_1\sigma_2 + \sigma_2\sigma_1 & 0 \\ 0 & \sigma_1\sigma_2 + \sigma_2\sigma_1 \end{pmatrix}$$

But

$$\sigma_1\sigma_2 + \sigma_2\sigma_1 = \begin{pmatrix} 0 & 1 \\ 1 & 0 \end{pmatrix}\begin{pmatrix} 0 & -i \\ i & 0 \end{pmatrix} + \begin{pmatrix} 0 & -i \\ i & 0 \end{pmatrix}\begin{pmatrix} 0 & 1 \\ 1 & 0 \end{pmatrix} = \begin{pmatrix} i & 0 \\ 0 & -i \end{pmatrix} + \begin{pmatrix} -i & 0 \\ 0 & i \end{pmatrix} = \begin{pmatrix} 0 & 0 \\ 0 & 0 \end{pmatrix} \equiv 0,$$

that is,

$$\gamma^1\gamma^2 + \gamma^2\gamma^1 = 0 = 2g^{12}1\,,$$

as required (0 and 1 here are the 4×4 null and unit matrices, respectively).

To take a second example,

$$\gamma^1\gamma^1 + \gamma^1\gamma^1 = 2\gamma^1\gamma^1 = 2\begin{pmatrix} 0 & \sigma_1 \\ -\sigma_1 & 0 \end{pmatrix}\begin{pmatrix} 0 & \sigma_1 \\ -\sigma_1 & 0 \end{pmatrix} = 2\begin{pmatrix} -\sigma_1^2 & 0 \\ 0 & -\sigma_1^2 \end{pmatrix} = 2\begin{pmatrix} -1 & 0 \\ 0 & -1 \end{pmatrix} = 2g^{11}1,$$

as required.

As a third example,

$$\gamma^0\gamma^0 + \gamma^0\gamma^0 = 2\gamma^0\gamma^0 = 2\begin{pmatrix} 1 & 0 \\ 0 & -1 \end{pmatrix}\begin{pmatrix} 1 & 0 \\ 0 & -1 \end{pmatrix} = 2\begin{pmatrix} 1 & 0 \\ 0 & 1 \end{pmatrix} = 2g^{00}1,$$

again as required.

It should be noted that applying a similarity transformation $\gamma'^i = U\gamma^i U^{-1}$ (U is a unitary 4×4 matrix) to the matrices γ^i gives a new set of matrices γ'^i that also satisfy the anticommutation relations (19.4). We shall use such an alternative set when we come to consider the property of the helicity of wave functions at the end of this chapter.

19.2 Lorentz and rotational covariance of the Dirac equation

An event has coordinates x^i in inertial frame K, and coordinates x'^i in inertial frame K', where

$$x^i = \alpha^i{}_j x'^j \tag{19.9}$$

(as usual, a repeated index implies a sum over that index). For example, in the case when frame K′ is moving with velocity V relative to frame K along the X axis, the matrix elements in (19.9) are those of the Lorentz-transformation matrix (see chapter 17 on the classical theory of fields)

$$(\alpha^i_{\ j})_{L_X(V)} = \begin{pmatrix} \dfrac{1}{\sqrt{1-V^2/c^2}} & \dfrac{V/c}{\sqrt{1-V^2/c^2}} & 0 & 0 \\ \dfrac{V/c}{\sqrt{1-V^2/c^2}} & \dfrac{1}{\sqrt{1-V^2/c^2}} & 0 & 0 \\ 0 & 0 & 1 & 0 \\ 0 & 0 & 0 & 1 \end{pmatrix} \equiv \begin{pmatrix} \cosh\omega & \sinh\omega & 0 & 0 \\ \sinh\omega & \cosh\omega & 0 & 0 \\ 0 & 0 & 1 & 0 \\ 0 & 0 & 0 & 1 \end{pmatrix}, \tag{19.10}$$

where we have written $V/c \equiv \tanh\omega$, so that $\sqrt{1-V^2/c^2} = \sqrt{1-\tanh^2\omega} = \mathrm{sech}\omega$.

Similarly, if K′ is stationary relative to K but has axes X' and Y' rotated through angle θ relative to X and Y, respectively, about the common Z axis, the matrix elements in (19.9) are those of the rotation matrix

$$(\alpha^i_{\ j})_{R_z(\theta)} = \begin{pmatrix} 1 & 0 & 0 & 0 \\ 0 & \cos\theta & -\sin\theta & 0 \\ 0 & \sin\theta & \cos\theta & 0 \\ 0 & 0 & 0 & 1 \end{pmatrix}. \tag{19.11}$$

If the four-component column-matrix wave function describing the state of a particle for an observer in frame K is $\Psi(x)$, the wave function $\Psi'(x')$ describing the state of this particle for an observer in frame K′ should obey a Dirac equation of **exactly the same form** as the Dirac equation in K, that is, **the Dirac equation should be Lorentz-covariant and rotationally covariant**.

The Dirac equation in K is

$$\left(i\hbar\gamma^i \frac{\partial}{\partial x^i} - 1mc\right)\Psi(x) = 0. \tag{19.12}$$

The wave function $\Psi'(x')$ describing the state of the particle in the frame K′ should satisfy a Dirac equation of the same form:

$$\left(i\hbar\gamma^i \frac{\partial}{\partial x'^i} - 1mc\right)\Psi'(x') = 0, \tag{19.13}$$

with, of course, the **same** Dirac matrices γ^i.

When written for the covariant rather than the contravariant coordinates, the Lorentz transformation (19.9) takes the form

$$x_i = \beta_i^{\ j} x'_j,$$

where the matrix β is related to the matrix α in (19.9) by the requirement that

$$x^i x_i = x'^i x'_i.$$

But

$$x^i x_i = \alpha^i{}_j \beta_i{}^k x'^j x'_k = x'^j x'_j$$

if and only if

$$\alpha^i{}_j \beta_i{}^k = \delta^k_j.$$

This can be rewritten as

$$\beta_i{}^k \alpha^i{}_j = (\beta^{\mathrm{T}})^k{}_i \alpha^i{}_j = \delta^k_j,$$

which shows that the transpose β^{T} of the matrix β is the inverse of the matrix α:

$$\alpha^{-1} = \beta^{\mathrm{T}}. \tag{19.14}$$

Since $\partial/\partial x^i$ is a covariant vector operator, we have

$$\frac{\partial}{\partial x^i} = \beta_i{}^j \frac{\partial}{\partial x'^j}.$$

We substitute this into equation (19.12), together with the relation

$$\Psi(x) = \mathrm{S}^{-1}(\alpha)\Psi'(x'),$$

where $\mathrm{S}(\alpha)$ is the 4×4 matrix [dependent on the elements $\alpha^j{}_i$ of the transformation matrix in equation (19.9)] that describes how the four-component column-matrix wave function $\Psi(x)$ transforms under (19.9):

$$\Psi'(x') = \mathrm{S}(\alpha)\Psi(x). \tag{19.15}$$

Then equation (19.12) becomes

$$\left(i\hbar\gamma^i \beta_i{}^j \frac{\partial}{\partial x'^j} - 1mc \right) \mathrm{S}^{-1}(\alpha)\Psi'(x') = 0.$$

If we multiply this on the left by $\mathrm{S}(\alpha)$ it takes the correct Dirac form (19.13) provided that

$$\mathrm{S}(\alpha)\gamma^i \mathrm{S}^{-1}(\alpha)\beta_i{}^j = \gamma^j.$$

This, after multiplication on the left by $\mathrm{S}^{-1}(\alpha)$ and on the right by $\mathrm{S}(\alpha)$, yields the equivalent matrix condition

$$\mathrm{S}^{-1}(\alpha)\gamma^j \mathrm{S}(\alpha) = \beta_i{}^j \gamma^i \tag{19.16}$$

(the $\beta_i{}^j$ here are, of course, just numerical coefficients).

To determine the dependence of the matrix S that satisfies (19.16) on the elements of the transformation matrix α it is convenient to use the matrix elements $a^i{}_j$ of the inverse of the transformation (19.9):

$$x'^i = a^i{}_j x^j = (\alpha^{-1})^i{}_j x^j = \beta_j{}^i x^j\,, \tag{19.17}$$

where we have used (19.14). Thus,

$$\beta_j{}^i = a^i{}_j, \tag{19.18}$$

and equation (19.16), written now as a condition on the matrix S as a function of the matrix a with elements $a^i{}_j$, becomes

$$\mathrm{S}^{-1}(\mathrm{a})\gamma^j \mathrm{S}(\mathrm{a}) = a^j{}_i \gamma^i\,. \tag{19.19}$$

If K′ and K are the same frame, that is, if in (19.19) we have $a^j{}_i = \delta^j_i$, then we must require

$$\Psi'(x') = \Psi(x), \text{ i.e., } \mathrm{S}(\mathrm{a}) = 1.$$

When the frame K′ differs from the frame K, the deviation of S(a) from the 4×4 unit matrix 1 will be determined by the deviation of $a^j{}_i$ from δ^j_i.

The first part of equation (19.17) can be rewritten as

$$x'^i = a^i{}_j x^j = a^i{}_j g^{jk} x_k \equiv a^{ik} x_k,$$

where we have introduced a new set of quantities a^{ik}, defined by

$$a^{ik} \equiv g^{jk} a^i{}_j. \tag{19.20}$$

For example, using (19.20), (19.18) and (19.14) we see that the matrix (a^{ij}) corresponding to the Lorentz-transformation matrix (19.10) is

$$\left(a^{ij}\right)_{\mathrm{L}x(V)} = \begin{pmatrix} \cosh\omega & \sinh\omega & 0 & 0 \\ -\sinh\omega & -\cosh\omega & 0 & 0 \\ 0 & 0 & -1 & 0 \\ 0 & 0 & 0 & -1 \end{pmatrix} \tag{19.21}$$

and the matrix (a^{ij}) corresponding to the rotation matrix (19.11) is

$$\left(a^{ij}\right)_{\mathrm{R}z(\theta)} = \begin{pmatrix} 1 & 0 & 0 & 0 \\ 0 & -\cos\theta & -\sin\theta & 0 \\ 0 & \sin\theta & -\cos\theta & 0 \\ 0 & 0 & 0 & -1 \end{pmatrix}. \tag{19.22}$$

When K′ and K are the same frame, that is, when $a^i{}_j = \delta^i_j$ (so that, as before, S(a) = 1), from (19.20) we have $a^{ik} = g^{ik}$ [equations (19.21) and (19.22) clearly exemplify this]. For **infinitesimally** different frames K and K′ we have

$$a^{ik} = g^{ik} + \delta a^{ik}. \tag{19.23}$$

For example, putting $V \to \delta V$ in (19.21), setting $\delta V/c \equiv \tanh \delta\omega$ and neglecting powers of $\delta\omega$ higher than the first, we have

$$\left(a^{ij}\right)_{\mathrm{L}x(\delta V)} = \begin{pmatrix} 1 & \delta\omega & 0 & 0 \\ -\delta\omega & -1 & 0 & 0 \\ 0 & 0 & -1 & 0 \\ 0 & 0 & 0 & -1 \end{pmatrix} = \left(g^{ij}\right) + \left(\delta a^{ij}\right)_{\mathrm{L}x(\delta V)}. \tag{19.24}$$

Similarly, putting $\theta \to \delta\theta$ in (19.22) and neglecting powers of $\delta\theta$ higher than the first, we have

$$\left(a^{ij}\right)_{\mathrm{R}z(\delta\theta)} = \begin{pmatrix} 1 & 0 & 0 & 0 \\ 0 & -1 & -\delta\theta & 0 \\ 0 & \delta\theta & -1 & 0 \\ 0 & 0 & 0 & -1 \end{pmatrix} = \left(g^{ij}\right) + \left(\delta a^{ij}\right)_{\mathrm{R}z(\delta\theta)}. \tag{19.25}$$

Thus, we can express the deviation of S(a) from 1 in terms of the deviations δa^{ik} of the elements a^{ik} from g^{ik}. These deviations are clearly antisymmetric in i and k, and so only elements with $k > i$ need be considered. We can then express the 4×4 matrix S(a) as

$$\mathrm{S(a)} = 1 - \frac{i}{4}\sigma_{ik}\delta a^{ik}, \tag{19.26}$$

where the σ_{ik} are a set of 4×4 matrices (not matrix elements!). Our task is to find the set of matrices σ_{ik} that ensure that (19.26) satisfies the condition (19.19), and because of the antisymmetry of the δa^{ik} we can restrict our search to matrices σ_{ik} such that $\sigma_{ki} = -\sigma_{ik}$.

To order δa^{ik} the inverse of the matrix (19.26) is clearly

$$\mathrm{S}^{-1}\mathrm{(a)} = 1 + \frac{i}{4}\sigma_{ik}\delta a^{ik}. \tag{19.27}$$

We rewrite (19.19) as

$$\mathrm{S}^{-1}\mathrm{(a)}\gamma^j \mathrm{S(a)} = a^j{}_i\gamma^i = g_{ik}a^{jk}\gamma^i \equiv a^{jk}\gamma_k = (g^{jk} + \delta a^{jk})\gamma_k, \tag{19.28}$$

where we have introduced the matrices

$$\gamma_k \equiv g_{ik}\gamma^i \quad \text{[i.e., } \gamma_0 = \gamma^0 \quad \text{and} \quad \gamma_\mu = -\gamma^\mu \quad (\mu = 1, 2, 3)\text{]}. \tag{19.29}$$

Putting (19.26), (19.27) and (19.23) into (19.28), we have

$$\gamma^j + \frac{i}{4}\left[\sigma_{ik}\delta a^{ik}\gamma^j - \gamma^j\sigma_{ik}\delta a^{ik}\right] = \gamma^j + \delta a^{jk}\gamma_k,$$

that is,

$$\frac{i}{4}\delta a^{ik}\left[\sigma_{ik}\gamma^j - \gamma^j\sigma_{ik}\right] = \delta a^{jk}\gamma_k. \tag{19.30}$$

A simple choice of 4×4 matrices σ_{ik} that satisfy $\sigma_{ki} = -\sigma_{ik}$ is provided by the commutators

$$\sigma_{ik} = \frac{i}{2}\left[\gamma_i, \gamma_k\right]. \tag{19.31}$$

Because γ_i and γ_k anticommute for $i \neq k$, we have

$$\sigma_{ik} = i\gamma_i\gamma_k \quad (i \neq k).$$

Thus, the left-hand side of (19.30) (in which, because of the antisymmetry of δa^{ik}, there are no terms with $i = k$) becomes

$$\begin{aligned}-\frac{1}{4}\delta a^{ik}[\gamma_i\gamma_k, \gamma^j] &= -\frac{1}{4}\delta a^{ik}\left\{\gamma_i[\gamma_k, \gamma^j] + [\gamma_i, \gamma^j]\gamma_k\right\}\\ &= -\frac{1}{4}\delta a^{ik}\left\{\gamma_i\left([\gamma_k, \gamma^j]_+ - 2\gamma^j\gamma_k\right) + \left([\gamma_i, \gamma^j]_+ - 2\gamma^j\gamma_i\right)\gamma_k\right\}.\end{aligned}$$

Substituting into this the anticommutator

$$[\gamma_k, \gamma^j]_+ = g_{ik}[\gamma^i, \gamma^j]_+ = 2g_{ik}g^{ij}1 = 2\delta_k^j 1,$$

we obtain

$$\text{L.H.S. of equation (19.30)} = -\frac{1}{4}\delta a^{ik}\left(2\gamma_i\delta_k^j - 2\gamma_i\gamma^j\gamma_k + 2\gamma_k\delta_i^j - 2\gamma^j\gamma_i\gamma_k\right).$$

The sum of the first and third terms in the brackets here is symmetric in *i* and *k*, and so gives zero contribution. The remaining two terms give

$$\begin{aligned}\text{L.H.S. of equation (19.30)} &= \frac{1}{2}\delta a^{ik}\left(\gamma_i\gamma^j\gamma_k + \gamma^j\gamma_i\gamma_k\right) = \frac{1}{2}\delta a^{ik}[\gamma_i, \gamma^j]_+\gamma_k\\ &= \delta a^{ik}\delta_i^j\gamma_k = \delta a^{jk}\gamma_k = \text{R.H.S. of equation (19.30)}.\end{aligned}$$

Thus, the choice (19.26) with (19.31) satisfies the condition (19.28), and so the 4×4 matrix S(a) in the law (19.15) of transformation of Dirac four-component column-matrix wave functions is given by

$$\mathrm{S(a)} = 1 - \frac{i}{4}\sigma_{ik}\delta a^{ik} = 1 + \frac{1}{8}[\gamma_i, \gamma_k]\delta a^{ik}. \tag{19.32}$$

In the case of a Lorentz transformation $L_X(\delta V)$ from inertial frame K to inertial frame K′ moving with infinitesimal velocity of magnitude δV relative to K along the X axis, the matrix S(a) is given by equation (19.32) in which the only nonzero δa^{ik} are $\delta a^{01} = \delta\omega$ and $\delta a^{10} = -\delta\omega$ [see equation (19.24)]. Thus, from (19.26),

$$\mathrm{S}_{\mathrm{L}_X(\delta V)} = 1 - \frac{i}{2}\sigma_{01}\delta\omega, \tag{19.33}$$

where, as before,

$$\delta V/c \equiv \tanh\delta\omega \approx \delta\omega. \tag{19.34}$$

In the case when the infinitesimal velocity $\delta\boldsymbol{V}$ of K relative to K′ has arbitrary direction, with direction cosines $\cos\alpha$, $\cos\beta$ and $\cos\gamma$, that is, when

$$\delta\boldsymbol{V} = \hat{\boldsymbol{i}}\delta V\cos\alpha + \hat{\boldsymbol{j}}\delta V\cos\beta + \hat{\boldsymbol{k}}\delta V\cos\gamma,$$

the only nonzero δa^{ik} [found from the appropriate modification of equation (19.24)] are δa^{01} $(= -\delta a^{10}) = \delta\omega\cos\alpha$, $\delta a^{02}(= -\delta a^{20}) = \delta\omega\cos\beta$, $\delta a^{03}(= -\delta a^{30}) = \delta\omega\cos\gamma$ (again, $\delta\omega = \delta V/c$), and equation (19.26) gives

$$\begin{aligned}\mathrm{S}_{\mathrm{L}(\delta\boldsymbol{V})} &= 1 - \frac{i}{2}\left(\sigma_{01}\delta a^{01} + \sigma_{02}\delta a^{02} + \sigma_{03}\delta a^{03}\right)\\ &= 1 - \frac{i}{2}\delta\omega\left(\sigma_{01}\cos\alpha + \sigma_{02}\cos\beta + \sigma_{03}\cos\gamma\right).\end{aligned} \tag{19.35}$$

But, because γ_0 and γ_i anticommute for $i \neq 0$, we have ($\mu = 1, 2, 3$)

$$\sigma_{0\mu} = \frac{i}{2}\left[\gamma_0, \gamma_\mu\right] = i\gamma_0\gamma_\mu \equiv -i\alpha_\mu. \tag{19.36}$$

[Here we have defined the 4×4 matrices

$$\alpha_\mu \equiv -\gamma_0\gamma_\mu = \gamma^0\gamma^\mu, \tag{19.37}$$

in which the latter form follows from (19.29).] Putting (19.36) into (19.35), we find

$$\mathrm{S}_{\mathrm{L}(\delta\boldsymbol{V})} = 1 - \frac{\delta\omega}{2}\left(\alpha_1\cos\alpha + \alpha_2\cos\beta + \alpha_3\cos\gamma\right). \tag{19.38}$$

Writing

$$\cos\alpha = \frac{\delta V_x}{|\delta\boldsymbol{V}|} \equiv \frac{\delta V_1}{|\delta\boldsymbol{V}|}, \text{ etc.,}$$

we bring this to the form

$$\mathrm{S}_{\mathrm{L}(\delta\boldsymbol{V})} = 1 - \frac{\delta\omega}{2}\frac{\boldsymbol{\alpha}\cdot\delta\boldsymbol{V}}{|\delta\boldsymbol{V}|}, \tag{19.39}$$

in which the vector 4×4 matrix $\boldsymbol{\alpha}$ is given by $\boldsymbol{\alpha} = \hat{\boldsymbol{\imath}}\alpha_1 + \hat{\boldsymbol{\jmath}}\alpha_2 + \hat{\boldsymbol{k}}\alpha_3$.

A finite Lorentz transformation from an inertial frame K to an inertial frame K′ moving relative to K with velocity $\boldsymbol{V}$ is described by a finite parameter ω, as in (19.10), and may be regarded as consisting of N ($\to \infty$) successive infinitesimal Lorentz transformations through $\delta\omega = \omega/N$. Thus, we can use (19.39) to find

$$\mathrm{S}_{\mathrm{L}(\boldsymbol{V})} = \lim_{N\to\infty}\left(1 - \frac{1}{2}\frac{\omega}{N}\frac{\boldsymbol{\alpha}\cdot\boldsymbol{V}}{|\boldsymbol{V}|}\right)^N = \exp\left(-\frac{\omega}{2}\frac{\boldsymbol{\alpha}\cdot\boldsymbol{V}}{|\boldsymbol{V}|}\right). \tag{19.40}$$

We now find the form of the matrix S corresponding to an infinitesimal rotation $R_Z(\delta\theta)$ of K′ relative to K about the Z axis. According to equation (19.25), the only nonzero δa^{ij} appearing in expression (19.32) for this case are $\delta a^{12} = -\delta\theta$ and $\delta a^{21} = \delta\theta$. Thus, for this case equation (19.32) gives

$$\mathrm{S}_{\mathrm{R}_Z(\delta\theta)} = 1 + \frac{i}{2}\sigma_{12}\delta\theta, \tag{19.41}$$

where

$$\sigma_{12} = \frac{i}{2}\left[\gamma_1, \gamma_2\right] = \begin{pmatrix} \sigma_3 & 0 \\ 0 & \sigma_3 \end{pmatrix}. \tag{19.42}$$

A finite rotation $R_Z(\theta)$ through angle θ about the Z axis can be regarded as N ($\to \infty$) successive rotations through $\delta\theta = \theta/N$, and so

$$\mathrm{S}_{\mathrm{R}_Z(\theta)} = \lim_{N\to\infty}\left(1 + \frac{i}{2}\sigma_{12}\frac{\theta}{N}\right)^N = \mathrm{e}^{i(\theta/2)\sigma_{12}}. \tag{19.43}$$

Note that because of the half-angle in (19.43) we have

$$\mathrm{S}_{\mathrm{R}_Z(\theta+2\pi)} \neq \mathrm{S}_{\mathrm{R}_Z(\theta)}.$$

In other words, a rotation through 4π from frame K to frame K′ is needed in order that the wave function in K′ be the same as that in K. Wave functions that transform in this manner under spatial rotations and that transform as described by (19.15) with (19.40) under Lorentz transformations are called **spinors**.

19.3 The current four-vector

We now multiply the Dirac equation (19.12) on the left by the matrix γ^0:

$$\left(i\hbar\gamma^0\gamma^i\frac{\partial}{\partial x^i} - \gamma^0 mc\right)\Psi(x) = 0. \tag{19.44}$$

Taking the adjoint of each term, we obtain the adjoint equation

$$-i\hbar\frac{\partial\Psi^\dagger}{\partial x^i}\gamma^{i\dagger}\gamma^{0\dagger} - mc\Psi^\dagger\gamma^{0\dagger} = 0, \tag{19.45}$$

where, by definition of the adjoint,

$$\Psi^\dagger(x) = \left(\psi^{(1)*}(x)\ \psi^{(2)*}(x)\ \psi^{(3)*}(x)\ \psi^{(4)*}(x) \right), \tag{19.46}$$

in which an asterisk denotes the complex conjugate. Using the facts that $\gamma^{0\dagger} = \gamma^0$, $\gamma^{\mu^\dagger} = -\gamma^\mu$ (since $\sigma_\mu^\dagger = \sigma_\mu$), and γ^0 commutes with itself but anticommutes with γ^μ, we can rewrite (19.45) as

$$-i\hbar \frac{\partial \Psi^\dagger}{\partial x^i} \gamma^0 \gamma^i - mc\Psi^\dagger \gamma^0 = 0. \tag{19.47}$$

Now multiply equation (19.44) on the left by the row-matrix wave function $\Psi^\dagger(x)$, multiply equation (19.47) on the right by the column-matrix wave function $\Psi(x)$, and take the difference of the resulting equations. This gives

$$i\hbar \frac{\partial}{\partial x^i} (\Psi^\dagger \gamma^0 \gamma^i \Psi) = 0, \tag{19.48}$$

which has the form of the **continuity equation** (see section 17.5.1)

$$\frac{\partial j^i}{\partial x^i} = 0 \tag{19.49}$$

with **probability four-current**

$$j^i = c\Psi^\dagger \gamma^0 \gamma^i \Psi \tag{19.50}$$

[recall that $\Psi^\dagger$ is a row matrix, the γ^i ($i = 0, 1, 2, 3$) are 4×4 matrices, and Ψ is a column matrix, so that each component j^i is a single (dimensional) number]. The constant in (19.50) has been chosen to be the velocity of light c in order that the $i = 0$ component be $c\rho$, where ρ is the probability density (see the discussion of the corresponding relation between the electric four-current and electric-charge density in chapter 17 on the classical theory of fields):

$$j^0 = c\Psi^\dagger \gamma^0 \gamma^0 \Psi = c\Psi^\dagger \Psi = c \sum_{r=1}^{4} \left| \psi^{(r)}(x) \right|^2 = c\rho, \tag{19.51}$$

that is, the probability four-current (19.50) is of the form

$$j^i = (c\rho, j_x, j_y, j_z), \tag{19.52}$$

in which we have identified the components with $i = 1, 2, 3$ as the components j_x, j_y, j_z of the probability-flux (probability-current density) three-vector.

19.4 Compact form of the Dirac equation

Since

$$\frac{\partial}{\partial x^i} = g_{ij}\frac{\partial}{\partial x_j}, \tag{19.53}$$

the Dirac equation (19.12) can also be written as

$$\left(i\hbar\gamma^i g_{ij}\frac{\partial}{\partial x_j} - 1mc \right)\Psi(x) = 0,$$

that is, as

$$\left(i\hbar\gamma_j\frac{\partial}{\partial x_j} - 1mc \right)\Psi(x) = 0,$$

where the matrices γ_j are **defined** as

$$\gamma_j \equiv g_{ij}\gamma^i. \tag{19.54}$$

But

$$\begin{aligned} i\hbar\frac{\partial}{\partial x_j} &= i\hbar\left(\frac{\partial}{\partial(ct)}, -\frac{\partial}{\partial x}, -\frac{\partial}{\partial y}, -\frac{\partial}{\partial z}\right) \\ &= \hat{p}^j = (\hat{p}^0, \hat{p}^1, \hat{p}^2, \hat{p}^3) = (\hat{E}/c, \hat{p}_x, \hat{p}_y, \hat{p}_z) \end{aligned}$$

is the quantum-mechanical operator of the energy and the momentum components. Thus, the Dirac equation (19.12) can be written as

$$\left(\gamma_j\hat{p}^j - 1mc\right)\Psi(x) = 0,$$

or, using (19.53) and (19.54), as

$$\left(\gamma^j\hat{p}_j - 1mc\right)\Psi(x) = 0. \tag{19.55}$$

We now introduce the Feynman slash notation

$$\not{p} \equiv \gamma_j\hat{p}^j = \gamma^j\hat{p}_j,$$

in which the Dirac equation takes the compact form

$$(\not{p} - 1mc)\,\Psi(x) = 0. \tag{19.56}$$

As explained in chapter 17, in the presence of an electromagnetic field described by the four-potential

$$A^i = (A^0, A^1, A^2, A^3) = (\varphi/c, A_x, A_y, A_z)$$

we must replace p^i by $p^i - qA^i = p^i + eA^i$, where q is the electric charge of the particle, assumed here to be an electron ($q = -e$). The Dirac equation then becomes

$$\left(\not{p} + e\not{A} - 1mc\right)\Psi(x) = 0. \tag{19.57}$$

19.5 Dirac wave function of a free particle

Consider a free particle at rest in frame K, that is, $p_x = p_y = p_z = 0$. Then (since $A^i = 0$ also) the Dirac equation reduces to

$$\left(\gamma^0 \hat{p}_0 - 1mc\right)\Psi(x) = 0,$$

that is,

$$\left(i\hbar\gamma^0 \frac{\partial}{\partial(ct)} - 1mc\right)\Psi(x) = 0.$$

Multiply this by the matrix γ^0 and use the result $(\gamma^0)^2 = 1$. The Dirac equation becomes

$$i\hbar\frac{\partial\Psi(x)}{\partial t} = \gamma^0 mc^2\Psi(x). \tag{19.58}$$

But

$$\gamma^0 = \begin{pmatrix} 1 & 0 & 0 & 0 \\ 0 & 1 & 0 & 0 \\ 0 & 0 & -1 & 0 \\ 0 & 0 & 0 & -1 \end{pmatrix}.$$

Therefore, we have four possible solutions (Dirac wave functions of a free particle at rest) of equation (19.58):

$$\Psi^{\text{I}} = \mathrm{e}^{-imc^2t/\hbar}\begin{pmatrix}1\\0\\0\\0\end{pmatrix},\quad \Psi^{\text{II}} = \mathrm{e}^{-imc^2t/\hbar}\begin{pmatrix}0\\1\\0\\0\end{pmatrix},\quad \Psi^{\text{III}} = \mathrm{e}^{imc^2t/\hbar}\begin{pmatrix}0\\0\\1\\0\end{pmatrix},\quad \Psi^{\text{IV}} = \mathrm{e}^{imc^2t/\hbar}\begin{pmatrix}0\\0\\0\\1\end{pmatrix}. \tag{19.59}$$

Since the time dependence of the wave function of a particle corresponding to definite energy is always $\mathrm{e}^{-i[\text{energy}\times t]/\hbar}$ (see the review of the basic principles of quantum mechanics in chapter 1), we see that Ψ^{I} and Ψ^{II} are **positive-energy solutions**, with energy equal to the rest energy mc^2 of the particle, and Ψ^{III} and Ψ^{IV} are **negative-energy solutions**, with energy equal to minus the rest energy $(-mc^2)$. (A physical interpretation of the negative-energy solutions is given later in this chapter.)

To obtain the spin of these solutions we use the result (19.43) that, upon rotation of the coordinate system through angle θ about the Z axis, the four-component column-matrix wave function of a particle in the new coordinates is obtained from that in the old coordinates by application of the 4×4 matrix

$$\mathrm{S}_{\mathrm{R}_Z(\theta)} = \mathrm{e}^{i(\theta/2)\sigma_{12}}, \tag{19.60}$$

in which the 4×4 matrix σ_{12} is given by

$$\sigma_{12} = \frac{i}{2}[\gamma_1, \gamma_2] = \begin{pmatrix} \sigma_3 & 0 \\ 0 & \sigma_3 \end{pmatrix}. \tag{19.61}$$

Thus, in equation (19.60) the "generator" of rotations about the Z axis (3 axis) is σ_{12}, and so σ_{12} can be identified with the matrix operator of the z component of the intrinsic angular momentum (**spin**) of the system. The form (19.61) fits neatly with this interpretation, if we recall that the matrix operator S_z in nonrelativistic quantum mechanics is given by $\mathrm{S}_z = \frac{1}{2}\hbar\sigma_z \equiv \frac{1}{2}\hbar\sigma_3$ (see the end of section 2.7).

Applying the matrix operator $\mathrm{S}_z = \frac{1}{2}\hbar\sigma_{12}$ to our four column-matrix wave functions (19.59) we see immediately that Ψ^{I} and Ψ^{III} are eigenstates of S_z with eigenvalue $\frac{1}{2}\hbar$, and Ψ^{II} and Ψ^{IV} are eigenstates of S_z with eigenvalue $-\frac{1}{2}\hbar$.

Thus, to summarize, Ψ^{I} is a spin-up state with energy $+mc^2$, Ψ^{II} is a spin-down state with energy $+mc^2$, Ψ^{III} is a spin-up state with energy $-mc^2$, and Ψ^{IV} is a spin-down state with energy $-mc^2$.

We now determine how these wave functions are modified when the particle has nonzero velocity $\boldsymbol{v}$. We have just shown that the four $\boldsymbol{v} = 0$ wave functions have the form

$$\Psi^r(x) = \mathrm{w}^r(0)\mathrm{e}^{-i\varepsilon_r mc^2 t/\hbar}, \tag{19.62}$$

where $\mathrm{w}^r(0)$ is the spinor $\mathrm{w}^r(\boldsymbol{p})$ for $\boldsymbol{p} = 0$:

$$\mathrm{w}^{\mathrm{I}}(0) = \begin{pmatrix} 1 \\ 0 \\ 0 \\ 0 \end{pmatrix}, \quad \mathrm{w}^{\mathrm{II}}(0) = \begin{pmatrix} 0 \\ 1 \\ 0 \\ 0 \end{pmatrix}, \quad \mathrm{w}^{\mathrm{III}}(0) = \begin{pmatrix} 0 \\ 0 \\ 1 \\ 0 \end{pmatrix}, \quad \mathrm{w}^{\mathrm{IV}}(0) = \begin{pmatrix} 0 \\ 0 \\ 0 \\ 1 \end{pmatrix}; \tag{19.63}$$

$\varepsilon_r = +1$ for $r = \mathrm{I, II}$, and $\varepsilon_r = -1$ for $r = \mathrm{III, IV}$.

We can get the four possible wave functions of a free particle with arbitrary velocity $\boldsymbol{v}$ relative to our frame K by taking the wave functions (19.59) of a particle at rest in the frame K and determining the form that these wave functions take when viewed from an inertial frame K′ moving with velocity $\boldsymbol{V} = -\boldsymbol{v}$ relative to K.

To determine this form we must find the result of such a Lorentz transformation on both the space–time part and the spinor part of (19.62).

The exponent in the space–time part of (19.62) contains the factor

$$mc^2 t = \frac{E}{c}ct = p_0 x^0 = p_i x^i,$$

where in the last step we have used the fact that equations (19.62) describe the possible states of a particle with zero momentum $\boldsymbol{p} = 0$. But $p_i x^i$ is manifestly a Lorentz scalar (see chapter 17 on the classical theory of fields), that is, $p_i x^i = p'_i x'^i$, and so the space–time part of the wave function in the coordinates x'^i of the particle, that is, as viewed from the frame K′ (relative to which the particle is moving with velocity $\boldsymbol{v}$ and has four-momentum p'^i), is

$$e^{-i\varepsilon_r p'_i x'^i/\hbar}. \tag{19.64}$$

To obtain the spinor parts of the wave functions of a particle moving with velocity

$$\boldsymbol{v} = v(\hat{\boldsymbol{i}}\cos\alpha + \hat{\boldsymbol{j}}\cos\beta + \hat{\boldsymbol{k}}\cos\gamma)$$

relative to K, we act on the spinor $\mathrm{w}^{(r)}(0)$ in (19.62) with the spinor-transformation matrix S(a) corresponding to a Lorentz transformation to a frame K′ moving with velocity $\boldsymbol{V} = -\boldsymbol{v}$ relative to K.

Thus, we have

$$\mathrm{w}^r(\boldsymbol{p}) = \mathrm{S(a)}\mathrm{w}^r(0), \tag{19.65}$$

with [see equation (19.40)]

$$\mathrm{S(a)} = \exp\left(-\frac{\omega}{2}\frac{\boldsymbol{\alpha}\cdot\boldsymbol{V}}{|\boldsymbol{V}|}\right) = \exp\left(\frac{\omega}{2}\frac{\boldsymbol{\alpha}\cdot\boldsymbol{v}}{|\boldsymbol{v}|}\right) = \cosh\left(\frac{\omega}{2}\frac{\boldsymbol{\alpha}\cdot\boldsymbol{v}}{|\boldsymbol{v}|}\right) + \sinh\left(\frac{\omega}{2}\frac{\boldsymbol{\alpha}\cdot\boldsymbol{v}}{|\boldsymbol{v}|}\right). \tag{19.66}$$

Here, the matrix

$$\frac{\boldsymbol{\alpha}\cdot\boldsymbol{v}}{|\boldsymbol{v}|} = \alpha_1\cos\alpha + \alpha_2\cos\beta + \alpha_3\cos\gamma, \tag{19.67}$$

where the matrices α_μ are defined by equation (19.37).

The expansion of cosh$x$ contains only even powers of x, and so we first calculate the square of (19.67). In this square, no cross terms of the type $\alpha_\mu\alpha_\nu$ occur, since the anticommutator of two distinct α matrices is zero:

$$\begin{aligned}\alpha_\mu\alpha_\nu + \alpha_\nu\alpha_\mu &= \gamma^0\gamma^\mu\gamma^0\gamma^\nu + \gamma^0\gamma^\nu\gamma^0\gamma^\mu \\ &= -\gamma^\mu\gamma^0\gamma^0\gamma^\nu - \gamma^\nu\gamma^0\gamma^0\gamma^\mu = -\gamma^\mu\gamma^\nu - \gamma^\nu\gamma^\mu = 0\end{aligned}$$

and so we have

$$\left(\frac{\boldsymbol{\alpha}\cdot\boldsymbol{v}}{|\boldsymbol{v}|}\right)^2 = \alpha_1^2\cos^2\alpha + \alpha_2^2\cos^2\beta + \alpha_3^2\cos^2\gamma. \tag{19.68}$$

But [see equation (19.4)]

$$\alpha_\mu^2 = \gamma^0\gamma^\mu\gamma^0\gamma^\mu = -\gamma^\mu\gamma^0\gamma^0\gamma^\mu = -\gamma^\mu\gamma^\mu = -\frac{1}{2}2g^{\mu\mu}1 = -(-1)1 = 1,$$

and so equation (19.68) becomes

$$\left(\frac{\boldsymbol{\alpha}\cdot\boldsymbol{v}}{|\boldsymbol{v}|}\right)^2 = 1(\cos^2\alpha + \cos^2\beta + \cos^2\gamma) = 1.$$

Therefore,

$$\left(\frac{\boldsymbol{\alpha}\cdot\boldsymbol{v}}{|\boldsymbol{v}|}\right)^{2n} = 1 \quad \text{and} \quad \left(\frac{\boldsymbol{\alpha}\cdot\boldsymbol{v}}{|\boldsymbol{v}|}\right)^{2n+1} = \alpha_1\cos\alpha + \alpha_2\cos\beta + \alpha_3\cos\gamma,$$

and so

$$\begin{aligned} \mathrm{S(a)} &= 1\cosh\frac{\omega}{2} + (\alpha_1\cos\alpha + \alpha_2\cos\beta + \alpha_3\cos\gamma)\sinh\frac{\omega}{2} \\ &= \cosh\frac{\omega}{2}\left[1 + (\alpha_1\cos\alpha + \alpha_2\cos\beta + \alpha_3\cos\gamma)\tanh\frac{\omega}{2}\right]. \end{aligned} \tag{19.69}$$

But

$$\begin{aligned} \tanh\frac{\omega}{2} &= \frac{\sinh(\omega/2)}{\cosh(\omega/2)} = \left(\frac{e^{\omega/2} - e^{-\omega/2}}{e^{\omega/2} + e^{-\omega/2}}\right)\left(\frac{e^{\omega/2} + e^{-\omega/2}}{e^{\omega/2} + e^{-\omega/2}}\right) = \frac{e^{\omega} - e^{-\omega}}{e^{\omega} + e^{-\omega} + 2} \\ &= \frac{\sinh\omega}{\cosh\omega + 1} = \frac{\tanh\omega}{1 + \mathrm{sech}\omega} = \frac{v/c}{1 + \sqrt{1 - v^2/c^2}} \\ &= \frac{mvc\Big/\sqrt{1 - v^2/c^2}}{\left(mc^2\Big/\sqrt{1 - v^2/c^2}\right) + mc^2} = \frac{pc}{E + mc^2}, \end{aligned}$$

and, as a consequence,

$$\begin{aligned} \cosh\frac{\omega}{2} &= \frac{1}{\mathrm{sech}(\omega/2)} = \frac{1}{\sqrt{1 - \tanh^2(\omega/2)}} = \frac{E + mc^2}{\sqrt{(E + mc^2)^2 - p^2c^2}} \\ &= \frac{E + mc^2}{\sqrt{E^2 + 2mc^2E + m^2c^4 - p^2c^2}} = \frac{E + mc^2}{\sqrt{2mc^2E + 2m^2c^4}} = \sqrt{\frac{E + mc^2}{2mc^2}}. \end{aligned}$$

Therefore,

$$\begin{aligned} \mathrm{S(a)} &= \sqrt{\frac{E + mc^2}{2mc^2}}\left[1 + \frac{c}{E + mc^2}(\alpha_1 p\cos\alpha + \alpha_2 p\cos\beta + \alpha_3 p\cos\gamma)\right] \\ &= \sqrt{\frac{E + mc^2}{2mc^2}}\left[1 + \frac{c}{E + mc^2}\left(\alpha_1 p_x + \alpha_2 p_y + \alpha_3 p_z\right)\right]. \end{aligned}$$

Using the standard forms of the matrices $\alpha_\mu = \gamma^0\gamma^\mu$ we find

$$\mathrm{S(a)} = \sqrt{\frac{E + mc^2}{2mc^2}}\begin{pmatrix} 1 & 0 & \dfrac{p_z c}{E + mc^2} & \dfrac{p_- c}{E + mc^2} \\ 0 & 1 & \dfrac{p_+ c}{E + mc^2} & -\dfrac{p_z c}{E + mc^2} \\ \dfrac{p_z c}{E + mc^2} & \dfrac{p_- c}{E + mc^2} & 1 & 0 \\ \dfrac{p_+ c}{E + mc^2} & -\dfrac{p_z c}{E + mc^2} & 0 & 1 \end{pmatrix}, \tag{19.70}$$

where the quantities $p_\pm$ are defined as

$$p_\pm = p_x \pm i p_y.$$

Thus, the four free-particle solutions for a particle with four-momentum p^i are

$$\Psi^r(x) = \mathrm{w}^r(\boldsymbol{p}) \mathrm{e}^{-i\varepsilon_r p_i x^i/\hbar} \quad (r = \mathrm{I, II, III, IV}) \tag{19.71}$$

where $\mathrm{w}^r(\boldsymbol{p})$ is given by column r of the matrix (19.70); $\varepsilon_r = +1$ for $r = \mathrm{I, II}$, and $\varepsilon_r = -1$ for $r = \mathrm{III, IV}$. (The common factor $\left[\left(E + mc^2\right)/2mc^2\right]^{1/2}$ affects only the normalization and can be omitted.)

19.6 Motion of an electron in an electromagnetic field

The Dirac equation (19.57) describing the motion of an electron in a magnetic field,

$$\left(\gamma_j(\hat{p}^j + eA^j) - 1mc\right)\Psi(x) = 0 \tag{19.72}$$

can be written as

$$\gamma_0 \hat{p}^0 \Psi(x) = [-\gamma_\mu(\hat{p}^\mu + eA^\mu) - \gamma_0 e A^0 + 1mc]\Psi(x).$$

Multiplying this on the left by γ_0 and using the definition (19.37) of α_μ and the fact that $(\gamma_0)^2 = 1$, we obtain

$$\hat{p}^0 \Psi(x) = [\alpha_\mu(\hat{p}^\mu + eA^\mu) - eA^0 1 + \gamma_0 mc]\Psi(x).$$

Putting $\hat{p}^0 \equiv i\hbar\partial/\partial(ct)$, using the fact that the time component A^0 is the scalar potential φ (divided by c) and the space components A^μ are the components of the vector potential $\boldsymbol{A}$, we find

$$\begin{aligned} i\hbar\frac{\partial\Psi(x)}{\partial t} &= [c\boldsymbol{\alpha}\cdot(\hat{\boldsymbol{p}} + e\boldsymbol{A}) - e\varphi 1 + \gamma_0 mc^2]\Psi(x) \\ &= [c\boldsymbol{\alpha}\cdot\hat{\boldsymbol{\pi}} - e\varphi 1 + \gamma_0 mc^2]\Psi(x), \end{aligned} \tag{19.73}$$

where $\hat{\boldsymbol{\pi}}$ is the operator of the kinetic momentum $\boldsymbol{\pi} = \boldsymbol{p} + e\boldsymbol{A}$ (see chapter 17). We now write the four-component column-matrix wave function $\Psi(x)$ as

$$\Psi(x) = \begin{pmatrix} \tilde{\phi}(x) \\ \tilde{\chi}(x) \end{pmatrix}, \tag{19.74}$$

where $\tilde{\phi}(x)$ and $\tilde{\chi}(x)$ are both two-component column-matrix wave functions. Then equation (19.73) takes the form

$$i\hbar\frac{\partial}{\partial t}\begin{pmatrix} \tilde{\phi} \\ \tilde{\chi} \end{pmatrix} = c\boldsymbol{\sigma}\cdot\hat{\boldsymbol{\pi}}\begin{pmatrix} \tilde{\chi} \\ \tilde{\phi} \end{pmatrix} - e\varphi\begin{pmatrix} \tilde{\phi} \\ \tilde{\chi} \end{pmatrix} + mc^2\begin{pmatrix} \tilde{\phi} \\ -\tilde{\chi} \end{pmatrix}, \tag{19.75}$$

where we have used (19.37) and the form (19.7) of the γ-matrices. [Equation (19.75) must be read as two equations – one for each of the two-component spinors $\tilde{\phi}(x)$ and $\tilde{\chi}(x)$.]

We now consider the case when the solution of equation (19.75) is a positive-energy solution, and consider the nonrelativistic limit of this equation, when the largest contribution to the energy of the electron is the rest energy mc^2. Since the time dependence of the wave function of a state with definite energy mc^2 is $\mathrm{e}^{-imc^2t/\hbar}$, we can write

$$\begin{pmatrix} \tilde{\phi} \\ \tilde{\chi} \end{pmatrix} = \mathrm{e}^{-imc^2t/\hbar} \begin{pmatrix} \phi \\ \chi \end{pmatrix}, \tag{19.76}$$

so that the residual time dependence in the two-component column-matrix wave functions ϕ and χ thus defined will be very weak, that is, ϕ and χ will be very slowly varying in time.

Substituting (19.76) into (19.75) we find

$$i\hbar\frac{\partial}{\partial t}\begin{pmatrix} \phi \\ \chi \end{pmatrix} = c\boldsymbol{\sigma}\cdot\hat{\boldsymbol{\pi}}\begin{pmatrix} \chi \\ \phi \end{pmatrix} - e\varphi\begin{pmatrix} \phi \\ \chi \end{pmatrix} - 2mc^2\begin{pmatrix} 0 \\ \chi \end{pmatrix}. \tag{19.77}$$

Since χ is slowly varying and $e\varphi \ll 2mc^2$, the second of the two two-component column-matrix equations in equation (19.77) can be approximated by

$$\chi = \frac{\boldsymbol{\sigma}\cdot\hat{\boldsymbol{\pi}}}{2mc}\phi. \tag{19.78}$$

Since the magnitude π of the kinetic momentum is $\approx mv$ in the nonrelativistic limit, equation (19.78) shows that the wave function χ is smaller than the wave function ϕ by a factor of order v/c in this limit, that is, χ are the "small components" of the wave function Ψ.

We now substitute (19.78) into the equation for ϕ in equation (19.77):

$$i\hbar\frac{\partial\phi(x)}{\partial t} = \left[\frac{(\boldsymbol{\sigma}\cdot\hat{\boldsymbol{\pi}})(\boldsymbol{\sigma}\cdot\hat{\boldsymbol{\pi}})}{2m} - e\varphi 1\right]\phi(x). \tag{19.79}$$

But, for any two three-vectors $\boldsymbol{a}$ and $\boldsymbol{b}$,

$$\begin{aligned}(\boldsymbol{\sigma}\cdot\boldsymbol{a})(\boldsymbol{\sigma}\cdot\boldsymbol{b}) = \sigma_\mu a_\mu \sigma_\nu b_\nu &= \frac{1}{2}\left[(\sigma_\mu\sigma_\nu + \sigma_\nu\sigma_\mu) + (\sigma_\mu\sigma_\nu - \sigma_\nu\sigma_\mu)\right]a_\mu b_\nu \\ &= \delta_{\mu\nu}1 a_\mu b_\nu + i\varepsilon_{\mu\nu\lambda}\sigma_\lambda a_\mu b_\nu = (\boldsymbol{a}\cdot\boldsymbol{b})1 + i\boldsymbol{\sigma}\cdot\boldsymbol{a}\times\boldsymbol{b},\end{aligned}$$

where we have used the expressions for the anticommutator and commutator of two Pauli matrices:

$$\sigma_\mu\sigma_\nu + \sigma_\nu\sigma_\mu = 2\delta_{\mu\nu}1 \quad \text{and} \quad \sigma_\mu\sigma_\nu - \sigma_\nu\sigma_\mu = 2i\varepsilon_{\mu\nu\lambda}\sigma_\lambda$$

(1 here is the 2×2 unit matrix). Thus, we can write

$$\begin{aligned}(\boldsymbol{\sigma}\cdot\hat{\boldsymbol{\pi}})(\boldsymbol{\sigma}\cdot\hat{\boldsymbol{\pi}})\phi &= \hat{\pi}^2\phi + i\boldsymbol{\sigma}\cdot\hat{\boldsymbol{\pi}}\times\hat{\boldsymbol{\pi}}\phi \\ &= \hat{\pi}^2\phi + i\boldsymbol{\sigma}\cdot(\hat{\boldsymbol{p}} + e\boldsymbol{A})\times(\hat{\boldsymbol{p}} + e\boldsymbol{A})\phi \\ &= \hat{\pi}^2\phi + i\boldsymbol{\sigma}\cdot(-i\hbar\boldsymbol{\nabla} + e\boldsymbol{A})\times(-i\hbar\boldsymbol{\nabla} + e\boldsymbol{A})\phi.\end{aligned}$$

But $\nabla \times \nabla\phi = 0$ and $A \times A\phi = 0$, and so this becomes

$$(\sigma \cdot \hat{\pi})(\sigma \cdot \hat{\pi})\phi = \hat{\pi}^2\phi + e\hbar\sigma \cdot [\nabla \times (A\phi) + A \times \nabla\phi]$$
$$= \hat{\pi}^2\phi + e\hbar\sigma \cdot (\nabla \times A)\phi = \hat{\pi}^2\phi + e\hbar\sigma \cdot B\phi$$

where we have used the standard vector-calculus identity for $\nabla \times (A\phi)$ and introduced the magnetic induction $B = \nabla \times A$. Thus, equation (19.79) becomes

$$i\hbar\frac{\partial\phi(x)}{\partial t} = \left[\frac{(\hat{p} + eA)^2}{2m}1 + \frac{e\hbar}{2m}\sigma \cdot \mathbf{B} - e\varphi 1\right]\phi(x), \tag{19.80}$$

which is known as the **Pauli equation**. We can describe a **uniform** magnetic induction B by the vector potential

$$A = \frac{1}{2}B \times r.$$

(This is checked easily as follows:

$$\nabla \times A = \frac{1}{2}\nabla \times (B \times r) = \frac{1}{2}[B(\nabla \cdot r) - r(\nabla \cdot B) + (r \cdot \nabla)B - (B \cdot \nabla)r] = \frac{1}{2}[3B - B] = B,$$

since the second term is zero by virtue of the Maxwell equation $\nabla \cdot B = 0$ and the third term is zero by virtue of the uniformity of B in the case under consideration.)

Keeping only terms linear in B in equation (19.80), we can write

$$(\hat{p} + eA)^2\phi = \left(-i\hbar\nabla + \frac{e}{2}B \times r\right) \cdot \left(-i\hbar\nabla + \frac{e}{2}B \times r\right)\phi$$
$$= \hat{p}^2\phi - \frac{i\hbar e}{2}B \times r \cdot \nabla\phi - \frac{i\hbar e}{2}\nabla \cdot (B \times r\phi)$$
$$= \hat{p}^2\phi - \frac{i\hbar e}{2}B \cdot r \times \nabla\phi - \frac{i\hbar e}{2}\nabla \cdot (B \times r\phi).$$

But

$$\nabla \cdot (B \times r\phi) = r\phi \cdot \nabla \times B - B \cdot \nabla \times (r\phi) = -B \cdot \nabla \times (r\phi)$$
$$= -B \cdot [(\nabla\phi) \times r + \phi\nabla \times r] = -B \cdot (\nabla\phi) \times r = B \cdot r \times \nabla\phi,$$

where we have again used the uniformity of B and standard vector-calculus identities. Thus, we have

$$(\hat{p} + eA)^2\phi = \hat{p}^2\phi - i\hbar eB \cdot r \times \nabla\phi = \hat{p}^2\phi + eB \cdot r \times \hat{p}\phi = \hat{p}^2\phi + eB \cdot \hat{L}\phi,$$

where $\hat{L} = r \times \hat{p}$ is the operator of the orbital angular momentum. Thus, to terms linear in B the Pauli equation (19.80) becomes

$$i\hbar\frac{\partial\phi(x)}{\partial t} = \left[\frac{\hat{p}^2}{2m}1 + \frac{e}{2m}B \cdot \left(\hat{L}1 + 2\mathbf{S}\right) - e\varphi 1\right]\phi(x), \tag{19.81}$$

where

$$\mathbf{S} = \frac{1}{2}\hbar\boldsymbol{\sigma} \tag{19.82}$$

is the 2×2 matrix operator of the **spin** vector, and, in order that all terms in the sum in the square brackets have the same matrix nature, the 2×2 unit matrices 1 have been inserted explicitly. The second term in the 2×2 matrix operator in the right-hand side of equation (19.81) takes the form of the operator $-\hat{\boldsymbol{\mu}} \cdot \boldsymbol{B}$ of the interaction energy of a magnetic moment μ with a magnetic field of induction $\boldsymbol{B}$, with the (2×2-matrix) **magnetic-moment operator** given by

$$\hat{\boldsymbol{\mu}} = \frac{\mu_{\mathrm{B}}}{\hbar}\left(\hat{\boldsymbol{L}}1 + 2\mathbf{S}\right), \tag{19.83}$$

where

$$\mu_{\mathrm{B}} = \frac{q_{\mathrm{e}}\hbar}{2m} = -\frac{e\hbar}{2m} \tag{19.84}$$

is the **Bohr magneton** of the electron (which has charge $q_{\mathrm{e}} = -e$). The factor $\hbar$ has been introduced into (19.83), and hence into the definition (19.84), because measurement of any component of the operator $\hat{\boldsymbol{L}}1 + 2\mathbf{S}$ can yield only an integer (positive, negative or zero) number of units of $\hbar$, so that measurement of any component of the magnetic moment will yield an integer (positive, negative or zero) number of Bohr magnetons. (The latter, therefore, may be regarded as the elementary unit of magnetic moment.)

Note that the magnetic-moment operator [see (19.83)] is proportional not to the operator $\hat{\boldsymbol{J}} = \hat{\boldsymbol{L}} + \hat{\boldsymbol{S}}$ of the total angular-momentum vector, but to the operator $\hat{\boldsymbol{L}} + 2\hat{\boldsymbol{S}}$. The factor 2 multiplying $\hat{\boldsymbol{S}}$ in this operator is called the **gyromagnetic ratio** of the electron.

19.7 Behavior of spinors under spatial inversion

In a coordinate transformation consisting of spatial inversion from a frame K to a frame K′ the space–time coordinates x'^i of an event in K′ are related to the coordinates x^i of the same event in K by [see (19.17)]

$$x'^i = a^i{}_j x^j \tag{19.85}$$

with

$$\left(a^i{}_j\right) = \begin{pmatrix} 1 & 0 & 0 & 0 \\ 0 & -1 & 0 & 0 \\ 0 & 0 & -1 & 0 \\ 0 & 0 & 0 & -1 \end{pmatrix}. \tag{19.86}$$

If in K a particle is described by a Dirac spinor wave function $\Psi(x)$ satisfying the Dirac equation in K, that is, satisfying

$$\left(i\hbar\gamma^i \frac{\partial}{\partial x^i} - 1mc\right)\Psi(x) = 0,$$

then for covariance of the Dirac equation under spatial inversion we must require that in K′ the corresponding wave function $\Psi'(x')$ satisfies

$$\left(i\hbar\gamma^i \frac{\partial}{\partial x'^i} - 1mc\right)\Psi'(x') = 0,$$

with, of course, the **same** Dirac matrices γ^i. By precisely the same arguments as in the case of a Lorentz transformation from K to K′, the spinor $\Psi'(x')$ must be related to the spinor $\Psi(x)$ by

$$\Psi'(x') = \mathrm{S(a)}\Psi(x),$$

where S(a) is the 4×4 spinor-transformation matrix corresponding to the spatial-coordinate inversion (19.85), (19.86), and [see equation (19.19)] must satisfy

$$\mathrm{S}^{-1}(\mathrm{a})\gamma^j \mathrm{S(a)} = a^j{}_i\gamma^i.$$

Using for the matrix $\mathrm{S}(a)$ the standard symbol P, and the elements of the matrix (19.86), we can write this as

$$\mathrm{P}^{-1}\gamma^0\mathrm{P} = \gamma^0 \quad \text{and} \quad \mathrm{P}^{-1}\gamma^\mu\mathrm{P} = -\gamma^\mu.$$

These are clearly satisfied by

$$\mathrm{P} = \mathrm{e}^{i\varphi}\gamma^0 \; (= \mathrm{e}^{i\varphi}\gamma_0) \tag{19.87}$$

(since $(\gamma^0)^{-1} = \gamma^0$ and, by the anticommutation relations for the γ-matrices, $\gamma^0\gamma^\mu\gamma^0 = -\gamma^\mu$).

If we require that four successive inversions, like a 4π spatial rotation, return a spinor to itself, the phase factor $\mathrm{e}^{i\varphi}$ can be any of the four fourth roots of unity ($\pm 1, \pm i$).

Because

$$\gamma^0 = \begin{pmatrix} 1 & 0 & 0 & 0 \\ 0 & 1 & 0 & 0 \\ 0 & 0 & -1 & 0 \\ 0 & 0 & 0 & -1 \end{pmatrix}$$

it is clear that the four possible spinor wave functions (19.59) of a particle at rest are all eigenstates of P (19.87) (the corresponding eigenvalues are called the **parities** of the states). The positive-energy states Ψ^{I} and Ψ^{II} have eigenvalue $\mathrm{e}^{i\varphi}$ and the negative-energy states Ψ^{III} and Ψ^{IV} have eigenvalue $-\mathrm{e}^{i\varphi}$, that is, for a particle at rest the positive-energy and negative-energy states have opposite parities. For a particle not at rest the possible spinor wave functions are given by (19.71), (19.70), and are clearly not eigenstates of P. It is clear, however, that for low velocities they are approximate eigenstates of P.

19.8 Unitarity properties of the spinor-transformation matrices

A matrix (or operator) S is said to be unitary if its adjoint is equal to its inverse, that is, if

$$\mathrm{S}^\dagger = \mathrm{S}^{-1}.$$

For a Lorentz transformation from frame K to frame K′ moving with velocity $\boldsymbol{V}$ relative to K we found [(19.40) and (19.36)]

$$\mathrm{S}_{\mathrm{L}(\boldsymbol{V})} = \exp\left(-\frac{\omega}{2}\frac{\boldsymbol{\alpha}\cdot\boldsymbol{V}}{|\boldsymbol{V}|}\right) = \exp\left(-\frac{\omega}{2}\frac{\alpha_\mu V_\mu}{|\boldsymbol{V}|}\right) = \exp\left(-i\frac{\omega}{2}\frac{\sigma_{0\mu}V_\mu}{|\boldsymbol{V}|}\right)$$

(here $\tanh\omega = V/c$), while for a spatial rotation through θ about the Z axis we found [see (19.43)]

$$\mathrm{S}_{\mathrm{R}_Z(\theta)} = \exp\left(i\frac{\theta}{2}\sigma_{12}\right).$$

To find the adjoints of these we need the adjoints of the matrices

$$\sigma_{ij} = \frac{i}{2}[\gamma_i, \gamma_j].$$

They are

$$\sigma_{ij}^\dagger = -\frac{i}{2}[\gamma_i, \gamma_j]^\dagger = -\frac{i}{2}[\gamma_j^\dagger, \gamma_i^\dagger] = \frac{i}{2}[\gamma_i^\dagger, \gamma_j^\dagger].$$

But

$$\gamma_0^\dagger = \gamma_0 \quad \text{and} \quad \gamma_\mu^\dagger = -\gamma_\mu,$$

and so

$$\sigma_{0\mu}^\dagger = -\sigma_{0\mu} \quad \text{and} \quad \sigma_{\mu\nu}^\dagger = \sigma_{\mu\nu}.$$

Hence,

$$\mathrm{S}_{\mathrm{L}(\boldsymbol{V})}^\dagger = \exp\left(i\frac{\omega}{2}\frac{\sigma_{0\mu}^\dagger V_\mu}{|\boldsymbol{V}|}\right) = \exp\left(-i\frac{\omega}{2}\frac{\sigma_{0\mu}V_\mu}{|\boldsymbol{V}|}\right) = \mathrm{S}_{\mathrm{L}(\boldsymbol{V})} \neq \mathrm{S}_{\mathrm{L}(\boldsymbol{V})}^{-1}$$

and

$$\mathrm{S}_{\mathrm{R}_Z(\theta)}^\dagger = \exp\left(-i\frac{\theta}{2}\sigma_{12}^\dagger\right) = \exp\left(-i\frac{\theta}{2}\sigma_{12}\right) = \mathrm{S}_{\mathrm{R}_Z(\theta)}^{-1}$$

that is, $\mathrm{S}_{\mathrm{L}(\boldsymbol{V})}$ is not unitary but $\mathrm{S}_{\mathrm{R}_Z(\theta)}$ is. Similarly, P is also unitary, since [from (19.87)]

$$\mathrm{P}^\dagger = \mathrm{e}^{-i\varphi}\gamma_0^\dagger = \mathrm{e}^{-i\varphi}\gamma_0 = \mathrm{P}^{-1}.$$

In fact, all three satisfy the "generalized unitarity" relation

$$\mathrm{S}^\dagger = \gamma_0\mathrm{S}^{-1}\gamma_0 \text{ (and so } \mathrm{S}^{-1} = \gamma_0\mathrm{S}^\dagger\gamma_0,\ \gamma_0\mathrm{S}^{-1} = \mathrm{S}^\dagger\gamma_0, \text{ etc.)} \tag{19.88}$$

(as can be seen by considering the power expansions of $S_{L(V)}$ and $S_{R_Z(\theta)}$ and using the fact that γ_0 commutes with σ_{12} but anticommutes with $\sigma_{0\mu}$ [see (19.31)]).

19.9 Proof that the four-current is a four-vector

Under the Lorentz transformation (19.85)

$$x'^i = a^i{}_j x^j,$$

for the four-current $j^i(x) = c\Psi^\dagger(x)\gamma^0\gamma^i\Psi(x)$ (19.50) to be a four-vector its components in frame K′ must be related to those in frame K by the same transformation law:

$$j'^i(x') = a^i{}_j j^j(x).$$

We now show that this is so:

$$\begin{aligned} j'^i(x') &= c\Psi'^\dagger(x')\gamma^0\gamma^i\Psi'(x') = c\Psi^\dagger(x)S^\dagger\gamma^0\gamma^i S\Psi(x) \\ &= c\Psi^\dagger(x)\gamma^0 S^{-1}\gamma^i S\Psi(x) = a^i{}_j c\Psi^\dagger(x)\gamma^0\gamma^j S\Psi(x) \\ &= a^i{}_j j^j(x), \end{aligned} \tag{19.89}$$

where in the second line we have used (19.88) and (19.19).

Similarly, we can show that the four-divergence of the four-current is a Lorentz invariant (as it must be, since it is zero by the continuity equation):

$$\frac{\partial j^j}{\partial x^j} = \frac{\partial x'^i}{\partial x^j}\frac{\partial j^j}{\partial x'^i} = a^i{}_j\frac{\partial j^j}{\partial x'^i} = \frac{\partial j'^i}{\partial x'^i},$$

where we have used (19.85) and (19.89).

The combination $\Psi^\dagger(x)\gamma^0$ occurs frequently and is called the **adjoint spinor**; we denote it by

$$\overline{\Psi}(x) = \Psi^\dagger(x)\gamma^0.$$

Under Lorentz transformations, spatial rotations or spatial inversion it transforms as follows:

$$\overline{\Psi}'(x') = \Psi'^\dagger(x')\gamma^0 = \Psi^\dagger(x)S^\dagger\gamma^0 = \Psi^\dagger(x)\gamma^0 S^{-1} = \overline{\Psi}(x)S^{-1}$$

where we have used (19.88). From this it follows that

$$\overline{\Psi}'(x')\Psi'(x') = \overline{\Psi}(x)S^{-1}S\Psi(x) = \overline{\Psi}(x)\Psi(x),$$

so that $\overline{\Psi}\Psi$ is a scalar under Lorentz transformations, spatial rotations and spatial inversion.

Later, as well as currents of the type $j^i = c\Psi^\dagger\gamma^0\gamma^i\Psi \equiv c\overline{\Psi}\gamma^i\Psi$ [see (19.50)], we shall consider "transition currents" $j^i = \overline{\Psi}(\mathrm{b})\gamma^i\Psi(\mathrm{a})$ describing a transition from state (or particle) a to state (or particle) b, and it is clear that the steps used in (19.89) demonstrate that this current too is a four-vector.

19.10 Interpretation of the negative-energy states

If we now include interaction with the radiation field (that is, the possibility of energy exchange with the radiation field), there appears to be nothing to stop an electron in a positive-energy state from giving up energy (in the form of a photon) to the radiation field and falling into a state described by a negative-energy solution of the Dirac equation.

All electrons, for example, should simply cascade downwards into oblivion! How can the theory be modified to avoid this catastrophe?

Dirac proposed that the **vacuum** is described by a state in which every one of the (infinite number of) spin-up and spin-down negative-energy states, with energies E in the range $-\infty < E \leq -mc^2$, is already occupied by an electron, and all the positive-energy states are empty. Since, by the **Pauli principle**, there can be no more than one fermion in any given single-particle state, any electrons added to the vacuum must go into positive-energy states, and transitions of such electrons into negative-energy states are blocked. For example, the state of an electron at rest (with energy $+mc^2$) is then stable.

Since, even in the presence of matter, there will always be unoccupied positive-energy single-particle states, it should be possible to knock an electron out of the infinite Dirac sea of negative-energy electrons that constitutes the vacuum into such a state, leaving a **hole** behind. All that is needed is a photon of the appropriate energy $\geq 2mc^2$ (see figure 19.1).

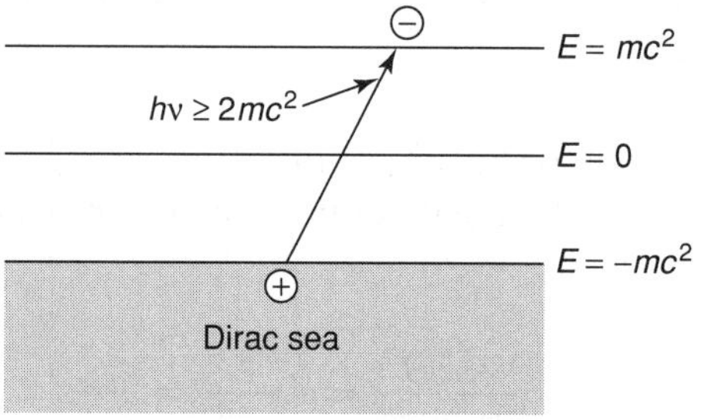

Figure 19.1 Creation of an electron–positron pair.

The resulting hole registers the absence from the vacuum of a negative-energy, negatively charged particle, and could be interpreted as the presence of a positive-energy, positively charged particle of the same mass m, that is, as a **positron**. Thus, figure 19.1 clearly provides a model of **electron–positron pair creation**.

Similarly, electron–positron pair annihilation is interpreted as resulting from a positive-energy electron falling into a negative-energy hole with emission of a photon (see figure 19.2).

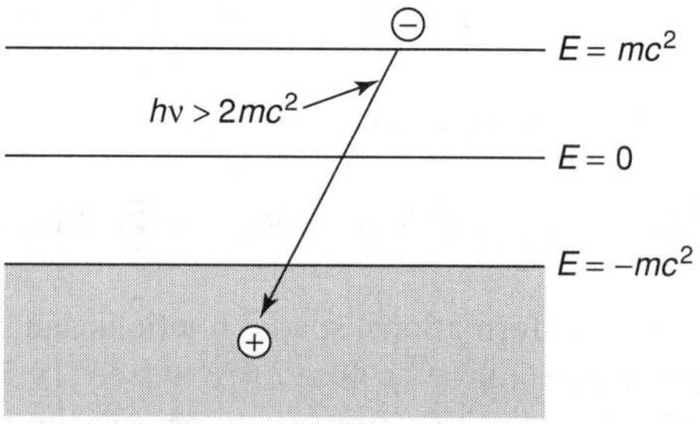

Figure 19.2 Annihilation of an electron–positron pair.

19.11 Charge conjugation

A positron of energy E and charge e ($e > 0$) can be interpreted **either** as the absence of an electron of negative energy $-E$ and charge $-e$, that is, of an electron corresponding to a negative-energy solution of the Dirac equation for the electron [see equations (19.57) and (19.55)]:

$$(i\hbar \not\nabla + e\not A - 1mc)\Psi = 0, \tag{19.90}$$

where

$$\not\nabla = \gamma^i \frac{\partial}{\partial x^i} = \gamma_i \frac{\partial}{\partial x_i},$$

or as a positive-energy solution of the positron Dirac equation

$$(i\hbar \not\nabla - e\not A - 1mc)\Psi_c = 0. \tag{19.91}$$

How can we construct Ψ_c from Ψ?
Take the complex conjugate of (19.90):

$$\left[\left(-i\hbar\frac{\partial}{\partial x^i} + eA_i\right)\gamma^{i*} - 1mc\right]\Psi^* = 0. \tag{19.92}$$

If we can find a nonsingular matrix (which we write as the product $C\gamma^0$), such that

$$(C\gamma^0)\gamma^{i^*}(C\gamma^0)^{-1} = -\gamma^i, \quad \text{i.e.,} \quad (C\gamma^0)\gamma^{i^*} = -\gamma^i(C\gamma^0), \tag{19.93}$$

then multiplying (19.92) on the left by $C\gamma^0$ we get

$$(i\hbar \not\nabla - e\not A - 1mc)(C\gamma^0\Psi^*) = 0,$$

which has the same form as (19.91) if

$$\Psi_c = C\gamma^0\Psi^*.$$

But the adjoint spinor $\overline{\Psi} = \Psi^\dagger\gamma^0$ has transpose

$$\overline{\Psi}^T = (\Psi^\dagger\gamma^0)^T = (\gamma^0)^T(\Psi^\dagger)^T = \gamma^0\Psi^*,$$

so that the positron wave function can be written as

$$\Psi_c = C\overline{\Psi}^T.$$

To find a matrix C satisfying the condition (19.93) we first write (19.93) in the form

$$C\gamma^0\gamma^{i^*}\gamma^0 C^{-1} = -\gamma^i. \tag{19.94}$$

But

$$\gamma^0\gamma^{i*}\gamma^0 = \begin{cases} \gamma^0\gamma^i\gamma^0 & (i=0,1,3) \\ -\gamma^0\gamma^i\gamma^0 & (i=2) \end{cases} = \begin{cases} \gamma^i & (i=0,2) \\ -\gamma^i & (i=1,3) \end{cases}$$

where in the first step we have used the fact that the nonzero elements of the matrix γ^i are all real for $i = 0$, 1, 3 and all pure-imaginary for $i = 2$, and in the second step we have used the anticommutation properties of the γ^i. Condition (19.94) then becomes

$$C\gamma^i C^{-1} = -\gamma^i \quad (i=0,2); \qquad -C\gamma^i C^{-1} = -\gamma^i \quad (i=1,3),$$

that is, C commutes with γ^1 and γ^3 and anticommutes with γ^0 and γ^2. A matrix that clearly has these properties is

$$C = i\gamma^2\gamma^0. \tag{19.95}$$

Therefore, the positron wave function is

$$\Psi_c = C\overline{\Psi}^T = C\gamma^0\Psi^* = i\gamma^2\gamma^0\gamma^0\Psi^* = i\gamma^2\Psi^*. \tag{19.95$'$}$$

For example, if we take for Ψ the solution of the electron Dirac equation for an electron at rest with negative energy $(-mc^2)$ and spin down:

$$\Psi = \Psi^{IV} = e^{imc^2t/\hbar}\begin{pmatrix}0\\0\\0\\1\end{pmatrix},$$

the wave function of the positron corresponding to absence of this electron from the Dirac sea is obtained from this by the following operation, called charge conjugation:

$$\Psi_c = i\gamma^2\Psi^{IV*} = i\gamma^2\begin{pmatrix}0\\0\\0\\1\end{pmatrix}e^{-imc^2t/\hbar} = i\begin{pmatrix}0&0&0&-i\\0&0&i&0\\0&i&0&0\\-i&0&0&0\end{pmatrix}\begin{pmatrix}0\\0\\0\\1\end{pmatrix}e^{-imc^2t/\hbar} = \begin{pmatrix}1\\0\\0\\0\end{pmatrix}e^{-imc^2t/\hbar} = \Psi^I,$$

which is the positive-energy, spin-up, zero-momentum (rest) solution of the Dirac equation, as required by our understanding of the nature of the positron.

19.12 Time reversal

The Dirac equation for an electron in an electromagnetic field is

$$(i\hbar\nabla\!\!\!/ + eA\!\!\!/ - 1mc)\Psi = 0, \tag{19.90}$$

where

$$\not{\nabla} = \gamma^i \frac{\partial}{\partial x^i} = \gamma_i \frac{\partial}{\partial x_i}.$$

To consider the effect of time reversal we multiply this throughout by $c\gamma^0$ (recalling that $(\gamma^0)^2 = 1$) and then separate out the time-derivative term. Equation (19.90) then takes the "hamiltonian form"

$$i\hbar \frac{\partial \Psi(\boldsymbol{r}, t)}{\partial t} = \mathrm{H}\Psi(\boldsymbol{r}, t) = \left[c\boldsymbol{\alpha} \cdot (-i\hbar \boldsymbol{\nabla} + e\boldsymbol{A}) + \beta mc^2 - e\varphi 1\right] \Psi(\boldsymbol{r}, t), \tag{19.96}$$

where we have used the conventional notation $\beta \equiv \gamma^0$ and $\alpha_\mu \equiv \gamma^0\gamma^\mu$ ($= -\gamma^\mu\gamma^0$) [see (19.37)].

If the time-reversed motion is also physically possible, there must be a corresponding time-reversed state described by a wave-function $\Psi'(\boldsymbol{r}, t') = \mathcal{T}\Psi(\boldsymbol{r}, t)$ $(t' = -t)$ satisfying the time-reversed Dirac equation

$$\begin{aligned} i\hbar \frac{\partial \Psi'(\boldsymbol{r}, t')}{\partial t'} &= \mathrm{H}'\Psi'(\boldsymbol{r}, t') = \left[c\boldsymbol{\alpha} \cdot (-i\hbar \boldsymbol{\nabla} + e\boldsymbol{A}') + \beta mc^2 - e\varphi' 1\right] \Psi'(\boldsymbol{r}, t') \\ &= \left[c\boldsymbol{\alpha} \cdot (-i\hbar \boldsymbol{\nabla} - e\boldsymbol{A}) + \beta mc^2 - e\varphi 1\right] \Psi'(\boldsymbol{r}, t') \end{aligned} \tag{19.97}$$

because $\boldsymbol{A}' = -\boldsymbol{A}$ (since $\boldsymbol{A}$ is generated by currents which reverse sign when the direction of time is reversed) and $\varphi' = \varphi$. In equation (19.96) we replace $\partial/\partial t$ by $-\partial/\partial t'$, put $\Psi(\boldsymbol{r}, t) = \mathcal{T}^{-1}\Psi'(\boldsymbol{r}, t')$, and act on both sides of the resulting equation with $\mathcal{T}$. The result is

$$\begin{aligned} -\mathcal{T}i\mathcal{T}^{-1}\hbar \frac{\partial \Psi'(\boldsymbol{r}, t')}{\partial t'} &= \mathcal{T}\mathrm{H}\mathcal{T}^{-1}\Psi'(\boldsymbol{r}, t') \\ &= \left[c\mathcal{T}\boldsymbol{\alpha} \cdot (-i\hbar \boldsymbol{\nabla} + e\boldsymbol{A})\mathcal{T}^{-1} + \mathcal{T}\beta mc^2\mathcal{T}^{-1} - e\mathcal{T}\varphi\mathcal{T}^{-1}\right] \Psi'(\boldsymbol{r}, t'). \end{aligned} \tag{19.98}$$

Equation (19.98) is the same as equation (19.97) if the operator $\mathcal{T}$ can be written in the form

$$\mathcal{T} = \text{take the complex conjugate and then multiply by a } 4 \times 4 \text{ constant matrix T} \tag{19.99}$$

[so that $\mathcal{T}i\mathcal{T}^{-1} = -i$], with a matrix T such that

$$\mathrm{T}\alpha^*\mathrm{T}^{-1} = -\alpha \quad \text{and} \quad \mathrm{T}\,\beta\,\mathrm{T}^{-1} = \beta.$$

Recalling that $\alpha_\mu \equiv \gamma^0\gamma^\mu$ and $\beta \equiv \gamma^0$, and that $\gamma^{i^*} = \gamma^i$ for $i = 0, 1, 3$ while $\gamma^{2^*} = -\gamma^2$, we clearly need a matrix T that commutes with γ^0 and γ^2 and anticommutes with γ^1 and γ^3. The matrix

$$\mathrm{T} = i\gamma^1\gamma^3 \tag{19.100}$$

is satisfactory (the phase factor i here is conventional).

To summarize, suppose that a movie film is taken of the motion of a particle and that the evolution of the state of the particle is described by a wave function $\Psi(\boldsymbol{r}, t)$. Then the wave function describing a particle in the state of motion observed by running the film backward (that is, in the fully time-reversed state of motion, including reversal of currents that give rise to the electromagnetic field) is

$$\Psi'(\boldsymbol{r}, t') = \mathcal{T}\Psi(\boldsymbol{r}, t) = \mathrm{T}\Psi^*(\boldsymbol{r}, t) = i\gamma^1\gamma^3\Psi^*(\boldsymbol{r}, t). \tag{19.101}$$

As an example, let $\Psi(\boldsymbol{r}, t)$ describe a positive-energy, spin-up free electron [see (19.71) and (19.70)]:

$$\Psi(\boldsymbol{r}, t) \propto \begin{pmatrix} 1 \\ 0 \\ \dfrac{p_z c}{E + mc^2} \\ \dfrac{p_+ c}{E + mc^2} \end{pmatrix} \mathrm{e}^{-ip_i x^i/\hbar}.$$

What is the nature of the particle undergoing the precisely time-reversed motion? Equation (19.101) gives the answer:

$$\Psi'(\boldsymbol{r}, t') = i\gamma^1\gamma^3\Psi^*(\boldsymbol{r}, t) \propto \gamma^1\gamma^3 \begin{pmatrix} 1 \\ 0 \\ \dfrac{p_z c}{E + mc^2} \\ \dfrac{p_- c}{E + mc^2} \end{pmatrix} \mathrm{e}^{ip_i x^i/\hbar}$$

$$= \begin{pmatrix} 0 & 0 & 0 & 1 \\ 0 & 0 & 1 & 0 \\ 0 & -1 & 0 & 0 \\ -1 & 0 & 0 & 0 \end{pmatrix} \begin{pmatrix} 0 & 0 & 1 & 0 \\ 0 & 0 & 0 & -1 \\ -1 & 0 & 0 & 0 \\ 0 & 1 & 0 & 0 \end{pmatrix} \begin{pmatrix} 1 \\ 0 \\ \dfrac{p_z c}{E + mc^2} \\ \dfrac{p_- c}{E + mc^2} \end{pmatrix} \mathrm{e}^{ip_i x^i/\hbar}$$

$$= \begin{pmatrix} 0 & 1 & 0 & 0 \\ -1 & 0 & 0 & 0 \\ 0 & 0 & 0 & 1 \\ 0 & 0 & -1 & 0 \end{pmatrix} \begin{pmatrix} 1 \\ 0 \\ \dfrac{p_z c}{E + mc^2} \\ \dfrac{p_- c}{E + mc^2} \end{pmatrix} \mathrm{e}^{ip_i x^i/\hbar}$$

$$= \begin{pmatrix} 0 \\ -1 \\ \dfrac{p_- c}{E + mc^2} \\ \dfrac{-p_z c}{E + mc^2} \end{pmatrix} \mathrm{e}^{ip_i x^i/\hbar} = -\begin{pmatrix} 0 \\ 1 \\ \dfrac{-p_- c}{E + mc^2} \\ \dfrac{p_z c}{E + mc^2} \end{pmatrix} \mathrm{e}^{i(Et - \boldsymbol{p}\cdot\boldsymbol{r})/\hbar} = -\begin{pmatrix} 0 \\ 1 \\ \dfrac{p'_- c}{E + mc^2} \\ \dfrac{-p'_z c}{E + mc^2} \end{pmatrix} \mathrm{e}^{-i[Et' - \boldsymbol{p}'\cdot\boldsymbol{r}]/\hbar}. \quad (19.102)$$

Equation (19.102) clearly corresponds to an electron travelling backward in time ($t' = -t$) and with spin and momentum ($\boldsymbol{p}' = -\boldsymbol{p}$) opposite to those of the electron described by $\Psi(\boldsymbol{r}, t)$. (Since an electron with this spin and momentum **can** exist, we could not be sure whether the movie we were watching was being run forward or backward, that is, time reversal appears to be a symmetry of nature at least as far as a free electron is concerned!)

Recall now the effect of spatial inversion on any spinor (the corresponding transformation matrix is $\mathrm{P} = \mathrm{e}^{i\varphi}\gamma^0$).

For example, the effect of P on the wave function of a negative-energy spin-down electron of momentum $\boldsymbol{p}$ is:

$$P\Psi_{p}^{\rm IV}(\boldsymbol{r},t) = e^{i\varphi}\gamma^0\Psi_{p}^{\rm IV}(\boldsymbol{r},t) = e^{i\varphi}\begin{pmatrix}1 & 0 & 0 & 0\\ 0 & 1 & 0 & 0\\ 0 & 0 & -1 & 0\\ 0 & 0 & 0 & -1\end{pmatrix}\begin{pmatrix}\dfrac{p_-c}{E+mc^2}\\ \dfrac{-p_zc}{E+mc^2}\\ 0\\ 1\end{pmatrix} e^{ip_ix^i/\hbar}$$

$$= e^{i\varphi}\begin{pmatrix}\dfrac{p_-c}{E+mc^2}\\ \dfrac{-p_zc}{E+mc^2}\\ 0\\ -1\end{pmatrix} e^{i(Et-\boldsymbol{p}\cdot\boldsymbol{r})/\hbar} = -e^{i\varphi}\begin{pmatrix}\dfrac{p'_-c}{E+mc^2}\\ \dfrac{-p'_zc}{E+mc^2}\\ 0\\ 1\end{pmatrix} e^{i(Et-\boldsymbol{p}'\cdot\boldsymbol{r}')/\hbar} = -e^{i\varphi}\Psi_{p'}^{\rm IV}(\boldsymbol{r}',t)$$

which is the wave function [in the new coordinates $\boldsymbol{r}'$ $(= -\boldsymbol{r})$, t] of a particle of reversed momentum $\boldsymbol{p}' = -\boldsymbol{p}$.

19.13 PCT symmetry

(1) The effect of time reversal on the wave function $\Psi(x)$ of an electron propagating backward in time is to give the wave function of an electron propagating forward in time with unchanged energy but reversed momentum and spin.

(2) The effect of charge conjugation on an electron wave function is to give a positron wave function with unchanged momentum but reversed sign of the energy and spin. [Note that "reversal of the sign of the energy" here means going from a negative-energy spinor of the type $\Psi^{\rm III}$ or $\Psi^{\rm IV}$ to a positive-energy spinor of the type $\Psi^{\rm I}$ or $\Psi^{\rm II}$ (or vice versa). The sign of the quantity E in these spinors does not change, since E was **defined** as $E = mc^2/\sqrt{1-v^2/c^2}$ with the positive square root; see the derivation of equation (19.70).]

(3) The effect of spatial inversion on an electron wave function is to give an electron wave function with unchanged energy and spin but reversed momentum.

Imagine now applying the above three operations, in the order given, to the backward (in time) evolution of a negative-energy electron with a given momentum and spin. The resulting wave function

$$\Psi_{\rm PCT}(x') \equiv \mathrm{P}\mathcal{C}\mathcal{T}\Psi(x) \tag{19.103}$$

$(x'^i = -x^i)$ should clearly be the wave function of a positive-energy positron **with the same momentum and spin** as those of the original (negative-energy) electron.

Thus, we take equation (19.103) (rather than Ψ_c) as the **definition** of the positron wave function.

Expanding equation (19.103) we get

$$\begin{aligned}\Psi_{\rm PCT}(x') &\equiv \mathrm{P}\mathcal{C}\mathcal{T}\Psi(x) = \mathrm{PCT}\Psi^*(x) = \mathrm{PC}\gamma^0[\mathrm{T}\Psi^*(x)]^* = \mathrm{PC}\gamma^0[i\gamma^1\gamma^3\Psi^*(x)]^*\\ &= -i\mathrm{PC}\gamma^0\gamma^1\gamma^3\Psi(x) = -i\mathrm{P}i\gamma^2\gamma^0\gamma^0\gamma^1\gamma^3\Psi(x) = \mathrm{P}\gamma^2\gamma^1\gamma^3\Psi(x)\\ &= e^{i\varphi}\gamma^0\gamma^2\gamma^1\gamma^3\Psi(x) = -e^{i\varphi}\gamma^0\gamma^1\gamma^2\gamma^3\Psi(x) \equiv ie^{i\varphi}\gamma_5\Psi(x),\end{aligned}$$

where the new matrix γ_5 is given by

$$\gamma_5 = i\gamma^0\gamma^1\gamma^2\gamma^3 = \begin{pmatrix} 0 & 0 & 1 & 0 \\ 0 & 0 & 0 & 1 \\ 1 & 0 & 0 & 0 \\ 0 & 1 & 0 & 0 \end{pmatrix}. \tag{19.104}$$

For example, take $\Psi(x)$ to be the wave function of a negative-energy spin-down electron of momentum $\boldsymbol{p}$ [see (19.71), (19.70)], moving backward in time:

$$\Psi(x) = \begin{pmatrix} \dfrac{p_-c}{E+mc^2} \\ \dfrac{-p_zc}{E+mc^2} \\ 0 \\ 1 \end{pmatrix} e^{ip_ix^i/\hbar}$$

(the coordinate $x^0 = ct$ in the space–time part increases in the direction of earlier times).

Then we can construct

$$\begin{aligned} \Psi_{\mathrm{PCT}}(x') &= ie^{i\varphi}\gamma_5\Psi(x) \\ &= ie^{i\varphi}\begin{pmatrix} 0 & 0 & 1 & 0 \\ 0 & 0 & 0 & 1 \\ 1 & 0 & 0 & 0 \\ 0 & 1 & 0 & 0 \end{pmatrix}\begin{pmatrix} \dfrac{p_-c}{E+mc^2} \\ \dfrac{-p_zc}{E+mc^2} \\ 0 \\ 1 \end{pmatrix} e^{ip_ix^i/\hbar} = ie^{i\varphi}\begin{pmatrix} 0 \\ 1 \\ \dfrac{p_-c}{E+mc^2} \\ \dfrac{-p_zc}{E+mc^2} \end{pmatrix} e^{-ip_ix'^i/\hbar}, \end{aligned}$$

which, as expected, is a positive-energy solution of the positron Dirac equation, with the same momentum $\boldsymbol{p}$ and spin as those of the electron described by $\Psi(x)$. (Note also that it is a wave function of the coordinates $x'^i = -x^i$, so that, in particular, $t' = -t$ is increasing in the direction of the future.)

This is the mathematical basis of Feynman's identification of a "positron as a negative-energy electron travelling backward in time".

Of course, this identification can be made only if PCT is an exact symmetry of nature, so that applying it to the wave function of any particle (whatever type of interaction the particle can participate in) gives a function that is an equally valid description of the state of that particle. That PCT is indeed an exact symmetry can be proved if we assume both Lorentz invariance of the theory and the **spin–statistics connection**, which states that particles of integer or zero spin obey Bose–Einstein statistics (that is, are bosons) and particles of half-odd-integer spin obey Fermi–Dirac statistics (that is, are fermions).

We now postulate that negative-energy states are described by wave functions that can propagate only backward in time. But, as we have just seen, the particle state described by a negative-energy, backward-propagating wave function $\Psi(x)$ is, by the PCT symmetry, the same as the antiparticle state described by the positive-energy, forward-propagating wave function $\Psi_{\mathrm{PCT}}(x')$. Thus, we have a re-interpretation of antiparticles (and a role for the negative-energy states) that does not lead to the necessity of a Dirac sea.

Is it true that P, C and T are symmetries of nature separately? Unlike the case of the product PCT, there is no **general** proof that they are, that is, there is no proof that they are for all types of particle and for all types of interaction.

Consider, in particular, the property of **helicity**, denoted by λ and defined as the projection of the spin along the direction of motion:

$$\lambda = \frac{\boldsymbol{S} \cdot \boldsymbol{p}}{|\boldsymbol{p}|}. \tag{19.105}$$

Spin-1/2 particles can clearly have helicity $\pm(1/2)\hbar$, corresponding to spin projection parallel and antiparallel to the direction of motion. States with positive helicity are said to be "right-handed", while those with negative helicity are "left-handed". Helicity is clearly a pseudoscalar quantity, since the parity operator P reverses the sign of the momentum $\boldsymbol{p}$ (a vector) but leaves the spin $\boldsymbol{S}$ (a pseudovector, or axial vector) unchanged:

$$\mathrm{P} : \lambda \to -\lambda. \tag{19.106}$$

Also, because the charge-conjugation operation $\mathcal{C}$ leaves the momentum $\boldsymbol{p}$ unchanged but reverses the sign of the spin $\boldsymbol{S}$, it too changes the helicity:

$$\mathcal{C} : \lambda \to -\lambda: \tag{19.107}$$

Consider the elementary **weak-interaction** processes $\mu^- \to e^-\bar{\nu}_e\nu_\mu$ (the decay of a muon to an electron, an electron anti-neutrino and a muon neutrino) and $\nu_e d \to u e^-$ (the interaction of an electron neutrino with a down quark to produce an up quark and an electron). All the particles involved here are elementary fermions: the fermions d and u are **hadrons**, defined as elementary fermions able to participate in ("feel") the **strong interaction**, while the others are **leptons**, defined as elementary fermions that are unable to feel the strong interaction.

It is an experimentally established fact that in all weak-interaction processes the fermion states involved, whether they be leptonic (electron, muon, electron neutrino, muon neutrino, etc.) or hadronic (up quark, down quark, etc.) are always left-handed, while the anti-fermion states involved, whether leptonic (positron, anti-muon, electron anti-neutrino, muon anti-neutrino, etc.) or hadronic (anti up quark, anti down quark, etc.) are always right-handed.

If parity P were a symmetry of the weak interaction, (19.106) would imply the existence of weak-interaction processes with right-handed neutrinos and left-handed anti-neutrinos. Such processes are never observed. Thus, parity P cannot be a symmetry of the weak interaction. In the standard terminology, parity is violated by the weak interaction.

Similarly, $\mathcal{C}$ is also violated in the weak interaction (as we have seen, it reverses the spin but leaves $\boldsymbol{p}$ unchanged), but in all but two cases the product P$\mathcal{C}$ is an exact symmetry (and hence so too is $\mathcal{T}$, since P$\mathcal{CT}$ is an exact symmetry). In these two cases (the weak decay of the neutral kaon K^0, and weak $b\bar{b}$ annihilation, where b is the bottom quark) the weak interaction is not invariant under P$\mathcal{C}$, and so, because P$\mathcal{CT}$ is exact, these processes cannot be invariant under $\mathcal{T}$ either. Thus, even though these processes are apparently elementary, only the "forward-movie" versions of them are observed in nature.

Because the fermion states involved in weak-interaction processes are all left-handed, and the anti-fermion states are all right-handed, to construct a theory of the weak interaction it will be helpful to find a way of constructing Dirac states of definite helicity. For this it is convenient to rewrite the free-particle Dirac equation (19.55) as

$$\left(\gamma'^j \hat{p}_j - 1mc\right) \Psi'(x) = 0 \tag{19.55a}$$

with a different set of gamma matrices $\gamma'^j = U\gamma^j U^{-1}$and a correspondingly different Dirac spinor $\Psi'(x) = U\Psi(x)$, where U is the 4×4 unitary matrix effecting a **similarity transformation** to new gamma matrices clearly satisfying the anticommutation relations (19.4) satisfied by the Dirac matrices γ^j (19.7):

$$\begin{aligned}\gamma'^j\gamma'^i + \gamma'^i\gamma'^j &= U\gamma^j U^{-1}U\gamma^i U^{-1} + U\gamma^i U^{-1}U\gamma^j U^{-1}.\\ &= U\gamma^j\gamma^i U^{-1} + U\gamma^i\gamma^j U^{-1} = 2g^{ij}1.\end{aligned}$$

The adjoint of the particular matrix

$$U = \frac{1}{\sqrt{2}}\begin{pmatrix} i\sigma_2 & i\sigma_2 \\ i\sigma_2 & -i\sigma_2 \end{pmatrix} = \frac{1}{\sqrt{2}}\begin{pmatrix} 0 & 1 & 0 & 1 \\ -1 & 0 & -1 & 0 \\ 0 & 1 & 0 & -1 \\ -1 & 0 & 1 & 0 \end{pmatrix},$$

is clearly $U^\dagger = U^T = -U = U^{-1}$ (so that this U is unitary), and the matrices $\gamma'^j = U\gamma^j U^{-1}$ obtained from the Dirac matrices γ^j (19.7) are easily found to be

$$\gamma'^0 = \begin{pmatrix} 0 & 1 \\ 1 & 0 \end{pmatrix}, \quad \gamma'^\mu = \gamma^\mu = \begin{pmatrix} 0 & \sigma_\mu \\ -\sigma_\mu & 0 \end{pmatrix} \quad (\mu = 1, 2, 3). \tag{19.108}$$

[The matrices (19.108) are said to form the **Weyl basis**, or **chiral basis**, and provide an acceptable alternative to the Dirac basis (19.7).]

For later use, the corresponding γ_5 matrix [see (19.104)] is

$$\begin{aligned}\gamma'_5 &= i\gamma'^0\gamma'^1\gamma'^2\gamma'^3 = i\gamma'^0\gamma'^1\begin{pmatrix} 0 & \sigma_2 \\ -\sigma_2 & 0 \end{pmatrix}\begin{pmatrix} 0 & \sigma_3 \\ -\sigma_3 & 0 \end{pmatrix} = i\gamma'^0\gamma'^1\begin{pmatrix} -\sigma_2\sigma_3 & 0 \\ 0 & -\sigma_2\sigma_3 \end{pmatrix}\\ &= i\gamma'^0\gamma'^1\begin{pmatrix} -i\sigma_1 & 0 \\ 0 & -i\sigma_1 \end{pmatrix} = \gamma'^0\begin{pmatrix} 0 & \sigma_1 \\ -\sigma_1 & 0 \end{pmatrix}\begin{pmatrix} \sigma_1 & 0 \\ 0 & \sigma_1 \end{pmatrix} = \gamma'^0\begin{pmatrix} 0 & \sigma_1\sigma_1 \\ -\sigma_1\sigma_1 & 0 \end{pmatrix} = \gamma'^0\begin{pmatrix} 0 & 1 \\ -1 & 0 \end{pmatrix}\\ &= \begin{pmatrix} 0 & 1 \\ 1 & 0 \end{pmatrix}\begin{pmatrix} 0 & 1 \\ -1 & 0 \end{pmatrix} = \begin{pmatrix} -1 & 0 \\ 0 & 1 \end{pmatrix}.\end{aligned} \tag{19.109}$$

Using the matrices (19.108), and multiplying (19.55a) above on the left by γ'^0, we get

$$\left(1\frac{E}{c} - \gamma'^0\gamma'^1\hat{p}_x - \gamma'^0\gamma'^2\hat{p}_y - \gamma'^0\gamma'^3\hat{p}_z - \gamma'^0 mc\right)\Psi'(x) = 0, \tag{19.110}$$

where we have used the fact that $(\gamma'_0)^2 = 1$ and replaced the operator $\hat{p}_0$ by its eigenvalue E/c for a stationary-state solution $\Psi'(x)$ with energy E. Using (19.108), and writing the wave function $\Psi'(x)$ in the form $\Psi'(x) = \begin{pmatrix} \phi(x) \\ \chi(x) \end{pmatrix}$, where $\phi(x)$ and $\chi(x)$ are two-component wave functions (Weyl spinors), we can rewrite (19.110) as

$$\left[\frac{E}{c}\begin{pmatrix} 1 & 0 \\ 0 & 1 \end{pmatrix} - \begin{pmatrix} -\sigma_\mu & 0 \\ 0 & \sigma_\mu \end{pmatrix}\hat{p}_\mu - mc\begin{pmatrix} 0 & 1 \\ 1 & 0 \end{pmatrix}\right]\begin{pmatrix} \phi(x) \\ \chi(x) \end{pmatrix} = 0.$$

This can now be written as two 2 × 2 matrix equations:

$$\frac{E}{c}\phi + \sigma \cdot \hat{\boldsymbol{p}}\phi - mc\chi = 0,$$
$$\frac{E}{c}\chi - \sigma \cdot \hat{\boldsymbol{p}}\chi - mc\phi = 0.$$

Because neutrinos are massless or nearly so, if in the above derivation $\Psi'(x)$ is the wave function of a neutrino or antineutrino we can set $m = 0$. We then find for the action of the operator $\boldsymbol{\sigma} \cdot \hat{\boldsymbol{p}}/|\boldsymbol{p}|$ on $\phi(x)$ and $\chi(x)$:

$$\frac{\sigma \cdot \hat{\boldsymbol{p}}}{|\boldsymbol{p}|}\phi = -\frac{E}{|\boldsymbol{p}|\,c}\phi = -\phi,$$
$$\frac{\sigma \cdot \hat{\boldsymbol{p}}}{|\boldsymbol{p}|}\chi = \frac{E}{|\boldsymbol{p}|\,c}\chi = \chi,$$

where we have used the fact that $E = pc$ for a massless particle. Thus, the part $\phi(x)$ is an eigenfunction of the helicity operator $\hat{\lambda} = \boldsymbol{S} \cdot \hat{\boldsymbol{p}}/|\boldsymbol{p}|$ with eigenvalue $-1/2\hbar$ and is, therefore, the negative-helicity (left-handed) part of the Dirac spinor $\Psi'_\nu(x)$ for a neutrino or antineutrino, while the part $\chi(x)$ is an eigenfunction of the helicity operator $\hat{\lambda}$ with eigenvalue $1/2\hbar$ and is, therefore, the positive-helicity (right-handed) part.

Because in weak-interaction processes the neutrinos detected are always left-handed, only the "upper part" $\phi(x)$ of a neutrino wave function $\Psi'_\nu(x)$ is relevant to the physics of weak interactions. It is clear that we can project it out from $\Psi'_\nu(x)$ by the action of the projection matrix $\mathrm{P_L} = \begin{pmatrix} 1 & 0 \\ 0 & 0 \end{pmatrix} = \frac{1}{2}(1 - \gamma'_5)$ [see (19.109)]. Thus, for a massless neutrino ν, the wave function $\frac{1}{2}(1 - \gamma'_5)\Psi'_\nu(x)$ contains just the left-handed component $\phi(x) \equiv \psi_{\mathrm{L}}(x)$ of $\Psi'_\nu(x)$.

Similarly, because in weak-interaction processes the antineutrinos detected are always right-handed, only the "lower part" $\chi(x)$ of an antineutrino wave function $\Psi'_{\bar{\nu}}(x)$ is relevant to the physics of weak interactions. Clearly, it can be projected out from $\Psi'_{\bar{\nu}}(x)$ by the action of the projection matrix $\mathrm{P_R} = \begin{pmatrix} 0 & 0 \\ 0 & 1 \end{pmatrix} = \frac{1}{2}(1 + \gamma'_5)$, and so, for a massless neutrino ν, the wave function $\frac{1}{2}(1 + \gamma'_5)\Psi'_\nu(x)$ contains only the right-handed component $\chi(x) \equiv \psi_{\mathrm{R}}(x)$ of $\Psi'_\nu(x)$.

(The above results for the effects of $\mathrm{P_L}$ and $\mathrm{P_R}$ are independent of the choice of gamma matrices used to derive them, and so we can drop the primes.)

At the end of section 19.9 we showed that the "transition current" $j^i = c\overline{\Psi}(\mathrm{b})\gamma^i\Psi(\mathrm{a})$ describing a transition from state (or particle) a to state (or particle) b is a four-vector. If, for example, $\Psi(\mathrm{a})$ here is a neutrino spinor (e.g., if a is ν_e), we shall want to consider just its left-handed part, that is, replace it by $\frac{1}{2}(1 - \gamma_5)\Psi(\nu_e)$, and so we shall need to consider currents of the type

$$j^i = c\overline{\Psi}(\mathrm{b})\gamma^i\gamma_5\Psi(\mathrm{a}). \tag{19.111}$$

Because, as was seen in section 19.13, $\gamma_5\Psi(\mathrm{a})$ represents a physically possible Dirac state if $\Psi(\mathrm{a})$ does (this is a statement of $\mathcal{PCT}$ symmetry), this j^i is also a four-vector under Lorentz transformations and spatial rotations. Under spatial inversion, however, Dirac states are acted upon by the matrix $\mathrm{S} = \mathrm{P} = e^{i\varphi}\gamma^0$ [see (19.87)], and we have $j^i(x) \to j'^i(x') = c\overline{\Psi}'(x')\gamma^i\gamma_5\Psi'(x') = c\overline{\Psi}(x)\mathrm{P}^{-1}\gamma^i\gamma_5\mathrm{P}\Psi(x) = c\overline{\Psi}(x)\gamma^0\gamma^i\gamma_5\gamma^0\Psi(x)$. For the $i = 0$ component this gives

$$j'^0(x') = c\overline{\Psi}(x)\gamma^0\gamma^0\gamma_5\gamma^0\Psi(x) = c\overline{\Psi}(x)\gamma_5\gamma^0\Psi(x) = -c\overline{\Psi}(x)\gamma^0\gamma_5\Psi(x) = -j^0(x)$$

while for the $i = \mu = 1, 2, 3$ components it gives

$$\begin{aligned} j'^\mu(x') &= c\overline{\Psi}(x)\gamma^0\gamma^\mu\gamma_5\gamma^0\Psi(x) \\ &= -c\overline{\Psi}(x)\gamma^\mu\gamma^0\gamma_5\gamma^0\Psi(x) = c\overline{\Psi}(x)\gamma^\mu\gamma^0\gamma^0\gamma_5\Psi(x) = j^\mu(x). \end{aligned}$$

Thus, under spatial inversion $(x^0, x^1, x^2, x^3 \to x^0, -x^1, -x^2, -x^3)$ we have

$$j^0, j^1, j^2, j^3 \to -j^0, j^1, j^2, j^3,$$

thereby demonstrating that $j^i = c\overline{\Psi}(\mathrm{b})\gamma^i\gamma_5\Psi(\mathrm{a})$ is an **axial** (pseudo) four-vector. Therefore, currents such as

$$j^i = c\overline{\Psi}(\mathrm{b})\gamma^i \tfrac{1}{2}(1 - \gamma_5)\Psi(\nu_\mathrm{e}),$$

which are introduced into weak-interaction theory to ensure the necessary parity violation, have the "$V - A$" structure $j^i = V^i - A^i$, with both a vector piece and an axial-vector piece.

19.14 Models of the weak interaction

We shall consider **current–current models** of the weak interaction, for example, nuclear β-decay, involving the process $\mathrm{d} \to \mathrm{u}\mathrm{e}^-\bar{\nu}_\mathrm{e}$. The corresponding "scattering" process is $\nu_\mathrm{e}\mathrm{d} \to \mathrm{u}\mathrm{e}^-$, which can be thought of as a leptonic transition $\nu_\mathrm{e} \to \mathrm{e}^-$ with an accompanying hadronic transition $\mathrm{d} \to \mathrm{u}$. The probability amplitude F for the process is postulated to be a Lorentz scalar involving the two corresponding transition currents, each of which is a Lorentz four-vector analogous to the four-current (19.50). The two four-currents (we write just the spinor parts of the corresponding particle wave functions) are $j_i(\nu_\mathrm{e} \to \mathrm{e}^-) = c\overline{\mathrm{w}}(\mathrm{e})\gamma_i\mathrm{w}(\nu_\mathrm{e})$ and $j_i(\mathrm{d} \to \mathrm{u}) = c\overline{\mathrm{w}}(\mathrm{u})\gamma_i\mathrm{w}(\mathrm{d})$, where $\overline{\mathrm{w}} \equiv \mathrm{w}^\dagger\gamma^0$ denotes the adjoint spinor corresponding to w (see section 19.9). [We shall incorporate the helicity-projection operator $\mathrm{P_L} = \frac{1}{2}(1 - \gamma_5)$ later.]

We shall consider two models of the coupling of these four-currents. In the first (the **Fermi model**), because of the low energy release in β-decay it is assumed that any momentum dependence of the probability amplitude F can be ignored, so that we have a point interaction with coupling constant G:

$$F = Gj_i(\nu_\mathrm{e} \to \mathrm{e}^-)j^i(\mathrm{d} \to \mathrm{u}) = Gc\overline{\mathrm{w}}(\mathrm{e})\gamma_i\mathrm{w}(\nu_\mathrm{e})g^{ik}c\overline{\mathrm{w}}(\mathrm{u})\gamma_k\mathrm{w}(\mathrm{d}). \tag{19.112}$$

This is the probability amplitude of a process corresponding to a point interaction, represented by the diagram shown in figure 19.3.

However, just as the electromagnetic interaction between two electric charges is not a point interaction but is mediated by the exchange of a virtual photon (the "carrier" of the electromagnetic interaction), we may postulate that the weak interaction is mediated by the exchange of a virtual boson. Unlike the electromagnetic currents, weak currents, such as the above transition currents $j_i(\nu_\mathrm{e} \to \mathrm{e}^-) = c\overline{\mathrm{w}}(\mathrm{e})\gamma_i\mathrm{w}(\nu_\mathrm{e})$ and $j_i(\mathrm{d} \to \mathrm{u}) = c\overline{\mathrm{w}}(\mathrm{u})\gamma_i\mathrm{w}(\mathrm{d})$, can be "charged", by which is meant charge-changing, by $+e$ or $-e$. There

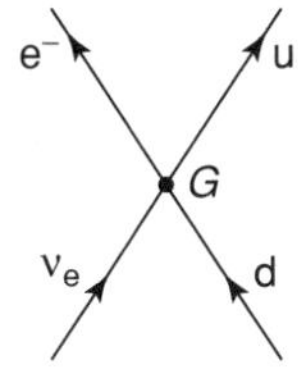

Figure 19.3 Point-interaction model (Fermi model) for the process $\nu_e d \to ue^-$.

also exist weak-interaction processes (such as neutrino–neutrino scattering) in which the transition currents are "neutral", meaning that the electric charge in the transition current remains unchanged.

It is therefore postulated that the carriers of the weak force are bosons (the so-called **intermediate vector bosons**, denoted by W^+, W^-, Z) carrying the corresponding charges, and that there is a universal weak-interaction coupling constant g that characterizes the coupling of any given leptonic or hadronic current to the relevant intermediate vector boson (in the same way that the coupling strength e couples an electromagnetic current to a photon). In this description the process $\nu_e d \to ue^-$ can be represented, to order g^2, by the diagram shown in figure 19.4.

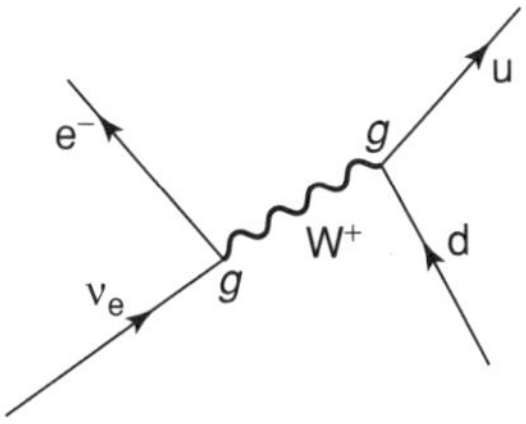

Figure 19.4 W^+-exchange graph in the process $\nu_e d \to ue^-$.

Consider, for example, the weak-interaction process $\mu^- \nu_e \to \nu_\mu e^-$, and in particular the process illustrated in figure 19.5.

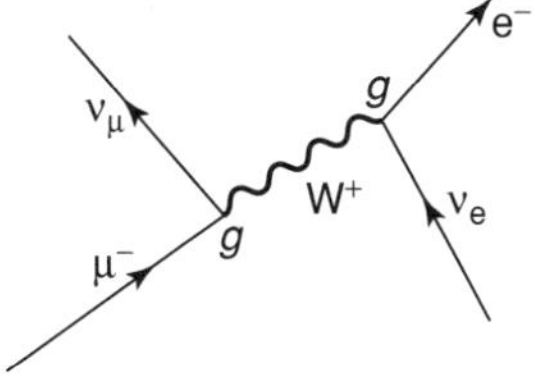

Figure 19.5 W^+-exchange graph in the process $\mu^- \nu_e \to \nu_\mu e^-$.

The amplitude F of the process in figure 19.5 is

$$F = \frac{g^2}{2} j_i(\mu^- \to \nu_\mu) G_W^{ik}(q) j_k(\nu_e \to e^-), \tag{19.113}$$

where

$$G_{\mathrm{W}}^{ik}(q) = \frac{-g^{ik} + \left(q^i q^k / M_{\mathrm{W}}^2 c^2\right)}{q^2 - M_{\mathrm{W}}^2 c^2} \tag{19.114}$$

is the W-boson propagator.

But the two currents in (19.113) need include only the left-handed parts of the electron-neutrino and muon-neutrino spinors, and so they can be expressed as:

$$j_k(\nu_\mathrm{e} \to \mathrm{e}^-) = c\overline{\mathrm{w}}(\mathrm{e})\gamma_k \mathrm{P_L} \mathrm{w}(\nu_\mathrm{e}) = c\overline{\mathrm{w}}(\mathrm{e})\gamma_k \frac{(1-\gamma_5)}{2}\mathrm{w}(\nu_\mathrm{e}) \tag{19.115}$$

and

$$\begin{aligned} j_i(\mu^- \to \nu_\mu) &= c\overline{\mathrm{P_L w}(\nu_\mu)}\gamma_i \mathrm{w}(\mu^-) = c\frac{1}{2}\overline{(1-\gamma_5)\mathrm{w}(\nu_\mu)}\gamma_i \mathrm{w}(\mu^-) \\ &= c\left[\frac{(1-\gamma_5)}{2}\mathrm{w}(\nu_\mu)\right]^\dagger \gamma^0\gamma_i \mathrm{w}(\mu^-) = c\mathrm{w}^\dagger(\nu_\mu)\frac{(1-\gamma_5)}{2}\gamma^0\gamma_i \mathrm{w}(\mu^-) \\ &= c\mathrm{w}^\dagger(\nu_\mu)\gamma^0\gamma_i\frac{(1-\gamma_5)}{2}\mathrm{w}(\mu^-) = c\overline{\mathrm{w}}(\nu_\mu)\gamma_i\frac{(1-\gamma_5)}{2}\mathrm{w}(\mu^-). \end{aligned} \tag{19.116}$$

where in deriving (19.116) we have used the definition [see section 19.9)] of the adjoint spinor (twice), the fact that $\gamma_5^\dagger = \gamma_5$, and also the fact that γ_5 commutes with the product $\gamma_0\gamma_i$. (As γ_i is passed successively through the four gamma matrices in γ_5 there are three changes of sign, and similarly when γ_0 is then passed through γ_5.) Note that (19.116) shows that the projection matrix $\mathrm{P_L}$ can be inserted into the same position in the expression for the current as in (19.115) even when the relevant neutrino spinor is described by its adjoint spinor.

Substituting (19.115) and (19.116) into (19.113), (19.114), we find

$$\begin{aligned} F &= \frac{g^2}{2} j_i(\mu^- \to \nu_\mu) G_{\mathrm{W}}^{ik}(q) j_k(\nu_\mathrm{e} \to \mathrm{e}^-) \\ &= \frac{g^2}{2} c\overline{\mathrm{w}}(\nu_\mu)\gamma_i\frac{(1-\gamma_5)}{2}\mathrm{w}(\mu^-)\frac{\left[-g^{ik} + \left(q^i q^k / M_{\mathrm{W}}^2 c^2\right)\right]}{q^2 - M_{\mathrm{W}}^2 c^2} c\overline{\mathrm{w}}(\mathrm{e})\gamma_k\frac{(1-\gamma_5)}{2}\mathrm{w}(\nu_\mathrm{e}). \end{aligned} \tag{19.117}$$

For energies such that the q^2 dependence of the propagator (19.114) can be neglected, that is, when $q^2 \ll M_{\mathrm{W}}^2 c^2$, the graph depicted in figure 19.3 reduces to the point-like approximation illustrated in figure 19.6.

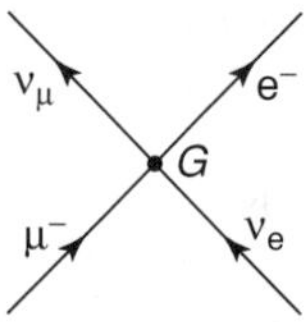

Figure 19.6 *Fermi-model (point-like) approximation to the W$^+$-exchange graph in figure 19.5.*

The amplitude (19.112) then becomes

$$F = F_{\text{Fermi}} = \frac{G_{\text{F}}}{\sqrt{2}}\overline{\text{w}}(\nu_{\mu})\gamma_i(1-\gamma_5)\text{w}(\mu^-)g^{ik}\overline{\text{w}}(\text{e})\gamma_k(1-\gamma_5)\text{w}(\nu_{\text{e}}) \tag{19.118}$$

The form (19.118) defines the Fermi constant G_{F}, and comparison of it with (19.117) in the case when $q^2 \ll M_{\text{W}}^2c^2$ yields the relation

$$\frac{G_{\text{F}}}{\sqrt{2}} = \frac{g^2}{8M_{\text{W}}^2}. \tag{19.119}$$

20

Gauge Theories of Quark and Lepton Interactions

In this chapter we shall show how the requirement that nature be symmetric (unchanged) under local changes of phase of wave functions compels us to introduce new fields (gauge fields) that may be viewed as responsible for the quark and lepton interactions.

20.1 Global phase invariance

Before looking at local changes of phase, we consider the effect of changing the phase of the wave function of a system by a constant phase factor:

$$\psi(x) \to \psi'(x) = \psi(x)\mathrm{e}^{i\alpha}. \tag{20.1}$$

The new function is still a solution of the wave equation satisfied by ψ (for example, if ψ was an energy eigenstate, then so is ψ', with the same energy), and the transformation also leaves all probability densities, expectation values and probabilities of transitions between states unchanged.

Because α is a constant (the same everywhere in space–time) the phase change (20.1) is said to be **global** and the corresponding symmetry is called **global phase invariance**.

20.2 Local phase invariance?

Is it possible that changing the phase of the wave function by an amount that differs from point to point in space–time will still leave the physics unchanged? In other words, is the transformation

$$\psi(\boldsymbol{r}, t) \to \psi'(\boldsymbol{r}, t) = \psi(\boldsymbol{r}, t)\mathrm{e}^{i\alpha(\boldsymbol{r},t)} \tag{20.2}$$

(with a phase angle α that depends on $\boldsymbol{r}$ and t) also a symmetry of nature?

A Course in Theoretical Physics, First Edition. P. J. Shepherd.

Suppose, for example, that $\psi(\boldsymbol{r}, t)$ satisfies the time-dependent Schrödinger equation for a free particle:

$$i h \frac{\partial \psi(\boldsymbol{r}, t)}{\partial t} = \frac{1}{2m}(-i\hbar\boldsymbol{\nabla})^2 \psi(\boldsymbol{r}, t). \tag{20.3}$$

Then, because α is a function of $\boldsymbol{r}$ and t, it is clear that the function $\psi'(\boldsymbol{r}, t)$ does not satisfy the time-dependent Schrödinger equation. In other words, the transformation (20.2) is not a symmetry of the free-particle Schrödinger equation.

But might it be a symmetry of some modified wave equation?

Consider, for example, the modified wave equation obtained from (20.3) by replacing the time and space derivatives as follows.

Using the relativistic notation

$$\partial^i \equiv (\partial^0, \partial^1, \partial^2, \partial^3) \equiv \left(\frac{\partial}{\partial x_0}, \frac{\partial}{\partial x_1}, \frac{\partial}{\partial x_2}, \frac{\partial}{\partial x_3}\right) = \left(\frac{\partial}{\partial (ct)}, -\frac{\partial}{\partial x}, -\frac{\partial}{\partial y}, -\frac{\partial}{\partial z}\right)$$

(which will be useful in considering the Dirac equation as well), we make in (20.3) the replacements

$$\partial^i \to D^i \equiv \partial^i + \frac{iqA^i}{\hbar}, \tag{20.4}$$

with A^i $(i = 0, 1, 2, 3)$ a contravariant four-vector

$$A^i = (A^0, A^1, A^2, A^3) \equiv (\varphi/c, A_x, A_y, A_z), \tag{20.5}$$

thereby obtaining a new wave equation

$$i\hbar c\left(\frac{\partial}{\partial(ct)} + \frac{iqA^0}{\hbar}\right)\psi(\boldsymbol{r}, t) = -\frac{\hbar^2}{2m}\left[\left(-\boldsymbol{\nabla} + \frac{iq\boldsymbol{A}}{\hbar}\right)\right]^2 \psi(\boldsymbol{r}, t) \tag{20.6}$$

(in which $\boldsymbol{A}$ is the three-vector $\boldsymbol{A} = \hat{\boldsymbol{\imath}} A_x + \hat{\boldsymbol{\jmath}} A_y + \hat{\boldsymbol{k}} A_z$).

Direct substitution now shows that the function $\psi'(\boldsymbol{r}, t)$ defined in (20.2), written in the form

$$\psi'(\boldsymbol{r}, t) = \psi(\boldsymbol{r}, t) e^{iq\chi(\boldsymbol{r},t)/\hbar} \tag{20.7}$$

with $\chi(\boldsymbol{r}, t)$ a Lorentz scalar, satisfies the equation

$$i\hbar c\left(\frac{\partial}{\partial(ct)} + \frac{iqA'^0}{\hbar}\right)\psi'(\boldsymbol{r}, t) = -\frac{\hbar^2}{2m}\left[\left(-\nabla + \frac{iq\boldsymbol{A}'}{\hbar}\right)\right]^2 \psi'(\boldsymbol{r}, t), \tag{20.8}$$

where

$$A'^0 = A^0 - \frac{\partial \chi}{\partial(ct)} \quad \text{and} \quad \boldsymbol{A}' = \boldsymbol{A} + \boldsymbol{\nabla}\chi. \tag{20.9}$$

Equation (20.6), obtained from the free-particle Schrödinger equation by the above modification, can be rewritten in the more familiar form

$$
\begin{aligned}
i\hbar\frac{\partial\psi(\boldsymbol{r},t)}{\partial t} &= \left[\frac{1}{2m}(-i\hbar\boldsymbol{\nabla} - q\boldsymbol{A})^2 + q\varphi\right]\psi(\boldsymbol{r},t) \\
&= \left[\frac{1}{2m}(\hat{\boldsymbol{p}} - q\boldsymbol{A})^2 + q\varphi\right]\psi(\boldsymbol{r},t),
\end{aligned}
\tag{20.10}
$$

which is recognized as the correct form of the Schrödinger equation for a particle of charge q in an **electromagnetic field** described by the four-potential (20.5), while (20.9) is the familiar set of so-called gauge transformations of the electromagnetic field (see section 17.4):

$$
\varphi' = \varphi - \frac{\partial\chi}{\partial t} \quad \text{and} \quad \boldsymbol{A}' = \boldsymbol{A} + \boldsymbol{\nabla}\chi. \tag{20.11}
$$

Because the electric-field intensity $\boldsymbol{E}$ and magnetic induction $\boldsymbol{B}$, which determine the forces on charges and are defined by

$$
\boldsymbol{E} = -\boldsymbol{\nabla}\varphi - \frac{\partial\boldsymbol{A}}{\partial t} \quad \text{and} \quad \boldsymbol{B} = \boldsymbol{\nabla}\times\boldsymbol{A},
$$

are clearly left unchanged by the gauge transformation (20.11), the four-potential A^i is called a **gauge field**.

Thus, we have shown that local phase invariance requires that there exist a gauge field A^i, that is, **the electromagnetic force itself is a consequence of local phase invariance of the wave function**.

20.3 Other global phase invariances

Experiment shows that, if we neglect the electromagnetic interaction, protons and neutrons behave so similarly that they can be regarded as **two states of the same particle** – the nucleon N. In particular, the short-range "strong" p–p, p–n and n–n forces are all the same, that is, the strong nucleon–nucleon force is "blind" in respect of the electric charge.

Thus, transformations of the proton and neutron wave functions, of the form

$$
\begin{aligned}
\psi_{\mathrm{p}} &\to \psi'_{\mathrm{p}} = \alpha\psi_{\mathrm{p}} + \beta\psi_{\mathrm{n}}, \\
\psi_{\mathrm{n}} &\to \psi'_{\mathrm{n}} = \gamma\psi_{\mathrm{p}} + \delta\psi_{\mathrm{n}},
\end{aligned}
\tag{20.12}
$$

with complex constants α, β, γ, δ, should leave the physics unchanged as far as the nucleon–nucleon force is concerned, that is, should be a **symmetry**.

The symmetry (20.12) can be represented as a global change of phase (albeit generalized to a matrix version), as can be seen below.

Write the nucleon wave function as a two-component wave function

$$
\Psi \equiv \begin{pmatrix} \psi_{\mathrm{p}} \\ \psi_{\mathrm{n}} \end{pmatrix}, \quad \text{i.e.,} \quad \Psi(x) \equiv \psi_{\mathrm{p}}(x)\begin{pmatrix} 1 \\ 0 \end{pmatrix} + \psi_{\mathrm{n}}(x)\begin{pmatrix} 0 \\ 1 \end{pmatrix}
$$

(a nucleon **isospinor**). Then we can rewrite (20.12) as

$$\Psi \to \Psi' = \mathrm{U}\Psi.$$

Thus, the transformed nucleon isospinor is

$$\begin{pmatrix} \psi'_{\mathrm{p}} \\ \psi'_{\mathrm{n}} \end{pmatrix} = \mathrm{U} \begin{pmatrix} \psi_{\mathrm{p}} \\ \psi_{\mathrm{n}} \end{pmatrix},$$

where U is a complex 2×2 matrix.

To preserve the normalization, U must be unitary, that is, $\mathrm{U}^{\dagger}\mathrm{U} = 1$, and so we must have

$$\det(\mathrm{U}\mathrm{U}^{\dagger}) = (\det \mathrm{U}^{\dagger})(\det \mathrm{U}) = (\det \mathrm{U})^{*}(\det \mathrm{U}) = |\det \mathrm{U}|^{2} = 1$$

so that

$$\det \mathrm{U} = \mathrm{e}^{i\theta},$$

where θ is a real number.

We can factor out an overall phase factor from the transformations (20.12) (since multiplying ψ_{p} and ψ_{n} by the same phase factor cannot change the physics), and thereby restrict ourselves to matrices U (**special** unitary matrices) that satisfy

$$\det \mathrm{U} = +1.$$

Such matrices constitute a **Lie group** called SU(2).

For Lie groups we can consider infinitesimal transformations, that is, with matrices U that differ from 1 by an infinitesimal matrix:

$$\mathrm{U} = 1 + i\xi,$$

where ξ is a 2×2 matrix whose elements are all infinitesimal quantities or zero. Taking the determinant of this, equating the result to 1, and neglecting terms of second order in the infinitesimal quantities in ξ, we find that ξ has zero trace:

$$\mathrm{Tr}\xi = 0.$$

The condition that U be unitary, that is, that

$$(1 + i\xi)(1 - i\xi^{\dagger}) = 1,$$

reduces (if we again ignore second-order quantities) to the condition

$$\xi^{\dagger} = \xi,$$

that is, ξ is a self-adjoint (hermitian) matrix.

A 2×2 hermitian traceless matrix has just three independent elements (as seen by counting the restrictions implied by $\mathrm{Tr}\xi = 0$ and $\xi^\dagger = \xi$) and can be written completely generally as

$$\xi = \boldsymbol{\varepsilon} \cdot \boldsymbol{\tau}/2,$$

where $\boldsymbol{\varepsilon}$ is an infinitesimal three-vector, with components $\varepsilon_1, \varepsilon_2, \varepsilon_3$, and $\boldsymbol{\tau}$ is the vector Pauli matrix, that is, with components [see (2.37)]

$$\tau_1 = \begin{pmatrix} 0 & 1 \\ 1 & 0 \end{pmatrix}, \quad \tau_2 = \begin{pmatrix} 0 & -i \\ i & 0 \end{pmatrix}, \quad \tau_3 = \begin{pmatrix} 1 & 0 \\ 0 & -1 \end{pmatrix}.$$

An infinitesimal transformation of the nucleon (p–n) doublet is therefore specified by

$$\begin{pmatrix} \psi'_{\mathrm{p}} \\ \psi'_{\mathrm{n}} \end{pmatrix} = \mathrm{U} \begin{pmatrix} \psi_{\mathrm{p}} \\ \psi_{\mathrm{n}} \end{pmatrix} = (1 + i\boldsymbol{\varepsilon} \cdot \boldsymbol{\tau}/2) \begin{pmatrix} \psi_{\mathrm{p}} \\ \psi_{\mathrm{n}} \end{pmatrix}. \tag{20.13}$$

We now express the dimensionless infinitesimal quantities $\varepsilon_1, \varepsilon_2, \varepsilon_3$ as finite phase angles $\alpha_1, \alpha_2, \alpha_3$ divided by N ($N \to \infty$) and build up a finite transformation by applying N infinitesimal transformations (20.13) and letting $N \to \infty$, using the identity

$$\mathrm{e}^{\mathrm{A}} = \lim_{N \to \infty} \left(1 + \frac{\mathrm{A}}{N}\right)^N.$$

Then

$$\begin{pmatrix} \psi'_{\mathrm{p}} \\ \psi'_{\mathrm{n}} \end{pmatrix} = \mathrm{e}^{i\boldsymbol{\alpha} \cdot \boldsymbol{\tau}/2} \begin{pmatrix} \psi_{\mathrm{p}} \\ \psi_{\mathrm{n}} \end{pmatrix}. \tag{20.14}$$

This is similar in nature to the global phase transformation considered previously:

$$\psi \to \psi' = \mathrm{e}^{i\alpha} \psi, \tag{20.15}$$

except that in (20.14) we now have three phase angles $\alpha_1, \alpha_2, \alpha_3$ and three (noncommuting) matrices τ_1, τ_2, τ_3. This noncommutation means that successive transformations of the type (20.14) do not, in general, commute. When the operations of a group do not commute, the group is said to be **non-abelian**, so that SU(2) is a non-abelian group. (All operations of the type (20.15) constitute the group U(1) and clearly commute, so that the group U(1) is **abelian**.)

The 2×2 matrices $\tau_r/2$ ($r = 1, 2, 3$) are called the **generators** (of infinitesimal transformations) in the **defining** (**fundamental**) matrix representation of the group SU(2). They obviously satisfy the commutation relations for angular-momentum components (see section 2.7):

$$[\tau_r/2, \tau_s/2] = i\varepsilon_{rst}\tau_t/2,$$

and we can think of them as acting in the space of two basis states – the proton ("isospin up") state and the neutron ("isospin down") state, forming the nucleon "isospin doublet".

The quantities ε_{rst} are the **structure constants** of SU(2).

We can find higher-dimensional matrices T_r ($r = 1, 2, 3$) satisfying the same commutation relations

$$[\mathrm{T}_r, \mathrm{T}_s] = i\varepsilon_{rst}\mathrm{T}_t, \tag{20.16}$$

that is, with the same structure constants, and these form higher-dimensional (that is, nonfundamental) irreducible matrix representations of the group SU(2). For example, there exist isospin triplets (e.g., Σ^-, Σ^0, Σ^+) and an isospin quartet (Δ^-, Δ^0, Δ^+, Δ^{++}) of particles, such that within each multiplet [as within the doublet (p, n)] the strong-interaction properties are the same. These are basis states of a three- and a four-dimensional irreducible matrix representation of SU(2), respectively.

The dimensionality of these matrices can be expressed as $2T + 1$, where T is the **total isospin quantum number** [analogous to the total-angular-momentum quantum number $j \equiv m_{\max}$ (2.31) in angular-momentum theory].

For example, in the three-dimensional irreducible representation of SU(2) the matrices T_r (and combinations of them such as $\mathrm{T}_- = \mathrm{T}_1 - i\mathrm{T}_2$ and $\mathrm{T}_+ = \mathrm{T}_1 + i\mathrm{T}_2$) can be thought of as acting in the space of the three "basis states" of an isospin-1 ($T = 1$) particle – states with values 1, 0, –1 of the 3-component of isospin (analogous to the angular-momentum states $|jm\rangle$ – see section 2.7 on the formal theory of angular momentum).

For example, the three 3×3 matrices T_r with elements

$$\left(\mathrm{T}_r^{(T=1)}\right)_{st} = -i\varepsilon_{rst} \tag{20.17}$$

satisfy (20.16), as is easily checked using the fourth index relation in subsection 17.1.6.

Acting in the space of a $(2T + 1)$-dimensional multiplet, a general SU(2) transformation has the form

$$\Psi^{(T)} \to \Psi'^{(T)} = \mathrm{e}^{i\boldsymbol{\alpha}\cdot\mathbf{T}^{(T)}}\Psi^{(T)} \equiv \mathrm{e}^{i(\alpha_1\mathrm{T}_1^{(T)}+\alpha_2\mathrm{T}_2^{(T)}+\alpha_3\mathrm{T}_3^{(T)})}\Psi^{(T)}, \tag{20.18}$$

with **three** angles α_i and **three** matrices T_r, of dimensionality $(2T + 1) \times (2T + 1)$.

If we do the same starting from SU(3) (the group of special unitary 3×3 matrices), then there are three basis states in the **defining** (alternatively, **fundamental**) representation, just as there were two (p and n) in the defining representation of SU(2). In the case of "flavor" SU(3) they are the up, down and strange states of the **quark**, so that an arbitrary state in this quark-flavor space can be written as

$$\Psi = \begin{pmatrix} \psi_\mathrm{u} \\ \psi_\mathrm{d} \\ \psi_\mathrm{s} \end{pmatrix}.$$

In the case of "color" SU(3 the three basis states are the "red", "green" and "blue" color states of the quark, so that an arbitrary state in this quark-color space can be written as

$$\Psi = \begin{pmatrix} \psi_\mathrm{r} \\ \psi_\mathrm{g} \\ \psi_\mathrm{b} \end{pmatrix}.$$

In either case, transformations in this space can be written in the form

$$\Psi \to \Psi' = \mathrm{U}\Psi \tag{20.19}$$

with

$$\mathrm{U} = \mathrm{e}^{i\boldsymbol{\alpha}\cdot\boldsymbol{\lambda}/2} \equiv \mathrm{e}^{i(\alpha_1\lambda_1/2+\alpha_2\lambda_2/2+\cdots+\alpha_8\lambda_8/2)}, \tag{20.20}$$

in which there are **eight** phase angles α_r and **eight** hermitian traceless 3×3 matrices λ_r [as compared with the **three** phase angles α_r and **three** hermitian traceless 2×2 matrices τ_r in the case of the fundamental representation of SU(2)].

In the case of color SU(3) the global phase transformations (20.20) are exact symmetries of the strong interaction, but in the case of flavor SU(3) they are only approximate symmetries of the strong interaction.

20.4 SU(2) local phase invariance (a non-abelian gauge theory)

Return to the **global** phase transformations (20.14) in the fundamental (doublet) representation of the group SU(2):

$$\Psi^{(T=1/2)} \to \Psi'^{(T=1/2)} = \mathrm{e}^{i\boldsymbol{\alpha}\cdot\boldsymbol{\tau}/2}\Psi^{(T=1/2)} \equiv \mathrm{e}^{i(\alpha_1\tau_1/2+\alpha_2\tau_2/2+\alpha_3\tau_3/2)}\Psi^{(T=1/2)},$$

but now let the three phase angles α_r depend on $x \equiv \boldsymbol{r}, t$, that is, make the transformations local:

$$\Psi^{(1/2)} \to \Psi'^{(1/2)} = \mathrm{e}^{ig\boldsymbol{\tau}\cdot\boldsymbol{\alpha}(x)/2}\Psi^{(1/2)} \tag{20.21}$$

(here we have inserted a dimensionless "coupling strength" g, analogous to the (dimensional) charge q that we inserted in the U(1) local phase transformation (20.7) $\psi \to \psi' = \mathrm{e}^{iq\chi(x)/\hbar}\psi$).

As in the U(1) case, (20.21) cannot be a symmetry of a free-particle wave equation, because the operators $\partial/\partial t$ and $\boldsymbol{\nabla}$ act on the angles $\alpha_i(x)$. But, just as, in the case of U(1), we replaced these operators by the covariant derivative

$$D^i = \partial^i + \frac{iqA^i(x)}{\hbar},$$

in the case of SU(2) we make the corresponding 2×2 matrix replacement

$$1\partial^i \to 1D^i = 1\partial^i + \frac{ig\boldsymbol{\tau}\cdot\boldsymbol{W}^i(x)}{2}, \tag{20.22}$$

that is, we have introduced **three** gauge fields $W_1^i(x)$, $W_2^i(x)$, $W_3^i(x)$ [each of which is a four-vector, like the four-potential $A_i(x)$]. They are called **SU(2) gauge fields** (the SU(2) instance of **Yang–Mills gauge fields**).

Then, if $\Psi^{(1/2)}$ satisfies some wave equation obtained from a free-particle wave equation in which the operators ∂^i have been replaced as in (20.22), we require that the function $\Psi'^{(1/2)}$ satisfies an equation of the same form but with transformed gauge fields $W_1'^i(x)$, $W_2'^i(x)$, $W_3'^i(x)$, that is, with the covariant derivatives

$$1D'^i = 1\partial^i + \frac{ig\boldsymbol{\tau}\cdot\boldsymbol{W}'^i(x)}{2}.$$

It is clear that this will be achieved if, even though the space–time-dependent phase factor in the wave function

$$\Psi'^{(1/2)} = e^{ig\boldsymbol{\tau}\cdot\boldsymbol{\alpha}(x)/2}\Psi^{(1/2)}$$

feels the action of the operators ∂^i, the fields $W_1'^i(x)$, $W_2'^i(x)$, $W_3'^i(x)$ are related to the fields $W_1^i(x)$, $W_2^i(x)$, $W_3^i(x)$ in such a way that this phase factor "passes through" the combined operator D'^i and "converts" it into D^i, that is, if

$$D'^i\Psi'^{(1/2)} = e^{ig\boldsymbol{\tau}\cdot\boldsymbol{\alpha}(x)/2}D^i\Psi^{(1/2)}. \tag{20.23}$$

If the wave equation is bilinear or higher-order in D^i, applying D'^i to (20.23) we can repeat the operation to find, for example,

$$D'^i D'^i\Psi'^{(1/2)} = e^{ig\boldsymbol{\tau}\cdot\boldsymbol{\alpha}(x)/2}D^i D^i\Psi^{(1/2)},$$

and so the wave equation involving D'^i and $\Psi'^{(1/2)}$ yields (after cancellation of $e^{ig\boldsymbol{\tau}\cdot\boldsymbol{\alpha}(x)/2}$ from every term) the original wave equation involving D^i and $\Psi^{(1/2)}$, and so must be equivalent to this wave equation.

The required relation (gauge transformation) between the fields $W_1'^i(x)$, $W_2'^i(x)$, $W_3'^i(x)$ and the fields $W_1^i(x)$, $W_2^i(x)$, $W_3^i(x)$ is simplest in the case of an infinitesimal transformation, that is, one in which the three phase angles $\alpha_r(x)$ are three infinitesimals $\eta_r(x)$. In this case the left-hand side of (20.23) becomes

$$\left(1\partial^i + \frac{ig\boldsymbol{\tau}\cdot\boldsymbol{W}'^i(x)}{2}\right)\left(1 + \frac{ig\boldsymbol{\tau}\cdot\boldsymbol{\eta}(x)}{2}\right)\Psi^{(1/2)}$$

and the right-hand side becomes

$$\left(1 + \frac{ig\boldsymbol{\tau}\cdot\boldsymbol{\eta}(x)}{2}\right)\left(1\partial^i + \frac{ig\boldsymbol{\tau}\cdot\boldsymbol{W}^i(x)}{2}\right)\Psi^{(1/2)}.$$

Equating these expressions, with $\boldsymbol{W}'^i$ written as $\mathbf{W}'^i = \boldsymbol{W}^i + \delta\boldsymbol{W}^i$ (since the above infinitesimal transformation would be expected to make an infinitesimal change to the fields $\boldsymbol{W}^i$), and neglecting products of infinitesimal quantities, we find (after cancelling terms)

$$\boldsymbol{\tau}\cdot\delta\boldsymbol{W}^i(x) = -\boldsymbol{\tau}\cdot\boldsymbol{\partial}^i\boldsymbol{\eta}(x) - g\boldsymbol{\tau}\cdot\boldsymbol{\eta}(x)\times\boldsymbol{W}^i(x) \tag{20.24}$$

where we have used the relation

$$\partial^i(\eta\Psi) = (\partial^i\eta)\Psi + \eta\partial^i\Psi$$

and the result (proved in section 19.6)

$$(\boldsymbol{\tau}\cdot\boldsymbol{a})(\boldsymbol{\tau}\cdot\boldsymbol{b}) = \boldsymbol{a}\cdot\boldsymbol{b} + i\boldsymbol{\tau}\cdot\boldsymbol{a}\times\boldsymbol{b}.$$

Then equation (20.24) gives

$$\delta\boldsymbol{W}^i(x) = -\partial^i\boldsymbol{\eta}(x) - g\boldsymbol{\eta}(x)\times\boldsymbol{W}^i(x) \tag{20.25}$$

which gives us the changes in the gauge fields $W_1^i(x)$, $W_2^i(x)$, $W_3^i(x)$ under the infinitesimal gauge transformation described by the "phase angles" $\eta_1(x)$, $\eta_2(x)$, $\eta_3(x)$.

The corresponding change in the electromagnetic four-potential A^i was $-\partial^i\chi(x)$, that is, there was no term equivalent to the last term in (20.25).

Equation (20.25) written for the changes $\delta W_r^i(x)$ ($r = 1, 2, 3$) in the three individual gauge fields is (by definition of the cross product)

$$\delta W_r^i(x) = -\partial^i \eta_r(x) - g\varepsilon_{rst}\eta_s(x)W_t^i(x) \tag{20.26}$$

What is the meaning of the last term in (20.26)?

Recall that a general unitary transformation in the $T = 1$ representation of SU(2) [that is, the $(2T + 1 = 3)$-dimensional representation] is of the form [see equation (20.18)]

$$\Psi^{(1)}(x) \to \Psi'^{(1)}(x) = e^{i(\alpha_1 T_1^{(1)} + \alpha_2 T_2^{(1)} + \alpha_3 T_3^{(1)})}\Psi^{(1)}(x),$$

the infinitesimal (and local) form of which gives

$$\begin{aligned}\delta\Psi^{(1)}(x) &= \Psi'^{(1)}(x) - \Psi^{(1)}(x)\\ &= ig(\eta_1(x)T_1^{(1)} + \eta_2(x)T_2^{(1)} + \eta_3(x)T_3^{(1)})\Psi^{(1)}(x) \equiv ig\eta_s(x)T_s^{(1)}\Psi^{(1)}(x),\end{aligned}$$

which, for the individual components of the isospin triplet, reads

$$\delta\psi_r^{(1)}(x) = ig\eta_s(x)\left(T_s^{(1)}\right)_{rt}\psi_t^{(1)}(x)$$

But [see (20.17)] the matrices $T_s^{(T=1)}$ have elements

$$\left(T_s^{(T=1)}\right)_{rt} = -i\varepsilon_{srt},$$

and so this becomes

$$\delta\psi_r^{(1)}(x) = g\varepsilon_{srt}\eta_s(x)\psi_t^{(1)}(x) = -g\varepsilon_{rst}\eta_s(x)\psi_t^{(1)}(x)$$

Thus, the extra (second) term in (20.26) is the change in $W_r^i(x)$ that we would expect from an infinitesimal SU(2) transformation [through "phase angles" $\eta_s(x)$] of the r-th member of an isospin triplet such as $\Psi^{(1)}(x)$.

Thus, the gauge fields $W_1^i(x)$, $W_2^i(x)$, $W_3^i(x)$ form an **isospin triplet**.

Since they form an isospin triplet, the wave equation that they satisfy will have the space–time derivatives ∂^i replaced as follows [compare with (20.22)]:

$$1\partial^i \to 1D^i = 1\partial^i + ig\mathbf{T}^{(1)}\cdot\mathbf{W}^i(x),$$

that is, the wave equation will have self-interaction terms bilinear and trilinear in the fields $\mathbf{W}^i(x)$ and proportional to the "charge" g and its square.

As we should expect (since the photon does not carry electric charge), such self-interaction terms do **not** appear in the wave equation (derived from the Maxwell equations) for the electromagnetic four-potential $A^i(x)$.

20.5 The "gauging" of color SU(3) (quantum chromodynamics)

In color SU(3), global transformations (in the fundamental, defining representation) are obtained by applying the 3×3 matrices U given by (20.20):

$$U = e^{i\boldsymbol{\alpha}\cdot\boldsymbol{\lambda}/2} \equiv e^{i(\alpha_1\lambda_1/2+\alpha_2\lambda_2/2+\cdots+\alpha_8\lambda_8/2)}$$

to the quark color triplet, and are an exact symmetry of the strong interaction.

Making this transformation local, that is, letting the eight "phase angles" α_r ($r = 1, 2, \ldots, 8$) depend on $x \equiv \boldsymbol{r}, t$ and introducing an SU(3) coupling constant g_s (analogous to the coupling constants q and g introduced in the U(1) and SU(2) cases), leads to unchanged physics only if the space–time derivatives ∂^i are replaced by the covariant derivatives

$$\begin{aligned} 1\partial^i \to 1D^i &= 1\partial^i + ig_s\boldsymbol{\lambda}\cdot\boldsymbol{G}^i(x) \\ &= 1\partial^i + ig_s\left(\lambda_1 G_1^i(x) + \lambda_2 G_2^i(x) + \cdots \lambda_8 G_8^i(x)\right), \end{aligned}$$

where the eight four-vector fields $G_r^i(x)$ ($r = 1, 2, \ldots, 8$) are gauge fields.

The gauge fields $G_r^i(x)$ (the "**gluon**" **fields**) form an SU(3) octet [just as the three gauge fields $W_r^i(x)$ form an SU(2) triplet], and their quanta (eight types of gluon) carry the strong force, that is, the force between the color charges of quarks. Unlike the photon, the gluon fields themselves carry nonzero charge ("color charge", in this case), and so the gluon wave equations contain gluon–gluon-interaction terms.

This theory (**quantum chromodynamics**) provides the basis of our current understanding of the **strong interaction**.

20.6 The weak interaction

Consider the weak-interaction process

$$n \to pe^-\bar{\nu}_e \text{ (}\beta\text{-decay)}$$

Since the quark structure of the neutron n is udd (an up quark and two down quarks), and the quark structure of the proton p is uud, the elementary process is

$$d \to ue^-\bar{\nu}_e,$$

equivalent to

$$\nu_e d \to ue^-.$$

This can be represented by a "two-current" process of the type in figure 20.1 in which (in the leptonic transition current) ν_e is converted to e^-, and (in the hadronic transition current) d is converted to u, and we have postulated the existence of a particle W^+ carrying the weak force responsible (see section 19.14).

This suggests that ν_e and e^- can be considered as two states of the same doublet, and similarly for the pair u and d.

We postulate that each pair forms an SU(2) doublet (a **weak-isospin** doublet).

For weak isospin we use the notation t (not T) for the total-isospin quantum number, and t_3 (not T_3) for the quantum number of the third component of isospin.

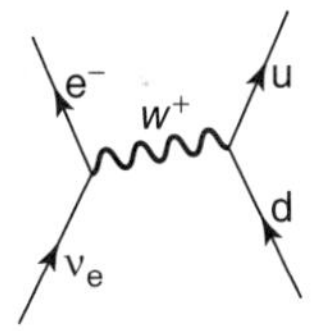

Figure 20.1 *Two-current representation of the process* $\nu_e d \to ue^-$.

In fact, what is assumed is that the following weak-isospin lepton doublets exist:

$$t = \tfrac{1}{2} \quad t_3 = \begin{cases} +\tfrac{1}{2} \\ -\tfrac{1}{2} \end{cases} \begin{pmatrix} \nu_e \\ e^- \end{pmatrix}_L \begin{pmatrix} \nu_\mu \\ \mu^- \end{pmatrix}_L \begin{pmatrix} \nu_\tau \\ \tau^- \end{pmatrix}_L$$

and the following weak-isospin quark doublets:

$$t = \tfrac{1}{2} \quad t_3 = \begin{cases} +\tfrac{1}{2} \\ -\tfrac{1}{2} \end{cases} \begin{pmatrix} u \\ d_C \end{pmatrix} \begin{pmatrix} c \\ s_C \end{pmatrix} \begin{pmatrix} t \\ b_C \end{pmatrix}.$$

The subscript L denotes that the components of these doublets are just the **left-handed** parts of the corresponding wave functions. This is because, in the weak interaction, neutrinos are detected only in the left-handed state (and antineutrinos only in the right-handed state) – see sections 19.13 and 19.14. The subscript C on d, s and b indicates that the relevant quark states are orthogonal linear superpositions (resulting from Cabibbo–Kobayashi–Maskawa mixing) of the down-quark, strange-quark and bottom-quark states. Each of the above doublets clearly belongs to the fundamental representation of the weak-isospin group SU. Because each doublet is left-handed, we shall call this group $SU(2)_L$.

Because right-handed (positive-helicity) neutrinos are never detected, that is, currents such as $e_R^- \leftrightarrow \nu_{e,R}$, $\mu_R^- \leftrightarrow \nu_{\mu,R}$ and $\tau_R^- \leftrightarrow \nu_{\tau,R}$ are not present, the right-handed (positive-helicity) states of the electron, muon and tauon must be singlet states ($t = 0$) of the weak-isospin $SU(2)_L$ group. So the theory contains the states

$$t = 0 \quad t_3 = 0: \quad e_R^- \quad \mu_R^- \quad \tau_R^-$$

and

$$t = 0 \quad t_3 = 0: \quad u_R \quad c_R \quad t_R \quad (d_C)_R \quad (s_C)_R \quad (b_C)_R$$

We may also consider neutral-current weak processes, such as neutrino–electron scattering (see figure 20.2). This must clearly be mediated by a neutral force carrier, which we may call Z.

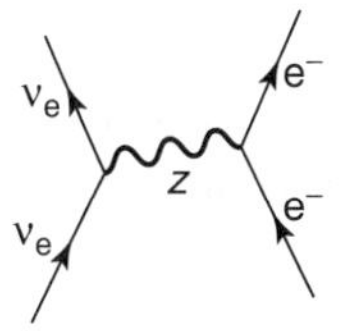

Figure 20.2 *A neutral-current weak process.*

Unlike the p and n states of the nucleon doublet (a hadronic-isospin doublet), the ν_e and e^- states of the first of the three leptonic weak-isospin doublets listed above have very different masses. So the weak-isospin $SU(2)_L$ symmetry, if it really exists, must be very strongly masked!

We assume that it exists, but is somehow hidden, and then "gauge the symmetry" (make it local) in the hope that the three gauge fields $W_1^i(x)$, $W_2^i(x)$, $W_3^i(x)$ that arise will have as their quanta the carriers of the weak force. (We shall return to the problem of the fermion masses in section 20.8.)

In fact, the covariant derivative used in the wave equation of an SU(2) doublet is

$$\begin{aligned} 1D^i &= 1\partial^i + \frac{ig\boldsymbol{\tau}\cdot\mathbf{W}^i(x)}{2} \\ &= 1\partial^i + \frac{ig\left(\tau_1 W_1^i(x) + \tau_2 W_2^i(x) + \tau_3 W_3^i(x)\right)}{2} \\ &= 1\partial^i + \frac{ig\left(\tau^+ W_-^i(x) + \tau^- W_+^i(x) + 2\tau_3 W_3^i(x)\right)}{4}, \end{aligned}$$

where (in analogy with the Pauli matrices in section 2.7) $\tau^{\pm} \equiv \tau_1 \pm i\tau_2$ are isospin raising and lowering operators (such as convert e– to ν_e and vice versa, respectively), and the associated fields $W_{\mp}^i(x) \equiv W_1^i(x) \mp iW_2^i(x)$ could have as their quanta the gauge bosons W^- and W^+ emitted in these transitions.

Similarly, since the matrix τ_3 does not change the isospin, and the isospin is not changed in neutral currents, we **might** expect that $W_3^i(x)$ will be the Z-boson field. But this turns out not to be the case.

Instead of considering gauged weak-isospin $SU(2)_L$ alone, Glashow, Weinberg and Salam considered its direct product with the gauged group U(1), that is, they considered the gauged group $SU(2)_L \times U(1)$.

In $SU(2)_L \times U(1)$, the $SU(2)_L$ gauge fields are the above weak-isospin triplet $\mathbf{W}^i(x)$ (that is, with $t = 1$, so that $t_3 = 1, 0, -1$) and the U(1) gauge field is an isospin-singlet $B^i(x)$ (that is, with $t = 0$, so that $t_3 = 0$). The field $B^i(x)$ is analogous to (but not the same as) the U(1) gauge field $A^i(x)$ of electromagnetism.

Each possible isospin multiplet has its own value of a property called the "**hypercharge**", which is defined as twice the average electric charge (in units of e) of the multiplet, for example,

$$y = -1 \text{ for } \begin{pmatrix} \nu_e \\ e^- \end{pmatrix}_L, \quad y = \frac{1}{3} \text{ for } \begin{pmatrix} u \\ d_C \end{pmatrix}_L$$

since the up and down quarks have electric charges $2e/3$ and $-e/3$, respectively.

In fact, the "charge" associated with the U(1) group in $SU(2)_L \times U(1)$ is this "**weak hypercharge**" y, always related to the electric charge $q = Qe$ of the particle by the "weak" **Gell-Mann–Nishijima relation**

$$Q = t_3 + \frac{y}{2},$$

so that $B^i(x)$ cannot be identified with $A^i(x)$.

Thus, in the $SU(2)_L \times U(1)$ gauge theory, the space–time derivatives ∂^i in the free-particle equations get replaced by covariant derivatives D^i with terms containing both the SU(2) and the U(1) gauge fields:

$$1\partial^i \to 1D^i = 1\partial^i + ig\mathbf{t}^{(t)}\cdot\mathbf{W}^i(x) + i1\left(\frac{yg'}{2}\right)B^i(x), \qquad (20.27)$$

where the choice of the matrices $\mathbf{t}^{(t)}$ depends on the value of t for the multiplet on which D^i is operating, and y is the hypercharge of this multiplet (g and g' are the weak "charges" associated with the $SU(2)_L$ and $U(1)_y$ parts of the gauge group, and the factor 1/2 is conventional). For example, on the (ν_e, e^-) doublet, which has $t = 1/2$ and $y = -1$, the correct covariant derivative to use is

$$1\partial^i \to 1D^i = 1\partial^i + ig\frac{\boldsymbol{\tau}\cdot \boldsymbol{W}^i(x)}{2} - i1\left(\frac{g'}{2}\right)B^i(x).$$

Recall that **the electromagnetic force is long-range** and the quanta of $A^i(x)$ – photons – are massless.

The weak force is short-range, and so the quanta of the corresponding gauge fields must have mass.

So the aim of the above theory is to show that it contains three massive gauge fields and one massless one (the photon).

Looking ahead, if the masses are generated by the **Higgs mechanism**, the Z field appears in the form of a linear combination of $W_3^i(x)$ and $B^i(x)$ (and, like the fields of the W^+ and W^- bosons, acquires a mass), while the photon field $A^i(x)$ appears as the linear combination of $W_3^i(x)$ and $B^i(x)$ orthogonal to the Z field and remains massless.

This unifies the weak force and electromagnetism. (It is the Weinberg–Salam "**electroweak theory**", or "**standard model**").

20.7 The Higgs mechanism

The **gauge bosons** of the weak interaction are believed to acquire their mass by the **Higgs mechanism.**

We shall first illustrate this mechanism for the simpler case of the "acquisition of mass" by the electromagnetic four-potential $A^i(x)$ (that is, by the photon!) in a superconductor.

Consider the Maxwell equations with sources (see section 17.4):

$$\frac{\partial F^{ik}}{\partial x^k} = -\frac{j^i}{\varepsilon_0 c^2}, \tag{20.28}$$

where F^{ik} is the electromagnetic-field tensor, defined by

$$F^{ik} = \frac{\partial A^k}{\partial x_i} - \frac{\partial A^i}{\partial x_k}.$$

It is easily checked [see (17.41)] that equations (20.28) are two of the Maxwell equations, namely,

$$i = 0 \Rightarrow \boldsymbol{\nabla}\cdot \boldsymbol{E} = \frac{\rho}{\varepsilon_0},$$
$$i = 1, 2, 3 \Rightarrow \boldsymbol{\nabla}\times \boldsymbol{B} = \frac{1}{c^2}\frac{\partial \boldsymbol{E}}{\partial t} + \frac{1}{\varepsilon_0 c^2}\boldsymbol{j}.$$

Written for the four-potential $A^i(x)$, equations (20.28) take the form

$$\partial_k(\partial^i A^k) - \partial_k\partial^k A^i = -\frac{1}{\varepsilon_0 c^2} j^i,$$

that is,

$$\partial^k \partial_k A^i - \partial^i (\partial_k A^k) = \frac{1}{\varepsilon_0 c^2} j^i. \tag{20.29}$$

This equation (the wave equation for A^i) is obviously invariant under the gauge transformation

$$A^i \to A'^i = A^i - \partial^i \chi.$$

In (20.29) the four-current $j^i = (c\rho, j_x, j_y, j_z)$ in the right-hand side can be regarded as the **source** of the electromagnetic four-potential A^i in the left-hand side.

In some physical situations **screening currents** exist, that is, currents that are themselves proportional to A^i. For example, in a superconductor, the spatial components of the screening current are given by

$$\boldsymbol{j}_{\rm sc} = -\frac{q^2}{m} |\psi|^2 \boldsymbol{A},$$

where ψ is the macroscopic wave function of the Cooper pairs responsible for the superconductivity and for the screening, $m = 2m_{\rm e}$ is the mass of a pair, and $q = -2e$ is the charge of a pair. If we put the four-vector analogue of this:

$$\frac{1}{\varepsilon_0 c^2} j^i = \frac{1}{\varepsilon_0 c^2} j^i_{\rm sc} = -\frac{M^2 c^2}{\hbar^2} A^i \tag{20.30}$$

(which defines a new quantity M^2) into (20.29), the wave equation for A^i becomes

$$\left(\partial^k \partial_k + \frac{M^2 c^2}{\hbar^2}\right) A^i - \partial^i (\partial_k A^k) = 0. \tag{20.31}$$

Taking the four-divergence of this, that is, contracting with ∂_i to the left, we find (after cancelling terms) that $\partial_i A^i = 0$, so that equation (20.31) becomes

$$\left(\hbar^2 \partial^k \partial_k + M^2 c^2\right) A^i = 0. \tag{20.32}$$

This can be thought of as describing a **photon** field with nonzero mass! [Compare it with the Klein–Gordon equation (19.2) for a free particle of mass M.]

To see the origin of the term "screening current", we consider (20.32) for a static (independent of t) field with just one-dimensional variation – say, with the coordinate x. Then (20.32) becomes

$$\hbar^2 \frac{\partial^2 A^i}{\partial x^2} = M^2 c^2 A^i,$$

which, if the superconductor occupies the half-space $x \geq 0$, has the solution

$$A^i(x) = A^i(0) \mathrm{e}^{-Mcx/\hbar}.$$

This states that in the superconductor the four-potential decays over a distance $l = \hbar/Mc$ (the **screening length**) to a value equal to e^{-1} of its value at the surface.

This accounts for the **Meissner effect** – the exclusion of magnetic flux $\boldsymbol{B} = \boldsymbol{\nabla} \times \boldsymbol{A}$ from the interior of a superconductor below its critical temperature.

Equation (20.32), because of the mass term, is no longer gauge-invariant, that is, the screening current arising from the Cooper-pair field has hidden the gauge invariance and given the photon a mass.

We now return to the gauge fields $W_r^i(x)$ $(r = 1, 2, 3)$ of the weak interaction, which are responsible for a short-range (that is, "screened") interaction (the weak force), and so must be massive.

How can they acquire mass from the **vacuum**? It is postulated that:

(i) the vacuum is pervaded by a field [the **Higgs field** $\phi(x)$] that has a nonzero expectation value $\langle\phi\rangle_0$ in the ground state (that is, in the vacuum);
(ii) the Higgs field couples to the gauge fields $W_r^i(x)$ through the covariant derivatives in the field equations for the $W_r^i(x)$, giving, in the latter, screening-current terms which, when the Higgs field is replaced by its nonzero (and constant) vacuum expectation value $\langle\phi\rangle_0$, are of the form $-\left(M^2c^2/\hbar^2\right) W_r^i(x)$.

Thus, the fields $W_r^i(x)$ acquire mass.

The Higgs field $\phi(x)$ is postulated to be a **weak-isospin doublet** ($t = 1/2$) of complex scalar fields:

$$\phi(x) = \frac{1}{\sqrt{2}}\begin{pmatrix} \phi_1(x) + i\phi_2(x) \\ \phi_3(x) + i\phi_4(x) \end{pmatrix}, \quad t_3 = \begin{cases} +\frac{1}{2} \\ -\frac{1}{2} \end{cases}$$

that has a nonzero expectation value $\langle\phi\rangle_0$ in the vacuum.

The Higgs field can clearly also be written in the form

$$\phi(x) = \mathrm{e}^{i(\boldsymbol{\tau}/2)\cdot\boldsymbol{\alpha}(x)}\begin{pmatrix} 0 \\ \chi(x) \end{pmatrix} \tag{20.33}$$

(which, with real $\boldsymbol{\alpha}(x)$ and $\chi(x)$, also has four independent functions of $x \equiv \boldsymbol{r}, t$). Without changing the physics, we can subject (20.33) to an isospin rotation (different at different points x) that brings the phase angles $\alpha_\beta(x)$ $(\beta = 1, 2, 3)$ to zero everywhere (provided, of course, that the space–time derivatives ∂^i are replaced by covariant derivatives containing the gauge fields $W_r^i(x)$ and $B^i(x)$ and that these fields undergo corresponding gauge transformations).

In this gauge the Higgs field takes the form

$$\phi(x) = \begin{pmatrix} 0 \\ \chi(x) \end{pmatrix}, \tag{20.34}$$

with $\chi(x)$ real. The isospin state (20.34) of the Higgs field ϕ clearly has $t_3 = -1/2$, and it is this state that is assumed to have a nonzero vacuum expectation value $\langle\phi\rangle_0$:

$$\langle\phi\rangle_0 = \begin{pmatrix} 0 \\ f/\sqrt{2} \end{pmatrix} \tag{20.35}$$

(f is a constant). In the true physical vacuum we do not allow a field with nonzero electric charge q to have a nonzero expectation value. Thus, for the isospin state (20.34) of the Higgs field the charge $q = Qe = 0$. The weak Gell-Mann–Nishijima relation $Q = t_3 + \frac{y}{2}$ (see section 20.6) then tells us that the Higgs field (regardless, of course, of its isospin state) has $y = 1$.

Since only the Higgs field has a nonzero vacuum expectation value, only it will contribute to the vacuum screening currents appearing as sources in the right-hand sides of the four wave equations for the four gauge fields $W_r^i(x)$ and $B^i(x)$.

Therefore, all we need is the form of the three weak-isospin currents $j_r^i(\phi)$ and weak-hypercharge current $j_y^i(\phi)$ (due to the Higgs field ϕ) that appear in these equations.

The structure of the weak-hypercharge current $j_y^i(\phi)$ in the wave equation for $B^i(x)$ will be analogous to that of the electromagnetic current $j_q^i(\phi)$ (due to a field ϕ) in the wave equation for $A^i(x)$. In the case when ϕ is a Klein–Gordon field $\phi(x)$ (appropriately normalized), $j_q^i(\phi)$ is the electric charge q of the field multiplied by the Klein–Gordon probability four-current:

$$j_q^i(\phi) = iq[\phi^*(\partial^i \phi) - (\partial^i \phi)^* \phi].$$

In the case when the source is an isospin doublet $\phi(x)$ this **electromagnetic current** generalizes to

$$j_q^i(\phi) = iq[\phi^\dagger(\partial^i \phi) - (\partial^i \phi)^\dagger \phi].$$

Thus, the correct form of the **weak-hypercharge current** as a source in the wave equation for the gauge field $B^i(x)$ is

$$j_y^i(\phi) = iy\frac{g'}{2}[\phi^\dagger(\partial^i \phi) - (\partial^i \phi)^\dagger \phi]. \tag{20.36}$$

The three **weak-isospin currents** $j_r^i(x)$ that are sources in the wave equations for the three gauge fields $W_r^i(x)$ are

$$j_1^i(\phi) = ig\left[\phi^\dagger \frac{\tau_1}{2}(\partial^i \phi) - (\partial^i \phi)^\dagger \frac{\tau_1}{2}\phi\right], \tag{20.37}$$

$$j_2^i(\phi) = ig\left[\phi^\dagger \frac{\tau_2}{2}(\partial^i \phi) - (\partial^i \phi)^\dagger \frac{\tau_2}{2}\phi\right], \tag{20.38}$$

$$j_3^i(\phi) = ig\left[\phi^\dagger \frac{\tau_3}{2}(\partial^i \phi) - (\partial^i \phi)^\dagger \frac{\tau_3}{2}\phi\right]. \tag{20.39}$$

In each of equations (20.36)–(20.39) we must replace ∂^i by the covariant derivative (20.27) in its correct form for the case $t = 1/2$, $y = 1$ (the quantum numbers of the Higgs field ϕ), that is, we put

$$1\partial^i \to 1D^i = 1\partial^i + i\frac{g}{2}\boldsymbol{\tau}\cdot\mathbf{W}^i(x) + i1\frac{g'}{2}B^i(x).$$

To get the **vacuum** screening currents, we also replace ϕ in (20.36)–(20.39) by the vacuum expectation value $\langle\phi\rangle_0$, given by (20.35), that is, we put

$$\phi \to \langle\phi\rangle_0 = \begin{pmatrix} 0 \\ f/\sqrt{2} \end{pmatrix} \tag{20.40}$$

and (because f is real)

$$\phi^\dagger \to \langle\phi\rangle_0^\dagger = \begin{pmatrix} 0 & f/\sqrt{2} \end{pmatrix}. \tag{20.41}$$

Then, because f is a constant, the terms in (20.36)–(20.39) that involve space–time derivatives ∂^i vanish. Thus,

$$\partial^i\phi \to D^i\phi = \left(i\frac{g}{2}\boldsymbol{\tau}\cdot\boldsymbol{W}^i(x) + i\frac{g'}{2}1B^i(x)\right)\begin{pmatrix}0\\ f/\sqrt{2}\end{pmatrix}$$
$$= i\frac{g}{2}W_1^i\begin{pmatrix}f/\sqrt{2}\\ 0\end{pmatrix} + \frac{g}{2}W_2^i\begin{pmatrix}f/\sqrt{2}\\ 0\end{pmatrix} - i\frac{g}{2}W_3^i\begin{pmatrix}0\\ f/\sqrt{2}\end{pmatrix} + i\frac{g'}{2}B^i\begin{pmatrix}0\\ f/\sqrt{2}\end{pmatrix}, \tag{20.42}$$

which implies the replacement

$$(\partial^i\phi)^\dagger \to -i\frac{g}{2}W_1^i\left(f/\sqrt{2}\ \ 0\right) + \frac{g}{2}W_2^i\left(f/\sqrt{2}\ \ 0\right) + i\frac{g}{2}W_3^i\left(0\ \ f/\sqrt{2}\right) - i\frac{g'}{2}B^i\left(0\ \ f/\sqrt{2}\right). \tag{20.43}$$

The first term in the four equations that result from equations (20.36)–(20.39) is of the form

$$\langle\phi\rangle_0^\dagger \mathrm{A} D^i\langle\phi\rangle_0 = \left(0\ \ f/\sqrt{2}\right)\mathrm{A}D^i\langle\phi\rangle_0$$

(with $\mathrm{A} = 1, \tau_1, \tau_2, \tau_3$, respectively), and, in this expression, only two terms of (20.42) give a nonzero result (the last two when $\mathrm{A} = 1$ or τ_3, and the first two when $\mathrm{A} = \tau_1$ or τ_2).

Similarly, the second term in the four equations that result from equations (20.36)–(20.39) is of the form

$$-(D^i\langle\phi\rangle_0)^\dagger\mathrm{A}\langle\phi\rangle_0 = -(D^i\langle\phi\rangle_0)^\dagger\mathrm{A}\begin{pmatrix}0\\ f/\sqrt{2}\end{pmatrix}$$

(with $\mathrm{A} = 1, \tau_1, \tau_2, \tau_3$, respectively), and, in this expression, only two terms of (20.43) give a nonzero result (the last two when $\mathrm{A} = 1$ or τ_3, and the first two when $\mathrm{A} = \tau_1$ or τ_2).

Thus, equation (20.36), with $y = 1$ (the hypercharge of the Higgs field), then gives

$$\left\langle j_y^i(\phi)\right\rangle_0 = i\frac{g'}{2}\left[\left(0\ \ f/\sqrt{2}\right)\left(-i\frac{g}{2}W_3^i + i\frac{g'}{2}B^i\right)\begin{pmatrix}0\\ f/\sqrt{2}\end{pmatrix} - \left(0\ \ f/\sqrt{2}\right)\left(i\frac{g}{2}W_3^i - i\frac{g'}{2}B^i\right)\begin{pmatrix}0\\ f/\sqrt{2}\end{pmatrix}\right]$$
$$= \frac{gg'f^2}{4}W_3^i - \frac{g'^2f^2}{4}B^i, \tag{20.36a}$$

while equation (20.37) gives

$$\left\langle j_1^i(\phi)\right\rangle_0 = ig\left[\left(0\ \ f/\sqrt{2}\right)\frac{\tau_1}{2}\left(i\frac{g}{2}W_1^i + \frac{g}{2}W_2^i\right)\begin{pmatrix}f/\sqrt{2}\\ 0\end{pmatrix} - \left(f/\sqrt{2}\ \ 0\right)\left(-i\frac{g}{2}W_1^i + \frac{g}{2}W_2^i\right)\frac{\tau_1}{2}\begin{pmatrix}0\\ f/\sqrt{2}\end{pmatrix}\right]$$
$$= -\frac{g^2f^2}{4}W_1^i, \tag{20.37a}$$

Equation (20.38) gives

$$\langle j_2^i(\phi)\rangle_0 = ig\left[\begin{pmatrix}0 & f/\sqrt{2}\end{pmatrix}\frac{\tau_2}{2}\left(i\frac{g}{2}W_1^i + \frac{g}{2}W_2^i\right)\begin{pmatrix}f/\sqrt{2}\\0\end{pmatrix} - \begin{pmatrix}f/\sqrt{2} & 0\end{pmatrix}\left(-i\frac{g}{2}W_1^i + \frac{g}{2}W_2^i\right)\frac{\tau_2}{2}\begin{pmatrix}0\\f/\sqrt{2}\end{pmatrix}\right]$$

$$= -\frac{g^2 f^2}{4}W_2^i, \tag{20.38a}$$

and equation (20.39) gives

$$\langle j_3^i(\phi)\rangle_0 = i\frac{g}{2}\left[\begin{pmatrix}0 & f/\sqrt{2}\end{pmatrix}\tau_3\left(-i\frac{g}{2}W_3^i + i\frac{g'}{2}B^i\right)\begin{pmatrix}0\\f/\sqrt{2}\end{pmatrix} - \begin{pmatrix}0 & f/\sqrt{2}\end{pmatrix}\left(i\frac{g}{2}W_3^i - i\frac{g'}{2}B^i\right)\tau_3\begin{pmatrix}0\\f/\sqrt{2}\end{pmatrix}\right]$$

$$= -\frac{g^2 f^2}{4}W_3^i + \frac{gg' f^2}{4}B^i. \tag{20.39a}$$

With the vacuum screening currents (20.36a)–(20.39a) as sources, the wave equations for the four gauge fields become

$$\partial^k\partial_k B^i - \partial^i(\partial_k B^k) = \frac{gg' f^2}{4}W_3^i - \frac{g'^2 f^2}{4}B^i, \tag{20.36b}$$

$$\partial^k\partial_k W_1^i - \partial^i(\partial_k W_1^k) = -\frac{g^2 f^2}{4}W_1^i, \tag{20.37b}$$

$$\partial^k\partial_k W_2^i - \partial^i(\partial_k W_2^k) = -\frac{g^2 f^2}{4}W_2^i, \tag{20.38b}$$

$$\partial^k\partial_k W_3^i - \partial^i(\partial_k W_3^k) = -\frac{g^2 f^2}{4}W_3^i + \frac{gg' f^2}{4}B^i. \tag{20.39b}$$

The gauge field $W_1^i(x)$ obeys a wave equation [equation (20.37b)] with a screening-current term, proportional to $W_1^i(x)$, that can be written in the form $-\left(M_{\mathrm{W}}^2c^2/\hbar^2\right)W_1^i(x)$, where M_{W}, to be interpreted as the mass of the gauge field $W_1^i(x)$ [see equation (20.31)], is given by

$$M_{\mathrm{W}} = \frac{\hbar g f}{2c}. \tag{20.44}$$

By the same argument, equation (20.38b) shows that the gauge field $W_2^i(x)$ also has this mass.

Equally, any linear combination of $W_1^i(x)$ and $W_2^i(x)$ obeys a wave equation with the same coefficient of that linear combination in the source term, that is, has the same mass M_{W}. In particular, the combinations

$$W_\pm^i(x) \equiv W_1^i(x) \pm iW_2^i(x)$$

clearly obey the wave equations

$$\partial^k\partial_k W_\pm^i - \partial^i(\partial_k W_\pm^k) = -\frac{g^2 f^2}{4}W_\pm^i,$$

and so both have the mass (20.44). Their quanta are the $W^{\pm}$ bosons.

Equations (20.36b) and (20.39b) clearly couple the gauge fields $B^i(x)$ and $W_3^i(x)$. Multiply equation (20.36b) by g' and subtract the resulting equation from that obtained by multiplying equation (20.39b) by g. The result is

$$\partial^k \partial_k Z^i - \partial^i(\partial_k Z^k) = -\frac{f^2}{4}\left(g^2 + g'^2\right) Z_0^i, \tag{20.45}$$

where we have introduced the linear combination

$$Z^i(x) = g W_3^i(x) - g' B^i(x). \tag{20.46}$$

Equation (20.45) is clearly the equation of a massive field $Z^i(x)$, given by (20.46), with mass

$$M_Z = \frac{\hbar f}{2c}\left(g^2 + g'^2\right)^{1/2}. \tag{20.47}$$

The quantum of this field is the Z boson.

Now, instead, multiply equation (20.36b) by g and add the resulting equation to that obtained by multiplying equation (20.39b) by g'. The result is

$$\partial^k \partial_k A^i - \partial^i(\partial_k A^k) = 0, \tag{20.48}$$

where we have introduced the linear combination

$$A^i(x) = g' W_3^i(x) + g B^i(x). \tag{20.49}$$

Equation (20.48) is clearly the equation of a **massless** field $A^i(x)$, given by (20.49), and, in fact, has the form of the Maxwell equations (20.29) with no sources (no currents).

Thus, the field (20.49) (apart from an appropriate dimensional factor) can be identified with the electromagnetic four-potential (the **photon** field), and so **both electromagnetism and the weak force** have emerged naturally from the Weinberg–Salam $SU(2)_L \times U(1)$ theory.

The fields (20.46) and (20.49) can be written equivalently as

$$\begin{aligned} Z^i(x) &= W_3^i(x)\cos\theta_W - B^i(x)\sin\theta_W, \\ A^i(x) &= W_3^i(x)\sin\theta_W + B^i(x)\cos\theta_W, \end{aligned} \tag{20.50}$$

where θ_W is the Glashow–Weinberg angle, defined in terms of the fundamental constants g and g' by

$$\tan\theta_W = \frac{g'}{g}, \quad \cos\theta_W = \frac{g}{\left(g^2 + g'^2\right)^{1/2}}, \quad \sin\theta_W = \frac{g'}{\left(g^2 + g'^2\right)^{1/2}}. \tag{20.51}$$

In a weak process involving the exchange of a $W^{\pm}$ boson, in the case when the magnitude q of the four-momentum exchanged between the participating particles is small compared with the $W^{\pm}$-boson mass M_W, the probability amplitude for the process is independent of q and contains an effective coupling constant

$G_F/\sqrt{2}$ (G_F is the **Fermi coupling constant**), related to the fundamental constant g and M_W by [see (19.119)]

$$\frac{G_F}{\sqrt{2}} = \frac{g^2}{8M_W^2}. \tag{20.52}$$

Putting $M_W = \hbar g f/2c$ (20.44) into (20.52), we can find f in terms of the experimentally known value of G_F ($G_F \approx 1.14 \times 10^{-5}$ [GeV/c^2]$^{-2}$), the result being

$$\frac{\hbar f}{c} = \frac{2M_W}{g} = \frac{1}{2^{1/4}\sqrt{G_F}} \approx 246 \text{ GeV}/c^2. \tag{20.53}$$

Similarly, the Glashow–Weinberg angle θ_W can be found experimentally by studying the relative probabilities of charged and neutral weak processes. The result is

$$\sin^2\theta_W \approx 0.23 \pm 0.01.$$

Finally, we take the general covariant derivative (20.27) in $\mathrm{SU(2)_L} \times \mathrm{U(1)}$ theory:

$$1\partial^i \rightarrow 1D^i = 1\partial^i + ig\mathbf{t}^{(t)}\cdot \boldsymbol{W}^i(x) + i1\left(\frac{yg'}{2}\right)B^i(x),$$

and re-express it in terms of the fields $Z^i(x)$ and $A^i(x)$ using the inverse of (20.50):

$$\begin{aligned} W_3^i(x) &= Z^i(x)\cos\theta_W + A^i(x)\sin\theta_W, \\ B^i(x) &= -Z^i(x)\sin\theta_W + A^i(x)\cos\theta_W. \end{aligned} \tag{20.54}$$

The charge-nonchanging part of the covariant derivative becomes

$$\begin{aligned} 1D^i &= 1\partial^i + ig\mathrm{t}_3^{(t)}W_3^i(x) + i1\left(\frac{yg'}{2}\right)B^i(x) \\ &= 1\partial^i + ig\mathrm{t}_3^{(t)}\left[Z^i(x)\cos\theta_W + A^i(x)\sin\theta_W\right] + i1\left(\frac{yg'}{2}\right)\left[-Z^i(x)\sin\theta_W + A^i(x)\cos\theta_W\right] \\ &= 1\partial^i + ig\sin\theta_W A^i(x)\left[\mathrm{t}_3^{(t)} + 1\frac{g'}{g}\frac{y}{2}\frac{\cos\theta_W}{\sin\theta_W}\right] + igZ^i(x)\left[\mathrm{t}_3^{(t)}\cos\theta_W - 1\frac{g'}{g}\frac{y}{2}\sin\theta_W\right] \\ &= 1\partial^i + ig\sin\theta_W A^i(x)\left[\mathrm{t}_3^{(t)} + 1\frac{y}{2}\right] + i\frac{g}{\cos\theta_W}Z^i(x)\left[\mathrm{t}_3^{(t)}\cos^2\theta_W - 1\frac{y}{2}\sin^2\theta_W\right], \end{aligned}$$

where we have used (20.51). But, for a particle of charge $q = Qe$, the Gell-Mann–Nishijima relation is

$$Q = t_3 + \frac{y}{2}, \tag{20.55}$$

which suggests that this covariant derivative contains the part found in the covariant derivative of electromagnetism [see (20.4)], provided that (after we have given the gauge field $A^i(x)$ the dimensions of the electromagnetic gauge field) we have the following relation between the fundamental constants e and g:

$$g \sin\theta_W \propto e. \tag{20.56}$$

From (20.44) and (20.47) for the masses of the W and Z gauge bosons, using (20.51) we have the relation

$$M_Z = \frac{M_W}{\cos\theta_W}.$$

The $W^{\pm}$ and Z bosons were produced in proton–antiproton collisions in the SPS (Super Proton Synchrotron) collider at CERN in 1983, and again in 1989–90, this time in electron–positron collisions, in the successor LEP (Large Electron Positron) collider. The masses determined in the latter experiment were

$$M_{W^{\pm}} \approx 80\,\mathrm{GeV}/c^2 \quad \text{and} \quad M_Z \approx 90\,\mathrm{GeV}/c^2, \tag{20.57}$$

and are consistent with the above experimental value for $\sin^2\theta_W$.

By analogy with the screening length $l = \hbar/Mc$ in the Meissner effect (see above), these masses suggest that the range of the weak interaction is of the order of 10^{-17} to 10^{-16} m.

The above explanation, based on the Higgs mechanism, of how the W and Z gauge bosons acquire mass, together with the eventual observation of W and Z bosons with the predicted masses (20.57), gave strong reason to believe in the existence of the particle known as the standard-model Higgs boson H (the lowest excitation of the Higgs field ϕ). In July 2012 it was announced that a new particle had been discovered, with mass close to $125\,\mathrm{GeV}/c^2$ (about 130 times the proton mass), in high-energy proton–antiproton collisions in the LHC (Large Hadron Collider) at CERN. The decisive evidence that this particle was indeed the Higgs boson was not its expected (and undoubtedly present) dominant decays to real quark–anti-quark pairs (of which the decay to $\mathrm{b\bar{b}}$ has the greatest probability [the bottom-quark mass m_b is $4.2\,\mathrm{GeV}/c^2$]), since the final products of such decays are not easily distinguished in the background of such products from other sources. Rather, the distinctive signature of the presence of the Higgs boson was primarily the decay to two photons, mediated by an initial decay to a virtual, very short-lived $\mathrm{t\bar{t}}$ pair. [The fact that the top-quark mass m_t is of the order of $173\,\mathrm{GeV}/c^2$ (see the next section) ensures, of course, that such a pair could only be virtual.] This particular decay to two photons represents a fraction 0.002 of all the Higgs-boson decays, but is sufficiently distinctive (and a sufficient number of such decays were observed) to ensure a probability of less than 6×10^{-7} that the relevant signal was arising from background statistical fluctuations in the data resulting from decays of already-familiar particles. Higgs-boson decays mediated by an initial decay to two virtual weak gauge bosons are also sufficiently distinctive to be observable against the background and, although rarer than the decays via a virtual top–anti-top pair, were also observed in the expected numbers and with the expected properties.

In **extensions** of the standard model there can be more than one type of Higgs particle. Further data and analysis are needed to determine whether the particle observed is the only Higgs particle or one of several such particles.

20.8 The fermion masses

The Dirac equation for a free electron (say) of mass m is [see (19.55)]

$$\left(i\hbar\gamma_i\partial^i - 1mc\right)\Psi(x) = 0. \tag{20.58}$$

In the $SU(2)_L \times U(1)$ gauge theory, the space–time derivatives ∂^i in the free-particle equations get replaced by covariant derivatives D^i with terms containing both the SU(2) and the U(1) gauge fields [see (20.27)]:

$$1\partial^i \to 1D^i = 1\partial^i + ig\mathbf{t}^{(t)}\cdot\mathbf{W}^i(x) + i1\left(\frac{yg'}{2}\right)B^i(x),$$

where the choice of the vector matrix $\mathbf{t}^{(t)}$ depends on the value of t for the multiplet on which D^i is operating and y is the hypercharge of this multiplet (g and g' are the weak "charges" associated with the $SU(2)_L$ and $U(1)_y$ parts of the gauge group, and the factor 1/2 is conventional). For example, on the $SU(2)_L$ doublet $(\nu_e, e^-)_L$, which has $t = 1/2$ and $y = -1$, the correct covariant derivative to use is

$$1\partial^i \to 1D^i = 1\partial^i + ig\frac{\boldsymbol{\tau}\cdot\mathbf{W}^i(x)}{2} - i1\left(\frac{g'}{2}\right)B^i(x),$$

while on the $SU(2)_L$ singlet e_R^-, which has $t = 0$ and $y = -2$, the covariant derivative to use is clearly

$$\partial^i \to D^i = \partial^i - ig'B^i(x).$$

Thus, the covariant derivatives are different for the left-handed ($SU(2)_L$-doublet) part

$$\Psi_L(x) = \frac{1}{2}(1-\gamma_5)\Psi(x)$$

and right-handed ($SU(2)_L$-singlet) part

$$\Psi_R(x) = \frac{1}{2}(1+\gamma_5)\Psi(x)$$

of the electron wave function $\Psi(x)$. This means that, if $m \neq 0$, it is not possible to make the Dirac equation (20.58) locally gauge-covariant. One way out of this problem is to proceed as follows. Apply the projection operator $P_R = \frac{1}{2}(1+\gamma_5)$ to equation (20.58) and use the fact that $\gamma_5\gamma_i = -\gamma_i\gamma_5$. We get

$$i\hbar\gamma_i\partial^i\Psi_L(x) = mc\Psi_R(x). \tag{20.59}$$

Similarly, applying $P_L = \frac{1}{2}(1-\gamma_5)$ to equation (20.58) we get

$$i\hbar\gamma_i\partial^i\Psi_R(x) = mc\Psi_L(x). \tag{20.60}$$

The right-hand sides here, which clearly destroy the local gauge covariance if $m \neq 0$, can be reproduced by starting from the corresponding massless equations $i\hbar\gamma_i\partial^i\Psi_L(x) = 0$ and $i\hbar\gamma_i\partial^i\Psi_R(x) = 0$ (which, with the appropriate covariant derivative, are each locally gauge-covariant) and introducing an interaction (of appropriate SU(2) matrix dimensionality) between the given wave function [$\Psi_L(x)$ or $\Psi_R(x)$] and a Higgs isodoublet with nonzero vacuum expectation value [see (20.40) and (20.41)]:

$$\phi \to \langle\phi\rangle_0 = \begin{pmatrix} 0 \\ f/\sqrt{2} \end{pmatrix}. \tag{20.61}$$

With this interaction (a Yukawa-type interaction) the massless equations become

$$i\hbar\gamma_i\partial^i\Psi_L(x) = g_f\hbar\phi\Psi_R(x) \tag{20.62}$$

and

$$i\hbar\gamma_i\partial^i\Psi_R(x) = g_f\hbar\phi^\dagger\Psi_L(x), \tag{20.63}$$

where g_f is the Yukawa fermion–Higgs coupling constant. Both sides of (20.62) have the structure of an SU(2) two-element column matrix (doublet), while both sides of (20.63) have the structure of an SU(2) singlet. Thus, we no longer have the problem of different terms (in either equation) transforming in different ways under $SU(2)_L$, and so each of these equations can be made locally covariant by introducing the appropriate covariant derivative. Substituting $\langle\phi\rangle_0$ (20.35) for ϕ we find that equations (20.62) and (20.63) yield the correct equations for a nonzero fermion mass [equations (20.59) and (20.60), respectively] if the fermion–Higgs coupling constant g_f is given by

$$g_f = \frac{\sqrt{2}mc}{\hbar f} = \frac{g}{\sqrt{2}}\left(\frac{m}{M_W}\right), \tag{20.64}$$

where, in deriving the second equality, we have used (20.44).

It is expected that there is a distinct Yukawa coupling constant g_f for the coupling of each of the elementary fermions f (the leptons and quarks) to the Higgs field. Efforts are being made to calculate the constants g_f as fixed points of renormalization-group equations (see the supplementary literature at the end of the book); the largest of them ($g_f = g_t$), believed to be of the order of 1, should yield the top-quark mass $m = m_t$ by (20.64).

The top quark was finally discovered in 1995, in proton–antiproton collisions in the CDF (Collider Detector at Fermilab), and was found to have a mass of the order of 173 GeV$/c^2$. This result is clearly consistent with the first of equations (20.64), if we use the value 246 GeV$/c^2$ for $\hbar f/c$ [equation (20.53)], where f is the vacuum expectation value of the Higgs field.

Appendices

A.1 Proof that the scattering states $|\phi+\rangle \equiv \Omega_+|\phi\rangle$ exist for all states $|\phi\rangle$ in the Hilbert space $\mathcal{H}$

The asymptotic condition for (4.12), that is, the existence of the states $|\phi+\rangle \equiv \Omega_+|\phi\rangle$ for **all** states $|\phi\rangle$ in the Hilbert space $\mathcal{H}$, is proved as follows. (The proof for (4.13) is almost identical.)

In (4.8), (4.9) we write $U^\dagger(t)\,U_0(t)$ as the integral of its derivative, using

$$\frac{\mathrm{d}}{\mathrm{d}t}(U^\dagger(t)U_0(t)) = \frac{\mathrm{d}}{\mathrm{d}t}(\mathrm{e}^{iHt/\hbar}\mathrm{e}^{-iH_0t/\hbar}) = \frac{i}{\hbar}\mathrm{e}^{iHt/\hbar}(H - H_0)\mathrm{e}^{-iH_0t\hbar} = \frac{i}{\hbar}U^\dagger(t)VU_0(t),$$

so that

$$U^\dagger(t)\,U_0(t)\,|\psi_{\text{in}}\rangle = |\psi_{\text{in}}\rangle + i\int_0^t \mathrm{d}\tau\, U^\dagger(\tau)\,VU_0(\tau)\,|\psi_{\text{in}}\rangle.$$

Therefore, $|\phi+\rangle$ exists if the integral here converges as $t \to -\infty$ [see (4.9)]. If we denote the norm of the ket $U^\dagger(\tau)\,VU_0(\tau)|\psi_{\text{in}}\rangle$ by $\|U^\dagger(\tau)VU_0(\tau)\psi_{\text{in}}\|$, a sufficient condition for this convergence is that

$$\int_{-\infty}^{0} \mathrm{d}\tau\,\|U^\dagger(\tau)\,VU_0(\tau)\,\psi_{\text{in}}\| < \infty,$$

Since U is unitary, this condition can be written as

$$\int_{-\infty}^{0} \mathrm{d}\tau\,\|VU_0(\tau)\,\psi_{\text{in}}\| < \infty,$$

and this must now be proved for any $|\psi_{\text{in}}\rangle$ in the Hilbert space $\mathcal{H}$. When expressed in the coordinate representation, any $|\psi_{\text{in}}\rangle$ in $\mathcal{H}$ can be arbitrarily closely approximated by a linear combination of a finite number of Gaussians, and so the convergence will be proved if we can prove it for a single Gaussian. We therefore take $|\psi_{\text{in}}\rangle$ to correspond to a Gaussian wave function

$$\langle \boldsymbol{r}|\psi_{\text{in}}\rangle = \mathrm{e}^{-(\boldsymbol{r}-\boldsymbol{r}_0)^2/2\xi^2},$$

A Course in Theoretical Physics, First Edition. P. J. Shepherd.

with centre $\boldsymbol{r}_0$ and width ξ. The effect of the free-evolution operator on this $|\psi_{\text{in}}\rangle$ is to "spread" the state, so that the squared modulus of the resulting position-space wave function increases in width and decreases in height with time:

$$|\langle \boldsymbol{r}|U_0(\tau)|\psi_{\text{in}}\rangle|^2 = \left(1+\frac{\hbar^2\tau^2}{m^2\xi^4}\right)^{-3/2} \exp\left[-\frac{|\boldsymbol{r}-\boldsymbol{r}_0|^2}{\xi^2+\hbar^2\tau^2/m^2\xi^2}\right].$$

Thus,

$$\begin{aligned}\|V U_0(\tau)\,\psi_{\text{in}}\|^2 &= \int \mathrm{d}^3\boldsymbol{r}|V(\boldsymbol{r})|^2\left(1+\frac{\hbar^2\tau^2}{m^2\xi^4}\right)^{-3/2} \exp\left[-\frac{|\boldsymbol{r}-\boldsymbol{r}_0|^2}{\xi^2+\hbar^2\tau^2/m^2\xi^2}\right]\\ &\le \left(1+\frac{\hbar^2\tau^2}{m^2\xi^4}\right)^{-3/2}\int \mathrm{d}^3\boldsymbol{r}|V(\boldsymbol{r})|^2.\end{aligned}$$

But all known physical scattering potentials $V(\boldsymbol{r})$ fall off sufficiently fast with r to ensure convergence of the spatial integral here, and so we find that

$$\int_{-\infty}^{0} \mathrm{d}\tau\|V U_0(\tau)\,\psi_{\text{in}}\| \le \left[\int \mathrm{d}^3\boldsymbol{r}|V(\boldsymbol{r})|^2\right]^{1/2} \int_{-\infty}^{0}\left(1+\frac{\hbar^2\tau^2}{m^2\xi^4}\right)^{-3/4} \mathrm{d}\tau < \infty.$$

This completes the proof of the existence of $|\phi+\rangle \equiv \Omega_+|\phi\rangle$ for all states $|\phi\rangle$ in $\mathcal{H}$. By the same procedure we can also prove the existence of $|\phi-\rangle \equiv \Omega_-|\phi\rangle$ for all states $|\phi\rangle$ in $\mathcal{H}$, and hence of the asymptotic condition.

A.2 The scattering matrix in momentum space

The relation

$$\langle \boldsymbol{p}'|S|\boldsymbol{p}\rangle = \delta^{(3)}(\boldsymbol{p}'-\boldsymbol{p}) - 2\pi i\delta(E_{p'}-E_p)\langle \boldsymbol{p}'|T(E_p+i0)|\boldsymbol{p}\rangle$$

[see (4.44) and (4.82)] is proved as follows.

For arbitrary $|\chi\rangle$ and $|\phi\rangle$ we have, from (4.20) with (4.9) and (4.11),

$$\langle\chi|S|\phi\rangle = \langle\chi|\Omega_-^\dagger\Omega_+|\phi\rangle = \lim_{\substack{t\to\infty\\ t'\to-\infty}} \langle\chi|(\mathrm{e}^{iH_0t/\hbar}\mathrm{e}^{-iHt/\hbar})(\mathrm{e}^{iHt'/\hbar}\mathrm{e}^{-iH_0t'/\hbar})|\phi\rangle.$$

Since the order of the limits is immaterial, we can put $t' = -t$ and let $t\to\infty$:

$$\langle\chi|S|\phi\rangle = \lim_{t\to\infty}\langle\chi|\mathrm{e}^{iH_0t/\hbar}\mathrm{e}^{-2iHt/\hbar}\mathrm{e}^{iH_0t/\hbar}|\phi\rangle.$$

We now write $e^{iH_0t/\hbar}e^{-2iHt/\hbar}e^{iH_0t/\hbar}$ as the integral of its derivative, using the expression

$$\begin{aligned}
&\frac{d}{dt}\left(e^{iH_0t/\hbar}e^{-2iHt/\hbar}e^{iH_0t/\hbar}\right)\\
&\quad=\frac{i}{\hbar}\left(e^{iH_0t/\hbar}(H_0-H)e^{-2iHt/\hbar}e^{iH_0t/\hbar}+e^{iH_0t/\hbar}e^{-2iHt/\hbar}(-H+H_0)e^{iH_0t/\hbar}\right)\\
&\quad=-\frac{i}{\hbar}\left(e^{iH_0t/\hbar}Ve^{-2iHt/\hbar}e^{iH_0t/\hbar}+e^{iH_0t/\hbar}e^{-2iHt/\hbar}Ve^{iH_0t/\hbar}\right),
\end{aligned}$$

and the fact that

$$e^{iH_0t/\hbar}e^{-2iHt/\hbar}e^{iH_0t/\hbar}=1 \text{ at } t=0.$$

Then,

$$\begin{aligned}
\langle\chi|S|\phi\rangle&=\langle\chi|\phi\rangle-\frac{i}{\hbar}\int_0^\infty dt\,\langle\chi|\left(e^{iH_0t/\hbar}Ve^{-2iHt/\hbar}e^{iH_0t/\hbar}+e^{iH_0t/\hbar}e^{-2iHt/\hbar}Ve^{iH_0t/\hbar}\right)|\phi\rangle\\
&=\langle\chi|\phi\rangle-\frac{i}{\hbar}\lim_{\varepsilon\downarrow 0}\int_0^\infty dt e^{-\varepsilon t/\hbar}\langle\chi|\left(e^{iH_0t/\hbar}Ve^{-2iHt/\hbar}e^{iH_0t/\hbar}+e^{iH_0t/\hbar}e^{-2iHt/\hbar}Ve^{iH_0t/\hbar}\right)|\phi\rangle.
\end{aligned}$$

For example, let $\langle\chi|=\langle\boldsymbol{p}'|$ and $|\phi\rangle=|\boldsymbol{p}\rangle$. Then, since

$$e^{iH_0t/\hbar}|\boldsymbol{p}\rangle=e^{iE_pt/\hbar}|\boldsymbol{p}\rangle \quad\text{and}\quad \langle\boldsymbol{p}'|e^{iH_0t/\hbar}=\langle\boldsymbol{p}'|e^{iE_{p'}t/\hbar},$$

we get

$$\begin{aligned}
\langle\boldsymbol{p}'|S|\boldsymbol{p}\rangle&=\delta^{(3)}(\boldsymbol{p}'-\boldsymbol{p})-\frac{i}{\hbar}\lim_{\varepsilon\downarrow 0}\int_0^\infty dt\,\langle\boldsymbol{p}'|\left(Ve^{i(E_{p'}+E_p+i\varepsilon-2H)t/\hbar}+e^{i(E_{p'}+E_p+i\varepsilon-2H)t/\hbar}V\right)|\boldsymbol{p}\rangle\\
&=\delta^{(3)}(\boldsymbol{p}'-\boldsymbol{p})-\lim_{\varepsilon\downarrow 0}\left(\langle\boldsymbol{p}'|V(-1)\left(E_{p'}+E_p+i\varepsilon-2H\right)^{-1}+(-1)\left(E_{p'}+E_p+i\varepsilon-2H\right)^{-1}V\right)|\boldsymbol{p}\rangle\\
&=\delta^{(3)}(\boldsymbol{p}'-\boldsymbol{p})+\frac{1}{2}\lim_{\varepsilon'\downarrow 0}\left(\langle\boldsymbol{p}'|V\left(\frac{E_{p'}+E_p}{2}+i\varepsilon'-H\right)^{-1}+\left(\frac{E_{p'}+E_p}{2}+i\varepsilon'-H\right)^{-1}V\right)|\boldsymbol{p}\rangle.
\end{aligned}$$

But, from (4.46),

$$\left(\frac{E_{p'}+E_p}{2}+i\varepsilon'-H\right)^{-1}=G\left(\frac{E_{p'}+E_p}{2}+i\varepsilon'\right)$$

and [see (4.66)] $VG(z)|\boldsymbol{p}\rangle=T(z)G^0(z)|\boldsymbol{p}\rangle$, which for this value of z is equal to

$$\begin{aligned}
&T\left(\frac{E_{p'}+E_p}{2}+i\varepsilon'\right)\left(\frac{E_{p'}+E_p}{2}+i\varepsilon'-H_0\right)^{-1}|\boldsymbol{p}\rangle\\
&\quad=T\left(\frac{E_{p'}+E_p}{2}+i\varepsilon'\right)\left(\frac{E_{p'}+E_p}{2}+i\varepsilon'-E_p\right)^{-1}|\boldsymbol{p}\rangle
\end{aligned}$$

Similarly [see (4.65)], $\langle \boldsymbol{p}'|G(z)V = \langle \boldsymbol{p}'|G^0(z)T(z)$, which for the same value of z is equal to

$$\langle \boldsymbol{p}'| \left(\frac{E_{p'} + E_p}{2} + i\varepsilon' - E_{p'} \right)^{-1} T \left(\frac{E_{p'} + E_p}{2} + i\varepsilon' \right)$$

Therefore,

$$\begin{aligned} \langle \boldsymbol{p}'|S|\boldsymbol{p}\rangle &= \delta^{(3)}(\boldsymbol{p}' - \boldsymbol{p}) + \lim_{\varepsilon \downarrow 0} \left(\frac{1}{E_{p'} - E_p + i\varepsilon} + \frac{1}{E_p - E_{p'} + i\varepsilon} \right) \langle \boldsymbol{p}'|T \left(\frac{E_{p'} + E_p}{2} + i\varepsilon \right) |\boldsymbol{p}\rangle \\ &= \delta^{(3)}(\boldsymbol{p}' - \boldsymbol{p}) - 2\pi i \delta(E_{p'} - E_p) \langle \boldsymbol{p}'|T\left(E_p + i0\right)|\boldsymbol{p}\rangle, \end{aligned}$$

where we have used (4.57), and the delta-function has enabled us to replace $(E_{p'} + E_p)/2$ by E_p.

A.3 Calculation of the free Green function $\langle \boldsymbol{r}|G^0(z)|\boldsymbol{r}'\rangle$

Here we evaluate the free-particle Green function (4.99):

$$\begin{aligned} \langle \boldsymbol{r}|G^0(z)|\boldsymbol{r}'\rangle &= \frac{1}{(2\pi\hbar)^3} \int \mathrm{d}^3\boldsymbol{p} \frac{\mathrm{e}^{i\boldsymbol{p}\cdot(\boldsymbol{r}-\boldsymbol{r}')/\hbar}}{z - p^2/2m} \\ &= \frac{2m}{(2\pi\hbar)^3} \int_0^{2\pi} \mathrm{d}\varphi \int_0^{\pi} \sin\theta \mathrm{d}\theta \int_0^{\infty} \mathrm{d}p p^2 \frac{\mathrm{e}^{ip|\boldsymbol{r}-\boldsymbol{r}'|\cos\theta/\hbar}}{2mz - p^2} \\ &= \frac{2m}{(2\pi)^2\hbar^3} \int_0^{\infty} \mathrm{d}p \frac{p^2}{2mz - p^2} \int_{-1}^{1} \mathrm{d}x \mathrm{e}^{ip|\boldsymbol{r}-\boldsymbol{r}'|x/\hbar} \\ &= -\frac{m}{2\pi^2 i\hbar^2|\boldsymbol{r}-\boldsymbol{r}'|} \int_0^{\infty} \mathrm{d}p \frac{p}{p^2 - 2mz} \left(\mathrm{e}^{ip|\boldsymbol{r}-\boldsymbol{r}'|/\hbar} - \mathrm{e}^{-ip|\boldsymbol{r}-\boldsymbol{r}'|/\hbar} \right). \end{aligned}$$

Now replace the integration variable p by $-p$ in the second term:

$$\langle \boldsymbol{r}|G^0(z)|\boldsymbol{r}'\rangle = \frac{im}{2\pi^2\hbar^2|\boldsymbol{r}-\boldsymbol{r}'|} \left(\int_0^{\infty} \mathrm{d}p \frac{p\mathrm{e}^{ip|\boldsymbol{r}-\boldsymbol{r}'|/\hbar}}{p^2 - 2mz} - \int_0^{-\infty} \mathrm{d}(-p) \frac{(-p)\mathrm{e}^{ip|\boldsymbol{r}-\boldsymbol{r}'|/\hbar}}{p^2 - 2mz} \right)$$

$$= \frac{im}{2\pi^2\hbar^2|\boldsymbol{r}-\boldsymbol{r}'|} \int_{-\infty}^{\infty} \mathrm{d}p \frac{p\mathrm{e}^{ip|\boldsymbol{r}-\boldsymbol{r}'|/\hbar}}{p^2 - 2mz} = \frac{im}{2\pi^2\hbar^2|\boldsymbol{r}-\boldsymbol{r}'|} \int_{-\infty}^{\infty} \mathrm{d}p \frac{p\mathrm{e}^{ip|\boldsymbol{r}-\boldsymbol{r}'|/\hbar}}{(p - \sqrt{2mz})(p + \sqrt{2mz})}.$$

We choose the root $\sqrt{2mz}$ to be the one with positive imaginary part, so that the pole at $p = \sqrt{2mz}$ lies in the upper half of the complex p-plane (see figure A.1).

We close the integration contour by drawing an infinite semicircle in the upper half-plane, as illustrated in figure A.1.

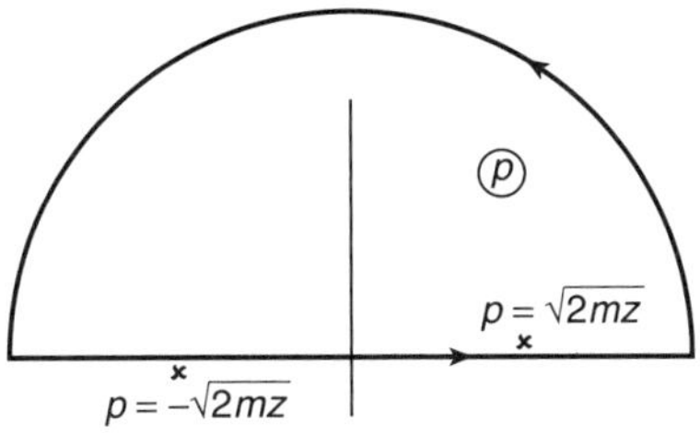

Figure A.1 *Poles of $\langle \boldsymbol{r}|G^0(z)|\boldsymbol{r}'\rangle$ in the complex p-plane.*

The integrand vanishes on the semicircle, and the residue of the integrand at the pole is

$$\frac{\sqrt{2mz}\,e^{i\sqrt{2mz}|\boldsymbol{r}-\boldsymbol{r}'|/\hbar}}{2\sqrt{2mz}}.$$

Therefore, from Cauchy's residue theorem we have

$$\langle \boldsymbol{r}|G^0(z)|\boldsymbol{r}'\rangle = -\frac{m}{2\pi\hbar^2}\frac{e^{i\sqrt{2mz}|\boldsymbol{r}-\boldsymbol{r}'|/\hbar}}{|\boldsymbol{r}-\boldsymbol{r}'|}.$$

If $z = E_p + i0 = (p^2/2m) + i0$ then

$$(2mz)^{1/2} = (p^2 + i0)^{1/2} = +(p + i0),$$

where the overall plus sign is chosen so as to locate the pole in the upper half-plane. Similarly, if $z = E_p - i0 = (p^2/2m) - i0$, then

$$(2mz)^{1/2} = (p^2 - i0)^{1/2} = -(p - i0),$$

where the overall negative sign is chosen for the same reason.

Therefore,

$$\langle \boldsymbol{r}|G^0(E_p \pm i0)|\boldsymbol{r}'\rangle = -\frac{m}{2\pi\hbar^2}\frac{e^{\pm ip|\boldsymbol{r}-\boldsymbol{r}'|/\hbar}}{|\boldsymbol{r}-\boldsymbol{r}'|}.$$

Where the overall negative sign is chosen for the same reason

Therefore

Supplementary Reading

For Module I: Nonrelativistic Quantum Mechanics

Schiff, L.I. (1968) *Quantum Mechanics*, 3rd edn. McGraw–Hill, New York.
Taylor, J.R. (2009) *Scattering Theory: The Quantum Theory of Nonrelativistic Collisions*, 2nd edn. Dover Publications, New York.

For Module II: Thermal and Statistical Physics

Callen, H.B. (1985) *Thermodynamics and an Introduction to Thermostatistics*, 2nd edn. John Wiley & Sons, Inc., New York.
Landau, L.D. and Lifshitz, E.M. (1980) *Statistical Physics*, 3rd edn. Butterworth–Heinemann, Oxford.
Waldram, J.R. (1985) *The Theory of Thermodynamics*. Cambridge University Press.

For Module III: Many-Body Theory

Nozières, P. (1997) *Theory of Interacting Fermi Systems*. Westview Press, Boulder, CO.
Pines, D. (1994) *Theory of Quantum Liquids, Volume I: Normal Fermi Liquids*. Westview Press, Boulder, CO.
Pines, D. (1999) *Elementary Excitations in Solids: Lectures on Phonons, Electrons, and Plasmons*. Westview Press, Boulder, CO.
Tinkham, M. (2004) *Introduction to Superconductivity*, 2nd edn. Dover Publications, New York.

For Module IV: Classical Field Theory and Relativity

Landau, L.D. and Lifshitz, E.M. (1980) *The Classical Theory of Fields*, 4th edn. Butterworth–Heinemann, Oxford.
Schutz, B. (2009) *A First Course in General Relativity*, 2nd edn. Cambridge University Press.

For Module V: Relativistic Quantum Mechanics and Gauge Theories

Aitchison, I.J.R. and Hey, A.J.G. (2003) *Gauge Theories in Particle Physics: Volume I: From Relativistic Quantum Mechanics to QED*, 3rd edn. Taylor and Francis, Abingdon, UK.
Aitchison, I.J.R. and Hey, A.J.G. (2003) *Gauge Theories in Particle Physics: Volume II: Non-Abelian Gauge Theories: QCD and the Electroweak Theory*, 3rd edn. Taylor and Francis, Abingdon, UK.
Bjorken, J.D. and Drell, S.D. (1965) *Relativistic Quantum Mechanics* McGraw–Hill, New York.

A Course in Theoretical Physics, First Edition. P. J. Shepherd.

Index

A Course in Theoretical Physics, First Edition. P. J. Shepherd.